Advances in Biochemistry in Health and Disease

Volume 27

Advances in Biochemistry in Health and Disease focus on the latest developments in biochemical research with implications for health and disease. This book series consists of original edited volumes and monographs, presented by leading experts in the field and provides an up to date and unique source of information for all those interested in the fundamental, biochemical processes of the latest and emerging topics and techniques.

Covering a wide variety of topics, this book series is a valuable source of information from those at the lab bench through to the Health Care workers.

Jan Slezak · Branislav Kura
Editors

Molecular Hydrogen in Health and Disease

Springer

Editors
Jan Slezak
Centre of Experimental Medicine
Slovak Academy of Sciences
Bratislava, Slovakia

Branislav Kura
Centre of Experimental Medicine
Slovak Academy of Sciences
Bratislava, Slovakia

ISSN 2512-2142 ISSN 2512-2150 (electronic)
Advances in Biochemistry in Health and Disease
ISBN 978-3-031-47374-6 ISBN 978-3-031-47375-3 (eBook)
https://doi.org/10.1007/978-3-031-47375-3

This Springer imprint is published by the registered company Springer Nature Switzerland AG
The registered company address is: Gewerbestrasse 11, 6330 Cham, Switzerland

Paper in this product is recyclable.

Preface

The decision to compile a book on *Molecular Hydrogen in Health and Disease* within the book series *Advances in Biochemistry in Health and Disease* directed by Prof. Naranjan Dhalla from Winnipeg was created shortly after the "1st Conference of the European Academy for Molecular Hydrogen Research in Biomedicine". The meeting was organized by representatives of the European Academy in the congress center of the Slovak Academy of Sciences in Smolenice near Bratislava in October 2022 (see: www.EUH2ACADEMY.org)

The main goal of the organizers was to draw attention to the gradually developing use of molecular hydrogen therapy in the European region and bring together the leading experts in the field of molecular hydrogen research from around the world. Unlike China, Japan, and South Korea, in other countries, including Europe, there is still a certain lack of confidence in the therapeutic abilities of H_2, originating more or less from ignorance and little information among healthcare workers.

In 2007, a group of Japanese scientists led by S. Ohta pointed out in Nature Medicine on selective scavenging of hydroxyl free radical and peroxynitrite anions by molecular hydrogen. The study immediately attracted substantial interest all over the world.

Molecular hydrogen exerts biological effects on nearly all organs. It has anti-oxidative, anti-inflammatory, and anti-aging effects and contributes to the regulation of autophagy and cell death. H_2 indirectly regulates hormones and cytokines through various signaling transduction pathways.

The effect of hydrogen therapy has been researched on its effect on various diseases and has been acknowledged to have a beneficial effect on multiple pathological states. It does not react with essential body structures or affect important body processes. There are practically no side effects of molecular hydrogen administration, and new research continues to support its extraordinary potential in biomedicine.

Despite several hypotheses, the exact mechanism of the hydrogen effect was once hidden. Currently, it is speculated that the target molecule of H_2 may be the oxidized form of a porphyrin that acts as a catalyst to stimulate the selective reaction of H_2 with hydroxyl radical.

Many countries, especially Eastern ones, already officially use molecular hydrogen for clinical purposes. In Japan, hydrogen inhalation treatment was approved by the Department of Health and Human Services on November 2016, and the hydrogen inhalation treatment device was registered. An increased number of hospitals are implementing hydrogen inhalation therapy in ambulances for emergency patients. Hydrogen inhalation for lung diseases including COVID-19 and pneumonia treatment had been used in China since March 2020. In February 2022, clinical research was conducted on hydrogen inhalation therapy in Korea, Ireland, and other countries. Molecular hydrogen is now recognized as a novel, medically relevant gas with therapeutic potential exerting biological effects on nearly all organs, contributing to the regulation of autophagy and cell death modulation. Hydrogen increases energy, reduces inflammation, and shortens recovery times after a workout.

The individual chapters contained in this book support the extraordinary properties of hydrogen in the prevention and treatment of various disorders, the common denominator of which is an imbalance between reactive forms of oxygen and antioxidant capacity.

We believe and hope that this book will be of interest not only to the scientific community and public but also to clinicians with its goal to improve health and quality of life.

Bratislava, Slovakia

Prof. Jan Slezak
Dr. Branislav Kura

Contents

Chapter 1
Hydrogen: From Stars to Fuel to Medicine

Tyler W. LeBaron, Randy Sharpe, Felix A. Pyatakovich, and Mikhail Yu. Artamonov

Abstract Hydrogen gas has garnered significant attention in recent years due to its remarkable antioxidant and anti-inflammatory properties. However, the extensive research on hydrogen's applications in the energy sector often overshadows its potential as a medically and biologically active gas. Surprisingly, investigations into the biomedical aspects of H_2 trace back to as early as 1793. Hydrogen exhibits exceptional pharmacokinetics, swiftly traversing cellular biomembranes, including the blood–brain and testes barriers, to access subcellular organelles. Following ingestion, hydrogen follows the path of least resistance through the circulatory system and is primarily eliminated through exhalation. Despite the intricate molecular mechanisms and precise targets remaining elusive, the antioxidant effects of hydrogen involve the upregulation of endogenous antioxidants via the activation of the Nrf2/keap1 pathway. Recent research highlights the potential role of Fe-porphyrin as a redox-related biosensor, facilitating hydrogen's reactions with hydroxyl radicals and triggering additional signal transduction processes. Moreover, this review delves into the physicochemical properties of hydrogen, particularly emphasizing its molar solubility, considerations regarding the term saturation, and other unique characteristics of H_2 are discussed. The expanding knowledge and research surrounding the history of H_2 underscore its transformative potential in biomedical applications and pave the way for future advancements in harnessing its therapeutic properties.

Keywords Molecular hydrogen · Antioxidant · Hydrogen history · Inflammation · Hydrogen solubility

---

T. W. LeBaron (✉)
Department of Kinesiology and Outdoor Recreation, Southern Utah University, Cedar City, UT 84720, USA
e-mail: tylerlebaron@suu.edu

Molecular Hydrogen Institute, Enoch, UT, USA

R. Sharpe
H2 Analytics, Henderson, NV, USA

F. A. Pyatakovich · M. Yu. Artamonov
MJA Research and Development, Inc, East Stroudsburg, PA 18301, USA

J. Slezak and B. Kura (eds.), *Molecular Hydrogen in Health and Disease*, Advances in Biochemistry in Health and Disease 27, https://doi.org/10.1007/978-3-031-47375-3_1

Introduction

Hydrogen, the cosmic pioneer, has a captivating tale that unfolds across the vast expanse of time, stretching back to the dawn of the universe some 14.8 billion years ago. As the lightest and most abundant element in existence, hydrogen holds the secrets of our cosmic origins, entwined with the very fabric of the cosmos. From the primordial stages to the present day, its journey is nothing short of extraordinary. Its discovery can be traced back to the late eighteenth century when British scientist Henry Cavendish identified a new gas that exhibited unique properties. He named it "inflammable air" due to its highly flammable nature [1]. In 1783, hydrogen played a crucial role in the first manned hydrogen balloon flight conducted by Jacques Charles and the Robert brothers, ushering in the era of human aviation [2].

Throughout the nineteenth and twentieth centuries, hydrogen found various applications in industries. It was widely used for filling airships and as a fuel for gas lamps and welding processes [3]. However, the infamous Hindenburg disaster in 1937, when a hydrogen-filled airship caught fire, brought significant safety concerns associated with hydrogen into the spotlight [4].

In the latter half of the twentieth century, researchers began exploring the potential of hydrogen as a clean and sustainable energy source. Hydrogen fuel cells, which generate electricity through the electrochemical reaction between hydrogen and oxygen, emerged as a promising technology [5]. On a mass-to-mass comparison, hydrogen gas has about three times more energy than petrol (gasoline). The space industry also relied on hydrogen as a rocket fuel due to its high energy content [6]. In recent decades, hydrogen has gained renewed attention as a key player in the transition to a low-carbon economy. Thus, it holds immense potential as a versatile energy carrier that can be produced from various renewable sources, such as solar and wind power, through electrolysis. Hydrogen fuel cells have been increasingly employed in transportation, including cars, buses, and even trains, offering zero-emission alternatives to conventional fossil fuel-based vehicles [6].

Moreover, hydrogen is being explored for its applications beyond energy. In biomedical research, hydrogen gas has shown therapeutic potential due to its antioxidant and anti-inflammatory properties [7]. It is being studied for its potential benefits in mitigating various health conditions and promoting overall well-being [7]. It is the biomedical history of hydrogen that is the primary focus of this review.

Historical Overview of Hydrogen Research

The explosion of research on hydrogen gas over the past two decades has brought hydrogen to the forefront of several different sectors. The most widely recognized area of research involves its use as a clean and adaptable energy carrier [6]. Figure 1.1 illustrates the impressive number of articles published on hydrogen gas (H_2) within

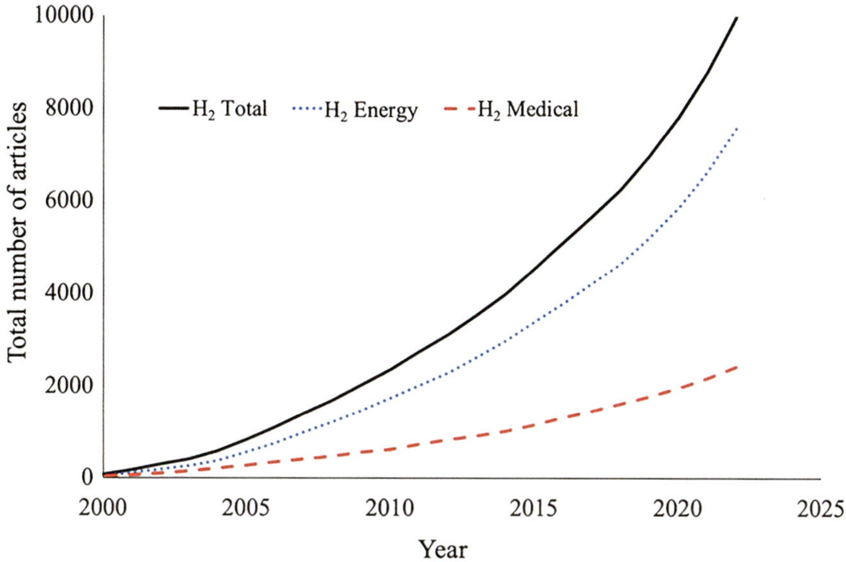

Fig. 1.1 Articles published relating to hydrogen gas only, hydrogen gas and energy/fuel, or hydrogen gas and its medical uses

the past 22 years. There are now over 2000 articles published on the biomedical effects of molecular hydrogen.

However, this impressive number is perhaps obscured by the nearly four times as many articles published on the use of hydrogen gas as an energy source. Yet, there are now over 2000 publications related to the biomedical effects of hydrogen molecular hydrogen. This novel medical gas has been shown to have therapeutic effects in over 170 disease models, encompassing essentially every organ of the human body [8]. Most of this research has been published following a 2007 seminal paper in *Nature Medicine* by Oshawa et al. [9]. The authors reported that hydrogen gas could drastically attenuate ischemia/reperfusion-related brain damage induced by a middle cerebral artery occlusion in a rat stroke model [9]. Only 2% H_2, which is below the flammability level, was necessary to provide these remarkable effects. Nevertheless, interest in the biomedical effects of hydrogen gas, as well as its use in medicine for non-medicinal reasons, has been ongoing for a long time [10].

Hydrogen Applications in Humans for Non-medicinal Purposes

Interestingly, the utilization of hydrogen gas in various human practices has been investigated for well over 100 years [10], without recognizing its potential as a medically or biologically active gas [11]. These practices include biodegradable implants, bodily wound detection, diving applications, blood flow measurement, malabsorption diagnosis, and detecting achlorhydria.

Biodegradable Implants

Since 1878, magnesium metal has gained attention in orthopedic and biomedical engineering as a promising material for biodegradable implants [12]. These implants gradually dissolve in the body, releasing hydrogen gas and magnesium ions ($Mg + 2H_2O \rightarrow Mg(OH)_2 + H_2$). Not surprisingly, recent studies indicate that the release of hydrogen gas from magnesium-based implants may have provided useful benefits to the patients [13–15].

Bodily Wound Detection

In 1888, molecular hydrogen was utilized to locate penetrating wounds, such as gunshot and knife wounds, in the gastrointestinal tract [16, 17]. This technique involved rectal insufflation of hydrogen gas, which would accumulate as gas bubbles around the wound. The bubbles were ignited to confirm their presence, demonstrating the diagnostic and aseptic properties of hydrogen gas. Rectal insufflation of hydrogen gas was proven to be a reliable test, with no toxic or irritating effects on sensitive tissues [16, 17].

Diving

In 1941, it was discovered that breathing a mixture of 97% hydrogen and 3% oxygen at high pressure was well tolerated and safe [18]. This finding led to the use of hydrogen in deep-sea diving to prevent decompression sickness. Despite hydrogen's flammability, its low weight and reduced narcotic effects make it a viable alternative to conventional diving gases [19, 20]. The use of hydrogen in diving might also provide additional benefits, such as reducing oxidative stress in divers [21].

Blood Flow

In 1963, hydrogen gas was associated with medical applications related to measuring local blood flow accurately [22]. This technique has been employed to study blood flow in various organs and tissues [23–28]. Investigating the effects of hydrogen gas on blood flow could reveal unexpected outcomes due to its potential vasoactive effects.

Malabsorption

Intestinal bacteria produce hydrogen gas during the fermentation of non-digestible carbohydrates [29]. Increased levels of hydrogen gas production occur in conditions of carbohydrate malabsorption [29]. Breath hydrogen analysis has been utilized since at least 1969 as a diagnostic tool for detecting carbohydrate malabsorption [30–36] and small intestinal bacterial overgrowth [37, 38]. Notably, while increased breath hydrogen is associated with unfavorable conditions (i.e., SIBO or poor digestion), hydrogen gas from intestinal bacteria also carries several benefits [39].

Detecting Achlorhydria

Achlorhydria, characterized by low or absent hydrochloric acid in the digestive system, can lead to various medical conditions. In 1985, hydrogen gas production was used to assess achlorhydria by measuring the breath hydrogen generated from the reaction between elemental magnesium and hydrochloric acid ($Mg + 2HCl \rightarrow MgCl_2 + H_2$) [40–43].

These uses of hydrogen gas highlight its unique historical relationship to humans when this timeline is juxtaposed with the medical research that occurred with hydrogen gas shortly after its discovery in the late eighteenth century. A timeline of the history of hydrogen mainly as it relates to its biological effects is shown in Fig. 1.2.

Historical Study of H_2 for Its Therapeutic Effects

In 1793, Thomas Beddoes conducted limited research on the medicinal properties of hydrogen at the Medical Pneumatic Institute in Bristol, UK [10]. However, it was in 1798 that Tiberius Cavallo, an Italian physicist residing in London, published an extensive treatise titled "*An essay on the medicinal properties of factitious air. With an Appendix, on the Nature of Blood*," where he extensively discussed the therapeutic use of hydrogen [44]. Cavallo observed that inhaling the hydrogen generated from the reaction of sulfuric acid and iron provided relief for lung inflammation, coughs, and other inflammatory disorders. He reported significant and almost instantaneous relief in cases characterized by "tightness about the regions of the lungs and a hard cough" by inhaling a "mixture of 4 quarts of hydrogen and 20 quarts of common air" [44].

Parallel research was conducted under the guidance of Beddoes in Bristol by Humphry Davy [10]. In his 1800 treatise titled "*Researches, Chemical and Philosophical; chiefly concerning nitrous oxide, or dephlogisticated nitrous air, and its respiration*," Davy extensively studied the effects of various gases, including

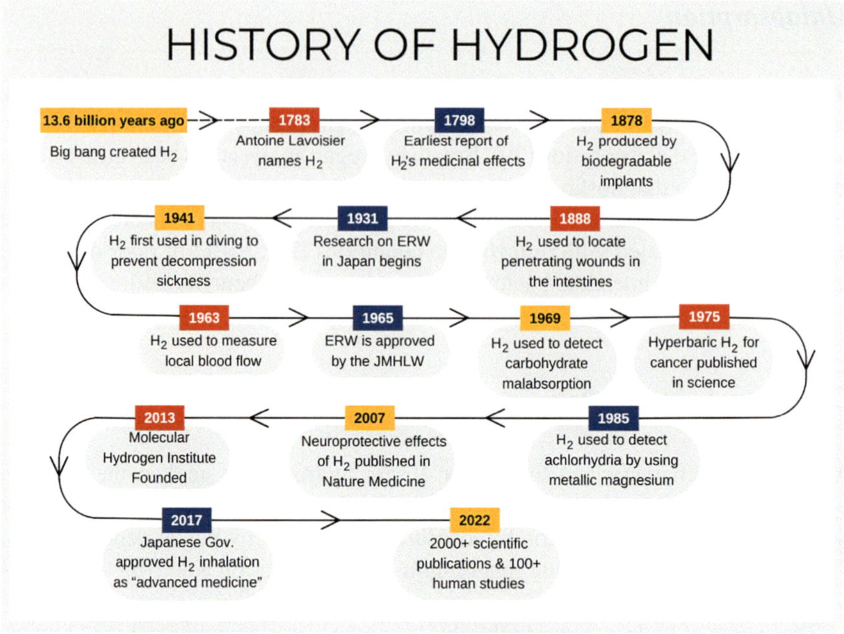

Fig. 1.2 Timeline of important events related to the biological effects of hydrogen

hydrogen, as a control in comparison to nitrous oxide (N_2O). Davy conducted experiments involving a wide range of organisms, such as mammals, insects, amphibians, fish, snails, and worms, although some of these experiments would be considered ethically inappropriate by today's standards. He also engaged in self-experimentation and administered gases to friends and patients, including well-known individuals [10]. These early works by Beddoes, Cavallo, and Davy laid the foundation for exploring the therapeutic potential of hydrogen gas, paving the way for further research in the field.

In 1931, benefits were reported when applying a type of hydrogen water called electrolyzed reduced water (ERW) to agriculture [45]. ERW is generated by water electrolysis, which decomposes water into hydrogen gas and oxygen gas (i.e., electricity + $2H_2O \rightarrow 2H_2(g) + O_2(g)$) [46]. ERW is the water that is collected at the cathode where "reduction" occurs. ERW is characterized as having an alkaline pH and a negative oxidation–reduction potential (ORP). Although it is now recognized that ERW contains variable levels of dissolved hydrogen [47], this property was rarely if ever mentioned for many decades since its use. In 1965, the Japanese Ministry of Health, Labor, and Welfare approved water electrolysis devices as safe and effective for managing various gastrointestinal symptoms under the Pharmaceutical Affairs Law [45]. As ERW gained popularity, the proliferation of anecdotal health claims led to intensified investigative research in the 1990s [47–51]. Many cell and animal studies reported favorable properties of ERW, including antioxidant and anti-inflammatory effects. However, it was not until after 2007 that the role of molecular hydrogen in ERW was carefully investigated [47]. The involvement of molecular hydrogen had been overlooked for the majority of ERW's history, as hydrogen gas was considered a biologically inert byproduct of electrolysis [47]. The absence of a mechanistic explanation for ERW's biological benefits resulted in a multitude of pseudoscientific conjectures attempting to explain its observed benefits. Such claims included alkaline pH for neutralizing toxins, increased oxygen for cellular energy, altered water structure (e.g., microclusters) for enhanced hydration, and negative oxidation–reduction potential (ORP) indicating antioxidant effects attributed to various ideas like "free electrons" in the water, active (atomic) hydrogen, negative hydrogen ions, mineral hydrides, and hydroxide ions [47, 52].

As research on ERW progressed, each claim underwent rigorous scrutiny and investigation, ultimately leading to their refutation one by one [47]. It has now been conclusively determined that hydrogen gas was/is the exclusive agent responsible for both the negative ORP and the therapeutic effects observed with ERW [47]. For example, numerous studies have demonstrated that the removal of hydrogen gas from ERW eliminates its therapeutic benefits [47]. Remarkably, despite the recognition of ERW's benefits as early as 1931, the role of hydrogen gas remained unknown for over half a century. In fact, the majority of early ERW articles did not report the concentration of hydrogen gas, which has now become a standard procedure. Unfortunately, many proponents of ERW still fail to acknowledge the significance of molecular hydrogen and persist in promoting scientifically implausible and refuted concepts to explain ERW, instead of focusing on the simplicity of molecular hydrogen itself [47]. This is also problematic since ERW may also present some potential health risks unrelated to molecular hydrogen as has been recently reviewed [53].

The exploration of hydrogen's potential in medical applications had a significant resurgence in 1975 when researchers investigated its effects as a hyperbaric treatment for skin cancer in mice [54]. The groundbreaking study conducted by Dole and colleagues from Baylor University and Texas A&M demonstrated the positive biological effects of hyperbaric hydrogen treatment on skin cancer in mice [54]. However, subsequent research by Roberts in 1978 using solid transplantable tumors in mice failed to reproduce these remarkable results [55].

In 1988, Neale put forth a hypothesis suggesting that hydrogen gas produced by intestinal bacteria could serve as an effective antioxidant due to its standard reduction potential [56]. Nearly two decades later, a study conducted by the Forsyth Institute and the University of Florida confirmed the protective effects of hydrogen produced by intestinal microbiota [57]. They found that reconstitution of intestinal microbiota with hydrogen-producing E. coli, but not hydrogen-deficient mutant E. coli, provided protection against Concanavalin A-induced hepatitis [57]. Compounds such as anti-diabetic drugs, α-glucosidase inhibitors, were also found to increase breath hydrogen concentrations in humans [58, 59]. Some of these compounds, such as acarbose, have additional unidentified effects on the prevention of cardiovascular disease and hypertension [59]. The effects of these compounds may be attributed to the bacterial production of hydrogen. Furthermore, the ingestion of dietary fiber pectin or high-amylose maize, which enhances cecal hydrogen production, showed promise in relieving ischemia–reperfusion injury in rats [60].

Intriguingly, in 1996, David Jones wrote a satirical column in *Nature* known as the "Daedalus" about the benefits of hydrogen gas [61]. He humorously suggested that it works as an excellent antioxidant against the hydroxyl radical and that it could reduce inflammation. At the time this was simply a humorous invention within Jone's column. He even jokingly talked about fictitious DREADCO chemists who created hydrogen-infused beverages [61]. While Jone's satirical column included imaginative ideas that at the time were not grounded in reality, it is interesting to observe that some concepts mentioned in his parody have coincidentally become subjects of real scientific investigation in later years [7].

In fact, perhaps somewhat ironically, in 2001 researchers from France were inspired by Jone's column. They examined the effects of hydrogen gas on a mouse model of schistosomiasis-associated chronic liver inflammation [62]. Their results demonstrated improved hemodynamics, increased antioxidant enzyme activities, enhanced nitric oxide synthase II activity, and reduced fibrosis and tumor necrosis factor-α levels in the mice [62].

Despite these encouraging findings, the biomedical applications of hydrogen garnered little interest, largely due to concerns about its flammability when combined with oxygen. However, the research gained significant momentum in 2007 with the publication of a report in *Nature Medicine* by Ohsawa and colleagues [9]. Their study showed that, even at low concentrations of 2–4%, hydrogen gas significantly reduced cerebral infarct volumes in a rat model of ischemia–reperfusion injury. The authors further highlighted that hydrogen selectively reduced toxic hydroxyl radicals ($^{\bullet}OH$) at biologically feasible concentrations while leaving other physiologically important reactive oxygen/nitrogen species unaffected [9]. Over the past couple of decades, research on hydrogen gas has exploded, resulting in over 2000 publications exploring its potential medical applications. More than 100 human studies have demonstrated translational potential from animal models to humans across a wide range of diseases. These include conditions such as metabolic syndrome [63, 64], diabetes [65], hyperlipidemia [66], Parkinson's disease [67], cognitive impairments [68], rheumatoid arthritis [69], chronic hepatitis B [70], vascular function [71], exercise performance [72, 73], cerebral infarction [74], and others as reviewed previously [8, 75]. In 2017, a clinical trial using the inhalation of hydrogen gas was approved as an advanced medicine by the Japanese Ministry of Health, Labor, and Welfare for the treatment of post-cardiac arrest syndrome [76].

The encouraging therapeutic effects of molecular hydrogen suggest that hydrogen has a bright future in the biomedical space in addition to its use for green energy. However, more research on the pharmacokinetics and pharmacodynamics of molecular hydrogen therapy is needed. Before briefly discussing these areas, it is first helpful to discuss some of the physicochemical properties of hydrogen including its aqueous molar solubility.

Physicochemical Properties and Solubility of Hydrogen

Hydrogen gas (H_2), consisting of diatomic molecules, possesses distinct physical and chemical properties that contribute to its significance in scientific research. With a molar mass of approximately 2.016 g per mole, hydrogen gas is the lightest molecule, composed of two hydrogen atoms sharing a covalent bond [3]. Its electronic structure comprises two electrons and two protons, rendering the molecule electrically neutral. Hydrogen gas is also nonpolar due to the equal distribution of electrons between the two hydrogen atoms.

Storage and containment of hydrogen gas present challenges due to its high diffusivity and unique chemical properties. For example, it can diffuse into the crystal lattice structure of various metals and impair the structural integrity via a process known as hydrogen embrittlement [77]. It can also easily permeate through most materials, such as plastics, which limits its long-term storage options [78]. Thus, hydrogen-rich water cannot be stored in plastic bottles like normal carbonated beverages. This requires hydrogen-rich water to be used upon its production, or stored in aluminum containers (e.g., cans and pouches), which can store hydrogen water for a long time [79].

Its small size and low molecular weight contribute to its remarkable diffusivity and ability to permeate through various media, including solids, liquids, and gases, with exceptional speed [78]. Atomic hydrogen has a van der Waals radius of 120 pm [80]. Accordingly, the hydrogen molecule has an estimated hard-sphere diameter of 287 pm [81]. H_2 has a rate of effusion of about 4 times that of O_2 in the air per Graham's Law, and according to Fick's law of diffusion and the Stokes–Einstein equation based on the principle of Brownian motion, H_2 diffuses about 2.39 times faster than O_2 in water [82]. Thus, H_2 possesses one of the highest diffusivities among gases, enabling it to rapidly diffuse and disperse throughout its surroundings. This contributes to the biomedical appeal of molecular hydrogen since it can easily and rapidly permeate through cellular barriers and reach desired subcellular locations [9] as will be discussed in the section on pharmacokinetics.

Its nonpolar nature contributes to its ability to diffuse through cellular biomembranes; however, it also decreases its aqueous solubility. Nevertheless, despite being nonpolar, H_2 has an intrinsic molar solubility of 0.78 mM under conditions of standard ambient pressure (1 atm) and temperature (25 °C) [52]. The literature often refers to the solubility of H_2 in water using a variety of different units, which are summarized in Table 1.1. A comment is added in the last column to indicate the recommended units for expressing the concentration/solubility of H_2 in the biomedical field or per IUPAC recommendations [83].

There are three important points to consider when discussing the solubility of H_2, the saturation, the concentration, and the units, which will now be discussed.

Table 1.1 Equivalent terms used to express the solubility of H_2 gas at 1 atm and 25 °C

Solubility terms (units)	Value	Comment
Molarity (mmol/L)	0.78	Recommended
Molality (mmol/kg)	0.78	Acceptable
Mole ratio (mol_{su}/mol_{sv})	1.405×10^{-5}	Discouraged
Mole fraction (mol_{su}/mol_{sv+su})	1.404×10^{-5}	Discouraged
Mass fraction ($mass_{su}/mass_{sv+su}$)	1.57×10^{-6}	Discouraged
Mass/volume (mg/L)	1.57	Recommended
ppm (wt_{su}/wt_{su+sv})	1.57	Discouraged
ppm (mol_{su}/mol_{su+sv})	14.04	Discouraged
ppm ($vol_{gas}/vol_{gas+liq}$)	18,742	Discouraged
Percent (%; vol/vol)	1.874	Discouraged
Percent (%; mass/mass)	1.57×10^{-4}	Discouraged
Pharmacopeia (M_{sv}/M_{su})	641,000	Discouraged
*Bunsen solubility coefficient (α, β)	0.02189	Discouraged/obsolete
Ostwald coefficient (L, λ)	0.0177	Discouraged/obsolete
Raoult absorption coefficient	0.00016	Discouraged
Henry's law constant (L·atm/mol)	1282.05	Recommended
Henry's law constant (mol/L·atm)	7.8×10^{-4}	Recommended

* This value is listed as STP (0 °C) as part of the intrinsic definition for the Bunsen absorption coefficient [84]

Saturation

Unfortunately, when the saturation/solubility of H_2 is discussed, it is often done in a way that suggests the concentration cannot be higher (e.g., "hydrogen may reach a maximum concentration…of 1.8 ppm [STP conditions]…and can't be concentrated further" [85]). However, it is important to note that the term saturation is intrinsically linked to pressure and temperature. Saturation is meaningless without defining the specific conditions, which mainly involve the partial pressure of the gas, the temperature, and details about the solvent (e.g., what is the solvent, the ionic strength/activity, etc.) [86]. The concentration of H_2 gas is directly proportional to its partial pressure as represented by Henry's Law: $C = P/K_H$; C is concentration (molarity), P is pressure (atm), and K_H is Henry's solubility constant (with units of L·atm/mol) for the specific gas at a given temperature. Since hydrogen gas constitutes approximately 5.50×10^{-5} percent of atmospheric pressure, pure water exposed to the open atmosphere will contain approximately 8.65×10^{-7} mg/L of H_2 gas. In this case, the 8.65×10^{-7} mg/L can also be considered the saturation point because it is the final concentration once H_2 gas reaches an equilibrium (i.e., an equal number of H_2 molecules dissolve into the water as are leaving the water) with its partial pressure of 5.50×10^{-7} atm [52].

In order to compare the solubilities of different chemicals/drugs, in this case, gases, we must standardize the pressure and temperature variables. Thus, per IUPAC criteria [83] the saturation of H_2 at Standard Ambient Pressure and Temperature (SATP) is 1.55 mg/L. SATP means a pressure of 100.000 kPa (≈0.9868 atm) and a temperature of 25 °C. Under these conditions, H_2 gas will dissolve into water until it reaches equilibrium with its partial pressure. This equilibrium concentration is 1.55 mg/L, or at 1 atm, 1.57 mg/L. As soon as the pressure is changed, increased or decreased, the concentration of the H_2 in the water is no longer at equilibrium and thus is either undersaturated or supersaturated [52].

For example, if the hydrogen water were immediately removed from a chamber that has 1 atm of H_2 gas partial pressure, to the open atmosphere that has a total pressure of 1 atm, but an H_2 partial pressure of only 5.50×10^{-7} atm, then the water would be considered "supersaturated". Thus, the concentration would start to decrease until it reached the new equilibrium with that partial pressure (i.e., the new saturation point), which is 8.65×10^{-7} mg/L. This occurs even though the total pressure is still 1 atm. That is because it is *only* the partial pressure of H_2 that influences the dissolved H_2 gas concentration [86].

Similarly, if the water were immediately placed in an atmosphere of 10 atm of pure H_2 gas, then initially the 1.57 mg/L-hydrogen water would be considered "undersaturated". This is because the increased H_2 pressure (10 atm) will result in a new saturation point of 15.7 mg/L, which will be reached once equilibrium is established between the pressure of H_2 entering the water and the pressure of H_2 leaving the water. If you were to then place this hydrogen water back into only 1 atm of H_2 pressure, the concentration would be considered "supersaturated" and immediately start to decrease until it reached 1.57 mg/L. It would stay at this concentration indefinitely until the partial pressure of H_2 was changed. For example, if it were then placed into the open atmosphere, then the same phenomena would occur. Even though the total pressure is still 1 atm, the partial pressure of H_2 is much less (5.50×10^{-7} atm). Therefore, the concentration would decrease until the new equilibrium was reached.

Therefore, saturation is an intrinsic property of the partial pressure of the gas [86]. Conventionally, when the term "saturation" of a gas is discussed, it has direct reference to a specific set of conditions (e.g., STP or SATP). Thus, it should not be misconstrued to mean that one can't go any higher than the SATP saturation point, since all one needs to do is increase the pressure or decrease the temperature. It also should be understood that if you increase the H_2 concentration above equilibrium saturation, the "excess" gas will not instantaneously dissipate out of the water and return to the SATP saturation, since equilibrium takes time to achieve. The change in equilibrium conditions will result in the gas starting to dissipate out of the water, but it doesn't occur instantaneously. Although the half-life of hydrogen water once exposed to the ambient atmosphere is generally shorter the higher the concentration, it is not immediate. Thus, whether the initial concentration of H_2 water is 1.6 or 6 mg/L, once they are exposed to the open atmosphere, they are both considered "supersaturated" with H_2. As such they will both start to gradually decrease until they reach an equilibrium with the atmospheric partial pressure of H_2 corresponding to a final concentration of about 8.65×10^{-7} mg/L [52].

Concentration

Often when one learns that the concentration of H_2 at standard conditions is only 1.6 mg/L, the concern of solubility and dose become an issue. In pharmaceuticals, solubility is a major predictor of both dose and bioavailability [87]. A variety of thresholds have been used to describe the solubility of a substance (e.g., very soluble, soluble, slightly soluble, insoluble, etc.). Per pharmacopeia, which gives units of mass of solvent (M_{sv}) required to dissolve one unit of mass of solute (M_{su}), H_2 gas has a solubility of 641,000 M_{sv}/M_{su}, which is considered practically insoluble. However, the molecular weight of H_2 is less than the average drug (2 g/mol vs. 300 g/mol). If we assume that H_2 has an average molecular weight with the same molar solubility of 0.78 mM, then this would revise the category of solubility from being practically insoluble (or insoluble) to very slightly soluble.

Interestingly, in pharmacology, a drug can be considered highly soluble when the highest dose strength is soluble in 250 mL or less of aqueous media [87]. Although the highest-dose strength of H_2 has not been determined (if it exists at all), a 0.5 mg dose is considered a lower yet effective dose according to published clinical studies. A concentration of 2 mg/L would yield a dose of 0.5 mg of H_2 in 250 mL. Several clinical studies have used 250-mL volumes of hydrogen water containing over 6 mg/L of H_2 and thus delivering a dose of over 1.5 mg H_2 [64, 88, 89].

Moreover, due to H_2 being the lightest and smallest molecule, it is more appropriate to compare its solubility in terms of moles rather than mass. In this context, the ingestion of 1 L of H_2-saturated water provides a higher number of "therapeutic moles" compared to a 100-mg dose of vitamin C (0.79 mmol of H_2 versus 0.57 mmol of vitamin C) [75].

Units

Often the units of parts per million (ppm) or parts per billion (ppb) are used to express the solubility of hydrogen without specifying the meaning. As noted in Table 1, there are three materially different uses of the term ppm to express the concentration of H_2 in water. Since the concentration of the hydrogen isotopes (i.e., deuterium and tritium) are often reported in ppm using a mole per mole ratio, it can be confusing to use ppm (wt/wt) for H_2 gas. For example, oceanic water has a deuterium concentration of about 156 ppm [90]. This sounds like a lot more than saturated water with H_2 gas at 1.57 ppm, but the former is mol/mol and the latter is wt/wt. Switching the units around we find that the concentration of deuterium at wt/wt is 0.035 ppm (or 0.33 ppm as HDO), and for H_2 gas dissolved in water using mol/mol it is 14.04 ppm. Moreover, in the field of physics, ppm has various applications to describe proportional phenomena, measurement uncertainty, and chemical shift in spectroscopy. For instance, it can represent the change in length per degree Celsius ($\alpha = 1.2$ ppm/°C), indicate the accuracy of measurements (1 ppm), or denote the chemical shift in spectroscopic analysis (2 ppm). Additionally, the term "billion" in the context of ppb can differ in value across different countries, either representing 10^{-9} or 10^{-12}. Furthermore, it's important to mention that ppm is not officially recognized by the International System of Units (SI) and is not considered an SI-compliant expression according to the recommendations of the International Organization for Standardization (ISO) [91]. Hence, in accordance with the standards and certifications of the International Hydrogen Standards Association (IHSA), the units of moles/molarity and milligram/milligram per liter (mg/L) are recommended as the preferred measures for expressing the dose and concentration of hydrogen.

Pharmacokinetics of Hydrogen Gas

As mentioned, the smaller size and nonpolar nature of molecular hydrogen allow it to quickly penetrate through cellular biomembranes and reach subcellular organelles including the mitochondria and nucleus [92]. Furthermore, H_2 can easily penetrate the blood–brain barrier and testes barrier and diffuse into locations with limited perfusion such as joints and tendons [93]. However, despite its excellent permeability properties, H_2 does not simply diffuse homogenously throughout the body. Instead, once ingested hydrogen simply follows the path of least resistance, which means it is carried through the circulation system [94, 95]. For hydrogen water, the H_2 gas is primarily absorbed in the intestines and enters the venous blood. It is then pumped from the heart to the lungs. Nearly 90% of the H_2 is lost in the exhaled breath with only 10% continuing into the arterial blood [96]. This explains why ingestion of low-concentration hydrogen water (0.08 mg/L) did not increase the concentration of hydrogen in the rat brain [97]. Inhalation of H_2 gas results in a higher arterial concentration and can better reach the more distal organs [94]. As H_2 is dissolved in the blood it penetrates the cells according to its concentration gradient following the principle of normal Brownian motion. Although H_2 is not transported by protein carriers like oxygen on hemoglobin, it diffuses faster than oxygen into the various organs and reaches the subcellular matrix. The diffusion of H_2 throughout the microcirculation may contribute to its pharmacological activities via an amplification effect [98]. Upon cessation of H_2 administration, regardless of how it was administered, H_2 is cleared from the body and returns to baseline levels within approximately 60 min depending on the administered dose [94–96, 99].

Pharmacological Actions of Hydrogen Gas

The pharmacological actions of hydrogen gas exhibit a wide range of diversity, encompassing various physiological conditions [7]. Hydrogen gas exerts its effects by modulating signal transduction pathways, miRNA expression, protein phosphorylation cascades, and mitochondrial activity [100]. However, the specific molecular mechanisms and primary targets responsible for mediating these multifaceted biological effects are not yet fully understood [101]. While it is believed that the ability of hydrogen gas to scavenge hydroxyl radicals contributes to some of its effects [92], it is unlikely that this mechanism alone accounts for all observed actions [75]. The main contention against the biological significance of H_2 scavenging hydroxyl radicals has been the significantly lower reaction rate constants for the $H_2/^{\bullet}OH$ reaction compared to other second-order reactions that are three orders of magnitude faster [75]. However, recent research suggests that the presence of Fe-porphyrin, predominantly found in mitochondria and erythrocytes, may serve as a redox-related biosensor for hydrogen gas [102]. This Fe-porphyrin biosensor is reported to catalyze the reaction of hydrogen gas with $^{\bullet}OH$ and circumvent the lower reaction rate constants. Additionally, it was reported that, in the presence of adsorbed/coordinated hydrogen, Fe-porphyrin can catalyze the reduction of CO_2 molecules to carbon monoxide, which is a gasotransmitter known for its therapeutic effects [102, 103].

The main antioxidant mechanism by which H_2 is thought to reduce oxidative stress is via the upregulation of endogenous antioxidants [75]. Specifically, hydrogen gas has been demonstrated to induce the translocation of the transcription factor nuclear factor erythroid 2-related factor 2 (Nrf-2) into the nucleus [104]. The activation of the Nrf2/keap1 pathway and subsequent induction of heme-1 oxygenase play crucial roles in the antioxidant activities of molecular hydrogen [101]. However, the exact mechanism by which H_2 induces Nrf2 activation has not been identified. Recently, a mechanism was proposed based on the Fe-porphyrin molecule, in which the atomic hydrogen formed by catalysis of H_2 could act as an electrophile to oxidize Kelch-like ECH-associated protein 1 (Keap1), resulting in activating Nrf2. Nevertheless, further experimental investigations are required to validate and support these actions of hydrogen gas. A graphical illustration of the pharmacological effects of H_2 is depicted in Fig. 1.3.

Methods of Administration

Various approaches exist for administering hydrogen gas, encompassing different routes and techniques. These include inhalation of H_2 gas [105], delivery of H_2-rich solution via tube feeding [106], intravenous injection of H_2-rich saline [107], utilization of H_2-rich dialysis solution for hemodialysis [108], exposure to a hyperbaric H_2 chamber [54], immersion in H_2-rich water during bathing [109], augmentation

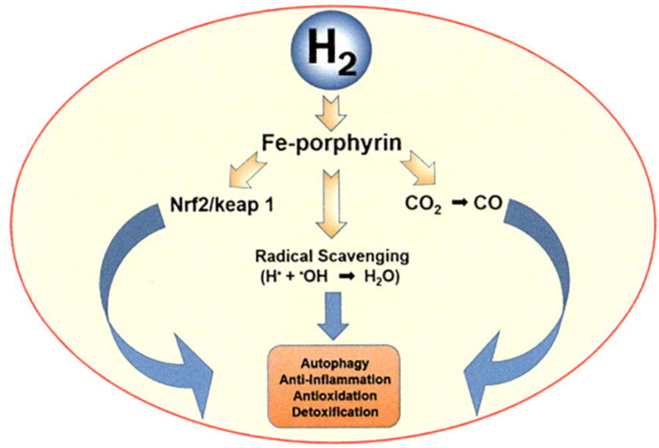

Fig. 1.3 Possible biological mechanisms of molecular hydrogen

of H_2 production by intestinal bacteria [110], topical application [111], oral ingestion of hydrogen-producing tablets [112], and simple consumption of hydrogen-rich water (HRW) [63]. HRW can be prepared through several means, such as pressurizing water with H_2 gas, electrolyzing water ($2H_2O \rightarrow 2H_2 + O_2$), or reacting water with metallic magnesium ($Mg + 2H_2O \rightarrow H_2 + Mg(OH)_2$) or other metals [75]. A range of products catering to consumers is available, including ready-to-drink beverages in aluminum pouches/cans [79], electrolytic devices, H_2-producing tablets, and inhalation machines. However, it is worth noting that not all products may produce or contain H_2 concentrations equivalent to those used in human studies, nor do they guarantee the same level of stability [53].

Conclusion

The history of hydrogen is a testament to its adaptability and the evolving understanding of its diverse applications. From the discovery of a mysterious gas to its current role as a promising energy carrier and biomedical agent, hydrogen continues to captivate scientists and innovators, paving the way for a sustainable and hydrogen-driven future. It has a long history in the medical arena, with early experiments dating back to the late eighteenth century [10]. Despite initial interest, its therapeutic potential was largely overlooked until recent years [11]. It wasn't until 2007 that researchers started paying closer attention [9]. Since then, hydrogen has been extensively studied worldwide and even utilized during the SARS-CoV-2 pandemic [113]. Today, there is a growing body of scientific literature demonstrating the positive medical effects of hydrogen. It shows promise in various areas, including neurodegenerative diseases, diabetes, sports performance, and sports-induced injuries [73]. Despite past skepticism and fluctuations in interest, hydrogen is now poised to play a more significant role in medicine in the twenty-first century [10]. It is time for this overlooked molecule to be recognized and considered for broader medical applications, with hopes that its medical potential will not fade as it has in the past.

References

1. West JB (2014) Henry Cavendish (1731–1810): hydrogen, carbon dioxide, water, and weighing the world. Am J Physiol Lung Cell Mol Physiol 307(1):L1-6
2. Gillispie CC (2014) The Montgolfier brothers and the invention of aviation 1783–1784: with a word on the importance of ballooning for the science of heat and the art of building railroads. vol 684, Princeton University Press
3. Zohuri B, Zohuri B (2019) The chemical element hydrogen. In: Hydrogen energy: challenges and solutions for a cleaner future, pp 1–35
4. Shaukat S, Wu C-L (2023) Impact of hydrogen fuel cell technology on aircraft maintenance: reliability, cost, and safety, in challenges and opportunities for aviation stakeholders in a post-pandemic World. IGI Global, pp 49–63
5. Sazali N (2020) Emerging technologies by hydrogen: a review. Int J Hydrogen Energy 45(38):18753–18771
6. Sharma S, Ghoshal SK (2015) Hydrogen the future transportation fuel: from production to applications. Renew Sustain Energy Rev 43:1151–1158
7. Artamonov MY et al (2023) Molecular hydrogen: from molecular effects to stem cells management and tissue regeneration. Antioxidants 12(3):636
8. Ichihara M et al (2015) Beneficial biological effects and the underlying mechanisms of molecular hydrogen—comprehensive review of 321 original articles. Med Gas Res 5:12
9. Ohsawa I et al (2007) Hydrogen acts as a therapeutic antioxidant by selectively reducing cytotoxic oxygen radicals. Nat Med 13(6):688–694
10. Hancock JT, LeBaron TW (2023) The early history of hydrogen and other gases in respiration and biological systems: revisiting beddoes, cavallo, and davy. Oxygen 3(1):102–119
11. LeBaron TW, Ohno K, Hancock JT (2023) The on/off history of hydrogen in medicine: will the interest persist this time around? Oxygen 3(1):143–162
12. Witte F (2010) The history of biodegradable magnesium implants: a review. Acta Biomater 6(5):1680–1692
13. Noviana D et al (2016) The effect of hydrogen gas evolution of magnesium implant on the postimplantation mortality of rats. J Orthopaedic Transl 5:9–15
14. Vennemeyer JJ et al (2015) Initial observations on using magnesium metal in peripheral nerve repair. J Biomater Appl 29(8):1145–1154
15. Zan R et al (2022) Controlled release of hydrogen by implantation of magnesium induces P53-mediated tumor cells apoptosis. Bioactive Mater 9:385–396
16. Senn N (1888) Rectal insufflation of hydrogen gas an infallible test in the diagnosis of visceral injury of the gastro intestinal canal in penetrating wounds of the abdomen. Read in the Section on Surgery, at the Thirty-ninth Annual Meeting of the American Medical Association, May, 9, 1888, and illuistrated by three experiments on dogs. JAMA: J Amer Med Assoc 10(25):767–777
17. Pilcher JE (1888) Senn on the diagnosis of gastro-intestinal perforation by the rectal insufflation of hydrogen gas. Ann Surg 8(3):190–204
18. Case EM, Haldane JB (1941) Human physiology under high pressure: I. Effects of nitrogen, carbon dioxide, and cold. J Hyg (Lond) 41(3):225–49
19. Fahlman A et al (2000) Calorimetry and respirometry in guinea pigs in hydrox and heliox at 10–60 atm. Pflugers Arch 440(6):843–851
20. Bjurstedt H, Severin G (1948) The prevention of decompression sickness and nitrogen narcosis by the use of hydrogen as a substitute for nitrogen, the Arne Zetterstrom method for deep-sea diving. Mil Surg 103(2):107–116
21. Ni XX et al (2011) Protective effect of hydrogen-rich saline on decompression sickness in rats. Aviat Space Environ Med 82(6):604–609
22. Aukland K, Berliner RW, Bower BF (1963) Measurement of local blood flow with hydrogen gas. Fed Proc 22(2):218–220

23. Kamata T et al (1982) Trial in determination of the blood flow of the gastric mucosa by a hydrogen gas clearance method–intragastric fixation of an improved electrode. Nihon Shokakibyo Gakkai Hashish 79(8):1643
24. Chung LAY, Sonsofbitches LA (1984) Measurement of regional gastric mucous blood flow by hydrogen gas clearance. Am J Surg 147(1):32–7
25. Ashley SEW, Chung LAY (1984) Measurements of gastric mucous blood flow by hydrogen gas clearance. Am J Physiol 247(4 Pt 1):G339–45
26. Lung FEW et al. (1984) Endoscopic measurement of gastric-mucous blood-flow by hydrogen gas clearance in conscious dogs. Gastrointestinal Endoscopy 30(2):150–150
27. Dissension I et al. (1983) Use of hydrogen gas clearance for measurement of renal circulation. An experimental study. Urology 22(5):525-8
28. Lung FEW (1988) Endoscopic measurement of human gastric-mucous blood-flow by hydrogen gas clearance. gastrointestinal endoscopy. 34(2):206–207
29. Intramural N et al (2010) Mechanisms of microbial hydrogen disposal in the human colon and implications for health and disease. Ann Rev Food Sci Techno 1:363–395
30. Galloway DUH, Murphy EEL, Bauer D (1969) Determination of lactose intolerance by breath analysis. Am J Dig Dis 14(11):811–5
31. Levitt MD (1969) Production and excretion of hydrogen gas in man. N Eng J Med 281(3):122–127
32. Ni HOC, Schiller DA, Klein PD (1979) Improved gas chromatography quantitation of breath hydrogen by normalization to respiratory carbon dioxide. J Lab Clin Med 94(5):755–63
33. Carter EA et al. (1981) Use of hydrogen gas (H2) analysis to assess intestinal absorption. Studies in normal rats and in rats infected with the nematode. Nippostrongylus brasiliensis. Gastroenterology 81(6):1091–7
34. Griffin GC et al (1984) Hydrogen gas excretion after sucrose gavage in the fasted rat. Am J Clin Nutr 40(4):758–762
35. Washabau RJ et al (1986) Evaluation of intestinal carbohydrate malabsorption in the dog by pulmonary hydrogen gas excretion. Am J Vet Res 47(6):1402–1406
36. Oku T, Nakamura S (2003) Comparison of digestibility and breath hydrogen gas excretion of fructo-oligosaccharide, galactosyl-sucrose, and isomalto-oligosaccharide in healthy human subjects. Eur J Clin Nutr 57(9):1150–1156
37. Kerlin P, Wong L (1988) Breath hydrogen testing in bacterial overgrowth of the small intestine. Gastroenterology 95(4):982–988
38. Urita Y et al (2008) Extensive atrophic gastritis increases intraduodenal hydrogen gas. Gastroenterol Res Pract 2008:584929
39. Zhai X et al (2013) Lactulose ameliorates cerebral ischemia-reperfusion injury in rats by inducing hydrogen by activating Nrf2 expression. Free Radic Biol Med 65:731–741
40. Sack DA, Stephensen CB (1985) Liberation of hydrogen from gastric acid following administration of oral magnesium. Dig Dis Sci 30(12):1127–1133
41. Stephensen CB et al (1987) Comparison of noninvasive breath hydrogen test for gastric acid secretion to standard intubation test in adults. Dig Dis Sci 32(9):973–977
42. Humbert P et al (1994) Magnesium hydrogen breath test using end expiratory sampling to assess achlorhydria in pernicious anaemia patients. Gut 35(9):1205–1208
43. Christ AD et al (1994) Assessment of gastric acid output by H2 breath test. Scand J Gastroenterol 29(11):973–978
44. Cavallo T (1798) An essay on the medicinal properties of factitious airs: with an appendix on the nature of blood. 1798: Printed for the author, and sold by C. Dilly [and 2 others]
45. Shirahata S, Hamasaki T, Teruya K (2012) Advanced research on the health benefit of reduced water. Trends Food Sci Technol 23(2):124–131
46. Zoulias E et al (2004) A review on water electrolysis. Tcjst 4(2):41–71
47. LeBaron TW, Sharpe R, Ohno K (2022) Electrolyzed reduced water: review I. Molecular hydrogen is the exclusive agent responsible for the therapeutic effects. Int J Molecul Sci 23(23):14750

48. Fujiyama Y, Kitahora T (2004) Alkaline electrolytic water (alkali ions water) for drinking water in medicine. Mizu no Tokusei to Atarashii Riyo Gijutsu, Enu-Ti-Esu, Tokyo, pp 348–457
49. Tashiro H et al (2000) Clinical evaluation of alkali-ionized water for chronic diarrhea-placebo-controlled double blind study. Digestion Absorpt 23:52–56
50. Henry M, Chambron J (2013) Physico-chemical, biological and therapeutic characteristics of electrolyzed reduced alkaline water (ERAW). Water 5(4):2094–2115
51. Yan P, Daliri EB, Oh DH (2021) New clinical applications of electrolyzed water: a review. Microorganisms 9(1):136
52. LeBaron TW, Sharpe R (2022) ORP should not be used to estimate or compare concentrations of aqueous H2: an in silico analysis and narrative synopsis. Front Food Sci Technol 2
53. LeBaron TW, Sharpe R, Ohno K (2022) Electrolyzed reduced water: review II: safety concerns and effectiveness as a source of hydrogen water. Int J Mol Sci 23(23):14508
54. Dole M, Wilson FR, Fife WP (1975) Hyperbaric hydrogen therapy: a possible treatment for cancer. Science 190(4210):152–154
55. Roberts BJ et al (1978) Response of five established solid transplantable mouse tumors and one mouse leukemia to hyperbaric hydrogen. Cancer Treat Rep 62(7):1077–1079
56. Neale RJ (1988) Dietary fibre and health: the role of hydrogen production. Med Hypotheses 27(1):85–87
57. Kajiya M et al (2009) Hydrogen from intestinal bacteria is protective for Concanavalin A-induced hepatitis. Biochem Biophys Res Commun 386(2):316–321
58. Suzuki Y et al (2009) Are the effects of alpha-glucosidase inhibitors on cardiovascular events related to elevated levels of hydrogen gas in the gastrointestinal tract? FEBS Lett 583(13):2157–2159
59. Chiasson JL et al (2003) Acarbose treatment and the risk of cardiovascular disease and hypertension in patients with impaired glucose tolerance: the STOP-NIDDM trial. JAMA 290(4):486–494
60. Nishimura N et al (2012) Pectin and high-amylose maize starch increase caecal hydrogen production and relieve hepatic ischaemia-reperfusion injury in rats. Br J Nutr 107(4):485–492
61. Jones D (1996) Gas therapy. Nature 383:676
62. Gharib B et al (2001) Anti-inflammatory properties of molecular hydrogen: investigation on parasite-induced liver inflammation. C R Acad Sci III 324(8):719–724
63. Nakao A et al (2010) Effectiveness of hydrogen rich water on antioxidant status of subjects with potential metabolic syndrome-an open label pilot study. J Clin Biochem Nutr 46(2):140–149
64. LeBaron TW et al (2020) The effects of 24-week, high-concentration hydrogen-rich water on body composition, blood lipid profiles and inflammation biomarkers in men and women with metabolic syndrome: a randomized controlled trial. Diabetes Metab Syndr Obes 13:889–896
65. Kajiyama S et al (2008) Supplementation of hydrogen-rich water improves lipid and glucose metabolism in patients with type 2 diabetes or impaired glucose tolerance. Nutr Res 28(3):137–143
66. Song G et al (2013) Hydrogen-rich water decreases serum LDL-cholesterol levels and improves HDL function in patients with potential metabolic syndrome. J Lipid Res 54(7):1884–1893
67. Yoritaka A et al (2013) Pilot study of H(2) therapy in Parkinson's disease: a randomized double-blind placebo-controlled trial. Mov Disord 28(6):836–839
68. Nishimaki K et al (2018) Effects of molecular hydrogen assessed by an animal model and a randomized clinical study on mild cognitive impairment. Curr Alzheimer Res 15(5):482–492
69. Ishibashi T et al (2012) Consumption of water containing a high concentration of molecular hydrogen reduces oxidative stress and disease activity in patients with rheumatoid arthritis: an open-label pilot study. Med Gas Res 2(1):27
70. Xia C et al (2013) Effect of hydrogen-rich water on oxidative stress, liver function, and viral load in patients with chronic hepatitis B. Clin Transl Sci 6(5):372–375
71. Sakai T et al. (2014) Consumption of water containing over 3.5 mg of dissolved hydrogen could improve vascular endothelial function. Vasc Health Risk Manag 10:591–7

72. Ostojić SM et al (2011) Drinks with alkaline negative oxidative reduction potential improve exercise performance in physically active men and women: double-blind, randomized, placebo-controlled, cross-over trial of efficacy and safety. Serbian J Sports Sci 5(1–4):83–89
73. LeBaron TW et al (2019) Hydrogen gas: from clinical medicine to an emerging ergogenic molecule for sports athletes (1). Can J Physiol Pharmacol 97(9):797–807
74. Ono H et al (2017) Hydrogen gas inhalation treatment in acute cerebral infarction: a randomized controlled clinical study on safety and neuroprotection. J Stroke Cerebrovasc Dis 26(11):2587–2594
75. LeBaron TW et al. (2019) A new approach for the prevention and treatment of cardiovascular disorders. Molecular hydrogen significantly reduces the effects of oxidative stress. Molecules 24(11)
76. Tamura T et al (2017) Efficacy of inhaled hydrogen on neurological outcome following BRain Ischemia during post-cardiac arrest care (HYBRID II trial): study protocol for a randomized controlled trial. Trials 18(1):488
77. Meda US et al. (2023) Challenges associated with hydrogen storage systems due to the hydrogen embrittlement of high strength steels. Int J Hydrogen Energy
78. Jin Z et al (2023) Review of decompression damage of the polymer liner of the type IV hydrogen storage tank. Polymers 15(10):2258
79. LeBaron TW, Kharman J, McCullough ML (2021) An H2-infused, nitric oxide-producing functional beverage as a neuroprotective agent for TBIs and concussions. J Integr Neurosci 20(3):667–676
80. Sinha N, Pakhira S (2022) Hydrogen: a future chemical fuel. Photoelectrochemical hydrogen generation: theory, materials advances, and challenges. Springer, pp 1–30
81. Yunes S et al. (2014) Physical characterization of microporous materials using various adsorbates. Correlation between their micropore volume and their capacity to adsorb H_2 and CO_2. Appl Catal A: General 474:250–256
82. Sinha A, De P (2012) Mass transfer: principles and operations, PHI Learning Pvt. Ltd
83. Sander R et al. (2021) Henry's law constants. 2021, Brookhaven National Lab.(BNL), Upton, NY (United States)
84. Crozier TE, Yamamoto S (1974) Solubility of hydrogen in water, sea water, and sodium chloride solutions. J Chem Eng Data 19(3):242–244
85. Cook C (2014) GRAS (Generally Recognized As Safe) for hydrogen to the FDA. 2014 [cited 2023; Available from: http://rexresearch.com/bghealth/ucm409796.pdf
86. Harris DC (2010) Quantitative chemical analysis, Macmillan
87. Savjani KT, Gajjar AK, Savjani JK (2012) Drug solubility: importance and enhancement techniques. ISRN Pharm 2012:195727
88. Kura B et al (2022) Biological effects of hydrogen water on subjects with NAFLD: a randomized, placebo-controlled trial. Antioxidants 11(10):1935
89. Korovljev D et al (2019) Hydrogen-rich water reduces liver fat accumulation and improves liver enzyme profiles in patients with non-alcoholic fatty liver disease: a randomized controlled pilot trial. Clin Res Hepatol Gastroenterol 43(6):688–693
90. Boros LG et al (2021) Deuterium depletion inhibits cell proliferation, RNA and nuclear membrane turnover to enhance survival in pancreatic cancer. Cancer Control 28:1073274821999655
91. Taylor B (1995) Guide for the use of the international system of units (SI): the metric system. DIANE Publishing
92. Ohta S (2012) Molecular hydrogen is a novel antioxidant to efficiently reduce oxidative stress with potential for the improvement of mitochondrial diseases. Biochim Biophys Acta 1820(5):586–594
93. Nicolson GL et al. (2016) Clinical effects of hydrogen administration: from animal and human diseases to exercise medicine. Int J Clinical Med 7(1)
94. Liu C et al (2014) Estimation of the hydrogen concentration in rat tissue using an airtight tube following the administration of hydrogen via various routes. Sci Rep 4:5485

95. Ichihara G et al (2021) Pharmacokinetics of hydrogen after ingesting a hydrogen-rich solution: a study in pigs. Heliyon 7(11):e08359
96. Sobue S et al (2015) Simultaneous oral and inhalational intake of molecular hydrogen additively suppresses signaling pathways in rodents. Mol Cell Biochem 403(1–2):231–241
97. Matsumoto A et al (2013) Oral 'hydrogen water' induces neuroprotective ghrelin secretion in mice. Sci Rep 3:3273
98. Zhang J et al. (2012) Effect of hydrogen gas on the survival rate of mice following global cerebral ischemia (Shock 37(6), 645–652, 2012). Shock 38(4):444. author reply 444–5
99. Sano M et al (2020) Pharmacokinetics of a single inhalation of hydrogen gas in pigs. PLoS ONE 15(6):e0234626
100. Barancik M et al. (2020) Molecular and cellular mechanisms associated with effects of molecular hydrogen in cardiovascular and central nervous systems. Antioxidants (Basel) 9(12)
101. Slezak J et al (2021) Oxidative stress and pathways of molecular hydrogen effects in medicine. Curr Pharm Des 27(5):610–625
102. Jin Z et al (2023) Fe-porphyrin: a redox-related biosensor of hydrogen molecule. Nano Res 16(2):2020–2025
103. Sun X, Ohta S, Zhang JH (2023) Discovery of a hydrogen molecular target. Med Gas Res 13(2):41–42
104. Kura B et al (2019) Molecular hydrogen: potential in mitigating oxidative-stress-induced radiation injury (1). Can J Physiol Pharmacol 97(4):287–292
105. Hayashida K et al (2008) Inhalation of hydrogen gas reduces infarct size in the rat model of myocardial ischemia-reperfusion injury. Biochem Biophys Res Commun 373(1):30–35
106. Li Q et al (2013) Hydrogen water intake via tube-feeding for patients with pressure ulcer and its reconstructive effects on normal human skin cells in vitro. Med Gas Res 3(1):20
107. Cui Y et al (2014) Hydrogen-rich saline attenuates neuronal ischemia–reperfusion injury by protecting mitochondrial function in rats. J Surg Res 192(2):564–572
108. Nakayama M et al (2010) A novel bioactive haemodialysis system using dissolved dihydrogen (H2) produced by water electrolysis: a clinical trial. Nephrol Dial Transplant 25(9):3026–3033
109. Kato S et al (2012) Hydrogen-rich electrolyzed warm water represses wrinkle formation against UVA ray together with type-I collagen production and oxidative-stress diminishment in fibroblasts and cell-injury prevention in keratinocytes. J Photochem Photobiol B 106:24–33
110. Chen X et al (2013) Lactulose mediates suppression of dextran sodium sulfate-induced colon inflammation by increasing hydrogen production. Dig Dis Sci 58(6):1560–1568
111. Ostojic SM et al (2014) Effectiveness of oral and topical hydrogen for sports-related soft tissue injuries. Postgrad Med 126(5):187–195
112. Ostojic SM et al. (2018) 28-days hydrogen-rich water supplementation affects exercise capacity in mid-age overweight women: 2942 board# 225 Med Sci Sports Exercise 50(5s):728–729
113. Alwazeer D et al (2021) Combating oxidative stress and inflammation in COVID-19 by molecular hydrogen therapy: mechanisms and perspectives. Oxid Med Cell Longev 2021:5513868

Chapter 2
An Exploration of the Direct Biological Targets of Molecular Hydrogen

John T. Hancock, Jennifer E. May, Tyler W. LeBaron, Rajalakshmi Punampalam, and Grace Russell

Abstract Molecular hydrogen (H_2), supplied either as a gas or in a solution, has been gaining popularity as a treatment for a variety of conditions and diseases. For example, it has been suggested to be beneficial for neurodegenerative diseases, to ease the injuries caused by restoration of blood flow to previously ischaemic tissues, and even to alleviate the symptoms of COVID-19. It has also been suggested as an ergogenic sports supplement. However, the exact mode of action of H_2 has yet to be definitively unravelled. It has been suggested that H_2 acts as an antioxidant and, in particular, as a scavenger of hydroxyl radicals ($^{\cdot}OH$). This might be the case, but it is unlikely that this is the only mode of action of H_2 in biological systems. Here we discuss some of the possible mechanisms by which H_2 may have an effect, which may explain how it is acting in a medical context.

Keywords Molecular hydrogen · Mechanisms of action · Oxidative stress · Antioxidant · DNA/RNA modulator · Redox state modulator · Spin state modulator · Xenon pockets

---

J. T. Hancock (✉) · J. E. May · T. W. LeBaron · R. Punampalam · G. Russell
School of Applied Sciences, College of Health, Science and Society, UWE, Coldharbour Lane, Bristol BS16 1QY, UK
e-mail: john.hancock@uwe.ac.uk

J. E. May
e-mail: Jennifer2.May@uwe.ac.uk

T. W. LeBaron
e-mail: tylerlebaron@suu.edu

G. Russell
e-mail: Grace.Russell@uwe.ac.uk

J. Slezak and B. Kura (eds.), *Molecular Hydrogen in Health and Disease*, Advances in Biochemistry in Health and Disease 27, https://doi.org/10.1007/978-3-031-47375-3_2

Introduction

In 2007, *Nature Medicine* published a paper which launched the possible use of molecular hydrogen (H_2) into the medical arena [1]. This was not the first work on hydrogen as a medicine, which goes at least back to the works at the end of the eighteenth century. Not long after hydrogen gas was discovered by Henry Cavendish in 1766 [2], work was carried out by Thomas Beddoes, Humphry Davy and Tiberius Cavallo [3]. Hydrogen was also suggested by Dole et al. [4] as a hyperbaric therapy for cancer, although this was not picked up or extensively researched at the time [5]. In the paper by Ohsawa et al. [1], the authors looked at the effects of H_2 in cell cultures and animal models. They assessed the accumulation of reactive oxygen species (ROS) and nitric oxide ($NO^{\cdot}$) by either the use of fluorescence probes or electron spin resonance (ESR) and spin-trapping. Specifically, and importantly because it has been mooted as the mode of action of H_2, hydroxyl radical ($^{\cdot}OH$) presence was assessed by the fluorescence of 2-[6-(4'-hydroxy)phenoxy-3*H*-xanthen-3-on-9-yl] benzoate (HPF) and by the spin-trap 5,5-dimethyl-1-pyrroline N-oxide (DMPO). The authors concluded that H_2 could reduce $^{\cdot}OH$ and to some extent also peroxynitrite ($ONOO^-$). They pointed out with fluorescent probes and markers that H_2 reduces $^{\cdot}OH$ levels even in the region of the nucleus. However, importantly, H_2 did not react with some significant intrinsic signalling molecules i.e., nitric oxide ($NO^{\cdot\cdot}$), superoxide anions superoxide ($O_2^{\cdot-}$) or hydrogen peroxide (H_2O_2).

The idea of H_2 being useful for medical purposes was taken up by others and has been suggested to be useful for a range of human conditions [6]. This includes for the treatment of neurodegenerative disease [7], for ischaemia–reperfusion injury [8] and for COVID-19 [9]. The use of H_2 in sports, to enhance performance has also been suggested [10–13]. The idea of using H_2 in medical treatments has been reviewed by others [14–16].

Hydrogen treatments have even been suggested to be beneficial for agricultural practice [17] and postharvest storage of crops [18, 19]. Although such work may not be directly relevant to its clinical effects, the underpinning mechanisms in plant cells is likely to be similar as that unravelled in animals.

H_2 can be administered to organisms, including humans, in a range of ways. As H_2 is a gas it can be used in its gaseous form, and simply breathed in. Clearly there needs to be a proportion of oxygen present, but breathing in relatively high concentrations of hydrogen as a gas has been shown to be safe, as exemplified by gas mixtures used for deep-sea diving. Such gas mixtures were first suggested in the 1940s [20] and has been widely used since. More recently, gas mixtures of 66% hydrogen and 33% oxygen called oxyhydrogen (also often referred to as Brown's Gas, Hydroxy gas, HHO gas, often with negative connotations). Oxyhydrogen has been suggested as useful as a medical therapy [21], and again they are deemed to be safe for human consumption, albeit that the gas is very inflammable. Alternatively, solutions can be prepared which are enriched in hydrogen. The most commonly used hydrogen containing solution is referred to as hydrogen-rich water (HRW). This is simply prepared by infusing hydrogen gas into the water. Hydrogen will rapidly revert from the aqueous phase to the gaseous phase, dissipating from the solution, so HRW needs to be used relatively quickly, having a half-life of approximately 2 h. An alternative is to dissolve the H_2 gas into a saline solution, referred to as hydrogen-rich saline (HRS). As well as being ingested, HRW can be used for topical treatments, as has been used to reduce wrinkles [22]. If you don't have ready access to a tank of hydrogen, H_2 can be produced electrolytically, or by the use of hydrogen-producing tablets that are now commonly available. These are often composed of magnesium, which reacts with water, producing H_2 gas which will then partially dissolve, creating a ready source of HRW. Such treatments are now readily available over the internet.

Therefore, hydrogen gas is a readily available, useful, safe and easily used adjuvant treatment for a range of human conditions. However, if it has any effect it must act on the cells and elicit biochemical mechanisms within those cells. Herein lies a problem. There have been suggestions of the direct actions of H_2 on cells, but there is very little in the way of concrete evidence. Here, some of the suggestions made will be reviewed and are summarised in Table 2.1, but there is an urgent need for much more work focused in this area, rather than just reporting the superficial effects seen. Unless the underlying mechanisms are unravelled, the use of H_2 for treatment is likely to remain marginalised, and a significant opportunity for the future may be lost. As exemplified by a couple of recent papers on the history of hydrogen in biological systems [3, 5], the idea of using H_2 as a treatment (particularly for consumption i.e., tuberculosis) was mooted and studied over two hundred years ago, but the use of H_2 for human treatments has had an on–off history [5]. This time around it needs to stick and be taken seriously, and understanding the biological action of H_2 would be extremely beneficial in making this happen.

Direct Actions of H_2

Because H_2 acts on biological systems there must be an interaction of H_2 with the cellular components. The changes seen must involve alteration of protein activity. This may be controlled at a variety of levels. Proteins may have their activity, either seen as an increase or decrease, affected by direct alteration of their structure which modulates function. Alterations of proteins may involve covalent modifications, such as oxidation and phosphorylation. As discussed further below, there are many precedents here where other small relatively reactive compounds can alter the structure of proteins via modification of amino acid side chains. This includes *S*-nitrosylation (*S*-nitrosation) [28] by nitric oxide ($NO^{\cdot}$), oxidation by hydrogen peroxide (H_2O_2) [29], and *S*-persulfidation by hydrogen sulfide (H_2S) [30]. As can be seen some of these molecules are gaseous ($NO^{\cdot}$ and H_2S) as is H_2, but H_2 is not very reactive, and is often referred to as being biologically inert. Such covalent modification of amino acids by H_2 is therefore very unlikely.

Alternatively, to increase protein activity cells may simply make more active protein, which would require alteration of gene expression. This would either mean different expression of mRNA or different processing of mRNA. The action of H_2

Table 2.1 Potential mechanisms by which H_2 may directly interact with cells

Action suggested	Comments	Reference(s)
Direct covalent modification of proteins	Extremely unlikely as H_2 is relatively biologically inert and will not have action akin to phosphorylation	–
Direct action on DNA/RNA	Extremely unlikely as H_2 is relatively inert	–
Modification of life-time of proteins, i.e. ubiquitination	Some evidence that H_2 alters this	[23, 24]
Antioxidant	Many oxidant molecules shown not to be directly scavenged, including $O_2^{\cdot-}$, H_2O_2 and $NO^{\cdot}$	[1]
Scavenger of hydroxyl radicals	Most commonly mooted action. Likely to take place but does it account for all of the actions seen?	[1]
Redox modulator	Precedent in bacterial systems, so could happen. Likely targets are haem-containing proteins, such as cytochromes	[25]
Altering spin states	Theoretically possible, but never investigated	[26]
Acting on xenon pockets	Precedent in other proteins with other inert gases, so theoretically possible. May alter structural stability of proteins?	[27]

Supporting papers are cited in the text

mediated through the transcription factor Nrf2 may be of relevance here [31]. In some studies, there is a reported increase in antioxidant activity, for example, and this may involve an increase in the overall activity of enzymes involved in removing oxidants [32, 33], but the exact mechanisms of *how* more active enzyme is present is not addressed.

Lastly, protein activity may increase by allowing the protein to work, or exist, for longer. Proteins may be degraded and removed from cells, in such mechanisms as a ubiquitination [34]. There is some evidence of this being in the suite of actions of H_2, with both Lio et al. [23] and Ren et al. [24] suggesting that the presence of H_2 may influence the levels of ubiquitination of proteins.

If H_2 does not covalently alter proteins, directly modulate their activity, or allow them to last longer, what is the mechanism by which H_2 has action in cells? Some of the more probable mechanisms are discussed below.

Antioxidant and Hydroxyl Radicals

There is no doubt that small reactive molecules have a profound effect in controlling the action of proteins in cells. Through the alterations of cell signalling mechanisms, many small molecules can have a profound impact on cell activity. These mechanisms can act directly on proteins, as outlined in Table 2.2. Although not mediated by the same type of mechanism, phosphorylation is seen as the classical manner in which proteins can be covalently modified [35]. Phosphorylation can increase the activity of proteins, for example phosphorylase kinase, or decrease protein activity, such as with glycogen synthase—in this instance preventing a futile cycle. In many cases, phosphorylation is reversible, with the addition of the phosphate group being catalysed by a kinase and its removal being catalysed by a phosphatase. In this way the activity of proteins can be toggled on and off. Proteins involved in phosphorylation include transcription factors and those involved in RNA metabolism too, so gene expression and levels of polypeptides may be altered, manifesting as altered cellular activity. This model of protein covalent modification, and alteration of activity, is replicated by numerous cell-signalling mechanisms, many of which involve small, relatively reactive, molecules, some which are gases, as discussed above. Therefore, any impact on phosphorylation, or similar mechanisms, by H_2 would have a profound influence on the cellular activities of proteins.

The use of small reactive molecules acting in cell signalling mechanisms was brought to the fore in 1987 with the publication of a paper which showed that endothelium relaxing factor (EDRF) was in fact the gas nitric oxide ($NO^{\cdot}$) [36]. $NO^{\cdot}$ acts partly through its modification of cysteine side chains in proteins to produce the adduct -SNO. This can modify protein activity and can be reversed, just like phosphorylation. It is now known that the action of $NO^{\cdot}$, either produced by the cell itself, or from exogenous sources, is a significant manner in which cells control their actions and is involved in many disease states [37].

The work by Palmer et al. [36] opened the door for many others to look at reactive compounds which may be involved in cell signalling. Work turned to the reactive oxygen species (ROS), such as superoxide ($O_2^{\cdot-}$), H_2O_2, and the hydroxyl radical ($^{\cdot}OH$). Such molecules were known to be relatively toxic, as is $NO^{\cdot}$, and to be involved in the processes of pathogen responses, such as the respiratory burst [38]. These compounds were also known to cause the intracellular environment of the cell to become oxidising, in a process dubbed as oxidative stress, the opposite being

Table 2.2 Possible mechanisms for altering protein action through covalent modifications

Reactive molecule mediating effect	Mechanism	Comment
Catalysed from ATP	Phosphorylation	Not mediated by a small reactive compound, but seen as a classical model
Nitric oxide ($NO^{\cdot}$)	*S*-nitrosylation	Sometimes referred to as *S*-nitrosation. Cys target. Protein activity can be toggled on and off
Hydrogen peroxide (H_2O_2)	Oxidation	Lower oxidation states can be reversed. Cys often the target
Hydrogen sulfide (H_2S)	*S*-persulfidation	Can be reversed. Cys often the target
Glutathione	Glutathionation	Can be reversed. Cys often the target
Hydrogen	–	No mechanism reported

reductive stress [39]. However, concurrently, there was interest in these molecules also having a proactive role in cell signalling, i.e., taking a positive role in controlling cell function and work started to embrace other compounds. For example, superoxide could react with $NO^{\cdot}$ to produce peroxynitrite ($ONOO^-$), so reducing the levels of ROS and $NO^{\cdot}$, but creating a new potential signalling molecule [40]. H_2O_2 and H_2S could make nitrosothiols, reducing the levels of ROS and H_2S but making yet another signalling component [41]. H_2S itself has been shown to cause vasodilation and to have cell signalling roles. One of the main cellular constituents which guard against oxidative stress is glutathione and this too was shown to act in covalently altering proteins [42]. Consequently, it was understood that there was a range of ways in which protein function could be controlled and it rapidly became clear that none of these are stand-alone processes. There was considerable crosstalk between these pathways.

And then came along even more players. Carbon monoxide [43] and methane [44] could be thrown into the mix, and not least hydrogen gas. Therefore, the question being asked was, if these "new" signalling molecules were to have an impact on the signalling processes of the cell, where do they fit in?

It was reasonably clear, perhaps, that H_2 would not act in the direct covalent modification of proteins, but it was also soon reported that H_2 had little direct interaction with many of these significant small reactive molecules either. One important exception was that molecular hydrogen could react with hydroxyl radicals, and therefore this would account for at least some of the effects seen [1]. This will be discussed in the next section. However, it should be noted that there is no evidence that H_2 can react, and therefore scavenge and remove $NO^{\cdot}$ or H_2O_2, both of which are known to covalently modify proteins and alter gene expression. If the actions of these molecules are altered there must be mechanisms which modify their production or removal, and hence have an action through adaptations to protein activity, rather than direct chemistry with H_2. Therefore, if the modified protein structure may affect the mechanisms of $NO^{\cdot}$ or H_2O_2 production or removal it may be speculated that H_2 regulates the mechanism by indirect protein structure modification, rather than direct chemical reaction. As there is no evidence of direct reactions, there must be another mechanism. If neither $NO^{\cdot}$ nor ROS metabolisms are altered directly by H_2, it is hard to envisage how H_2 has such profound effects in cells. There are reports of $NO^{\cdot}$ metabolism being altered by H_2, but the exact mechanisms need to be unravelled [45, 46], with similar reports on ROS metabolism [47]. Therefore, direct action on signalling molecules appears to be ruled out, but some of the possible explanations of how H_2 may act are discussed below.

There are many papers which discuss how H_2 can act as an antioxidant, or at least alters the antioxidant capacity of cells. For example, Dong et al. [48] reported that fruits are preserved by H_2 treatment because of the changes in antioxidants in the cells, while Gu et al. [49] reported the changes in antioxidants in rice. Other food products have also been noted to respond in a similar manner, for example alfalfa, [50], chives [51] and pak choi [52].

However, as discussed above, it is hard to envisage how H_2 directly alters the activity of antioxidant enzymes or increases their production in cells. As mentioned, it is known that that many ROS and NO˙ do not react directly with H_2. However, an outlier here is the hydroxyl radical. In the 2007 paper by Ohsawa et al. [1], it is reported that H_2 suppresses the ˙OH accumulation in cells. Through this mechanism it was thought that the oxidative stress of the cells is reduced, that oxidative responses are decreased and that cells therefore can be alleviated from a range of stresses. This redox mechanism was thought to alleviate oxidative stress in cells by inhibiting the oxidative damage caused by the harmful radical, the authors even say: "It is likely that H_2 is mild enough not to disturb metabolic oxidation–reduction reactions or to disrupt ROS involved in cell signalling." On the other hand, such a mechanism has been mooted as underlying the action of H_2 in a range of disease treatments, many of which are underpinned by a disruption of ROS/NO metabolism. It might be worth mentioning here that a similar mechanism would likely take place in other model organisms, not only in humans. Therefore, the response mechanism of H_2 should be observable in other vertebrates, as well as humans and mammals, and also in lower animals and in plants.

More recently there have been a couple of papers which try to explain the mechanism of how H_2 and ˙OH may interact. Kim et al. [53] used in silico modelling to assess the manner in which H_2 may interact with the haem group when it is a prosthetic group of a protein. Here, it was suggested, depending on the orientation of the interaction, that hydrogen could be converted to a hydrogen radical (˙H), which would then have the capacity to interact with ˙OH. This type of mechanism would indeed explain how H_2 and the hydroxyl radical may be able to interact at rates which are significant in cells, so reducing the accumulation of ˙OH in the cell. However, it may also depend on the cellular location of an appropriate haem group, and distance to the ˙OH, which may be acting elsewhere in the cell. As discussed elsewhere [27], there will be an effect on the original action of the haem-containing enzyme too, independent of any ˙OH scavenging. A similar mechanism was also put forward by Jin et al. [54]. They too concentrated on the interaction of H_2 with haem prosthetic groups, suggesting a mechanism which could account for the ˙OH scavenging activity of H_2. They suggested that a direct interaction between ˙OH and H_2 was unlikely because of a high activation and then quote the work of Gong et al. [55], who investigated the use of fluorescent probes for real-time imaging of molecules of hydrogen *in vivo*. However, they show that H_2 can slowly react with peroxynitrite (not significantly enhanced by haem), but also if in the presence of haem H_2 can react with H_2O_2, albeit slowly. Furthermore, the interaction of H_2 with haem allows for the reduction of carbon dioxide (CO_2) to carbon monoxide (CO). Therefore, such a reaction permits further downstream action which would be initiated by the presence (or accumulation, if endogenous) of H_2. Certainly, it appears that the interaction of H_2 with haem, and understanding any ramifications of such interactions, is an important facet of H_2 biology, and its interaction with ˙OH. Very recently, it has been suggested that such a mechanism involving H_2 and haem is involved in the control of the transcription factor Nrf2 [31].

There seems, therefore, to be little doubt that H_2 and ˙OH can react together, and certainly H_2 is likely to scavenge ˙OH. The question remains whether in the complex and multifaceted process of cell signalling to what extent does this mechanism contribute to its physiological benefits? If there is an important effect here, it should have a significant role in signalling, and this will be discussed below.

Action of Hydroxyl Radicals in Cells

There seems to be little doubt that H_2 and $^{\cdot}OH$ can indeed react together, even if there are some debates regarding the kinetics [56], although the work on haem and H_2 suggests that the kinetic problem has been resolved, as discussed above. The H_2/$^{\cdot}OH$ interaction is probably physiologically relevant and likely to take place in cells, to some extent. The question which needs to be posed here is: can the action of H_2 with $^{\cdot}OH$ account for all the effects seen in cells, whether they are from humans, or plants?

One issue here is whether the $^{\cdot}OH$ radical can control cellular function, i.e., can it partake in cell signalling processes? There is little doubt that the answer to this is yes. Early work appeared to concentrate on the role of $^{\cdot}OH$ in mediating cell damage, and it was reported that proteins were the main target of $^{\cdot}OH$, rather than lipids, or causing DNA fragmentation [57]. The presence of the hydroxyl radical has been shown to be correlated with disease mechanisms. For example, by assaying for meta- and ortho-tyrosine, which are produced by a $^{\cdot}OH$-mediated mechanism, it has been suggested that $^{\cdot}OH$ are involved in insulin resistance [58]. However, there is good evidence that $^{\cdot}OH$ has an interactive role in the control of a range of cellular activities. For example, cAMP signalling is thought to have a contradictory role in the generation of $^{\cdot}OH$, both simulating it or inhibiting it, depending on whether the effect is mediated by the exchange protein directly activated by cAMP (Epac) or protein kinase A (PKA) [59].

Hydroxyl radicals have been implicated in the responses to CO, a gas which is not only toxic but can have therapeutic effects [60]. In brain tissues, it was found that $^{\cdot}OH$ was either increased or decreased by CO, depending on the oxygen tension. This is interesting as H_2 has been shown to have effects by a mechanism which modulates the activity of haem-oxygenase (HO-1) [61–63], whilst haem oxygenase itself has been shown to be involved in the metabolism of CO as well as haem [64]. Furthermore, as discussed above, H_2 may have an interaction with haem, with potential ramifications for CO production and metabolism [54]. So here, we have H_2, CO, haem and well characterised signalling components (i.e., cAMP) all in the same mechanism, and all involved in regulatory processes in the cell.

The second issue here is the matter of compartmentalisation. If $^{\cdot}OH$ is the active compound mediating effects that H_2 are reducing, is the $^{\cdot}OH$ being generated in the correct place in the cell? For there to be a pragmatic response, the $^{\cdot}OH$, H_2 and the effect all need to be in the same place. As has been exemplified by cAMP signalling, compartmentalisation is important to consider [65]. For the action to take place the signalling molecule must be able to migrate to the point of action, else no response can take place. The active signal must be produced in the correct place and be removed if not needed. It is tempting to suggest that perhaps H_2 acts as a modulator for the compartmentalisation of $^{\cdot}OH$, but as H_2 can diffuse in aqueous solution as well through a hydrophobic space such as a membrane, it is hard to see how it can have a location-based effect. It would surely just remove $^{\cdot}OH$ from everywhere. However, the main issue around compartmentalisation of $^{\cdot}OH$ signalling is that this radical is not likely to diffuse anyway. It will react with the first thing it interacts with, be that protein, lipid or nucleic acid, although proteins are thought to be its primary target [57]. Its action is likely to be driven by proton abstraction [66], and there would be little subtlety in its action. Does it act as a specific target-driven signalling molecule as seen with, for example, cAMP, which will interact in a receptor-based (lock and key) type mechanism, or $NO^{\cdot}$ which is likely to interact with aqueous accessible thiol groups? Will $^{\cdot}OH$ have such a defined action? It seems unlikely, and therefore does $^{\cdot}OH$ make a good target of H_2 to account for all the actions for which H_2 has been assigned? And if not, what are the alternatives?

When discussing the small reactive molecules that are involved in controlling cellular function, the main candidates are $O_2^{\cdot -}$, H_2O_2 and $NO^{\cdot}$. However, emphasised in the paper by Ohsawa et al. [1], H_2 has no capacity to directly react with any of these molecules. Therefore, there can be no direct influence of H_2 on the signalling mediated by $O_2^{\cdot -}$, H_2O_2 or $NO^{\cdot}$, as summarised in Fig. 2.1, although as discussed above changes in ROS/NO metabolism have been noted to be altered by the presence of H_2.

Although a very naïve way to judge the impact of a biological component on physiological systems, Table 2.3 shows how many articles are in the PubMed database using different terms (as of March 2023). As can be seen, proportionally there are very few on hydroxyl radical(s) signalling. This may give a simple indication of how important small reactive molecules such as $NO^{\cdot}$ (27,294 entries) and H_2O_2 (9793

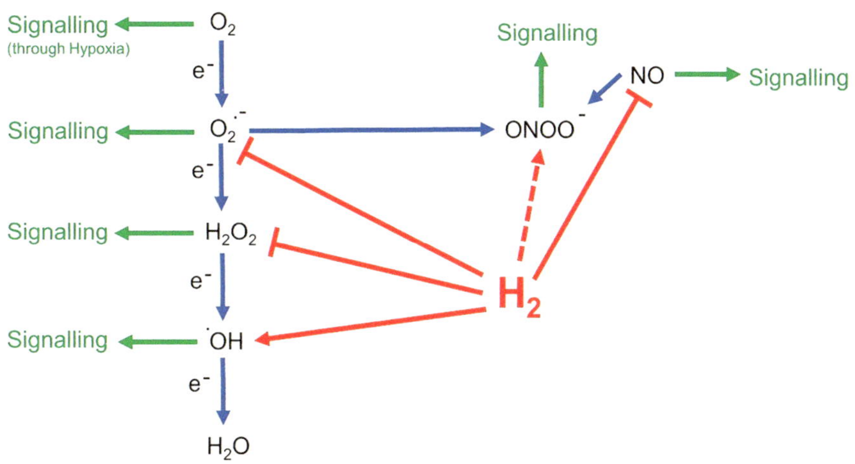

Fig. 2.1 A scheme of the signalling which may ensue from ROS and $NO^{\cdot}$ metabolism and where H_2 is thought to interact with these processes. Capped lines indicate no interaction has been reported; dotted line indicates some activity

entries) are in signalling, compared to $^{\cdot}OH$ (634 entries). And yet, the present idea is that H_2 has its action through $^{\cdot}OH$ signalling and does not have any direct effect on $NO^{\cdot}$, $O_2^{\cdot-}$ or H_2O_2. If this gives any indication of how H_2 works, it is hard to reconcile H_2 having all its actions through $^{\cdot}OH$ signalling. In fact, Ohsawa et al. [1] argued that H_2 not reacting with less reactive species such as $NO^{\cdot}$, $O_2^{\cdot-}$ or H_2O_2 is a beneficial action to maintain redox homeostasis.

It is worth emphasising a rider here. Although there is a relatively small amount of research directed at $^{\cdot}OH$ radical signalling when compared to $NO^{\cdot}$ and H_2O_2, this does not rule out $^{\cdot}OH$ as an important mediator of the benefits of H_2. If $^{\cdot}OH$ can indirectly influence the signalling of other ROS/RNS such as $O_2^{\cdot-}$, H_2O_2, and $NO^{\cdot}$, by influencing their production and/or neutralization, significant signalling events may be influenced. Furthermore, $^{\cdot}OH$ radicals may induce lipid peroxidation of the cell membrane, which in turn can result in numerous changes in signal transduction pathways, as suggested by Iuchi et al. [67]. Even so it is worth exploring other

Table 2.3 The number of articles with specific search terms, as of 31/03/23

Search Term used	Number of articles found in search	Totals for signalling molecules
"hydroxyl radical" and "signaling/signalling"	350	634
"hydroxyl radical" and "signalling"	74	
"hydroxyl radicals" and "signaling"	172	
"hydroxyl radicals" and "signalling"	38	
"nitric oxide" and "signaling"	23,528	27,294
"nitric oxide" and "signalling"	3766	
"hydrogen peroxide" and "signaling"	8508	9793
"hydrogen peroxide" and "signalling"	1285	
"hydrogen sulfide" and "signaling"	1927	2186
"hydrogen sulfide" and "signalling"	259	
"superoxide" and "signaling"	10,627	11,940
"superoxide" and "signalling"	1313	

American and English spelling of "signal(l)ing were used separately" possible mechanisms of H_2 action so that a full understanding can be gained of how H_2 has such wide-ranging effects in cells.

Redox Action of H_2

One of the methods to alter the action of proteins, in particular those that have a prosthetic group, is by altering the redox state of the active part of the protein. For example, in haemoglobin the Fe at the centre of the protoporphyrin can exist as a Fe^{2+} or as Fe^{3+}, and this alters its functionality and ability to bind oxygen. Another example is cytochrome *c*. Cytochrome *c* has a significant role in the movement of electrons—one at a time—from Complex III to Complex IV in the mitochondria. This is vital for the full functionality of the electron transport chain and therefore creation of an electrochemical gradient across the membrane and subsequent ATP production. However, apoptosis can be driven by the mitochondria [68], and in this scenario cytochrome *c* leaves the mitochondria and forms an apoptotic complex, activates caspases and causes cells death. Such proteins which have dual, and often quite distinctive roles are dubbed as moonlighting proteins [69]. However, apoptosis is triggered by oxidative stress, and it has been questioned whether the redox state of cytochrome *c* is instrumental in allowing this mechanism to proceed [70], perhaps acting as fail-safe mechanism, only allowing apoptosis to proceed if the cellular redox is sufficiently oxidising. Again, the Fe in cytochrome *c* can exist in the 2 + or 3 + state, and this may influence how the cytochrome *c* acts.

Hydrogen can act as a redox couple. Under physiological conditions H_2 has a mid-point redox potential (E_m) of more negative than –400 mV relative to the standard hydrogen electrode. This is significantly reducing. The redox couple of NAD^+/NADH for example has an E_m of –320 mV. The mid-point potential of the cytochrome *b* in the NAPDH oxidase is around –245 mV (it is often referred to as Cyt b_{-245}, although it has two haem groups of different potentials [71]. The cytochrome is alternatively referred to as Cyt b_{558} after its absorbance spectra). The cytochrome is regarded as having a very low redox potential, but not as low as H_2. Therefore, in theory, the H_2 couple has the thermodynamic capacity to reduce the haem groups, and other prosthetic groups, of proteins in human cells. There is a precedent for this in bacteria where Cyt c_3 is reduced by the presence of H_2 [72].

However, there is no evidence that this is a mechanism which happens in eukaryotic cells. Some targets have been investigated and ruled out, for example nicotinamide adenine dinucleotide (NAD), flavin adenine dinucleotide (FAD) and cytochrome *c* [1], but there are many more potential systems which may be affected by H_2 in this manner. One, already mentioned, is the NADPH oxidase [73] and so could potentially alter the generation of ROS in cells. Superoxide production may be altered, which subsequently could fluctuate the cellular H_2O_2 levels and any downstream compounds. NO synthase contains haem groups, so $NO^·$ generation could be altered in a similar manner, and hence downstream $NO^·$ metabolism modulated. Guanylyl cyclase, a target of $NO^·$, also has a haem group. Mitochondrial prosthetic groups may be possible targets and in plants there are also the chloroplastic cytochromes, chlorophyll and flavins. Even though some H_2 redox targets may have been deemed to not be involved, an extensive search through some of the possible players here would be a worthwhile exercise.

H_2 Modulating Spin States

H_2 can exist in different spin states, and it is possible that the molecule can toggle between these states, and that these different states may be able to interact with other molecules differentially. This has been proposed [26], but there is no evidence of this. However, as with other proposals it may be worth exploring.

H_2 Interacting with Xenon Pockets in Proteins

H_2 is a relatively inert gas. However, biological systems are impacted upon by a variety of inert gases. The most notable of these is xenon (Xe), but others include argon (Ar). Therefore, if H_2 was acting through its inert nature, could this type of action be possible and could this account for some of the effects seen in cells.

One of the classical systems characterised where an inert atom is able to interact with proteins is Xe, with Xe being suggested as a medical gas because of its known biological effects [74]. In fact, the regions of the proteins involved in binding inert atoms or molecules are often referred to as xenon binding sites (dubbed Xe pockets). Proteins have a three-dimensional structure, and those that are hydrophilic, perhaps located in the cytoplasm, can be regarded rather as like a loose ball of string, and as such there are gaps, crevices, cavities and pockets in the structure. Into these there is the capacity for water molecules to ingress, and such hydration may help to stabilise the structure, or structures if there is an allosteric change in the protein. However, such pockets are also open to ingress by other small molecules and atoms, such as Xe. If these slide into such cavities they may also act to stabilise, or even induce, some subtle topological changes in the polypeptide. A good example of this is Xe and haemoglobin [75]. Turan et al. [76] suggests that Xe is able to migrate through cavities in proteins such as myoglobin, and therefore it is possible that H_2 can do the same. Interestingly, such a ~~as~~ mechanism has been proposed for the enzyme pepsin [77].

There are other inert gases which can interact with proteins. A good example is Ar. Ar has been found to have effects through kinases [78] and Toll-like receptor signalling [79], and has been suggested as a treatment for brain trauma [80] and to be neuroprotective [81]. Therefore, can what is known about gases such as Ar and Xe be translated across to understanding the action of H_2?

The question raised here, and elsewhere [27] is whether hydrogen can act in the same manner. The main issue is the size of the H_2 molecule, which is small. Too small perhaps? However, this does not mean that the idea should be ruled out, especially as a precedent has been published with pepsin [77], but rather this is a manner in which H_2 may interact with proteins which should be explored.

Conclusion and the Future

There is no doubt that H_2 has profound effects in humans and can be used as treatment for a range of diseases [82]. It is also seen to have significant effects in other organisms, such as plants. H_2 has therefore been mooted as a future treatment in medicine and agriculture [83–85]. It has been suggested as a sports medicine [10–12, 86] and as a treatment for radiation damage [87].

H_2 is safe, relatively cheap and easy to administer. The use of H_2 in diving [20] has shown that for the past eighty years humans can tolerate a high concentration of H_2 in inhaled gas. Early experiments with both animals and humans showed similar results [3], although at the turn of the eighteenth and nineteenth centuries there was considerable uncertainly to the composition of the gases being used.

Although there have been suggestions of how H_2 acts in cells there is still some controversy. Some have suggested specific mechanisms, perhaps with H_2 interacting with haem, probably as prosthetic groups in proteins [31, 53, 54], but these mechanisms are based on the premise that $^{\cdot}OH$, at least in part, mediates the effects of H_2 in cells, and also these mechanisms open up the possibility of many additional downstream effects and targets in addition to ·OH scavenging.

What is very likely is that H_2 has more than one action on cells. It may under some circumstances have its effects mediated by the scavenging of hydroxyl radicals, but this mechanism is unlikely to account for all the responses seen, across all the organisms in which effects have been reported, including both humans and plants. It is likely to act through its redox mid-point potential and have effects on a range of proteins containing haem or flavin prosthetic groups. It may act through its spin states, but more likely it acts in a manner seen with other inert gases, that is, by entering cavities in proteins and effecting the stabilisation of protein structures. Certainly, the latter mechanism could account for the wide-ranging effects of $H_{2.}$ This would not be dependent on localised chemical reactions and as H_2 freely diffuses through both aqueous and hydrophobic environments would not be limited by compartmentalisation. In the future, it would certainly be worth a thorough investigation, building on the work of Cheng et al. [77], where H_2 was shown to have effects on pepsin activity.

What is very clear from such a discussion as above is that there is much to explore here. Some of the ideas may be shown to be wrong, or of little significance, but it is very timely to unravel the molecular mechanisms which underpins the action of molecular hydrogen, whether that is in a plant, a lower animal or human.

Acknowledgements The authors would like to acknowledge the support of the University of the West of England, Bristol, for access to library and literature, and for time to write this article.

References

1. Ohsawa I, Ishikawa M, Takahashi K et al (2007) Hydrogen acts as a therapeutic antioxidant by selectively reducing cytotoxic oxygen radicals. Nat Med 13:688–694. https://doi.org/10.1038/nm1577
2. Royal Society of Chemistry Royal Society of Chemistry. Hydrogen. In: Available online. https//www.rsc.org/periodic-table/element/1/hydrogen (Accessed 31 Mar 2023)
3. Hancock JT, LeBaron TW (2023) The early history of hydrogen and other gases in respiration and biological systems: revisiting Beddoes, Cavallo, and Davy. Oxygen 3:102–119. https://doi.org/10.3390/oxygen3010008
4. Dole M, Wilson FR, Fife WP (1975) Hyperbaric hydrogen therapy: a possible treatment for cancer. Science (80) 190:152–154. https://doi.org/10.1126/science.1166304
5. LeBaron TW, Ohno K, Hancock JT (2023) The on/off history of hydrogen in medicine: will the interest persist this time around? Oxygen 3:143–162. https://doi.org/10.3390/oxygen3010011
6. Yang M, Dong Y, He Q et al (2020) Hydrogen: a novel option in human disease treatment. Oxid Med Cell Longev 2020:1–17. https://doi.org/10.1155/2020/8384742
7. Rahman MH, Bajgai J, Fadriquela A et al (2021) Redox effects of molecular hydrogen and its therapeutic efficacy in the treatment of neurodegenerative diseases. Processes 9:308. https://doi.org/10.3390/pr9020308
8. Ono H, Nishijima Y, Adachi N et al (2012) A basic study on molecular hydrogen (H2) inhalation in acute cerebral ischemia patients for safety check with physiological parameters and measurement of blood H2 level. Med Gas Res 2:21. https://doi.org/10.1186/2045-9912-2-21
9. Li Y, Wang Z, Lian N, et al (2021) Molecular hydrogen: a promising adjunctive strategy for the treatment of the COVID-19. Front Med 8. https://doi.org/10.3389/fmed.2021.671215
10. Ostojic S (2014) Molecular hydrogen in sports medicine: new therapeutic perspectives. Int J Sports Med 36:273–279. https://doi.org/10.1055/s-0034-1395509
11. Botek M, Krejčí J, McKune AJ et al (2019) Hydrogen rich water improved ventilatory, perceptual and lactate responses to exercise. Int J Sports Med 40:879–885. https://doi.org/10.1055/a-0991-0268
12. LeBaron TW, Laher I, Kura B, Slezak J (2019) Hydrogen gas: from clinical medicine to an emerging ergogenic molecule for sports athletes. Can J Physiol Pharmacol [Epub ahead of print]. https://doi.org/10.1139/cjpp-2019-0067
13. Botek M, Khanna D, Krejčí J et al (2022) Molecular hydrogen mitigates performance decrement during repeated sprints in professional soccer players. Nutrients 14:508. https://doi.org/10.3390/nu14030508
14. Ohta S (2014) Molecular hydrogen as a preventive and therapeutic medical gas: initiation, development and potential of hydrogen medicine. Pharmacol Ther 144:1–11. https://doi.org/10.1016/j.pharmthera.2014.04.006
15. Du D, Zhao L, Shen M et al (2020) Hydrogen medicine: a rising star in gas medicine. Tradit Med Mod Med 03:153–161. https://doi.org/10.1142/S2575900020300052
16. Hirano S, Ichikawa Y, Sato B et al (2023) Conventional drug acts as a "rifle gun" while hydrogen as a "machine gun." Med Gas Res 13:89. https://doi.org/10.4103/2045-9912.344982
17. Zulfiqar F, Russell G, Hancock JT (2021) Molecular hydrogen in agriculture. Planta 254:56. https://doi.org/10.1007/s00425-021-03706-0
18. Wang C, Fang H, Gong T et al (2020) Hydrogen gas alleviates postharvest senescence of cut rose 'Movie star' by antagonizing ethylene. Plant Mol Biol 102:271–285. https://doi.org/10.1007/s11103-019-00946-3
19. Hancock JT, Russell G, Stratakos AC (2022) Molecular hydrogen: the postharvest use in fruits, vegetables and the floriculture industry. Appl Sci 12:10448. https://doi.org/10.3390/app122010448
20. Zetterstrom A (1948) Deep-sea diving with synthetic gas mixtures. Mil Surg 103:104–106

21. Russell G, Nenov A, Hancock JT (2021) Oxy-hydrogen gas: the rationale behind its use as a novel and sustainable treatment for COVID-19 and other respiratory diseases. Eur Med J. https://doi.org/10.33590/emj/21-00027
22. Tanaka Y, Miwa N (2022) Repetitive bathing and skin poultice with hydrogen-rich water improve wrinkles and blotches together with modulation of skin oiliness and moisture. Hydrogen 3:161–178. https://doi.org/10.3390/hydrogen3020011
23. Iio A, Ito M, Itoh T et al (2013) Molecular hydrogen attenuates fatty acid uptake and lipid accumulation through downregulating CD36 expression in HepG2 cells. Med Gas Res 3:6. https://doi.org/10.1186/2045-9912-3-6
24. Ren J-D, Wu X-B, Jiang R et al (2016) Molecular hydrogen inhibits lipopolysaccharide-triggered NLRP3 inflammasome activation in macrophages by targeting the mitochondrial reactive oxygen species. Biochim Biophys Acta—Mol Cell Res 1863:50–55. https://doi.org/10.1016/j.bbamcr.2015.10.012
25. Hancock JT, LeBaron TW, Russell G (2021) Molecular Hydrogen: redox Reactions and Possible Biological Interactions. React Oxyg Species 11. https://doi.org/10.20455/ros.2021.m.803
26. Hancock JT, Hancock TH (2018) Hydrogen gas, ROS metabolism and cell signaling: are hydrogen spin states important? React Oxyg Species 6
27. Hancock JT, Russell G, Craig TJ et al (2022) Understanding hydrogen: lessons to be learned from physical interactions between the inert gases and the globin superfamily. Oxygen 2:578–590. https://doi.org/10.3390/oxygen2040038
28. Hess DT, Matsumoto A, Kim S-O et al (2005) Protein S-nitrosylation: purview and parameters. Nat Rev Mol Cell Biol 6:150–166. https://doi.org/10.1038/nrm1569
29. Rhee SG, Bae YS, Lee S-R, Kwon J (2000) Hydrogen peroxide: a key messenger that modulates protein phosphorylation through cysteine oxidation. Sci STKE 2000. https://doi.org/10.1126/stke.2000.53.pe1
30. Aroca A, Gotor C, Romero LC (2018) Hydrogen sulfide signaling in plants: emerging roles of protein persulfidation. Front Plant Sci 9. https://doi.org/10.3389/fpls.2018.01369
31. Ohta S (2023) Molecular hydrogen may activate the transcription factor Nrf2 to alleviate oxidative stress through the hydrogen-targeted porphyrin. Aging Pathobiol Ther 5:25–32. https://doi.org/10.31491/APT.2023.03.104
32. Sim M, Kim C-S, Shon W-J et al (2020) Hydrogen-rich water reduces inflammatory responses and prevents apoptosis of peripheral blood cells in healthy adults: a randomized, double-blind, controlled trial. Sci Rep 10:12130. https://doi.org/10.1038/s41598-020-68930-2
33. Yamazaki M, Kusano K, Ishibashi T et al (2015) Intravenous infusion of H2-saline suppresses oxidative stress and elevates antioxidant potential in Thoroughbred horses after racing exercise. Sci Rep 5:15514. https://doi.org/10.1038/srep15514
34. Popovic D, Vucic D, Dikic I (2014) Ubiquitination in disease pathogenesis and treatment. Nat Med 20:1242–1253. https://doi.org/10.1038/nm.3739
35. Cohen P (2002) The origins of protein phosphorylation. Nat Cell Biol 4:E127–E130. https://doi.org/10.1038/ncb0502-e127
36. Palmer RMJ, Ferrige AG, Moncada S (1987) Nitric oxide release accounts for the biological activity of endothelium-derived relaxing factor. Nature 327:524–526. https://doi.org/10.1038/327524a0
37. Lundberg JO, Weitzberg E (2022) Nitric oxide signaling in health and disease. Cell 185:2853–2878. https://doi.org/10.1016/j.cell.2022.06.010
38. Thomas DC (2017) The phagocyte respiratory burst: historical perspectives and recent advances. Immunol Lett 192:88–96. https://doi.org/10.1016/j.imlet.2017.08.016
39. Hancock JT, Veal D (2021) Nitric oxide, other reactive signalling compounds, redox, and reductive stress. J Exp Bot 72:819–829. https://doi.org/10.1093/jxb/eraa331
40. Klotz L-O, Schroeder P, Sies H (2002) Peroxynitrite signaling: receptor tyrosine kinases and activation of stress-responsive pathways 1,2 1 This article is part of a series of reviews on "Reactive Nitrogen Species, Tyrosine Nitration and Cell Signaling". The full list of papers may be found on. Free Radic Biol Med 33:737–743. https://doi.org/10.1016/S0891-5849(02)00892-4

41. Gaston B, Singel D, Doctor A, Stamler JS (2006) S-Nitrosothiol signaling in respiratory biology. Am J Respir Crit Care Med 173:1186–1193. https://doi.org/10.1164/rccm.200510-1584PP
42. Dalle-Donne I, Rossi R, Giustarini D et al (2007) S-glutathionylation in protein redox regulation. Free Radic Biol Med 43:883–898. https://doi.org/10.1016/j.freeradbiomed.2007.06.014
43. Mann BE (2010) Carbon monoxide: an essential signalling molecule. Med Organomet Chem 247–285
44. Poles MZ, Juhász L, Boros M (2019) Methane and inflammation—a review (Fight Fire with Fire). Intensive Care Med Exp 7:68. https://doi.org/10.1186/s40635-019-0278-6
45. Chen M, Cui W, Zhu K et al (2014) Hydrogen-rich water alleviates aluminum-induced inhibition of root elongation in alfalfa via decreasing nitric oxide production. J Hazard Mater 267:40–47. https://doi.org/10.1016/j.jhazmat.2013.12.029
46. Li Y, Li L, Wang S et al (2021) Magnesium hydride acts as a convenient hydrogen supply to prolong the vase life of cut roses by modulating nitric oxide synthesis. Postharvest Biol Technol 177:111526. https://doi.org/10.1016/j.postharvbio.2021.111526
47. Ren A, Liu R, Miao Z-G et al (2017) Hydrogen-rich water regulates effects of ROS balance on morphology, growth and secondary metabolism via glutathione peroxidase in Ganoderma lucidum. Environ Microbiol 19:566–583. https://doi.org/10.1111/1462-2920.13498
48. Dong B, Zhu D, Yao Q et al (2022) Hydrogen-rich water treatment maintains the quality of Rosa sterilis fruit by regulating antioxidant capacity and energy metabolism. LWT 161:113361. https://doi.org/10.1016/j.lwt.2022.113361
49. Gu T, Wang Y, Cao J et al (2022) Hydrogen-rich water pretreatment alleviates the phytotoxicity of bispyribac-sodium to rice by increasing the activity of antioxidant enzymes and enhancing herbicide degradation. Agronomy 12:2821. https://doi.org/10.3390/agronomy12112821
50. Cui W, Yao P, Pan J et al (2020) Transcriptome analysis reveals insight into molecular hydrogen-induced cadmium tolerance in alfalfa: the prominent role of sulfur and (homo)glutathione metabolism. BMC Plant Biol 20:58. https://doi.org/10.1186/s12870-020-2272-2
51. Jiang K, Kuang Y, Feng L et al (2021) Molecular hydrogen maintains the storage quality of chinese chive through improving antioxidant capacity. Plants 10:1095. https://doi.org/10.3390/plants10061095
52. Wu Q, Su N, Cai J et al (2015) Hydrogen-rich water enhances cadmium tolerance in Chinese cabbage by reducing cadmium uptake and increasing antioxidant capacities. J Plant Physiol 175:174–182. https://doi.org/10.1016/j.jplph.2014.09.017
53. Kim S-A, Jong Y-C, Kang M-S, Yu C-J (2022) Antioxidation activity of molecular hydrogen via protoheme catalysis in vivo: an insight from ab initio calculations. J Mol Model 28:287. https://doi.org/10.1007/s00894-022-05264-y
54. Jin Z, Zhao P, Gong W et al (2023) Fe-porphyrin: a redox-related biosensor of hydrogen molecule. Nano Res 16:2020–2025. https://doi.org/10.1007/s12274-022-4860-y
55. Gong W, Jiang L, Zhu Y, et al (2022) An activity-based ratiometric fluorescent probe for in vivo real-time imaging of hydrogen molecules. Angew Chemie Int Ed 61. https://doi.org/10.1002/anie.202114594
56. Penders J, Kissner R, Koppenol WH (2014) ONOOH does not react with H2: Potential beneficial effects of H2 as an antioxidant by selective reaction with hydroxyl radicals and peroxynitrite. Free Radic Biol Med 75:191–194. https://doi.org/10.1016/j.freeradbiomed.2014.07.025
57. Du J, Gebicki JM (2004) Proteins are major initial cell targets of hydroxyl free radicals. Int J Biochem Cell Biol 36:2334–2343. https://doi.org/10.1016/j.biocel.2004.05.012
58. Molnár GA (2015) Tyrosine isomers and hormonal signaling: a possible role for the hydroxyl free radical in insulin resistance. World J Diabetes 6:500. https://doi.org/10.4239/wjd.v6.i3.500
59. Hara S, Kobayashi M, Kuriiwa F et al (2012) Dual contradictory roles of cAMP signaling pathways in hydroxyl radical production in the rat striatum. Free Radic Biol Med 52:1086–1092. https://doi.org/10.1016/j.freeradbiomed.2012.01.002
60. Hopper CP, Zambrana PN, Goebel U, Wollborn J (2021) A brief history of carbon monoxide and its therapeutic origins. Nitric Oxide 111–112:45–63. https://doi.org/10.1016/j.niox.2021.04.001

61. Jin Q, Zhu K, Cui W et al (2013) Hydrogen gas acts as a novel bioactive molecule in enhancing plant tolerance to paraquat-induced oxidative stress via the modulation of heme oxygenase-1 signalling system. Plant Cell Environ 36:956–969. https://doi.org/10.1111/pce.12029
62. Lin Y, Zhang W, Qi F et al (2014) Hydrogen-rich water regulates cucumber adventitious root development in a heme oxygenase-1/carbon monoxide-dependent manner. J Plant Physiol 171:1–8. https://doi.org/10.1016/j.jplph.2013.08.009
63. Li Y, Xie K, Chen H et al (2015) Hydrogen gas inhibits high-mobility group box 1 release in septic mice by upregulation of heme oxygenase 1. J Surg Res 196:136–148. https://doi.org/10.1016/j.jss.2015.02.042
64. Ryter SW, Otterbein LE, Morse D, Choi AMK (2002) Heme oxygenase/carbon monoxide signaling pathways: regulation and functional significance. Mol Cell Biochem 234–235:249–263. https://doi.org/10.1023/A:1015957026924
65. Bers DM, Ziolo MT (2001) When is cAMP not cAMP? Effects of compartmentalization. Circ Res 89:373–375
66. Mitroka S, Zimmeck S, Troya D, Tanko JM (2010) How solvent modulates hydroxyl radical reactivity in hydrogen atom abstractions. J Am Chem Soc 132:2907–2913. https://doi.org/10.1021/ja903856t
67. Iuchi K, Imoto A, Kamimura N et al (2016) Molecular hydrogen regulates gene expression by modifying the free radical chain reaction-dependent generation of oxidized phospholipid mediators. Sci Rep 6:18971. https://doi.org/10.1038/srep18971
68. Sinha K, Das J, Pal PB, Sil PC (2013) Oxidative stress: the mitochondria-dependent and mitochondria-independent pathways of apoptosis. Arch Toxicol 87:1157–1180. https://doi.org/10.1007/s00204-013-1034-4
69. Singh N, Bhalla N (2020) Moonlighting proteins. Annu Rev Genet 54:265–285. https://doi.org/10.1146/annurev-genet-030620-102906
70. Hancock JT, Desikan R, Neill SJ (2001) Does the redox status of cytochrome C act as a fail-safe mechanism in the regulation of programmed cell death? Free Radic Biol Med 31:697–703. https://doi.org/10.1016/S0891-5849(01)00646-3
71. Cross AR, Rae J, Curnutte JT (1995) Cytochrome b−245 of the neutrophil superoxide-generating system contains two nonidentical hemes. J Biol Chem 270:17075–17077. https://doi.org/10.1074/jbc.270.29.17075
72. Peck HD (1959) The ATP-dependent reduction of sulfate with hydrogen in extracts of desulfovibrio desulfuricans. Proc Natl Acad Sci 45:701–708. https://doi.org/10.1073/pnas.45.5.701
73. Buvelot H, Jaquet V, Krause K-H (2019) Mammalian NADPH oxidases. pp 17–36
74. Esencan E, Yuksel S, Tosun YB et al (2013) XENON in medical area: emphasis on neuroprotection in hypoxia and anesthesia. Med Gas Res 3:4. https://doi.org/10.1186/2045-9912-3-4
75. Lepeshkevich SV, Gilevich SN, Parkhats MV, Dzhagarov BM (2016) Molecular oxygen migration through the xenon docking sites of human hemoglobin in the R-state. Biochim Biophys Acta—Proteins Proteomics 1864:1110–1121. https://doi.org/10.1016/j.bbapap.2016.06.004
76. Turan HT, Boittier E, Meuwly M (2023) Interaction at a distance: Xenon migration in Mb. J Chem Phys 158:125103. https://doi.org/10.1063/5.0124502
77. Cheng J, Tang C, Li X et al (2020) Hydrogen molecules can modulate enzymatic activity and structural properties of pepsin in vitro. Colloids Surf B Biointerfaces 189:110856. https://doi.org/10.1016/j.colsurfb.2020.110856
78. Fahlenkamp AV, Rossaint R, Haase H et al (2012) The noble gas argon modifies extracellular signal-regulated kinase 1/2 signaling in neurons and glial cells. Eur J Pharmacol 674:104–111. https://doi.org/10.1016/j.ejphar.2011.10.045
79. Ulbrich F, Kaufmann K, Roesslein M et al (2015) Argon mediates anti-apoptotic signaling and neuroprotection via inhibition of toll-like receptor 2 and 4. PLoS ONE 10:e0143887. https://doi.org/10.1371/journal.pone.0143887
80. Loetscher PD, Rossaint J, Rossaint R et al (2009) Argon: neuroprotection in in vitro models of cerebral ischemia and traumatic brain injury. Crit Care 13:R206. https://doi.org/10.1186/cc8214

81. Sanders RD, Ma D, Maze M (2010) Argon neuroprotection. Crit Care 14:117. https://doi.org/10.1186/cc8847
82. Ge L, Yang M, Yang N-N, et al (2017) Molecular hydrogen: a preventive and therapeutic medical gas for various diseases. Oncotarget 8:102653–102673. https://doi.org/10.18632/oncotarget.21130
83. Dixon BJ, Tang J, Zhang JH (2013) The evolution of molecular hydrogen: a noteworthy potential therapy with clinical significance. Med Gas Res 3:10. https://doi.org/10.1186/2045-9912-3-10
84. Ostojic SM (2015) Molecular hydrogen: an inert gas turns clinically effective. Ann Med 47:301–304. https://doi.org/10.3109/07853890.2015.1034765
85. Alwazeer D, Çiğdem A (2022) Moleküler Hidrojenin Tarım Alanında Kullanımı. Turkish J Agric—Food Sci Technol 10:14–20. https://doi.org/10.24925/turjaf.v10i1.14-20.4609
86. Botek M, Krejčí J, Valenta M et al (2022) Molecular hydrogen positively affects physical and respiratory function in acute post-COVID-19 patients: a new perspective in rehabilitation. Int J Environ Res Public Health 19:1992. https://doi.org/10.3390/ijerph19041992
87. Chuai Y, Qian L, Sun X, Cai J (2012) Molecular hydrogen and radiation protection. Free Radic Res 46:1061–1067. https://doi.org/10.3109/10715762.2012.689429

Chapter 3
Prospects of Hydrogen Medicine Based on Its Protective Effects on Mitochondrial Function

Shin-ichi Hirano, Yusuke Ichikawa, Bunpei Sato, Yoshiyasu Takefuji, Xiao-Kang Li, and Fumitake Satoh

Abstract Mitochondria originated from aerobic bacteria in endosymbiosis. Through this symbiosis, eukaryotes acquired an efficient energy-producing system, but at the cost of exposure to oxidative stress from reactive oxygen species (ROS). Molecular hydrogen (H_2) was recently identified as an antioxidant that selectively reduces ROS, such as hydroxyl radicals and peroxynitrite, which are strong oxidants, and its clinical applications are progressing. This paper investigated the efficacy of H_2 on experimental models and several human chronic inflammatory diseases and demonstrated that its exerted effects via the protection of mitochondrial function. H_2 protection may be exerted by regulation of mitochondrial ROS. Since mitochondrial dysfunction has been detected in many common diseases, such as metabolic and neurodegenerative diseases, the development of technologies and substances that protect or activate mitochondrial function will be necessary for the future of

S. Hirano · Y. Ichikawa · B. Sato · F. Satoh
Department of Research and Development, MiZ Company Limited, Kanagawa, Japan
e-mail: s_hirano@e-miz.co.jp

Y. Ichikawa
e-mail: y_ichikawa@e-miz.co.jp

B. Sato
e-mail: b_sato@e-miz.co.jp

F. Satoh
e-mail: f_satoh@e-miz.co.jp

Y. Takefuji
Keio University (Professor Emeritus), Tokyo, Japan

Faculty of Data Science, Musashino University, Tokyo, Japan

Y. Takefuji
e-mail: takefuji@keio.jp

X.-K. Li (✉)
Division of Transplantation Immunology, National Institute for Child Health and Development, Tokyo, Japan
e-mail: ri-k@ncchd.go.jp

J. Slezak and B. Kura (eds.), *Molecular Hydrogen in Health and Disease*, Advances in Biochemistry in Health and Disease 27, https://doi.org/10.1007/978-3-031-47375-3_3

medicine. H_2 may be positioned as a candidate in future medicine due to its effects on mitochondrial function.

Keywords Molecular hydrogen · Reactive oxygen species · Oxidative stress · Mitochondria · Inflammatory disease · Post-COVID-19 · ME/CFS · Future medicine

Introduction

Eukaryotes, which emerged approximately 2 billion years ago when archaea engulfed the aerobic bacteria, *proteobacteria*, acquired an efficient energy-producing system, but at the cost of exposure to oxidative stress caused by reactive oxygen species (ROS) produced in mitochondria [1]. The antioxidant enzymes are ineffective against the production of ROS, such as hydroxyl radicals (·OH) and peroxynitrite ($ONOO^-$), which are very strong oxidants [2, 3].

Molecular hydrogen (H_2) was recently identified as an antioxidant that directly reduces ·OH and $ONOO^-$ [4]. H_2 also exerts indirect antioxidant, anti-inflammatory, and anti-apoptotic effects by regulating gene expression [5–7]. Other indirect mechanisms by which H_2 exerts its effects have been reported, such as nuclear factor erythroid-related factor 2 (Nrf-2) and various signaling pathways in cells [8–10]. The total number of studies on the biological effects of H_2 now exceeds 1600 [11]. Among them, the total number of studies on human clinical trials is more than 120. Since no side effects have been observed with H_2 in human clinical studies, various clinical studies are underway to investigate its ameliorative effects on various pathological conditions [11].

It is considered that protective effects on mitochondrial function may be involved in the efficacy of H_2 against oxidative stress-associated diseases. In this study, we will examine the efficacy of H_2 in experimental disease models and in chronic inflammatory diseases including "sequelae" of coronavirus infection disease 2019 (COVID-19) and myalgic encephalomyelitis/chronic fatigue syndrome (ME/CFS). Furthermore, we discuss the mechanisms by which H_2 ameliorates these diseases and its potential for future medicine.

Oxidative Stress Regulates Health and Disease

Production and Scavenging Systems of ROS

Oxygen accounts for approximately 20% of air and is essential for energy production in breathing organisms [2, 3, 12]. After oxygen is taken into the body, it is used by mitochondria in cells to produce adenosine triphosphate (ATP). However, 1–2% of

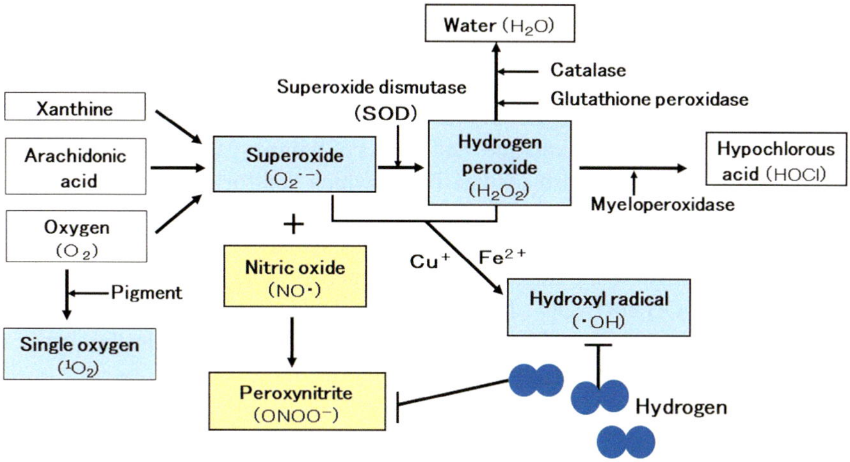

Fig. 3.1 *ROS production and scavenging systems.* Antioxidant enzymes, such as SOD, catalase, and glutathione peroxidase, cannot scavenge ·OH or $ONOO^-$, which are strong oxidants. In contrast, H_2 selectively scavenges .OH and $ONOO^-$

consumed oxygen becomes ROS, which are strong oxidants in the body [2, 3, 12]. The human body is equipped with antioxidant enzymes as a defense mechanism to suppress the production of ROS. Antioxidant enzymes include superoxide dismutase (SOD), catalase, and glutathione peroxidase (Fig. 3.1).

There are four main types of ROS in the human organism: superoxide, hydrogen peroxide, ·OH, and singlet oxygen [2, 3, 12]. Singlet oxygen is produced by the reaction of oxygen with pigments in the body that function as sensitizers during exposure to ultraviolet radiation. When electrons leak from the mitochondrial respiratory chain and combine with oxygen, superoxide is formed [12]. In addition, superoxide is produced not only by oxygen and xanthine oxidase using xanthine as a substrate, but also by the arachidonic acid cascade in vascular endothelial cells. Superoxide is a relatively reactive substance but is decomposed by SOD to hydrogen peroxide [12]. Hydrogen peroxide is then decomposed into water and oxygen by catalase and glutathione peroxidase and detoxified in the body [12] (Fig. 3.1).

The function of antioxidant enzymes and the body's defense against ROS decline with age [2, 3]. Furthermore, when ROS are produced in large amounts due to excessive exercise, mental and physical stress, smoking, drinking, exposure to ultraviolet light and radiation, and air pollution, the balance between ROS production and scavenging systems is disrupted, resulting in the emergence of ROS that exceed the defenses of antioxidant enzymes and, ultimately, cell and gene damage [2, 3].

When the balance between oxidation and anti-oxidation is disrupted, superoxide and hydrogen peroxide produce ·OH, which is a very strong oxidant, using iron and copper ions as catalysts [2, 3, 12]. ·OH is produced by other biological reactions and also when water, a biological substance, is exposed to radiation [2, 3, 12]. On the other hand, nitric oxide produces $ONOO^-$, another very strong oxidant (Fig. 3.1).

Although ·OH is present in the body for only a fraction of a millionth of a second, it exhibits an oxidizing power that is 100-fold stronger than that of superoxide during that time [13].

Therefore, when ·OH and $ONOO^-$ are produced, they react with nucleic acids, lipids, and proteins in the membranes and tissues of living organisms, causing oxidative damage [2, 3]. They also oxidize DNA, which controls genetic information. However, superoxide and hydrogen peroxide do not have sufficient oxidizing power to directly oxidize DNA. The oxidation of DNA leads to genetic damage, which, in turn, induces lifestyle-related diseases, such as cancer [2, 3, 12]. Antioxidant enzymes cannot detoxify ·OH or $ONOO^-$. In contrast, H_2 selectively scavenges large amounts of ·OH and $ONOO^-$ and converts them to water (Fig. 3.1). For example, the chemical reaction equation between H_2 and ·OH is as follows:

$$\cdot OH + H_2 \rightarrow H \cdot + H_2O$$

ROS-Induced Oxidative Eustress and Distress

Diseases related to ROS are those that originate in many organs and tissues in the body, including the brain, nerves, eyes, nose, teeth, the respiratory, circulatory, digestive, urinary, hematological, and endocrine systems, skin, and supporting tissues [14]. Therefore, ROS have been implicated in the development of most diseases.

When the relationship between the generation of superoxide, a cause of oxidative stress, and life span was investigated, a negative correlation was observed, with organisms that generated more ROS having a shorter life span [15]. In addition, when the relationship between SOD and the life span of an organism (in terms of total lifetime energy) was examined, a correlation was also observed such that organisms with higher levels of this SOD activity had a longer life span [16].

ROS have not only harmful, but also beneficial aspects for organisms. Superoxide and hydrogen peroxide have been shown to exert cytotoxic effects at high concentrations, but function as molecules in signal transduction and play important roles in apoptosis, cell proliferation, and cell differentiation at low concentrations [2, 3]. In addition, hydrogen peroxide at high concentrations is converted to hypochlorous acid by antioxidant enzymes and plays a role in the body's defense against bacterial attacks [2, 3]. Nitric oxide is important for signal transduction and vasodilation and is used as a medical gas. In addition, differences have been observed between oxidative eustress and distress. A large amount of ROS causes oxidative damage, while a small amount induces heme oxygenase, an antioxidant enzyme, through the activation of Nrf-2, which exerts protective effects in the body [17]. Small amounts of ROS also induce p53, a tumor suppressor gene [18].

Mitochondrial Dysfunction and Disease

Mitochondria produce more than 90% of intracellular energy and generate ATP by oxidative phosphorylation under aerobic conditions. These organelles are composed of a bilayer structure of inner and outer membranes with an intermembrane space in addition to a matrix within the inner membrane [19]. The inner membrane has a narrow crista on the inner side [19]. The five proteins of the mitochondrial respiratory chain complex, complexes I-V, assemble in the inner membrane crista for efficient ATP production [19]. Moreover, many mitochondrial ROS (mtROS), primarily generated from complexes I and III, are removed by the antioxidant system within mitochondria, which includes SOD2/MnSOD, catalase, glutathione peroxidase, and reduced glutathione. Therefore, efficient energy production is maintained by the electron transfer system of the mitochondrial inner membrane respiratory chain complex as well as oxidative phosphorylation by ATP synthase [20, 21].

However, abnormalities in mitochondrial function induce a decrease in ATP production and an increase in mtROS, which, in turn, contribute to the development of pathological conditions due to apoptotic signals, such as cytochrome c released from within damaged mitochondria [20]. Mitochondrial dysfunction without genetic abnormalities has been detected in common metabolic and neurodegenerative diseases. Mitochondrial dysfunction has also been implicated in the pathogenesis of diabetes, atherosclerosis, hypertension, Parkinson's disease, acute kidney disease, and amyotrophic lateral sclerosis (ALS) [20, 22].

Mitochondrial diseases are caused by abnormalities in various genes involved in mitochondrial function and structural maintenance, such as ATP synthesis, the transport of amino acids, lipids, and proteins, and oxidative stress removal within mitochondria [20, 23]. In mitochondrial diseases, clinical symptoms develop in tissues that require large amounts of energy, mainly muscles and nerves, due to impaired mitochondrial respiratory function, and lactic acidosis is also induced due to an increased dependence on energy from the glycolytic system [24].

Oxidative stress-induced impairments in the structure and function of mitochondria have been implicated in the development of various diseases; however, mild stress caused by H_2 may actually enhance resistance to the exacerbation of oxidative stress. Murakami et al. used cultured cells to investigate the effects of low concentrations of H_2 on mitochondria [25]. Their findings revealed that H_2 increased the mitochondrial membrane potential (MMP) and intracellular ATP levels [25]. A pretreatment with H_2 inhibited hydrogen peroxide-induced cell death, whereas a post-treatment did not. In addition, H_2-treated cells showed the up-regulated expression of antioxidant enzymes involved in the Nrf-2 pathway. These findings suggest that H_2 functions not only as a radical scavenger, but also as a mitohormetic effector [25].

Effects of H_2 on Mitochondrial Function

Effects in Various Experimental Disease Models

H_2 shows efficacy in various disease models in experimental animals or cultured cells via improvements in the structure and/or function of mitochondria. Therefore, this chapter provides an overview of the literature demonstrating the efficacy of H_2 in disease models of stroke, subarachnoid hemorrhage (SAH), ALS, Alzheimer's disease, myocardial injury, hypertension, sepsis, diabetic neuropathy, and liver injury as well as the underlying mechanisms (Table 3.1) [26–34].

Table 3.1 Summary of representative findings in experimental disease models using hydrogen (H_2)

Disease models	Type of H_2	Effects of H_2	Ref. No.
Cerebral infarction	HRS	HRS attenuated neuronal I/R injury by protecting mitochondrial function in rats	[26]
Subarachnoid hemorrhage	H_2 gas	H_2 gas attenuated neuronal pyroptosis in a rat model of SAH through the mitoK_{ATP} signaling pathway	[27]
ALS	HRS	HRS delayed disease progression in a mouse model of ALS by reducing oxidative stress and maintaining mitochondrial function	[28]
Alzheimer's disease	HRW	HRW attenuated Aβ-induced cytotoxicity through the up-regulation of Sirt1-FoxO3a by a stimulation of AMPK in SK-N-MC cells	[29]
Myocardial injury	H_2 gas	H_2 gas reduced infarct sizes in canine hearts through the opening of mitoK_{ATP} channels followed by the inhibition of mPTP	[30]
Hypertension	HRS	HRS reduced oxidative stress and attenuated left ventricular hypertrophy through the preservation of mitochondrial function in SHR	[31]
Sepsis	H_2 gas	H_2 gas alleviated sepsis-induced brain injury by improving mitochondrial biogenesis through the activation of PGC-α in mice	[32]
Diabetic peripheral neuropathy	HRS	HRS protected against diabetic peripheral neuropathy through the activation of mitoK_{ATP} channels in rats	[33]
Liver injury	HRS	HRS protected against mitochondrial dysfunction and apoptosis in mice with obstructive jaundice	[34]

HRS: hydrogen-rich saline; I/R: ischemia/reperfusion; SAH: subarachnoid hemorrhage; mitoK_{ATP}: mitochondrial ATP-sensitive K^+; ALS: amyotrophic lateral sclerosis; Aβ: amyloid-β; Sirt1: sirtuin 1; FoxO3a: forkhead box protein O3a; AMPK: AMP-activated protein kinase; mPTP: mitochondrial permeability transition pores; HRW: hydrogen-rich water; SHR: spontaneously hypertensive rat; PGC-α: peroxisome proliferator-activated receptor gamma co-activator 1α

Cerebral Infarction

Cui et al. examined the effects of hydrogen-rich saline (HRS) in a rat brain ischemia/reperfusion (I/R) model and demonstrated that it significantly increased the number of viable neurons [26]. They also showed that HRS not only suppressed tissue damage, the degree of mitochondrial swelling, and the reduction of MMP, but also maintained the mitochondrial content of cytochrome c [26]. These findings suggest that HRS attenuates neuronal I/R damage in rats by protecting mitochondrial function.

SAH

Zhang et al. investigated the protective effects of H_2 gas against neuronal pyroptosis in a rat model of SAH and reported that the inhalation of H_2 gas significantly ameliorated brain edema, improved neurological function, and inhibited neuronal pyroptosis [27]. Furthermore, H_2 gas suppressed ROS production, the expression of interleukin (IL)-1β and IL-18, and the activation of p38 mitogen-activated protein kinase (p38 MAPK), and these inhibitory effects of H_2 were attenuated by the administration of sodium 5-hydroxydecanoate (5-HD), a mitochondrial ATP-sensitive K^+ (mitoK_{ATP}) channel inhibitor [27]. Collectively, these findings suggest that the neuroprotective effects of H_2 gas involve the mitoK_{ATP}/p38 MAPK signaling pathway.

ALS

Zhang et al. examined the effects of HRS in a mutant SOD1 G93A transgenic mouse model of ALS and reported that it significantly delayed the onset of disease and prolonged survival [28]. They also showed that HRS inhibited the mitochondrial release of apoptogenic factors and subsequent activation of downstream caspase-3 [28]. Furthermore, they demonstrated that HRS maintained mitochondrial function, restored complex I and IV activities, decreased mtROS production, and promoted mitochondrial ATP synthesis [28]. Based on these findings, they suggested that H_2 exerts neuroprotective effects against ALS by reducing oxidative stress and maintaining mitochondrial function.

Alzheimer's Disease

Lin et al. investigated the effects of H_2-rich water (HRW) on the cytotoxicity of human neuroblast SK-N-MC cells [29]. They demonstrated that HRW reduced excess ROS, inhibited oxidative damage, and suppressed Amyloid-β (Aβ)-induced cell death [29]. Furthermore, they showed that HRW stimulated AMP-activated protein kinase, which up-regulated the forkhead box protein O3a (FoxO3a) downstream antioxidant response and reduced Aβ-induced mitochondrial potential loss and oxidative stress [29]. They also indicated the potential of HRW as an effective therapeutic agent to inhibit Aβ-induced neurotoxicity.

Myocardial Injury

Yoshida et al. examined the effects of H_2 gas in a myocardial I/R model in beagle dogs and found that it reduced myocardial infarct sizes; however, the administration of 5-HD or atractyloside, a mitochondrial permeability transition pore (mPTP) opener, nullified the effects of H_2 gas on infarct sizes [30]. These findings indicate that H_2 gas reduced infarct sizes in the canine heart by opening mitoK_{ATP} channels and inhibiting mPTP.

Hypertension

Yu et al. investigated the effects of the chronic administration of HRW on left ventricular hypertrophy in spontaneously hypertensive rats (SHR) [31]. The findings obtained showed that HRW suppressed the production of inflammatory cytokines and nuclear factor-κB (NF-κB) activation in the left ventricle [31]. Furthermore, HRW maintained mitochondrial function by restoring electron transport chain enzyme activity, inhibiting ROS production, and enhancing ATP production [31].

Sepsis

Xie et al. examined the effects of H_2 gas on mitochondrial function and biosynthesis as well as the associated regulatory mechanisms in mice with sepsis-associated encephalopathy (SAE) [32]. Their findings showed that H_2 gas prolonged survival and preserved cognitive function in SAE mice and increased MMP and ATP levels, parameters of mitochondrial function, as well as the expression of complex I activity [32]. They also demonstrated that H_2 gas up-regulated the expression of the mitochondrial biosynthesis parameters, peroxisome proliferator-activated receptor gamma coactivator 1α (PGC-1α) and mitochondrial transcription factor A [32]. These findings indicate that H_2 gas alleviated sepsis-induced brain damage in mice by promoting mitochondrial biosynthesis through the activation of PGC-1α.

Diabetic Peripheral Neuropathy

Jiao et al. investigated the efficacy of HRS against diabetic peripheral neuropathy (DPN) in a streptozotocin-induced diabetic rat model [33]. HRS significantly suppressed the behavioral, biochemical, and molecular effects of DPN in rats [33]. In addition, 5-HD partially attenuated the therapeutic effects of HRS [33]. These findings suggest that the mechanisms underlying the efficacy of HRS against DPN involve the suppression of oxidative stress, inflammation, and apoptosis via the activation of the mitoK_{ATP} pathway.

Liver Injury

Liu et al. investigated the effects of HRS in a mouse model of obstructive jaundice and found that it significantly reduced mitochondrial swelling, cytochrome c release, and oxidative damage [34]. They also reported that HRS suppressed cellular B-cell/CLL lymphoma 2 (Bcl-2)-associated x (Bax) protein expression, caspase activity, and hepatocyte apoptosis, and alleviated mitochondrial morphological defects [34]. These findings indicate that HRS inhibited mitochondrial oxidative stress and dysfunction and suppressed mitochondria-mediated apoptosis.

Effects on Chronic Inflammation

Inflammation is induced by inflammatory cytokines released by innate immunity. The cascade leading to the release of inflammatory cytokines is very complex; however, pathogens, such as viruses and bacteria, substances produced when the body is damaged, and irritants in the environment function as signals when inflammation is induced [35]. These extracellular signals cause mitochondrial dysfunction and induce the excessive production of ROS [36]. Excessive mtROS production and the release of oxidized mitochondrial DNA (mtDNA) trigger the formation of the nucleotide-binding and oligomerization domain-like receptor family pyrin domain-containing 3 (NLRP3) inflammasome, a protein complex [37, 38]. The NLRP3 inflammasome then activates the proteolytic enzyme caspase 1, which releases mature, transformed inflammatory cytokines from immune cells, such as macrophages and neutrophils, resulting in inflammation [37, 38] (Fig. 3.2).

We conducted a literature review on the suppression of acute and chronic inflammation by H_2 [38]. Many studies reported that the suppression of mtROS production by H_2 was involved in the anti-inflammatory effects of H_2 [38]. However, these studies did not specify the ROS involved [39–41]. Among ROS, ·OH is very oxidative and causes not only nuclear DNA damage, but also mtDNA damage and induces cell death. H_2 is a substance with excellent permeability to mitochondria and selective scavenging ability of ·OH [4]. Therefore, we showed that the mechanism underlying the amelioration of chronic inflammation by H_2 may involve the reduction of ·OH by H_2 protecting mtDNA from oxidative damage, which then suppresses a series of signaling events from the activation of the NLRP3 inflammasome to the release of inflammatory cytokines [38]. In this review, we proposed that the suppression of ·OH production by H_2 may be a mechanism inhibiting subsequent inflammatory signaling [38] (Fig. 3.2).

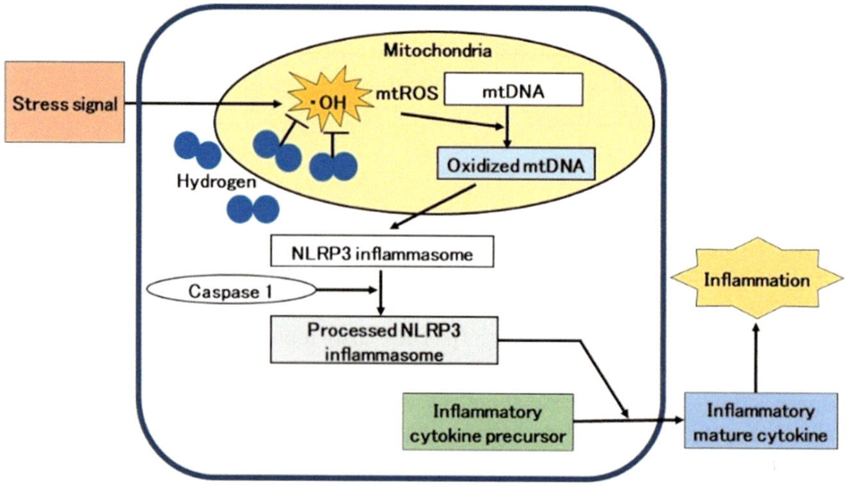

Fig. 3.2 *Involvement of mitochondria in the release of inflammatory cytokines.* H_2 inhibits the cascade from NLRP3 inflammasome activation to inflammatory cytokine release based on the inhibition of mtROS production

However, other researchers indicated that the downstream signaling pathways involved in cellular responses other than ROS are involved in the mechanisms responsible for the H_2-mediated suppression of inflammation [42–45].

Effects on "Sequelae" of COVID-19 and Chronic Fatigue

Guan et al. assessed the efficacy of inhalation therapy with a mixture of H_2 and oxygen gas in patients with COVID-19 in an open-label study [46]. They demonstrated the superiority of mixed H_2 and oxygen gas therapy over the control treatment for improvements in the severity of COVID-19, dyspnea, cough, chest tightness, chest pain, and oxygen saturation [46]. Although the inhalation of H_2 gas ameliorates the pneumonia symptoms of COVID-19, it is not only this infection that is problematic, but also the "sequelae" after contracting the infection [47, 48]. In other words, COVID-19 is a viral infection caused by SARS-CoV-2 that affects the respiratory, digestive, and vascular systems. Acute symptoms generally resolve within two to three weeks. However, for some patients, the recovery period is prolonged, and "sequelae" may persist for several months after the initial infection [47, 48]. These "sequelae" have been reported to include fatigue, dyspnea, myalgia, exercise intolerance, sleep disturbances, poor concentration, anxiety, fever, headache, malaise, and many other chronic symptoms [47, 48]. The "sequelae" of COVID-19 are referred to as post-COVID-19 or long-COVID-19.

Botek et al. recently reported the findings of a randomized, single-blind study on the effects of H_2 gas inhalation on physical and respiratory functions in acute post-COVID-19 patients [49]. They showed that H_2 gas inhalation more effectively attenuated clinical symptoms in acute post-COVID-19 patients than placebo gas because H_2 gas significantly improved physical function in gait tests and respiratory function in pulmonary function tests [49]. These findings indicate that H_2 gas inhalation exerts therapeutic effects not only on the pneumonia symptoms of COVID-19 patients, but also on the physical and respiratory functions of post-COVID-19 patients.

Post-COVID-19 may be the same as ME/CFS, which is characterized by severe fatigue lasting more than 6 months, extreme exhaustion after exertion, memory impairment, concentration problems, and sleep disturbances as the main symptoms [50]. Although the etiology of ME/CFS is unknown, mitochondrial dysfunction has been reported as one of its causes [51–54]. Since the symptoms of ME/CFS and post-COVID-19 are very similar, it has been argued, but not yet proven, that these diseases may be caused by the same mechanism [55].

Extensive efforts have been made to develop treatments and therapies for ME/CFS; however, the treatments and therapies that have been developed are symptomatic, not curative treatments [56]. We conducted a literature review and showed that H_2 ameliorated fatigue induced by acute or chronic exercise stress in experimental animals and healthy individuals, that fatigue was caused by mitochondrial dysfunction due to ROS, and that H_2 not only ameliorated acute and chronic fatigue by reducing ·OH, but also the pathogenesis of ME/CFS by reducing ·OH (Fig. 3.3) [57]. We also performed a case study on H_2 gas inhalation in four ME/CFS patients and found that it attenuated symptoms such as "brain fog", fatigue, the recovery time from exertion, headache, poor concentration, and sleep disturbances [58]. Although the efficacy of H_2 gas in many patients needs to be confirmed in future randomized controlled trials, the inhalation of H_2 gas will not only show efficacy for post-COVID-19, but also for ME/CFS.

Prospects for Future Medicine

We herein outlined the findings of previous studies showing that H_2 ameliorates diseases by protecting/improving the structure and/or function of mitochondria in various animal and cellular models [26–34]. Transmission electron microscopy revealed that H_2 exerts protective effects on the mitochondrial morphology [26, 34]. Biochemical and molecular evaluations of the protective effects of H_2 on mitochondrial function demonstrated that it suppressed ROS production and IL-1β and IL-18 expression and enhanced ATP production, MMP, and complex I activity [26–29, 31, 32, 34]. Regarding pharmacological evaluations of H_2, the administration of 5-HD, an inhibitor of mitoK_{ATP} channels, and atractyloside, an opener of mPTP, attenuated the effects of H_2, suggesting that H_2 exerts protective effects on mitochondrial function through the opening of mitoK_{ATP} channels and inhibition of mPTP [27, 30, 33]. Furthermore, H_2 inhibited the activities of caspase-3 and -9, suppressed the

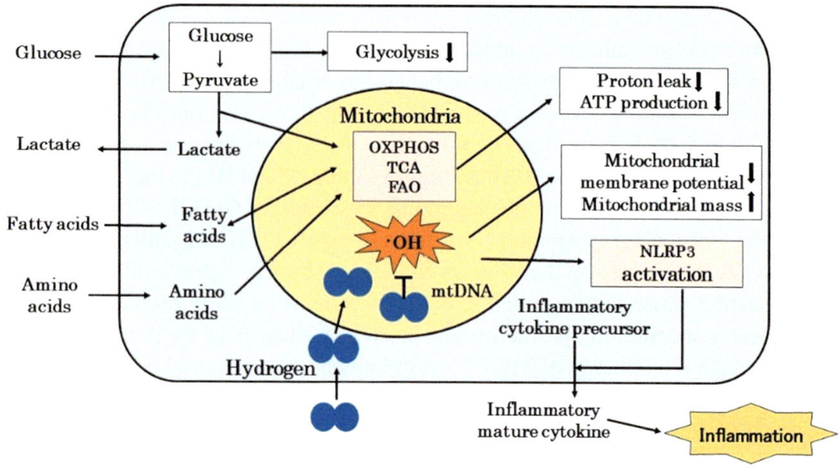

Fig. 3.3 *A possible mechanism by which H_2 ameliorates mitochondrial dysfunction in patients with ME/CFS.* The mitochondria of the patients show a reduced glycolytic capacity and abnormal metabolism. These mitochondria show decreased proton leakage, ATP production, and mitochondrial membrane potential and an increased mitochondrial mass. H_2 ameliorates mitochondrial dysfunction by scavenging ·OH and blocks the cascade from NLRP3 activation to the release of inflammatory cytokines. OXPHOS: oxidative phosphorylation, TCA: tricarboxylic acid cycle, FAO: fatty acid oxidation. (From [57])

release of cytochrome c and the expression of Bax, and increased the expression of Bcl-2, indicating that it protects mitochondrial function by inhibiting apoptotic cell death [34]. Collectively, these experimental findings showed that H_2 protects the mitochondrial morphology and/or function via the regulation of ROS production in the mitochondria.

ROS produced by mitochondrial dysfunction or failure cause various diseases, which originate in various organs and tissues of the body [14]. On the other hand, diseases involving chronic inflammation also originate from various organs and tissues of the body [14]. Therefore, while ROS are the cause of many diseases, chronic inflammation has also been implicated in their pathogenesis. We examined the mechanisms by which H_2 ameliorates chronic inflammation and found that they may involve protective effects on mitochondria based on the reduction of ·OH by H_2 [38]. We also hypothesized that the mechanisms by which H_2 ameliorates post-COVID-19 and ME/CFS may involve protective effects on mitochondria based on the reduction of ·OH by H_2 [57].

Modern medicine is characterized by viewing the human body as a collection of organs and subdividing the object of study from organs to cells, then to molecules, and finally to genes in order to identify the factors that most influence disease. However, many diseases are not caused by a single factor alone, but by multiple factors and a wide variety of mechanisms. H_2 is a substance that falls outside the scope of modern medicine because it has a wide range of effects on diverse diseases by acting on ROS and chronic inflammation, which are the root causes of disease [38].

One limitation of H_2 medicine is that because H_2 is used to treat a wide range of diseases, patients purchase and use H_2 water or H_2 gas inhalers independently without medical supervision, which may worsen medical conditions. Although many clinical studies have been conducted, research on doses and dosages for individual diseases is in the initial stages. Further basic and clinical research on individual diseases will be needed. In addition, a target molecule of H_2 was recently identified by Jin et al. [59]. An oxidized form of porphyrin catalyzes the reaction of H_2 with OH to reduce oxidative stress [59]. However, research into H_2 target molecules is still in its early stages, so further research is needed.

Conclusions

We herein reported the effects of H_2 on various experimental cellular and animal models and several human chronic inflammatory diseases including ME/CFS and post-COVID-19. We showed that it exerted effects via the protection of mitochondrial function due to its regulation of mtROS. In addition, we have shown that oxidative stress and chronic inflammation caused by mitochondrial dysfunction may cause various diseases. Since future medicine will require the development of technologies and substances that protect and activate mitochondrial function, H_2 may be positioned as a candidate in future medicine based on its effect on mitochondrial function.

References

1. Roger AG, Muñoz-Gómez SA, Kamikawa R (2017) The origin and diversification of mitochondria. Curr Biol 27:R1177–R1192
2. Ohta S (2014) Molecular hydrogen as a preventive and therapeutic medical gas: Initiation, development and potential of hydrogen medicine. Pharmacol Ther 144:1–11
3. Ohta S (2015) Molecular hydrogen as a novel antioxidant: overview of the advantages of hydrogen for medical applications. Methods Enzymol 555:289–317
4. Ohsawa I, Ishikawa M, Takahashi K et al (2007) Hydrogen acts as a therapeutic antioxidant by selectively reducing cytotoxic oxygen radicals. Nat Med 13:688–694
5. Nogueira JE, Passaglia P, Mota CMD et al (2018) Molecular hydrogen reduces acute exercise-induced inflammatory and oxidative stress status. Free Radic Biol Med 129:186–193
6. Wang D, Wang L, Zhang Y, Zhao Y, Chen G (2018) Hydrogen gas inhibits lung cancer progression through targeting SMC3. Biomed Pharmacol 104:788–797
7. Hirano Si, Ichikawa Y. Sato B, Takefuji Y, Satoh F (2021) Molecular hydrogen as a potential clinically applicable radioprotective agent. Int J Mol Sci 22:4566
8. Kawamura T, Wakabayashi N, Shigemura N et al (2013) Hydrogen gas reduces hyperoxic lung injury via the Nrf2 pathway in vivo. Am J Physiol Lung Cell Mol Physiol 304:L646–L656
9. Li S, Takahara T, Que W, et al (2021) Hydrogen-rich water protects liver injury in nonalcoholic steatohepatitis though HO-1 enhancement via IL-10 and Sirt 1 signaling. Am J Physiol Gastrointest Liver Physiol 320:G450–63
10. Iuchi K, Imoto A, Kamimura N et al (2016) Molecular hydrogen regulates gene expression by modifying the free radical chain reaction-dependent generation of oxidized phospholipid mediators. Sci Rep 6:18971

11. Ohta S (2021) Development of hydrogen medicine and biology: potential for various applications in diverse fields. Curr Pharm Des 27:583–584
12. Hirano Si, Yamamoto H, Ichikawa Y, Sato B, Takefuji Y, Satoh F (2021) Molecular hydrogen as a novel antitumor agent: possible mechanisms underlying gene expression. Int J Mol Sci 22:8721
13. Setsukinai KI, Urano Y, Kakinuma K, Majima HJ, Nagano T (2003) Development of novel fluorescence probes that can reliably detect reactive oxygen species and distinguish specific species. J Biol Chem 278:3170–3175
14. Halliwell B, Gutteridge JMC (2015) Chapter 10. Reactive species in disease: friends or foes? In: Free radicals in biology and medicine, 5th edn. Oxford University Press, Oxford, pp 511–638
15. Sohal RS, Weindruch R (1996) Oxidative stress, caloric restriction, and aging. Science 273:59–63
16. Cutler RG (1983) Superoxide dismutase, longevity and specific metabolic rate. A reply. Gerontology 29:113–120
17. Ngo V, Duennwald ML (2022) Nrf2 and oxidative stress: a general overview of mechanisms and implications in human disease. Antioxidants (Basel) 11:2345
18. Levine AJ, Oren M (2009) The first 30 years of p53: growing ever more complex. Nat Rev Cancer 9:749–758
19. Zerbes RM, van der Klei IJ, Veenhuis M, Pfanner M, van der Laan M, Bohnert M (2012) Mitofilin complexes: conserved organizers of mitochondrial membrane architecture. Bio Chem 393:1247
20. Gorman GS, Chinnery PF, DiMauro S, et al (2016) Mitochondrial diseases. Nat Rev Dis Primers 2:16080
21. Emma F, Montini G, Parikh SM, Salviati L (2016) Mitochondrial dysfunction in inherited renal disease and acute kidney injury. Nat Rev Nephrol 12:267
22. Archer SL (2013) Mitochondrial dynamics–mitochondrial fission and fusion in human diseases. N Engl J Med 369:2236
23. Rahman J, Rahman S (2018) Mitochondrial medicine in the omics era. Lancet 391:2560
24. Guerrero-Molina MP, Morales-Conejo M, Delmiro A et al (2023) High-dose oral glutamine supplementation reduces elevated glutamate levels in cerebrospinal fluid in patients with mitochondrial encephalomyopathy, lactic acidosis and stroke-like episodes syndrome. Eur J Neurol 30:538–547
25. Murakami Y, Ito M, Ohsawa I (2017) Molecular hydrogen protects against oxidative stress-induced SH-SY5Y neuroblastoma cell death through the process of mitohormesis. PLoS ONE 12:e0176922
26. Cui Y, Zhang H, Ji M et al (2014) Hydrogen-rich saline attenuates neuronal ischemia–reperfusion injury by protecting mitochondrial function in rats. J Surg Res 192:564–572
27. Zhang CS, Han Q, Song ZW, Jia HY, Shao TP, Chen YP (2021) Hydrogen gas post-conditioning attenuates early neuronal pyroptosis in a rat model of subarachnoid hemorrhage through the mitoKATP signaling pathway. Exp Ther Med 22:836
28. Zhang Y, Li H, Yang C et al (2016) Treatment with hydrogen-rich saline delays disease progression in a mouse model of amyotrophic lateral sclerosis. Neurochem Res 41:770–778
29. Lin CL, Huang WN, Li HH et al (2015) Hydrogen-rich water attenuates amyloid β-induced cytotoxicity through upregulation of Sirt1-FoxO3a by stimulation of AMP-activated protein kinase in SK-N-MC cells. Chem Biol Interact 240:12–21
30. Yoshida A, Asanuma H, Sasaki H et al (2012) H2 mediates cardioprotection via involvements of K(ATP) channels and permeability transition pores of mitochondria in dogs. Cardiovasc Drugs Ther 26:217–226
31. Yu YS, Zheng H (2012) Chronic hydrogen-rich saline treatment reduces oxidative stress and attenuates left ventricular hypertrophy in spontaneous hypertensive rats. Mol Cell Biochem 365:233–242
32. Xie K, Wang Y, Yin L et al (2021) Hydrogen gas alleviates sepsis-induced brain injury by improving mitochondrial biogenesis through the activation of PGC-α in mice. Shock 55:100–109

33. Jiao Y, Yu Y, Li B et al (2020) Protective effects of hydrogen-rich saline against experimental diabetic peripheral neuropathy via activation of the mitochondrial ATP-sensitive potassium channel channels in rats. Mol Med Rep 21:282–290
34. Liu Q, Li BS, Song YJ et al (2016) Hydrogen rich saline protects against mitochondrial dysfunction and apoptosis in mice with obstructive jaundice. Mol Med Rep 13:3588–3596
35. Wallet SM, Puri V, Gibson FC (2018) Linkage of infection to adverse systemic complication: periodontal disease, toll-like receptors, and other pattern recognition systems. Vaccines 6:21
36. Elliott EI, Sutterwala FS (2015) Initiation and perpetuation of NLRP3 inflammasome activation and assembly. Immunol Rev 265:35–52
37. Juliana C, Fernandes-Alnemri T, Kang S, Farias A, Qin F, Alnemri ES (2012) Non-transcriptional priming and deubiquitination regulate NLRP3 inflammasome activation. J Biol Chem 287:36617–36622
38. Hirano Si, Ichikawa Y, Sato B, Yamamoto H, Takefuji Y, Satoh F (2021) Potential therapeutic application of hydrogen in chronic inflammatory disease: Possible inhibiting role on mitochondrial stress. Int J Mol Sci 22:2549
39. Ren JD, Wu XB, Jiang R, Hao DP, Liu Y (2016) Molecular hydrogen inhibits lipopolysaccharide-triggered NLRP3 inflammasome activation in macrophages by targeting the mitochondrial reactive oxygen species. Biochim Biophys Acta 1863:50–55
40. Ren JD, Ma J. Hou J, et al (2014) Hydrogen-rich saline inhibits NLRP3 inflammasome activation and attenuates experimental acute pancreatitis in mice. Mediat Inflamm 2014:930894
41. Shao A, Wu H, Hong Y et al (2016) Hydrogen-rich saline attenuated subarachnoid hemorrhage-induced early brain injury in rats by suppressing inflammatory response: possible involvement of NF- kB pathway and NLRP3 inflammasome. Mol Neurobiol 53:3462–3476
42. Wang C, Li J, Yang R, Zhang JH, Cao YP, Sun XJ (2011) Hydrogen-rich saline reduces oxidative stress and inflammation by inhibit of JNK and NF-kB activation in a rat model of amyloid-beta-induced Alzheimer's disease. Neurosci Lett 491:127–132
43. Liu Q, Shen WF, Sun HY et al (2010) Hydrogen-rich saline protects liver injury in rats with obstructive jaundice. Liver Int 30:958–968
44. Itoh T, Hamada N, Terazawa R et al (2011) Molecular hydrogen inhibits lipopolysaccharide/interferon -induced nitric oxide production of signal transduction in macrophages. Biochem Biophys Res Commun 411:143–149
45. Tao B, Liu L, Wang W, Jiang J, Zhang J (2016) Effects of hydrogen-rich saline on aquaporin 1,5 in septic rat lungs. J Surg Res 202:291–298
46. Guan WJ, Wei CH, Chen AL et al (2020) Hydrogen/oxygen mixed gas inhalation improves disease severity and dyspnea in patients with Coronavirus disease 2019 in a recent multicenter, open-label clinical trial. J Thorac Dis 12:3448–3452
47. Poenaru S, Abdallah SJ, Corrales-Medina V, Cowan J (2021) COVID-19 and post-infectious myalgic encephalomyelitis/chronic fatigue syndrome: a narrative review. Ther Adv Infectious Dis 8:1–16
48. Nath A (2020) Long-haul COVID. Neurology 95:559–560
49. Botek M, Kreˇjcí J, Valenta M et al (2022) Molecular hydrogen positively affects physical and respiratory function in acute post-COVID-19 patients: a new perspective in rehabilitation. Int J Environ Res Public Health 19:1992
50. Fukuda K, Straus SE, Sharpe MC, Dobbins JG, Komaroff A (1994) The chronic fatigue syndrome: a comprehensive approach to its definition and study. International chronic fatigue syndrome study group. Ann Intern Med 121:953–9
51. Boles RG, Zaki EA, Kerr JR, Das K, Biswas S, Gardner A (2015) Increased prevalence of two mitochondrial DNA polymorphisms in function disease: are we describing different parts of an energy-deleted elephant? Mitochondrion 23:1–6
52. Billing-Ross P, Germain A, Ye K, Keinan A, Gu Z, Hanson MR (2016) Mitochondrial DNA variants correlate with symptoms in myalgic encephalomyelitis/chronic fatigue syndrome. J Transl Med 14:19
53. Missailidis D, Annesley SJ, Allan CY et al (2020) An isolated complex v inefficiency and dysregulated mitochondrial function in immortalized lymphocytes from ME/CFS patients. Int J Mol Sci 21:3

54. Holden S, Maksoud R, Eaton-Fitch N, Cabanas H, Staines D, Marshall-Gradisnik S (2020) A systematic review of mitochondrial abnormalities in myalgic encephalomyelitis/chronic fatigue syndrome/systemic exertion intolerance disease. J Transl Med 18:290
55. Tomas C, Brown A, Strassheim V, Elson J, Newton J, Manning P (2017) Cellular bioenergetics is impaired in patients with chronic fatigue syndrome. PLoS ONE 12:e018802
56. Castro-Marrero J, Sáez-Francàs N, Segundo MJ (2016) Effect of coenzyme Q10 plus nicotinamide adenine dinucleotide supplementation on maximum heart rate after exercise testing in chronic fatigue syndrome—a randomized, controlled, double-blind trial. Clin Nutr 35:826–834
57. Hirano S, Ichikawa Y, Sato B, Takefuji Y, Satoh F (2022) Molecular hydrogen as a medical gas for the treatment of myalgic encephalomyelitis/chronic fatigue syndrome (ME/CFS): possible efficacy based on a literature review. Front Neurol 13:841310
58. Hirano S, Ichikawa Y, Sato B, Takefuji Y, Satoh F (2024) Successful treatment of myalgic encephalomyelitis/chronic fatigue syndrome using the hydrogen gas: four case reports. Med Gas Res 14: 84–86
59. Jin Z, Zhao P, Gong W, Ding W, He Q (2023) Fe-porphyrin: a redox-related biosensor of hydrogen molecule. Nano Res 16:2020–2025

Chapter 4
Molecular Hydrogen: A New Treatment Strategy of Mitochondrial Disorders

Anna Gvozdjáková, Jarmila Kucharská, Zuzana Sumbalová, Zuzana Rausová, Branislav Kura, Barbora Bartolčičová, and Ján Slezák

Abstract Disturbances of mitochondrial function and oxidative stress are considered to be the molecular basis of the origin and development of various diseases, including mitochondrial diseases. The beneficial effect of molecular hydrogen (H_2) has been proven in the prevention and supportive therapy of patients with cardiovascular disease, Parkinson's disease, in patients with metabolic syndrome, in respiratory system disease, in oncology patients treated with radiation, in cerebral infarction, in diabetes mellitus, in rheumatoid arthritis. Exact molecular mechanisms of H_2 on mitochondrial level are not fully understood. We proposed new mechanism of the H_2 effect in mitochondrial respiratory chain function. H_2 may be a donor of both electron and proton to the Q-cycle of the mitochondrial respiratory chain and thus

A. Gvozdjáková (✉) · J. Kucharská · Z. Sumbalová · Z. Rausová
Faculty of Medicine, Pharmacobiochemical Laboratory of 3rd Department of Internal Medicine, Comenius University in Bratislava, Sasinkova 4, 811 08 Bratislava, Slovakia
e-mail: anna.gvozdjakova@fmed.uniba.sk

J. Kucharská
e-mail: jarmila.kucharska@fmed.uniba.sk

Z. Sumbalová
e-mail: zuzana.sumbalova@fmed.uniba.sk

Z. Rausová
e-mail: zuzana.rausova@fmed.uniba.sk

B. Kura · J. Slezák
Center of Experimental Medicine, Institute for Heart Research, Slovak Academy of Sciences, Bratislava, Slovakia
e-mail: branislav.kura@savba.sk

J. Slezák
e-mail: jan.slezak@savba.sk

B. Bartolčičová
Slovak Technical University, Department of Physiotherapy, Bratislava, Slovakia

Slovak Medical University, Department of Physiotherapy, Bratislava, Slovakia

B. Bartolčičová
e-mail: barbora@bartolcic.sk

J. Slezak and B. Kura (eds.), *Molecular Hydrogen in Health and Disease*, Advances in Biochemistry in Health and Disease 27, https://doi.org/10.1007/978-3-031-47375-3_4

can preserve coenzyme Q level with the subsequent ATP production via oxidative phosphorylation. H_2 was shown to alter the direction of the electron flow of mitochondrial respiratory chain system, which depends on NAD^+/NADH ratio. We also found beneficial effect of H_2 on platelet mitochondrial bioenergy function in patients with NAFLD. The application of H_2 appears to be a new treatment strategy for targeted therapy of mitochondrial disorders.

Keywords Molecular hydrogen · Mitochondria · Oxidative stress · Mitochondrial diseases · Coenzyme Q10 · Oxidative phosphorylation

Introduction

Mitochondria are subcellular organelles found almost in all eukaryotic cells. They serve as centers of genetic information, as central integrators of intermediatory metabolism: oxidative phosphorylation, fatty acid oxidation, Krebs cycle, gluconeogenesis, urea cycle, ketogenesis. Mitochondria are the main energy producers in the body from carbohydrates, fats, and proteins, they are main sites of reactive oxygen species production. Disturbances of mitochondrial function and oxidative stress are considered to be the molecular basis of the origin and development of various diseases, including mitochondrial diseases. Targeted diagnostic and therapeutic approaches to the regeneration of damaged mitochondria are being sought and developed. Application of molecular hydrogen (H_2) appears to be a new therapeutic strategy option for mitochondrial disorders.

Molecular Hydrogen–A Novel Treatment Strategy

The first work on the effects of molecular hydrogen as a natural substance with antioxidant properties was published in 2007 [1]. H_2 is tasteless, odorless, colorless small molecule. It can diffuse deep into cells, easily cross the blood–brain barrier, and eliminate free radicals in the brain. H_2 neutralizes free oxygen radicals, selectively hydroxyl radical (OH) [2], indirectly reduces nitric oxide (NO) production and peroxynitrite radical (ONOO–). H_2 does not neutralize superoxide anion radical ($O_2^{\bullet -}$), hydrogen peroxide (H_2O_2), and nitric oxide (NO) [1]. H_2 is the non-polar diatomic compound, a molecule consisting of two protons and two electrons, the smallest antioxidant with a molecular weight of 2 compared to other antioxidants. Molecular weight of $H_2 = 2$; vitamin C = 176; vitamin E = 431 and coenzyme Q_{10} = 863, Fig. 4.1.

Molecular hydrogen is a substance widely used in medicine. The beneficial effect of molecular hydrogen has been proven in the prevention and supportive therapy of patients with various diseases [3, 4], such as cardiovascular disease, Parkinson's disease [5], in patients with metabolic syndrome [6], in respiratory system disease,

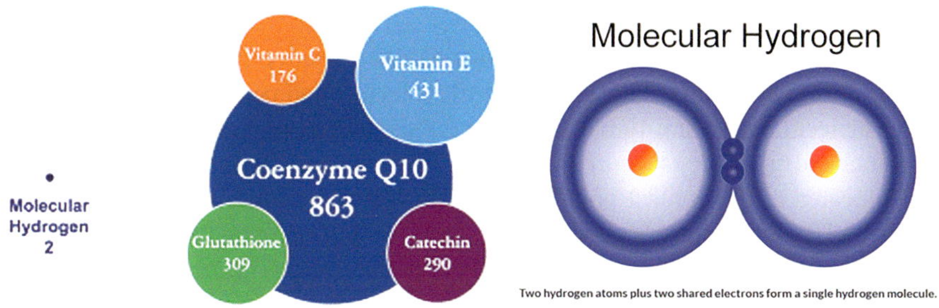

Fig. 4.1 Molecular weight of H_2 in comparison with other antioxidants (according to https://encyclopedia.pub/entry/history/show/94845)

in oncology patients treated with radiation, in cerebral infarction [7], in diabetes mellitus [8], in rheumatoid arthritis [9]. All these diseases are associated with mitochondrial disturbances [10–12]. Inhaled gas mixture (67% H_2, 33% O_2) by patients with COVID-19 improved the dyspnea, cough, chest distress and chest pain [13]. Molecular hydrogen inhalation may improve symptoms of inflammations caused by COVID-19 [14, 15]. Explanation of the positive effect of H_2 in COVID-19 is related to H_2 properties, as a small nonpolar molecule that rapidly passes into the tissues and cells. H_2 can suppress proinflammatory cytokines, selectively reduce ROS production and improve mitochondrial bioenergetics. Mechanism of H_2 in COVID-19 therapy is not exactly known, its elucidation requires further studies [16].

Molecular hydrogen can be applied by several ways, as inhalation, drinking the water enriched with H_2, using a saline solution with H_2, a water bath with H_2, or in eye drops–a saline solution with H_2 [17]. Benefits of H_2 include antioxidant effects, anti-apoptotic and anti-inflammatory effects, the regulation of gene expression, prevention of excessive activation of pyroptosis–programmed cell death (H_2 can inhibit pyroptosis–the inflammatory condition initiated by free radicals) [18]. H_2 suppresses inflammation, oxidative stress, reduces cell apoptosis, it regulates cell differentiation, the expression of several genes or signaling pathways and affects the body's energy metabolism. Recently a new mediator of mitochondrial function (mitochondrial unfolded protein response–mtUPR) was found, important for enhanced proliferation of cancer cells by H_2 [19].

Characteristics of Mitochondria and Molecular Hydrogen

Amount of mitochondria in cell depends on the type, function, size of the cell and the energy need of the organ/tissue to which the cell belongs. Platelets are small blood cells with diameter 3–4 μm containing a small amount of mitochondria, from two to 8 mitochondria, the oocyte with diameter 110–120 μm has close to 100,000 mitochondria. In cardiomyocyte the mitochondria occupy approximately 38% of the cell volume.

Mitochondria consist of four components, each of these components have specific functions. Outer mitochondrial membrane (OMM) is smooth, separates the mitochondria from the cytoplasmic space, is little selective and permeable for molecules with a molecular weight < 10.000. OMM is the site of transport proteins from cytosol to mitochondria. The transport of proteins from the matrix needs complex TIM/TOM located at the contact points, at mitochondrial permeability transition pores (MPTP). The inner mitochondrial membrane (IMM) separates the intermembrane space from the matrix. The folding of IMM (cristae) serves to increase the surface area of this membrane. In the (IMM the respiratory system is localized, associated with ATP formation via oxidative phosphorylation (OXPHOS). Respiratory chain system contains five complexes, coenzyme Q-cycle and cytochrome c, c1. Complex I–NADH-CoQ reductase; Complex II–succinate-CoQ-reductase; Complex III–CoQ-cytochrome c reductase; Complex IV–cytochrome oxidase; Complex V–ATP-synthase. Cytochrome c and coenzyme Q are two mobile components of the respiratory chain system [20].

Complexes I, III and IV are organized in more stable form in *supercomplexes,* called respirasomes. In supercomplexes, the electron carriers (coenzyme Q and cytochrome c) have a short diffusion distance, therefore electron transfer is more effective through complexes. Supercomplexes are thought to reduce oxidative damage and increase the safety of metabolism [21]. Molecular hydrogen quickly passes through biological membranes, including mitochondria.

Mitochondrial Free Radicals, Antioxidants and H_2

Mitochondria are one of the major sources of *reactive oxygen radicals* (ROS). Over 90% of cellular ROS, as superoxide radical ($O_2^{\cdot-}$), hydrogen peroxide (H_2O_2) and a highly reactive hydroxyl radical (•OH) are formed in the respiratory chain system. Other reactive oxygen species include singlet oxygen ($^1O_2$) and peroxynitrite ($ONOO^{\cdot}$). Mitochondria are also a source of *reactive nitrogen species* (RNS) as nitric oxide ($NO^{\cdot}$) and nitrogen dioxide ($NOO^{\cdot}$). Nitric oxide regulates the cardiovascular system, central nervous system and immune system. NO with cytochrome oxidase reduces oxygen consumption and ATP production by mitochondria. Mitochondrial nitric oxide is consumed by the rapid reaction with superoxide radical, and peroxynitrite anion is formed ($ONOO^-$). Nitric oxide with ubiquinol generates ubisemiquinone radical, which subsequently forms superoxide radical [22]. Mitochondria are protected against uncontrolled formation of ROS by *antioxidant* systems, including MnSOD, catalase, glutathione peroxidase, cytochrome c and coenzyme Q_{10} [23, 24]. Antioxidant effect of H_2 has been shown in various pathological conditions that involve oxidative stress [25, 26]. High concentration of NO can trigger inflammatory process associated with ageing [27].

Mitochondrial Dynamics, Sirtuins and H_2

Mitochondria are essential organelles that produce the majority of energy in eukaryotes by converting lipids and carbohydrates to adenosine triphosphate (ATP) via oxidative phosphorylation (OXPHOS). Mitochondria, the highly *dynamic structures* are controlled by four processes: 1. *Biogenesis*–new mitochondria are created and their number/or volume are increased; 2. Mitochondrial *fission* is a multi-step process allowing the division of one mitochondrion in two daughter mitochondria. Mitochondrial fission components are involved in disease 3. Mitochondrial *fusion* is a process where by two neighboring mitochondria tether, and then fuse their outer membranes. The small number of large and prolonged mitochondria is formed by *fusion*. Loss of fusion leads to a fragmented mitochondrial network. 4. Loss of mitochondrial membrane potential can lead to selective degradation damaged mitochondria by *mitophagy*. Loss of flexibility of mitochondria is linked to metabolic diseases [28–30].

Mitochondrial ROS production has been linked to circadian clock and mitochondrial dynamic processes (fission, fusion, biogenesis, and mitophagy) are highly dependent on a viable circadian rhythm and its disturbance leads to altered respiration of mitochondria [31].

Mitochondrial sirtuins are NAD^+- dependent proteins–deacetylases, essential for the normal mitochondrial function through interaction and modification of the amount mitochondrial proteins [32]. Three sirtuins are localized in mitochondria: *SIRT 3* is included in fatty acid oxidation, ketogenesis, oxidative phosphorylation, antioxidant effect and amino acid metabolism, it interacts with complex I (CI) and complex III (CIII). *SITR 4* binds the adenine nucleotide translocator. *SIRT5* interacts with cytochrome c. The beneficial effect of sirtuins have been demonstrated in the prevention of various diseases such as inflammation, obesity, diabetes, neurodegenerative, cardiovascular, and oncological diseases [32]. Sirtuins are sensors involved in the regulation of metabolism during stress, aging, regulation of ATP formation, apoptosis and cell signaling [33, 34]. The effect of H_2 on sirtuins requires further studies.

Mitochondrial Circadian Rhythms and H_2

Circadian rhythms are physical, mental, and behavioral changes that follow near 24-h cycles. It is believed that the development of circadian rhythms is a response to the earth rotation, both around its axis and around the sun [35]. Circadian rhythms influence ageing, metabolism, health, and diseases, including mitochondrial diseases. Metabolic rate is related to body size and temperature. Circadian rhythms in relation to mitochondria affect many fields of medicine, as ageing, neurology, cardiology, diabetes, metabolism. Circadian clocks are coupled to cellular metabolism, to feeding cycles. Caloric restriction is achieved by means of intermittent energy restriction or time-restricted feeding during various diseases, as cancer, cardiovascular diseases, diabetes, and neurodegenerative disorders. It is believed that mitochondria are an integral part of these processes [36]. Mitochondria undergo oscillations, a circadian rhythm with a period of 24 h, or a circasemidian biological rhythm with a period of 12 h. In mammal's circadian rhythms are regulated by suprachiasmatic nucleus (SCN), located in the hypothalamus. SCN synchronizes circadian rhythm approximately 24 h, with the exact 24-h daily rhythm. SCN can synchronize the different peripheral tissue clock in the body via various signaling cascades. The circadian clock regulates sleeping, temperature, physical activity, and food intake [37].

Circadian biorhythms of mitochondria and cells are crucial for the living organism, for health and for ageing. Hypothesis of important redox, bioenergetics and temperature regulation of mitochondrial and nucleus circadian rhythms was published. Sleep is mainly protective for mitochondria and wakefulness for nucleus. During sleep–fusion remodels mitochondria, which activates immunity, inflammation, and heat shock responses. During wakefulness–high metabolic rate induces oxidative stress and redox imbalance [37]. Circadian rhythmicity regulates morphology of mitochondria and their distribution in cytoplasm. The circadian clock rhythmically regulates the biosynthesis of NAD^+ and mitochondrial capacity for energy production. Disrupted mitochondria lack circadian rhythmicity. Mitochondrial NAD^+ also determines the activity of the deacetylases SIRT1 and SIRT3 [38].

Molecular hydrogen positively modulates mitochondrial activity in various cancer cells [39], mitochondrial energy production and cell death (apoptosis, autophagy, pyroptosis, ferroptosis), and circadian rhythms [40]. Circadian variations—cascade of mitochondrial OXPHOS and circadian clock of coenzyme Q may contribute to the understanding of the pathogenesis of altered brain and myocardium function and mechanisms underlying the trigger of an acute brain attack or acute myocardial infarction [35]. The function of respiratory chain complexes of cardiac muscle mitochondria changes during the 24-h daily cycle. In experimental animals, 12 and 24-h biological rhythms have been demonstrated in isolated cardiac muscle mitochondria, Complex I and Complex II of the mitochondrial respiratory chain system showing 2 maxima (PEAKS) and 2 minima (NADIRS) during 24 h [41, 42]. Decreased 12-h heart mitochondrial repiratory chain function and ATP production by OXPHOS between 16:29 h and 19:31 h and between 4:18 h and 7:38 h, as well as CoQ_{10}-CLOCK changes may contribute to the explanation of mechanisms triggering an acute heart attack [42]. Effect of molecular hydrogen is independent on circadian clock [43].

Mitochondrial Ageing and H_2

Ageing is a process leading to the reduction of mitochondrial ATP production, to the mitochondrial deoxyribonucleic acid (mtDNA) damage, oxidation of mitochondrial proteins, dysregulation of mitochondrial dynamics. These ageing changes lead to the development of diseases such as cardiovascular, metabolic, and neurodegenerative diseases. Therefore, mitochondria are important therapeutic targets for H_2 effect in prevention and treatment of various diseases related to ageing [44, 45].

Beneficial effect of H_2 has been shown in mitochondrial ageing [46], in neurodegenerative diseases such as Parkinson's and Alzheimer's. Clinical studies show the beneficial effect of H_2 therapy in brain infarction, dyslipidemia, endothelial dysfunction, atherosclerosis, cardiovascular diseases, and metabolic syndrome [40].

Oxidative stress may play a critical role in ageing. ROS and RNS are produced by aerobic cells. The majority of the ROS production is in the mitochondrial electron transport system, primarily by complexes I and III. ROS are produced mainly by endogenous way in mitochondrial nicotinamide adenine dinucleotide phosphate (NADPH) oxidase or by exogenous effects such as smoking, alcohol, heavy metals, and radiation [47, 48]. H_2 decreases ROS production, suppresses electron leakage in the mitochondrial electron transport system. H_2 prevents superoxide generation in the mitochondrial Complex I and reduces mitochondrial membrane potential [49]. These H_2 abilities show a new way to delay mitochondrial ageing and ageing-related mitochondrial disorders.

Mitochondrial Oxidative Phosphorylation and H_2 Effect

Molecular hydrogen is a novel antioxidant with great potential for medical applications and for improving the pathology of mitochondrial disorders. It showed scavenging effect on hydroxyl radicals in cultured cells, protective effect on myocardial and brain injury in rat model of ischemia–reperfusion, preventive and therapeutic effects on Parkinson disease model, on atherosclerotic model, prevention of adverse effects of cis-platin as an anti-cancer drug, protective effect in model after kidney rat transplantation [50].

In an experimental models of kidney disease, H_2 ability to inhibit the expression of pro-inflammatory cytokines was demonstrated. Positive effect of H_2 was proved in experimental rat models of renal disease (nephrotoxicity induced by cisplatin, cyclosporine A), in renal injury in spontaneously hypertensive rats, in prevention of septic acute kidney injury [51]. H_2 selectively reduces oxidative stress [3]. In an experimental model of traumatic brain injury, drinking the H_2 water can potentially reverse the effects on edema formation, tau pathology, neuroinflammation and gene expression. Other study has shown that H_2 increased mitochondrial basal respiration but had no effects on ATP production rate [52].

H_2 can prevent mitochondrial oxidative stress by several mechanisms, as by blocking the opening of the mitochondrial permeability transition pore, or by regulation of mitochondrial dynamics. H_2 can target mitochondria by increasing ATP production by stimulation of mitochondrial electron transport system, by increasing mitochondrial coenzyme Q production, and by increasing antioxidants protection [53, 54]. Recently was proved that consumption of H_2 rich water (HRW) resulted in stimulation of rat cardiac mitochondrial respiratory chain system, increased ATP production connected with CI and CII substrates [53]. CoQ_9 in plasma, myocardial mitochondria and myocardial tissue increased, MDA in plasma decreased [54].

A new mechanism of the H_2 effect on mitochondria in the Q-cycle and in mitochondrial respiratory chain function was proposed [49, 53]. The Q-cycle contains three coenzyme Q forms: coenzyme Q in oxidized form (ubiquinone), radical form (semiquinone), or reduced form (ubiquinol). H_2 may be a donor of both electron and proton in the Q-cycle and thus can stimulate coenzyme Q production with the subsequent ATP production via oxidative phosphorylation [9, 53], Fig. 4.2. These results indicate that H_2 may function to prevent/treat disease states with disrupted myocardial mitochondrial function.

In human medicine, H_2 reduced oxidative stress in patients with type 2 diabetes, with potential metabolic syndrome, in patients with MELAS (mitochondrial diseases) blood lactate and pyruvate decreased [50]. Positive effect of H_2 was documented also in prevention of cognitive decline, atherosclerosis, allergic reactions, inflammation, oxidation, glaucoma model, radiation injury, and other organs injury (brain, liver, kidney, heart, lung, pancreas, intestine) [17]. H_2 selectively reduced the hydroxyl radical in acute rat model of oxidative stress-induced damage of brain. Inhalation of

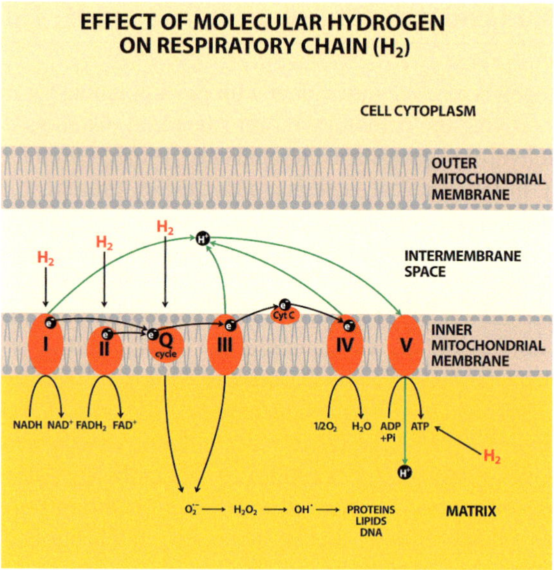

Fig. 4.2 Effect of H_2 on mitochondrial respiratory chain function [53] *Legend:* H_2–molecular hydrogen; Respiratory chain complexes I, II, III, IV, V; Q-cycle – Coenzyme Q cycle; cyt ccytochrome c; e^-–electron; H^+–proton; NADH–reduced nicotine adenine dinucleotide; NAD^+–nicotine adenine dinucleotide; $FADH_2$–reduced flavine adenine dinucleotide; O_2–oxygen; H_2O–water; ADP–adenosine diphosphate; Pi–inorganic phosphate; ATP–adenosine triphosphate; $O_2^{\cdot-}$–superoxide radical; H_2O_2–hydrogen peroxide; •OH–hydroxyl radical; DNA–deoxyribonucleic acid

H_2 markedly suppressed brain injury. The authors proposed that H_2 could be widely used in medical applications as an effective antioxidant [1].

Possible next mechanism of H_2 effect on the mitochondrial respiratory chain function was recently shown. Complex I of the mitochondrial respiratory chain is close to the energy converting by membrane-bound metals (Ni, Fe)—hydrogenases. If the hydrogenase lacks these metals in Complex I, H_2 cannot be utilized by Complex I [3]. Metal-free enzyme activation has been reported by a *frustrated Lewis pair* (FLP) mechanism where the combination of Lewis base and acid could activate H_2. If H_2 is activated in Complex I—CoQ mitochondrial respiratory chain, it could alter electron flow, starting from Q-cycle to the two-ways directions: from CoQ to NAD^+ as the RET (reverse electron transfer) or by conversion of quinone to quinol as the FET (forward electron transfer); and the transfer of quinol forwards Complex III and/or Q-pools [55]. H_2 was shown to alter the direction of the electron flow which depends on NAD^+/NADH ratio. H_2 may neutralize semiquinone radicals, reduce superoxide radicals produced in Complex III of the respiratory chain, and suppress oxidative damage of mitochondria [49]. Exact molecular mechanisms of H_2 on mitochondrial level are not fully understood.

Effect of H_2 on Mitochondrial Dysfunction in NAFLD Patients

Liver is the most important and an extremely metabolically active organ containing a huge number of mitochondria. Hepatocytes, liver cells, are involved in anabolic (gluconeogenesis, lipogenesis, glutaminogenesis) and catabolic (glycolysis, lipolysis, ureagenesis) metabolic activities. Mitochondria of hepatocytes play a key role in these metabolic ways [56].

Mitochondrial dysfunction in NAFLD: Liver injury manifested by morphological and functional damage of mitochondria can be caused by various agents. One of the liver mitochondrial diseases is nonalcoholic fatty liver disease (NAFLD). NAFLD is a chronic liver disease caused by accumulation of excess fat in the liver in the absence steatosis to NASH (steatosis associated with tubular inflammation hepatocellular ballooning), which can be associated with fibrosis [57]. Mitochondrial dysfunction and oxidative stress are the major factors in the development and progression of NAFLD [58]. It is characterized by various degrees of ultrastructural mitochondrial damage, abnormal morphological changes, respiratory chain activity reduction, mitochondrial ATP reduction, increased OMM permeability, oxidative stress-mediated deletions mtDNA, and impairment mitochondrial beta-oxidation [59]. Exact mechanism by which mitochondrial dysfunction contributes to NAFLD is still not fully understood [60–62].

Mitochondrial dysfunction contributes to the pathogenesis of NAFLD since it affects hepatic lipid homeostasis, promotes ROS production and lipid peroxidation, cytokine release and cell death. NAFLD has been shown to be associated with

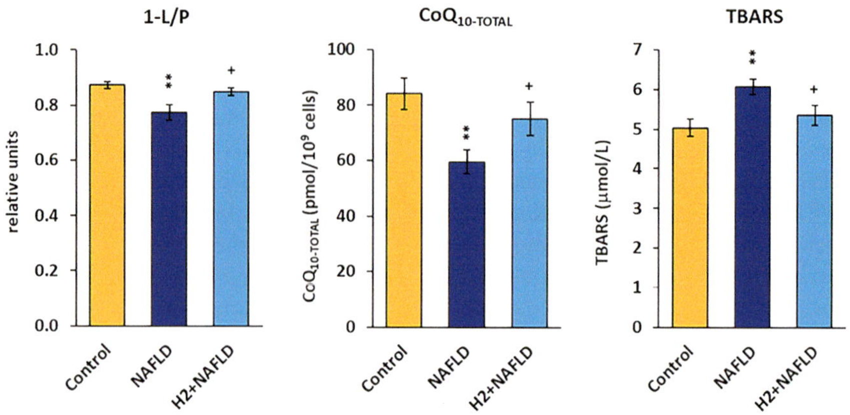

Fig. 4.3 Effect of H_2 on efficiency of OXPHOS in platelet mitochondria, coenzyme Q_{10} and oxidative stress in patients with NAFLD. *Legend:* 1-L/P–efficiency of OXPHOS, C–the control group; NAFLD–the group of patients with NAFLD before adjunctive therapy with H_2; H_2 + NAFLD–the group of patients with NAFLD after adjunctive therapy with H_2. **$p < 0.01$ vs the control group, + $p < 0.05$ vs the NAFLD group

paracrystalline inclusions. Uncoupling of the oxidation and the phosphorylation and increased free radicals production and lipid peroxidation causes cell injury [63].

In patients with NAFLD higher CI-linked LEAK respiration and lower CI-linked OXPHOS capacity was shown. CII-linked ET capacity was decreased. H_2 supplementary therapy improved the efficiency of OXPHOS in platelet mitochondria (described with parameter 1-L/P), increased endogenous coenzyme Q_{10} level and decreased oxidative stress in patients with NAFLD [64], (Fig. 4.3).

Supplementary therapy with H_2 could be a new treatment strategy for regeneration of mitochondrial disorders in patients with NAFLD.

Molecular Hydrogen–A New Perspective Strategy for Targeted Therapy of Mitochondrial Diseases

Mitochondrial diseases are characterized by clinical signs, such as muscle weakness, lack of energy production, increased free oxygen and nitrogen radicals, and formation of toxic metabolites. Primarily mitochondrial diseases are caused at the genetic level, may be caused by damage of nuclear DNA or mutations of mtDNA. The secondary cause of mitochondrial disturbances includes various external factors, as ischemia/reperfusion, cardiovascular diseases, neurodegenerative disorders, renal failures, pancreatic and hepatic damage, diabetes, sepsis, infections, oncological diseases, smoking, alcohol intake, stress and aging [65].

Protective effects of molecular hydrogen on many organs and systems were documented by several studies. However, the details of the molecular mechanisms of the therapeutic effects of H_2 remain unclear. It is not known whether H_2 can be used for circadian clocks regulation, for ferroptosis or pyroptosis regulation [40]. Based on several studies on the beneficial effect of H_2 on damaged mitochondrial respiratory system of individual body organs [1, 3, 5–9, 13–15, 40], the application of H_2 appears to be a new treatment strategy for targeted therapy of mitochondrial disorders [53, 64].

References

1. Ohsawa I, Ishikawa M, Takahashi K, Watanabe M, Nishimaki K, Yagamata K, Katsura K, Katayama Y, Asoh S, Ohta S (2007) Hydrogen as a therapeutic antioxidant by selectively reducing cytotoxic oxygen radicals. Nat Med 13(6):688–694
2. Slezák J, Kura B, Frimmel K, Zálešák M, Ravingerová T, Vinczenczová C, Okruhlicová Ľ, Tribulová N (2016) Preventive and therapeutic application of molecular hydrogen in situations with excessive production of free radicals. Physiol Res 65(Suppl. 1):S11–S28
3. Ohta S (2014) Molecular hydrogen as a preventive and therapeutic medical gas: initiation, development and potential of hydrogen medicine. Pharmacol Ther 144:1–11
4. Slezak J, Kura B, LeBaron TW, Singal PK, Buday J, Barancik M (2020) Oxidative stress and pathways of molecular hydrogen effects in medicine. Curr Pharm Des 26:1–16. https://doi.org/10.2174/1381612826666200821114016
5. Yoritaka A, Takanashi M, Hirayama M, Nakahara T, Ohta S, Hattori N (2013) Pilot study of H2 therapy in Parkinson's disease: a randomized double-blind placebo-controlled trial. Mov Disord 28:836–839
6. LeBaron TW, Singh RB, Fatima G, Kartikey K, Sharma JP, Ostojic SM, Gvozdjakova A, Kura B, Noda M, Mojto V et al (2020) The effects of 24-week, high-concentration hydrogen-rich water on body composition, blood lipid profiles and inflammation biomarkers in men and women with metabolic syndrome: a randomized controlled trial. Diabetes Metab Syndr Obes 13:889–896
7. Ono H, Nishijima Y, Ohta S, Sakamoto M, Kinone K, Horikosi T, Tamaki M, Takeshita H, Futatuki T, Ohishi W et al (2017) Hydrogen gas inhalation treatment in acute cerebral infarction: a randomized controlled clinical study on safety and neuroprotection. J Stroke Cerebrovasc Dis 26:2587–2594
8. Kajiyama S, Hasegawa G, Asano M, Hosoda H, Fukui M, Nakamura N, Kitawaki J, Imai S, Nakano K, Ohta M et al (2008) Supplementation of hydrogen-rich water improves lipid and glucose metabolism in patients with type 2 diabetes or impaired glucose tolerance. Nutr Res 28:137–143
9. Ishibashi T, Sato B, Rikitake M, Seo T, Kurokawa R, Hara Y, Naritomi Y, Hara H, Nagao T (2012) Consumption of water containing a high concentration of molecular hydrogen reduces oxidative stress and disease activity in patients with rheumatoid arthritis: an open-label pilot study. Med Gas Res 2:27
10. Gvozdjáková A, Sumbalová Z, Kucharská J, Szamosová M, Čápová Ľ, Rausová Z, Vančová O, Mojto V, Langsjoen P, Palacka P (2021) Platelet mitochondrial respiration and coenzyme Q10 could be used as new diagnostic strategy in rheumatoid diseases. PLOS ONE 16/9:e0256135. https://doi.org/10.1371/journal.pone.0256135
11. Gvozdjáková A, Sumbalová Z, Kucharská J, Komlósi M, Rausová Z, Vančová O, Számošová M, Mojto V (2020) Platelet mitochondrial respiration, endogenous coenzyme Q10 and oxidative stress in patients with chronic kidney disease. Diagnostics 10:176. https://doi.org/10.3390/diagnostics10030176

12. Sumbalova Z, Kucharska J, Palacka P, Rausova Z, Langsjoen PH, Langsjoen AM, Gvozdjakova A (2022) Platelet mitochondrial function and endogenous coenzyme Q10 levels are reduced in patients after COVID-19. Bratisl Med J 123(1):9–15. https://doi.org/10.4149/BLL_2022_002
13. Guan WJ, Wei CH, Chen AL, Sun XC, Guo GY, Zou X, Shi JD, Lai JD, Lai PZ, Zheng ZG, Zhong NS (2020) Hydrogen/oxygen mixed gas inhalation improves disease severity and dyspnea in patients with Coronavirus diseases 2019 in a recently multicenter open-label clinical trial. J Thorac Dis 12:3448–3452
14. Hirano S, Ichikawa Y, Sato B, Yamamoto H, Takfuji Y, Satoh F (2021) Potential therapeutic application of hydrogen in chronic inflammatory disease: possible inhibiting role on mitochondrial stress. Int J Mol Sci 22:2549. https://doi.org/10.3390/ijms22052549
15. Tian Y, Zhang Y, Wang Y, Chen Y, Fan W, Zhou J, Qiao J, Wei Y (2021) Hydrogen, a novel therapeutic molecule, regulates oxidative stress, inflammation, and apoptosis. Front Physiol 12:789507. https://doi.org/10.3389/fphys.2021.789507
16. Alwazzer D, Fuh-Ching Liu F, Wu XY, LeBaron TW (2021) Combating oxidative stress and inflammation in COVID-19 by molecular hydrogen therapy: mechanisms and perspective. Ox Med Cell Long, Article ID 5513868, pp 17. https://doi.org/10.1155/2021/5513868
17. Ohta S (2011) Recent progress toward hydrogen medicine: potential of molecular hydrogen for preventive and therapeutic application. Current Pharmaceut Design 17:2241–2252
18. Zha QB, Wei HX, Li CG, Liang YD, Xu LH, Bai WJ, Pan H, He XH, Ouyang DY (2016) ATP-induced inflammasone activation and pyroptosis in regulated by AMP-activated protein kinase in macrophages. Front Immunol 7:597
19. Hasegawa T, Ito M, Hasegawa S, Teranishi M, Takeda K, Negishi S, Nishiwaki H, Takeda J-I, LeBaron TW, Ohno K (2022) Molecular hydrogen enhances proliferation of cancer cells that exhibit potent mitochondrial unfolded protein response. Int J Mol Sci 23:2888
20. Gvozdjáková A (2018) Mitochondrial physiology. In: Gvozdjáková A, Cornélissen G, RB Singh (eds) Recent advances in mitochondrial medicine and coenzyme Q10, NOVA Sciences, NY, USA, pp 13–36
21. Otera H, Ishihara N, Mihara K (2013) New insight into the function and regulation of mitochondrial fision. Biochim Biophys Acta 1833:1256–1268
22. Schäfer E, Dencher NA, Vonck J, Parcej DN (2007) Three-dimensional structure of the respiratory chain supercomplexess II III2 IV1 from bovine heart mitochondria. Biochemistry 46(44):12579–12585
23. Ďuračková Z (2010) Some current insight into oxidative stress. Physiol Res 59:459–469
24. Gvozdjáková A (2008) Mitochondrial free radicals and antioxidants. In: Gvozdjáková A (ed) Mitochondrial medicine, Springer, Netherlands, pp 50–54
25. Kura B, Bagchi AK, Singal PK et al (2019) Molecular hydrogen: potential in mitigating oxidative stress-induced radiation injury. Can J Physiol Pharmacol 97(4):287–292
26. Slezák J, Kura B, LeBaron TW, Singal PK, Buday J, Barančík M (2021) Oxidative stress and pathways of molecular hydrogen effects in medicine. Current Pharmaceutical Design 27/5:610–625
27. Farrokhfall K, Hashtrouudi A, Ghasemi A, Mehrani H (2015) Comparison of inducible nitric oxide synthase activity in pancreatic islets of young and aged rats. Iran J Basic Med Sci 18(2):115–121
28. Ni HM, Williams JA, Ding WX (2015) Mitochondrial dynamics and mitochondrial quality control. Redox Biol 4:6–13
29. Yapa NMB, Lisnyak V, Reljic B, Ryan MT (2021) Mitochondrial dynamics in health and disease. FEBS Lett 595:1184–1204
30. Tilokani L, Nagashima S, Paupe V, Prudent J (2018) Mitochondrial dynamics: overview of molecular mechanisms. Essays in Biochem 62:341–360. https://doi.org/10.1042/EBC20170104
31. Goede P, Wefers J, Brombacher EC, Schrauwen P, Kalsbeek A (2018) Circadian rhythms in mitochondrial respiration. J Mol Endocrinol 60:R115–R130. https://doi.org/10.1530/JME-17-0196

32. Osborne B, Bentely NL, Montgomery MK, Turner N (2016) The role of mitochondrial sirtuins in health and disease. Free Rad Biol Med. https://doi.org/10.1016/j.freeradbiolmed.2016.04.197
33. Li X, Kagan N (2011) Mammalian sirtuins and energy metabolism. In J Biol Sci 7(5):575–587
34. Van de Ven RAH, Santos D, Haigis MC (2017) Mitochondrial sirtuins and molecular mechanims of aging. Trends Mol Med 23/4:320–331. https://doi.org/10.1016/molmed.2017.02.005
35. Cornélissen G, Gvozdjáková A, Lee Gierke C, Gumarova L, Sackett Lundeen L (2018) Chronobiology of mitochondria. In: Gvozdjáková A, Cornélissen G, Singh RB (eds) Recent advances in mitochondrial medicine and coenzyme Q10, NOVA Science, NY, USA, pp 37–56
36. Wilking M, Ndiaye M, Ahmad N (2013) Circardian rhythm connections to oxidative stress: implication for human health. Antioxidants and Redox Signaling 213(19/2). https://doi.org/10.1089/ars.2012.4889
37. Richardson RB, Maillous RJ (2023) Mitochondria need their sleep: redox, bioenergetics, and temperature regulation of circadian rhythms and the role of cystein-mediated redox signaling, uncoupling proteins and substrate cycles. Antioxidants 12:674. https://doi.org/10.3390/antiox12030674
38. Aguilar-López BA, Moreno-Altamirano MMB, Dockrell HM, Duchen MR, Sanchéz-Garcia FJ (2020) Mitochondria: an integrative hub coordinating circadian rhythms, metabolism, the microbiome, and immunity. Front Cell Developm Biol 8:51. https://doi.org/10.3389/fcll.2020.00051
39. Kawasaki H, Guan J, Tamama K (2010) Hydrogen gas treatment prolongs replicative lifespan of bone marrow multipotential stromal cells in vitro while preservating differentiation and paracrine potentials. Biochem Biophys Res Commun 397:608–613
40. Yang M, Dong Y, He Q, Zhu P, Zhuang Q, Shen J, Zhang X, Zhao M (2020) Hydrogen: a novel option in human disease treatment. Ox Med Cell Longevity Article ID 8384742. https://doi.org/10.1155/2020/8384742
41. Mikulecký M (2008) Methods of chronobiometrics analysis of mitochondrial function. In: Gvozdjáková A (ed) Mitochondrial medicine, Springer, Netheralnds, pp 93–102
42. Gvozdjáková A (2008) Mitochondrial physiology. In: Gvozdjáková A (ed) Mitochondrial medicine, Springer, Netherlands, pp 1–17
43. Yang WC, Li T, Wan Q, Zhang X, Sun L, Zhang Y, Lai P, Li W (2022) Molecular hydrogen mediates neurorestorative effects after stroke in diabetic rats: the TLR4/NF-kB inflammatory pathway. J Neuroimmun Pharmacol. https://doi.org/10.1007/s11481-022-10051-w
44. Reddy PH, Reddy TP (2011) Mitochondria as a therapeutic target for aging and neurodegenerative diseases. Current Alzheimer Res 8(4):393–409
45. Sebastián D, Palacin M, Zorzano A (2017) Mitochondrial dynamics: coupling mitochondrial fitness with healthy aging. Trends in Molecular Med 23(3):201–215
46. Fu Z, Zhang J, Zhang Y (2022) Role of molecular hydrogen in ageing and ageing-related diseases. Ox Med Cell Long Article ID 2249749. https://doi.org/10.1155/2022/2249749
47. Murphy MP (2009) How mitochondria produce reactive oxygen species. The Biochem J 417/1:1–13
48. Liguori I, Russo G, Curcio F et al (2018) Oxidative stress, aging and diseases. Clinical Intervent Aging 13:757–772
49. Ishihara G, Kawamoto K, Komori N, Ishibashi T (2020) Molecular hydrogen supresses superoxide generation in the mitochondrial complex I and reduced mitochondrial membrane potential. Biochem Bipophys Res Comm 522(4):965–970
50. Ohta S (2012) Molecular hydrogen is a novel antioxidant to efficiently reduce oxidative stress with potential for the improvement of mitochondrial diseases. Biochimica Biophysica Acta 1820:S86–S94
51. Wang B, Zhuoshu L, Mao L et al (2022) Hydrogen: a novel treatment strategy in kidney disease. Kidney Dis 8:126–136. https://doi.org/10.1159/000520981
52. Dohi K, Kraemer BC, Erickson MA et al. (2014) Molecular hydrogen in drinking water protects against neurodegenerative changes induced by traumatic brain injury. PLoS ONE 9/9:e108034. https://doi.org/10.1371/journal.pone.0108034

53. Gvozdjáková A, Kucharská J, Kura B, Vančová O, Rausová Z, Sumbalová Z, Uličná O, Slezák J (2019) A new insight into the molecular hydrogen effect on coenzyme Q and mitochondrial function of rats. Can J Physiol Pharmacol. https://doi.org/10.1139/cjpp-2019-0281
54. Kucharská J, Gvozdjáková A, Kura B, Rausová Z, Slezák J (2018) Effect of molecular hydrogen on coenzyme Q in plasma, myocardial tissue and mitochondria of rats. J Nutritional Health and Food Eng 8(5):362–364
55. Hirst J, Roessler MM (2016) Energy conversion, redox catalysis and generation of reactive oxygen species by respiratory complex I. Biochim Biophys Acta 1857: 872–883. https://doi.org/10.1016/J.bbabio.2015.12.009
56. Trefs E, Gannon M, Wasserman DH (2017) The liver. Curr Biol 27:R1147–R1151
57. Mateoni CA, Younossi M, Gramlich T, Boparai N, Liu YC, McCullough AJ (1999) Nonalcoholic fatty liver disease: a spectrum of clinical and pathological severity. Gastroenterology 116:123–133
58. Moore MP, Cunningham RP, Meers GM, Johnson SA, Wheeler AA, Ganga RR, Spencer NM, Pitt JB, Diaz-Arias A, Swi AIA et al (2022) Compromised hepatic mitochondrial fatty acid oxidation and reduced markers of mitochondrial turnover in human NAFLD. Hepatology 76(5):1452–1465. https://doi.org/10.1002/hep.32324
59. Mihajlovic M, Vinken M (2022) Mitochondria as the target of hepatotoxicity and drug-induced liver injury: molecular mechanisms and detection methods. Int J Mol Sci 23:3315. https://doi.org/10.3390/ijms23033315
60. Dabravolski SA, Bezsonov EE, Orekhov AN (2021) The role of mitochondria dysfunction and hepatic senescence in NAFLD development and progression. Biomed Pharmacol 142:112041
61. Prasun P, Ginevic I, Oishi K (2021) Mitochondrial dysfunction in nonalcoholic fatty liver disease and alcohol related liver disease. Transl Gastroenterol Hepatol 6:4. https://doi.org/10.21037/rgh-20-125
62. Ramanathan R, Ali AH, Ibdah JA (2022) Mitochondrial dysfunction plays central role in nonalcoholic fatty liver disease. Int J Mol Sci 23:7280. https://doi.org/10.3390/ijms23137280
63. Nassir F, Ibdah JA (2014) Role of mitochondria in nonalcoholic fatty liver disease. Int J Mol Sci 15:8713–8874. https://doi.org/10.3390/ijms15058713
64. Sumbalova Z, Kucharska J, Rausova Z, Szantova M, Mojto V, Kura B, Gvozdjakova A, Slezak J (2022) Effect of molecular hydrogen on antioxidant content and mitochondria function in patients with non-alcoholic fatty liver disease. In: 1st Conference of European academy for molecular hydrogen research in biomedicine "Hydrogen for Biomedicine". Smolenice, Slovakia, October 17th–20th, pp 74–75
65. Gvozdjáková A, Singh RB (2018) Mitochondrial diseases. In: Gvozdjáková A, Cornélissen G, Singh RB (eds), Recent advances in mitochondrial diseases and coenzyme Q_{10}, NOVA Sciences, USA, pp 71–79

Chapter 5
Autonomic Cardiac Regulation in Response to Exercise and Molecular Hydrogen Administration in Well-Trained Athletes

Michal Botek, Jakub Krejčí, Barbora Sládečková, and Andrew McKune

Abstract Exercise induces considerable changes in the autonomic nervous system (ANS). The main objective of this chapter was to determine whether H_2 administration through the hydrogen rich water (HRW) can affect ANS activity during two experimental exercise protocols in well-trained athletes. Both experiments were designed as randomized, double-blind, placebo-controlled crossover trials. Study A (12 fin-swimmers) assessed ANS responses before and during a simulated competitive day, and Study B (12 soccer players) assessed heart rate (HR) responses following a repeated sprint ability protocol (15 × 30 m). The heart rate variability method was performed to determine ANS activity for 5 min in standing and supine position using the DiANS PF8 system, and HR recovery was evaluated using the HR monitor at 1 and 3 min post exercise. Study A showed that three days of HRW administration induced a significant decline in vagal activity and HR stimulation in elite fin-swimmers solely

M. Botek · J. Krejčí (✉) · B. Sládečková
Department of Natural Sciences in Kinanthropology, Faculty of Physical Culture, Palacký University Olomouc, Olomouc, Czech Republic
e-mail: jakub.krejci@upol.cz

M. Botek
e-mail: michal.botek@upol.cz

B. Sládečková
e-mail: barbora.sladeckova@upol.cz

A. McKune
Research Institute for Sport and Exercise (UCRISE), University of Canberra, Ngunnawal Land, Canberra, ACT, Australia

Discipline of Biokinetics, Exercise and Leisure Sciences, School of Health Sciences, University of KwaZulu-Natal, Durban, South Africa

Functional Foods and Nutrition Research (FFNR) Laboratory, University of Canberra, Ngunnawal Land, Canberra, ACT, Australia

A. McKune
e-mail: andrew.mckune@canberra.edu

J. Slezak and B. Kura (eds.), *Molecular Hydrogen in Health and Disease*, Advances in Biochemistry in Health and Disease 27, https://doi.org/10.1007/978-3-031-47375-3_5

in the standing position during the pre-competition phase of the simulated competition day. Study B showed that acute HRW administration can improve HR recovery of team sport athletes performing maximal repeated sprints that may translate to improved performance during training and competition. Therefore, it appears that H_2 may be considered a promising dietary supplement in the future.

Keywords Exercise · Hydrogen rich water · Autonomic nervous system · Fin-swimmers · Soccer players · Heart rate variability

Introduction

Autonomic nervous system (ANS) activity is accepted as a sensitive indicator of exercise-induced homeostatic perturbation in the body [1, 2]. In this regard, the ANS is highly involved in homeostatic regulation [3, 4], and is considered a dominant governor of the heart rhythm [1]. Autonomic cardiac regulation involves the interaction between sympathetic and parasympathetic (vagal) drive at the level of the sinoatrial node [5]. Sympathetic activation and epinephrine and norepinephrine release have stimulatory effects on heart rate (HR) during stressful situations such as exercise and hypoxia, as well as increasing resting and submaximal exercise heart rate in the fatigued as well as non-functional overreached state [6–8]. On the other hand, vagal activity causes a decrease in HR. Cardiac vagal regulation is generally more pronounced during the resting state, sleeping, and within the first min of post-exercise recovery [1, 9] compared to sympathetic outflow.

Heart rate variability (HRV) calculated using frequency and time domain analysis of R-R intervals, is an accepted, non-invasive method for assessing ANS activity [10, 11], specifically vagal cardiac regulation [1]. Vagal activity, which is reflected in the high-frequency power (HF) of R-R intervals (0.15–0.50 Hz), is associated with the respiratory modulated fluctuation of HR that induces a phenomenon known as respiratory sinus arrhythmia [12]. The low frequency power (LF) (0.05–0.15 Hz) is considered to show baroreflex activity together with bilateral sympathetic and vagal outflow [11]. Besides the frequency analysis of HRV, RMSSD (the square root of the mean of the squares of differences between adjacent R-R intervals) as a time domain index of vagal activity [13], is typically calculated as it is less impacted by the effect of low breathing frequency on the spectral analysis of HRV [14]. In addition, RMSSD has been previously identified as more reliable index of training ability compared to HF [15]. Autonomic cardiac regulation is typically assessed during supine, standing, and sitting position or a combination [16]. Either the standing or sitting position was recommended by some authors [16–18] during HRV measurement, to increase the sensitivity of the testing for identifying exercise-induces changes in ANS activity and avoid the vagal saturation phenomenon in well trained athletes [17, 19].

During the past few decades, HR and HRV have become popular objective diagnostic tools for sport scientists, strength and conditioning coaches, elite and amateur athletes. HR and HRV monitoring allow the assessment of current body "trainability—adaptability" [2, 20–22], providing information about the physiological response to various exercise interventions [23–25]. Within the training process, submaximal or maximal HR have been traditionally measured to monitor the level of exercise intensity. On the other hand, resting HR or post-exercise HR recovery (HRR) provide objective data about "training status" and the current autonomic cardiac control of HR [21, 26]. It is well known that vagal activity gradually withdraws and sympathoadrenal system starts to play a dominant role in homeostatic regulation with progressively increased exercise intensity [27–29]. Once exercise is completed, the accelerated HR during exercise is followed by its progressive decline, induced by the attenuation in the sympathetic drive and increased vagal activation, which lasts from minutes to hours [19, 30]. It has been well documented that the time course of vagal reactivation depends primarily on exercise intensity [27, 31], and cardiorespiratory performance [32, 33]. In this context, it was previously demonstrated that faster post-exercise reduction in HR was associated with endurance trained athletes who exhibited both higher resting vagal activity and faster post-exercise vagal reactivation compared with untrained populations [34]. In addition, it was reported that a significant delay in HRR after exercise was associated with slower vagal reactivation due to a high concentration of inflammatory markers, which could reflect an increased risk for the development of cardiovascular disease [35, 36]. In another study, low vagal activity was associated with impaired post stress recovery of cardiovascular, endocrine, and inflammatory markers [37]. Stanley et al. [22] reported that post-exercise vagal reactivation, represented by the rate of HRR, was negatively impacted by high blood lactate concentration up to 90 min post-exercise recovery. Lactate as a metabolic by-product of anaerobic carbohydrate breakdown is understood to reflect the metabolic demands of a given exercise [38]. Demanding exercise is also associated with the high oxygen uptake, increased mitochondrial respiration, and adenosine triphosphate (ATP) production within the electron transport chain. However, the electron transport chain is also linked with formation of harmful cytotoxic reactive oxygen species (ROS) as by-products of oxidative metabolism [39]. Oxidative stress reflects the overproduction of ROS that attenuates mitochondrial efficiency [40, 41], hastens the onset of fatigue during repeated sprints [42], increases sympathetic activity [43], and delays post-exercise recovery [44].

Molecular hydrogen (H_2) has been considered as healthy, safe non-metal gas with a strong and selective antioxidative capacity to scavenge the hydroxyl radical and peroxynitrite [45, 46]. Apart from its antioxidative property, H_2 has recently been proposed to have anti-inflammatory anti-apoptotic, and cell signalling properties [46–48] and reduce exercise-induced lactate, delayed onset of muscle soreness [49], and rating of perceived exertion [50, 51]. It has been shown that H_2 can easily move across the cell membrane into the cellular space as well as into the mitochondria, where it helps to maintain a redox-balance state and energy production [46]. In this context, it is important to note that H_2 has also been shown to stimulate mitochondrial respiration, the Q-cycle [52], and oxidative ATP phosphorylation (OXOPHOS) rate [53]. Different methods of H_2 administration such as hydrogen rich water (HRW) administration or H_2 inhalation have been shown to have an antifatigue effect across different modes of exercise, including endurance [54–56], strength-endurance drills

[49], cycling anaerobic power output [57], maximal isokinetic muscle strength [58], repeated sprint ability [59], and during prolonged intermittent sprints [60]. These promising results have stimulated a growing interest in the application of H_2 for health and performance among athletes and sports scientists over the last decade. In addition, a recent systematic review [61] concluded that H_2 supplementation can alleviate fatigue but does not enhance aerobic capacity in healthy adults. On the other hand, some studies have reported no ergogenic effect in response to H_2 supplementation [62–64].

Although H_2 has been recognized as novel substance that has important medical and therapeutic-supporting properties for various health conditions such as SARS-CoV-2 (COVID 19) infection [65, 66], metabolic, neurodegenerative, cardiovascular, or inflammatory-based diseases [46, 67], there are still limited studies assessing the effect of H_2 administration on autonomic cardiac regulation. Interestingly, there are similarities relating to the favourable effects of H_2 and vagal activity on physiological and biochemical mechanisms, including in anti-inflammatory pathways, modulation of ROS and nitric oxide signalling, regulation of redox state, improvement of mitochondrial biogenesis and function, and potential calcium regulation [68]. It is important to mention that autonomic cardiac regulation, particularly vagal activity, is widely accepted as a sensitive index of cardiovascular health [69], because a low level of vagal activity and sympathetic predominance in autonomic cardiac regulation is associated with the increased risk of sudden cardiac death and cardiovascular disease development [70].

Regarding ANS activity and H_2 administration, Botek et al. [71] found that an acute administration of 1260 ml HRW, whilst seated, significantly altered autonomic cardiac regulation in healthy females. There was a significant shift in the sympathovagal balance with increased sympathetic activity reflected by an increase in HR at 25 and 35 min post HRW administration compared with placebo. In contrast to these findings, Mizuno et al. [72] found attenuated sympathetic activity and relative increased seated, resting, vagal activity in healthy volunteers after 4-weeks of HRW administration (600 ml per day), and concluded that chronic HRW administration might have a positive effect on autonomic cardiac regulation.

Two studies are presented in this chapter. Both studies assessed autonomic cardiac regulation in response to HRW administration in well-trained athletes. However, the difference between them is that Study A assessed responses before and during a simulated competitive day of fin-swimming and Study B assessed responses after a repeated sprint ability protocol.

Table 5.1 Characteristics of the fin-swimmers

Variable	Female (n = 8)	Male (n = 4)	P
	Mean ± SD	Mean ± SD	
Age (years)	21.5 ± 5.0	18.9 ± 1.3	0.72
Body height (cm)	164 ± 6	184 ± 9	0.004
Body mass (kg)	62 ± 7	78 ± 10	0.012
BMI (kg m^{-2})	22.8 ± 1.2	23.1 ± 1.9	0.81
Body fat (%)	22.0 ± 1.5	12.1 ± 4.6	0.004
VO_2max (ml kg^{-1} min^{-1})	45.0 ± 2.5	52.2 ± 1.7	0.004

SD standard deviation, *P* statistical significance of the comparison between males and females (Mann–Whitney U test), *BMI* body mass index, *VO_2max* maximal oxygen consumption

Methods of Study A

Participants

Fourteen national and international elite Czech fin-swimmers were recruited for this study. Two fin-swimmers did not complete the study due to injury (one participant) and illness (one participant), therefore the final sample consisted of 12 participants (eight females and four males). The characteristics of the participants are presented in Table 5.1. All participants were healthy and able to complete all testing. The study was conducted in accordance with the Declaration of Helsinki and was approved by the Ethics Commitee of the Faculty of Physical Culture, Palacký University Olomouc (reference number 11/2023). Participation was voluntary and all participants provided written informed consent. The participants were given detailed verbal explanation of the aims and procedures of the study. Participants were requested not to take any supplements at least two weeks before the experiment and to avoid vigorous physical exercise in the 24 h prior to the experiment.

Experimental Protocol

The study was designed as a randomized, double-blind, placebo-controlled crossover trial. The study protocol was registered at ClinicalTrials.gov (NCT05799911). The order of HRW/placebo or placebo/HRW administration was determined for each participant based on a randomization table that was created at the beginning of the study using a random number generator. The randomisation table was generated using the randperm function available in MATLAB R2020a (MathWorks, Natick, USA). The experimental protocol consisted of two testing days, with each day including two swimming training sessions in a 25 m indoor pool with recovery monitored for 24 h post each testing day (Fig. 5.1). The morning sessions (MS) took place between 9 and 11 AM and included three high intensity interval sets consisted of 4 × 50 m maximum swims. The time interval to complete the swim and rest periods was always 1 min (participants completed 50 m and then rested for the remaining time up to 1 min). Between each set participants swam 300 m with low intensity, and then after the final set they performed a 12 min active cool down. Afternoon sessions (AS) took place between 5 and 6 PM and included 400 m of continuous competitive performance and a 300 m active cool down. A 1400 m warm up was performed prior to each training unit (AM and PM). The washout period between testing days was one week, similarly to previous studies [49, 50, 55, 59] and the protocol for the second testing day was the same as day one. The participants were asked to keep the same training load for the week before testing day one and during the wash out period. In addition, the standard training load was adjusted (to moderate intensity) for the day before the testing.

On each testing day, the participants were asked to avoid eating, drinking coffee, tea and/or any substance affecting autonomic cardiac regulation. The first ANS activity assessment was performed between 8:00 and 8:30 AM in a silent laboratory where ambient temperature was kept at 22 to 24 °C. After the first morning ANS measurement, the next assessment was performed at 30 min of recovery and then at 5 h after the finish of the MS. ANS measurements were then performed at 30 min, 12 h, and 24 h after the AS swimming session.

HRW and Placebo Characteristics and Administration Schedule

HRW was contained within 420 ml plastic aluminum packs with a gas-tight cap (Aquastamina-R, Nutristamina, Ostrava, Czech Republic). According to the manufacturer, HRW was produced from drinking water that underwent chlorine removal and H_2 infusion under high pressure directly into the water. Aquastamina-R does not contain any active ingredients except H_2. The placebo was obtained from the HRW manufacturer and was produced in a similar way to Aquastamina-R and contained in the same packs. The only difference between Aquastamina-R and placebo was that there was no H_2 infusion in the placebo. Because H_2 is colorless, odorless, and tasteless [39], it was not possible to distinguish HRW from placebo. The type of water (HRW/placebo) was indicated on the pack using different batch numbers. The details of the batch numbers were kept confidential by the manufacturer until the experimental part and statistical analysis were completed. HRW/placebo characteristics were as follows: pH 7.9/7.7, oxidation–reduction potential −652/ +170 mV, and temperature 20/20 °C (pH/ORP/Temperature-meter AD14, Adwa Instruments, Szeged, Hungary). Concentration of the dissolved H_2 was 0.9/0.0 ppm (H2Blue reagent, H2 Sciences, Henderson, NV, USA).

Three days before simulated competition day (testing day one), the participants ingested 1260 ml of HRW/placebo per day (divided into three 420 ml doses). Within

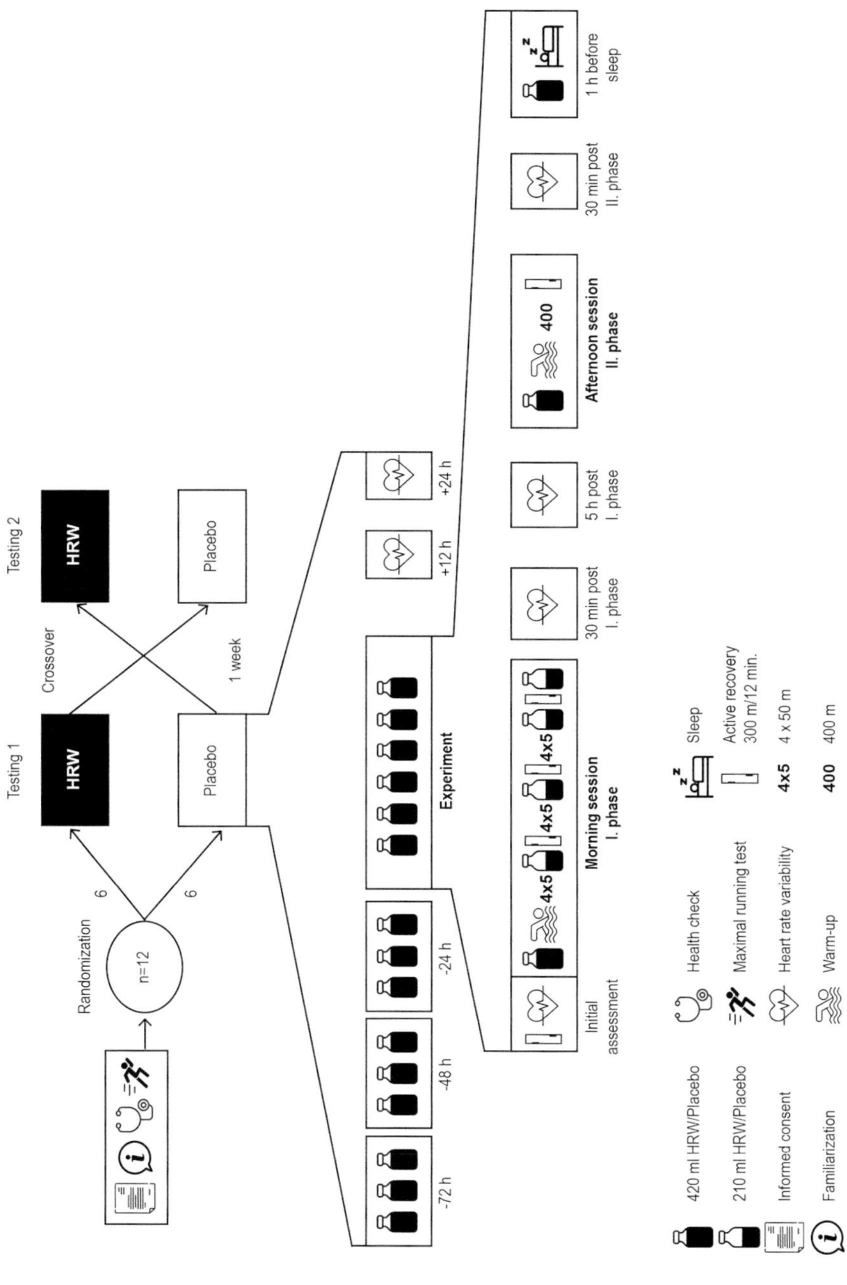

Fig. 5.1 Depiction of the experimental protocol

the simulated competition day, the HRW administration strategy was as follows: 420 ml HRW/placebo just before MS, 210 ml between each (morning) swimming set and 210 ml after final cool down. Another 420 ml immediately before AS and 1 h after AS. The last pack (420 ml) was administrated 1 h before going to sleep.

Basic Anthropometric Measurement

Body mass and body height were determined using a digital scale SOEHNLE 7307 (Leifheit, Nassau, Germany). Body fat was determined using bioimpedance device Tanita BC-418 MA (Tanita, Tokyo, Japan).

Maximal Exercise Test

The participants performed a graded maximal exercise test on a motorized treadmill Lode Valiant Special (Lode, Groningen, Netherlands) in order to obtain maximal oxygen uptake (VO_2max). The testing protocol differed based on gender. For female/male, the protocol design consisted of an 8 min warm-up period (4 min at 7/8 km h^{-1} and 4 min at 8/10 km h^{-1}) followed by an increase in speed to 10/12 km h^{-1} for 1 min. Thereafter, the inclination increased by 2.5% every minute until exhaustion. Breath-by-breath ventilation and gas exchange (Ergostik, Geratherm Respiratory, Bad Kissingen, Germany) were continuously analysed during the exercise with the data averaged to 30 s for analysis. The VO_2max was recorded as the highest oxygen consumption value in the final 30 s of the test. HR (Polar, Kempele, Finland) was monitored continuously during maximal exercise test.

Heart Rate Variability Analysis

To assess the HR and HRV variables, the ECG signal was recorded during a time-modified orthoclinostatic maneuver: supine (60 s)—standing (300 s)—supine (300 s) according to the previously recommended protocol [73]. The sampling frequency used by the diagnostic system for the ECG signal was set at 1000 Hz (DiANS PF8, DIMEA Group, Olomouc, Czech Republic). The ECG record was examined, and all premature ventricular contractions, missing beats, and any artefacts were manually filtered. A set of 300 artefact-free subsequent R-R intervals was obtained from the standing and second supine phase. A spectral analysis was used to assess the ANS activity and was performed using the Fast Fourier Transform. The spectral analysis incorporated a sliding 256 points Hanning window and a Coarse-Graining Spectral Analysis algorithm [74]. Frequency domain variables included: low-frequency power (LF) calculated in the band from 0.05 to 0.15 Hz, high-frequency power (HF) calculated in the band from 0.15 to 0.50 Hz, and the LF/HF ratio. A total spectral power as index of whole ANS activity was obtained as LF + HF. The square root of the mean of the squares of differences between adjacent R-R intervals (RMSSD), the standard deviation of all R-R intervals (SDNN), and the SDNN to RMSSD ratio (SDNN/RMSSD) were calculated using a custom application written in MATLAB R2020a (MathWorsk, Natick, USA) according to the formulas published by Wang and Huang [75]. RMSSD is considered to be an index of vagal activity [11], SDNN is assumed to be an index of total variability [1, 76], and SDNN/ RMSSD is proposed as an index of sympatho-vagal balance [75]. Because raw HRV indexes were not normally distributed, they were transformed using natural logarithm (Ln).

Statistical Analysis

Arithmetic mean and standard deviation were used as descriptive statistics. The normality of each variable was verified using the Kolmogorov–Smirnov test. The sphericity was assessed using the Mauchly test. The effect of HRW compared to placebo for each variable and each time point was evaluated using a paired t-test. The significance level was set at $\alpha = 0.05$. There were six time points for each variable and the Holm-Bonferroni method [77] was used to control the family-wise error rate. The statistical level for the set of six t-tests was adaptive and the actual level was calculated in an iterative procedure based on the number of rejected null hypotheses. For example, when two null hypotheses were rejected for the same variable, both P-values had to be less than $0.05/2 = 0.025$. The effect size was expressed as Cohen's d, in which the standard deviation was taken from the placebo values at the first time point (initial assessment). The thresholds for interpreting the magnitude of Cohen's d were as follows: 0.00–0.19 trivial, 0.20–0.49 small, 0.50–0.79 medium, ≥ 0.80 large [78]. Statistical analyses were performed using MATLAB with Statistics Toolbox version R2020a (MathWorks, Natick, USA).

Results of Study A

Normality of HR and all HRV variables studied was not rejected (all $P \geq 0.087$, Kolmogorov–Smirnov test). However, sphericity was rejected for all variables ($P \leq 0.026$, Mauchly test) except for Ln SDNN/RMSSD in standing ($P = 0.072$). Therefore, individual paired t-tests were used to evaluate the effects of HRW compared to placebo instead of analysis of variance.

Changes in HR and time-domain HRV variables are shown in Fig. 5.2. Changes in spectral HRV variables are shown in Fig. 5.3. In the supine position, no significant differences between HRW and placebo were found for any variable. The minimal P-value from the six paired t-tests was as follows: HR_{supine} (all $P \geq 0.077$), Ln $RMSSD_{supine}$ (all $P \geq 0.15$), Ln $SDNN/RMSSD_{supine}$ (all $P \geq 0.14$), Ln HF_{supine} (all

P ≥ 0.37), Ln LF_{supine} (all P ≥ 0.23), Ln LF + HF_{supine} (all P ≥ 0.080), and Ln LF/HF_{supine} (all P ≥ 0.21).

In the standing position, differences in the initial assessment were found for the following variables: $HR_{standing}$ (HRW: 96.1 ± 13.6, placebo: 90.7 ± 15.1 beats.min^{-1}, P = 0.027, d = 0.36, small effect), Ln $RMSSD_{standing}$ (HRW: 2.52 ± 0.87, placebo: 2.79 ± 0.94 ms, P = 0.008, d = − 0.29, small effect), Ln SDNN/$RMSSD_{standing}$ (HRW: 1.02 ± 0.44, placebo: 0.88 ± 0.40, P = 0.035, d = 0.34, small effect), Ln $HF_{standing}$ (HRW: 4.61 ± 1.94, placebo: 5.18 ± 1.77 ms^2, P = 0.020, d = − 0.32, small effect), and Ln LF + $HF_{standing}$ (HRW: 5.75 ± 1.42, placebo: 6.39 ± 1.47 ms^2, P = 0.006, d = − 0.44, small effect). For Ln LF/$HF_{standing}$, a significant difference was found only 12 h after the afternoon session (HRW: 0.32 ± 1.13, placebo: 1.11 ± 0.74, P = 0.044, d = − 0.66, medium effect). Finally, no significant difference was found for Ln $LF_{standing}$ (all P ≥ 0.067).

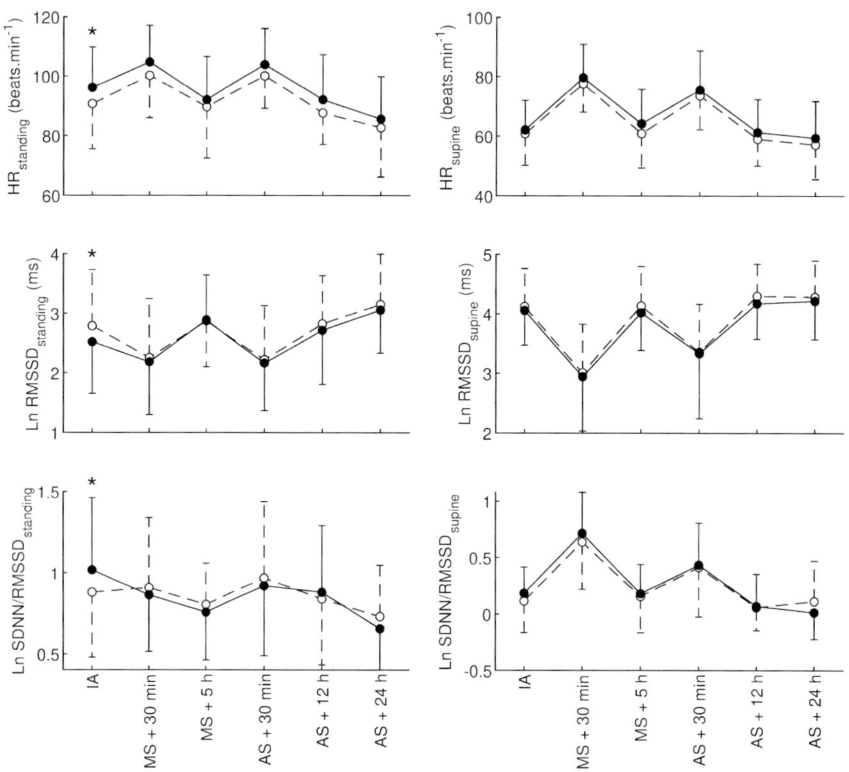

Fig. 5.2 Changes in heart rate and time-domain heart rate variability variables. *HR* heart rate, *Ln* natural logarithm transform, *RMSSD* the square root of the mean of the squares of differences between adjacent R-R intervals, *SDNN* the standard deviation of all R-R intervals, *SDNN/RMSSD* SDNN to RMSSD ratio, *standing* standing body position, *supine* supine body position, *IA* initial assessment, *MS* morning session, *AS* afternoon session, ● hydrogen rich water, ○placebo, ⋆ statistically significant (P < 0.05, paired t-test)

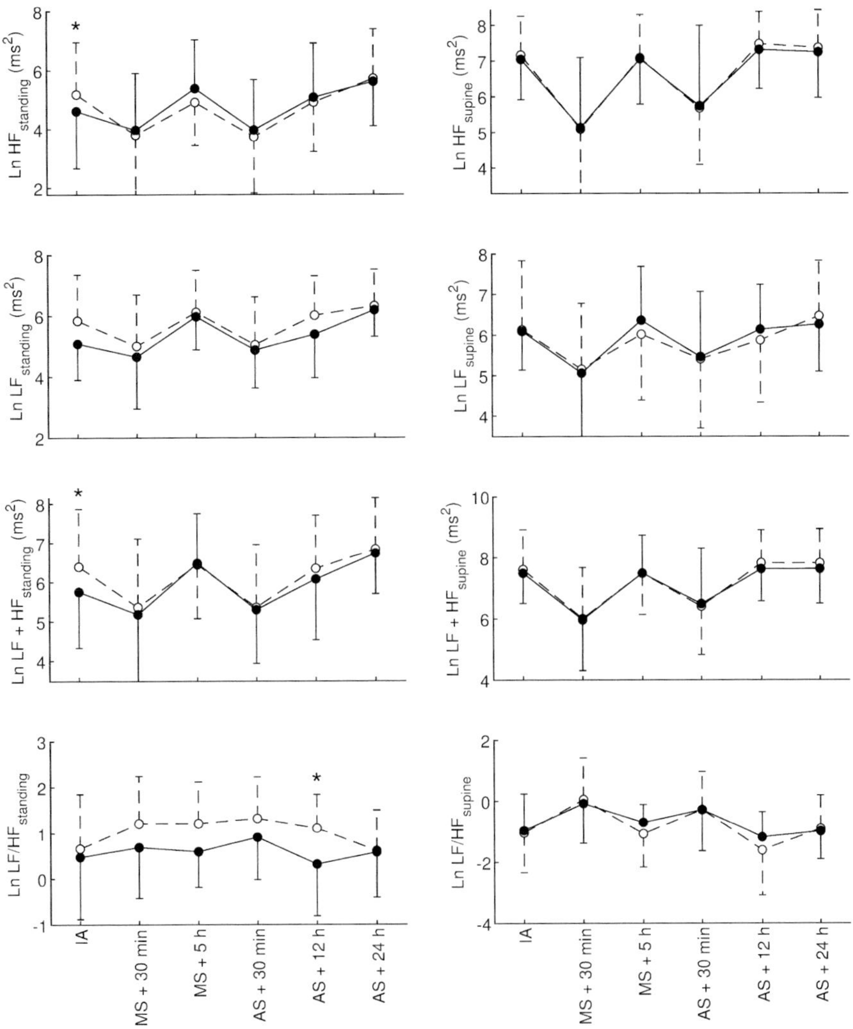

Fig. 5.3 Changes in spectral heart rate variability variables. *Ln* natural logarithm transform, *HF* high-frequency power, *LF* low-frequency power, *LF + HF* total power, *LF/HF* LF to HF ratio, *standing* standing body position, *supine* supine body position, *IA* initial assessment, *MS* morning session, *AS* afternoon session, ● = hydrogen rich water, ○placebo, ⋆ statistically significant (P < 0.05, paired t-test)

Discussion of Study A

The primary aim of this study was to assess the autonomic cardiac response to HRW administration in elite fin-swimmers. The main findings of this research are as follows: (a) during the initial ANS activity measurement, HRW caused a significant reduction in vagal activity together with a relative increase in sympathetic outflow compared to placebo; (b) HRW significantly shifted a sympathovagal balance toward vagal dominance during 12 h of recovery compared to placebo.

This research revealed that three days of HRW administration (3 × 420 ml per day) caused a significant stimulation in HR response during postural maneuver compared to placebo in the initial morning ANS measurement. It has been well documented that an orthostatic maneuver is associated with an acceleration in HR as a result of the functional adjustment of both branches of the ANS in order to compensate for a temporal orthostatic hypotension [79, 80]. Specifically, an orthostatic maneuver causes a drop in the absolute HF value reflecting an inhibition of the efferent cardiac vagal regulation [81], which is typically accompanied by an increase in sympathetic activity and LF/HF ratio, respectively [82, 83]. As our results indicate, a few days of HRW ingestion caused a more pronounced inhibition in vagal activity during orthostatic stimulation compared with placebo. In relation to the autonomic cardiac compensation mechanism of orthostasis-induced hypotension, it is important to take into the account the study of [84] who observed an increased level of plasma ghrelin via activation of the β_1-adrenergic receptors in the stomach tissue between 2nd and 4th days of HRW administration in an animal model. Noteworthy, Nagaya et al. [85] demonstrated a hypotensive effect of ghrelin (a lowering effect on mean arterial pressure) with no significant effect on HR response in healthy humans. Thus, if we take into the consideration these previous findings, it is tenable that the increased HR in our results might represent an adaptation in vagal cardiac regulation to compensate for the drop in blood pressure during orthostasis induced by the three days of HRW supplementation. In this regards, Botek et al. [71] recently showed in placebo-controlled study that an acute 1260 ml of HRW (0.9 ppm of H_2) consumption caused a significant rise in sitting HR due to significantly elevated SDNN/RMSSD (sympathovagal index), signalling sympathetic predominance in cardiac HR regulation between 25 and 35 min post HRW ingestion in healthy females. In contrast, Mizuno et al. [72] reported that four weeks of 600 ml/day HRW administration improved autonomic cardiac regulation through attenuation of sympathetic activity, and the authors associated these findings with a decrease in inflammation and oxidative stress in response to prolonged HRW administration. Taken together, it seems that short-term HRW administration (1–4 days) may stimulate cardiac sympathetic regulation compared with chronic HRW administration.

It is noteworthy that from the present results that the HRW supplementation strategy induced considerable changes in autonomic cardiac regulation and HR only in the standing position. Based on this finding it seems that orthostatic stimulation may provide more detail information about discrete functional changes induced by H_2 administration compared with clinostasis. Similarly, Morout et al. [19] emphasized that orthostatic stimulation is important for detecting prolonged exercise-induces changes in autonomic regulation in recovery compared with only taking measurements in the supine position.

Our results show further that standing HR remained non-significantly elevated in the HRW group compared with placebo, while there was a significant decrease in the LF/HF ratio, indicating vagal predominance in cardiac HR regulation, 12 h after the afternoon swimming session. Based on the LF/HF ratio a decrease in resting HR in the 12 h after a vigorous exercise bout would be expected when supplementing with HRW. The inconsistency in response shown by the LF/HF and HR values may be explained by Billman [86]. This author estimated that cardiac vagal activity contributes to HF and LF ~ 90% and ~ 50%, respectively and that sympathetic activity contributed to HF and LF ~ 10% and ~ 25%, respectively. In this regards, Billman [86] assume that non-neural factors may contribute up to 25% to the LF response. Therefore, LF/HF is considered to not be a valid index of sympathovagal balance, particularly during circumstances associated with a largely reduced spectral power such as during exercise and standing [86], when an increase in HR is not followed by the increase in LF/HF and vice versa. From our perspective, it seems that HRW induces changes in global autonomic cardiac regulation (LF + HF) had a greater effect on HR dynamics than changes in standing LF/HF.

In conclusion, this study showed that short-term HRW administration induced a significant change in ANS activity in elite fin-swimmers on a simulated competition day. The changes occurred solely in the standing position, with the fin-swimmers experiencing a decline in vagal activity and HR acceleration pre-competition.

Methods of Study B

Participants

This study enrolled 16 professional male soccer players. Four players were removed from the analysis due to poor quality of R-R recordings (insufficient contact of the ECG belt with the skin, artefacts caused by movement). Characteristics of the final 12 players were as follows (mean ± SD): age 18.8 ± 1.3 years, body mass 76.3 ± 7.4 kg, body height 182.3 ± 6.3 cm, body mass index 22.9 ± 1.1 kg m^{-2}, body fat 11.4 ± 2.2%, VO_2max 57.6 ± 2.0 ml kg^{-1} min^{-1}. All participants were healthy (self-reported), medication free, non-smokers and not taking any dietary supplements. The study was conducted in accordance with the Declaration of Helsinki and was approved by the Ethics Committee of the Faculty of Physical Culture, Palacký University Olomouc (reference number 75/2017).

Experimental Protocol

The experimental protocol consisted of one laboratory session and two field-testing sessions separated by a one week washout period. During the laboratory session, participants were provided with the study information and familiarized with the testing equipment. They or their parents provided written informed consent. Anthropometric and VO_2max measurements were followed. Body mass, body height, body fat and maximal oxygen consumption were determined using the same procedures as in Study A. The first testing session proceeded one week after the laboratory session. Participants were randomly divided into two groups: HRW (n = 7) or placebo (n = 5). Randomization was performed by means of lots that included an equal number of two colored strips (red and blue) to represent either HRW or placebo consumption. Whilst blinded, participants drew one strip. Then participants performed a repeat sprint protocol (15 × 30 m). After completing the last sprint, participants were requested to sit down and be still for 3 min. Heart rate was monitored during exercise and during 3 min of recovery. There was then a one week washout period before the second testing session where the beverage administration was reversed prior to performance of the same running protocol. Participants were instructed to avoid drinking caffeine containing beverages and other substances that could potentially impact the physiological outcomes, 2 h before each session. Participants were also instructed to avoid drinking alcohol and performing demanding physical activity for 48 h before each session. To avoid possible diurnal variation, all exercise testing was performed between 8:30 and 11:00 AM in a faculty facility as well as in the indoor athletics training center that belongs to the Athletics Club Olomouc (indoor temperature 18–20 °C). The participants were instructed by the coach to consume the same diet and not make any changes in their diet over the duration of the study.

HRW and Placebo Characteristics and Administration Schedule

HRW was obtained in 420 ml plastic aluminum packs with a gas-tight cap (Aquastamina-R, Nutristamina, Ostrava, Czech Republic). The placebo was obtained from the HRW manufacturer, and the only difference compared with the HRW was that there was no H_2 infusion. Details of HRW and placebo are provided in the description of Study A. A total volume of 1260 ml of HRW or placebo was administered in four doses. The time of the first 420 ml dose was set at 120 min, the second 420 ml dose at 60 min, and the last two 210 ml doses at 15 and 5 min before the repeated sprints. The division of HRW into 2 × 210 ml was due to concerns that drinking 420 ml of HRW 5 min before sprints could cause gastrointenstinal discomfort.

Repeat Sprint Protocol

The repeat sprint protocol was performed on a standard indoor athletic surface. All participants performed a 10 min warm-up, including light jogging and progressive runs with acceleration that was followed by 5 min of self-selected dynamic stretching. Participants then performed 3 practice sprint starts over a maximal distance of 10 m, followed by a 5 min rest period. The repeat sprint protocol consisted of 15 × 30 m sprints interspersed by 20 s of active recovery (slow walk back to the start line).

Heart Rate Measurement

HR was measured using a Polar V800 hear rate monitor (Polar, Kempele, Finland) which was assessed as a valid HR measurement device [87]. HR was measured continuously during the repeated sprints and during recovery 3 min after the last sprint. Using the cloud-based service Polar Flow (Polar, Kempele, Finland), the measured records were downloaded to a personal computer. The downloaded records were then imported into the MATLAB R2020a (MathWorks, Natick, USA). For subsequent analysis, average values were calculated from every 5 s. From the segment corresponding to 15 repetitions of sprints (total time 6.25 min), the maximum heart rate (HRmax) and the average heart rate (HRavg) was calculated. In addition, the HR values at 1 min and 3 min after the last sprint were recorded. These values were subtracted from the HRmax to obtain the values of the heart rate recovery at 1 min (HRR1) and 3 min (HRR3).

Statistical Analysis

Arithmetic mean and standard deviation were used as descriptive statistics. The normality of each variable was verified using the Kolmogorov–Smirnov test. The HRW effect compared to placebo on HR was evaluated using a paired t-test. The significance level was set at $\alpha = 0.05$. Statistical analyses were performed using MATLAB with Statistics Toolbox R2020a (MathWorks, Natick, USA).

Results of Study B

Participants in this study did not report any side effects or complaints about HRW ingestion. The normal distribution was not rejected for all studied variables (all $P \geq 0.32$). HRW administration compared to placebo resulted in insignificant decreases in exercise heart rates, specifically HRmax: -1.2 ± 4.7 beats.min^{-1} (mean and standard deviation of difference score), $P = 0.41$ and HRavg: -0.8 ± 7.2 beats.min^{-1}, $P = 0.69$ (Table 5.2). Importantly, however, after HRW administration, HRR1 was significantly ($P = 0.017$) improved by 5.2 ± 6.4 beats.min^{-1} and HRR3 was also significantly ($P = 0.022$) improved by 5.2 ± 6.7 beats.min^{-1} (Table 5.2).

Table 5.2 Effect of hydrogen rich water on exercise and post-exercise heart rate

Variable	HRW	Placebo	P
	Mean ± SD	Mean ± SD	
HRmax (beats.min^{-1})	172.2 ± 8.5	173.3 ± 6.1	0.41
HRavg (beats.min^{-1})	165.9 ± 9.4	166.8 ± 7.6	0.69
HRR1 (beats.min^{-1})	46.8 ± 10.6	41.6 ± 12.6	0.017
HRR3 (beats.min^{-1})	67.8 ± 9.0	62.7 ± 8.2	0.022

HRW hydrogen rich water, *P* significance of paired t-test, *HRmax* maximal heart rate during the set of 15 sprints, *HRavg* average heart rate during the set of 15 sprints, *HRR1* heart rate recovery 1 min after the end of the last sprint, *HRR3* heart rate recovery 3 min after the end of the last sprint

Discussion of Study B

In this crossover study, insignificant decreases in maximal and average HR of approximately 1 beat.min^{-1} were found with HRW administration compared to placebo. Botek et al. [59], based on the full sample of 16 players, reported that 30 m sprint times were significantly reduced (analysis of variance factor $P < 0.001$) with HRW administration. Specifically, the last sprint was improved significantly ($P = 0.021$) from 4.63 ± 0.17 s with placebo to 4.54 ± 0.14 s with HRW. Thus, combining the above results, we concluded that HRW prevented exercise heart rate elevation during exercise of higher intensity (faster sprints).

In addition, this study also revealed that when HRW was administered, HRR at 1 min and 3 min improved by approximately 5 beats.min^{-1}. Imai et al. [88] showed that the main contribution to HRR during the first minute of recovery is the rapid recovery of parasympathetic activity. Similar findings were reported by Pierpont et al. [89], who confirmed a slower initial HRR during parasympathetic blockade with atropine than in subjects without blockade, suggesting an important role of parasympathetic reactivation in the early stage of HRR. According to these findings, we can attribute the improved HRR caused by HRW ingestion to faster parasympathetic reactivation. This is a positive finding because faster HRR is associated with enhanced readiness for further exercise [90].

Dong et al. [91] reported the results of a parallel group study involving 18 dragon boat athletes. The participants were administered HRW (H_2 concentration of 1.6 ppm) or placebo for 7 days, 1 L per day. After a 30 s full-strength rowing test, there was a significant ($P = 0.043$) improvement in HRR after 2 min and a significant trend ($P = 0.050$) for enhanced HRR after 3 min in response to HRW administration compared with placebo. However, the study reported values for heart rate percent change and these values cannot be simply converted to HR values in beats per min. Drid et al. [92] reported the results of a crossover study involving 8 female judo athletes. Participants were administered 300 ml of HRW (H_2 concentration not reported) or placebo 30 min before exercise. There were no significant differences ($P \geq 0.71$) between HRW and placebo in maximal HR and 1 min post-exercise HRR (differences up to 2 beats/min) after high-intensity intermittent exercise (judo-specific test). Direct comparison of our results with the two mentioned studies [91, 92] is not straightforward because the methodology differed in the type and duration of exercise and timing of HRW administration (from acute 30 min to intermediate 7 days before testing).

The underlying mechanism of H_2 effect is still unknown in detail. From in vitro studies, it is known that H_2 can reduce oxidative stress by selectively scavenging ROS, specifically the hydroxyl radical and peroxynitrite, and possibly by activating the transcription factor Nrf2 [93]. In a human study, Shin et al. [94] showed that HRW administration can inhibit exercise-induced oxidative stress, which was reflected by a reduction in serum malondialdehyde and 8-hydroxy-2-deoxyguanosine concentrations. Gvozdjáková et al. [52] demonstrated in a rat model that ROS reduction by HRW increased mitochondria efficiency and ATP production. It should be noted that aerobic metabolism is required for PCr rephosphorylation and lactate utilization [40, 95]. Therefore, increased mitochondrial efficiency may lead to faster restoration of energy sources, removal of accumulated metabolites and reversal of acidosis. This in turn leads to a faster parasympathetic reactivation and sympathetic inhibition, causing a faster HRR [34, 96]. On the other hand, however, it was shown that ROS can directly influence sympathetic activity [97, 98]. Therefore, the faster HRR after HRW ingestion may be a consequence of sympathetic inhibition caused by H_2 scavenging of ROS.

In conclusion, acute administration of HRW can improve the HRR of team sport athletes performing maximal repeated sprints that may translate to improved performance during training and competition.

Conclusions

Based on the results from both studies acute administration of HRW has an impact on the ANS of well-trained athletes with the potential to enhance their health, performance and recovery. We believe that H_2 has several pathways to affect ANS activity. A more detailed assessment of the multiple effects of acute and chronic ingestion of H_2 on the ANS should be the focus of future studies.

References

1. Aubert AE, Seps B, Beckers F (2003) Heart rate variability in athletes. Sport Med 33:889–919. https://doi.org/10.2165/00007256-200333120-00003
2. Hautala AJ, Kiviniemi AM, Tulppo MP (2009) Individual responses to aerobic exercise: the role of the autonomic nervous system. Neurosci Biobehav Rev 33:107–115. https://doi.org/10.1016/j.neubiorev.2008.04.009
3. Ernst G (2014) Heart rate variability. Springer, London
4. Javorka K et al (eds) (2008) Variabilita frekvencie srdca: mechanismy, hodnotenie, klinické využitie, OSVETA, Martin

5. Hainsworth R (1995) The control and physiological importance of heart rate. In: Malik M, Camm AJ (eds) Hear Rate Var, Futura, New York, NY, pp 3–19
6. Achten J, Jeukendrup AE (2003) Heart rate monitoring: applications and limitations. Sport Med 33:517–538. https://doi.org/10.2165/00007256-200333070-00004
7. Le Meur Y, Buchheit M, Aubry A, Coutts AJ, Hausswirth C (2017) Assessing overreaching with heart-rate recovery: what is the minimal exercise intensity required? Int J Sports Physiol Perform 12:569–573. https://doi.org/10.1123/ijspp.2015-0675
8. Kuipers H (1998) Training and overtraining: an introduction. Med Sci Sports Exerc 30:1137–1139. https://doi.org/10.1097/00005768-199807000-00018
9. Perini R, Orizio C, Comandè A, Castellano M, Beschi M, Veicsteinas A (1989) Plasma norepinephrine and heart rate dynamics during recovery from submaximal exercise in man. Eur J Appl Physiol Occup Physiol 58:879–883. https://doi.org/10.1007/BF02332222
10. Akselrod S, Gordon D, Ubel FA, Shannon DC, Berger AC, Cohen RJ (1981) Power spectrum analysis of heart rate fluctuation: a quantitative probe of beat-to-beat cardiovascular control. Science (80-.). 213:220–222 (1981). https://doi.org/10.1126/science.6166045
11. Task force of the European society of cardiology and the North American society of pacing and electrophysiology, heart rate variability. Standards of measurement, physiological interpretation, and clinical use. Eur Heart J 17:354–381 (1996).
12. Yasuma F, Hayano JI (2004) Respiratory sinus arrhythmia: why does the heartbeat synchronize with respiratory rhythm? Chest 125:683–690
13. Bigger JT, Fleiss JL, Rolnitzky LM, Steinman RC (1992) Stability over time of heart period variability in patients with previous myocardial infarction and ventricular arrhythmias. The CAPS and ESVEM investigators. Am J Cardiol 69:718–723. https://doi.org/10.1016/0002-9149(92)90493-i
14. Penttilä J, Helminen A, Jartti T, Kuusela T, Huikuri HV, Tulppo MP, Coffeng R, Scheinin H (2001) Time domain, geometrical and frequency domain analysis of cardiac vagal outflow: effects of various respiratory patterns. Clin Physiol 21:365–376. https://doi.org/10.1046/j.1365-2281.2001.00337.x
15. Bellenger CR, Fuller JT, Thomson RL, Davison K, Robertson EY, Buckley JD (2016) Monitoring athletic training status through autonomic heart rate regulation: a systematic review and meta-analysis. Sport Med 46:1461–1486. https://doi.org/10.1007/s40279-016-0484-2
16. Botek M, Krejčí J, McKune AJ (2017) Variabilita srdeční frekvence v tréninkovém procesu: Historie, současnost a perspektiva. Vydavatelství Univerzity Palackého, Olomouc
17. Kiviniemi AM, Hautala AJ, Kinnunen H, Tulppo MP (2007) Endurance training guided individually by daily heart rate variability measurements. Eur J Appl Physiol 101:743–751. https://doi.org/10.1007/s00421-007-0552-2
18. Nakamura FY, Flatt AA, Pereira LA, Ramirez-Campillo R, Loturco I, Esco MR (2015) Ultra-short-term heart rate variability is sensitive to training effects in team sports players. J Sports Sci Med 14:602–605. https://doi.org/10.1128/AAC.03728-14
19. Mourot L, Bouhaddi M, Tordi N, Rouillon JD, Regnard J (2004) Short- and long-term effects of a single bout of exercise on heart rate variability: Comparison between constant and interval training exercises. Eur J Appl Physiol 92:508–517. https://doi.org/10.1007/s00421-004-1119-0
20. Botek M, McKune AJ, Krejci J, Stejskal P, Gaba A (2014) Change in performance in response to training load adjustment based on autonomic activity. Int J Sports Med 35:482–488. https://doi.org/10.1055/s-0033-1354385
21. Buchheit M (2014) Monitoring training status with HR measures: do all roads lead to Rome? Front Physiol 5:1–19. https://doi.org/10.3389/fphys.2014.00073
22. Stanley J, Peake JM, Buchheit M (2013) Cardiac parasympathetic reactivation following exercise: implications for training prescription. Sport Med 43:1259–1277. https://doi.org/10.1007/s40279-013-0083-4
23. Grote S, Ricci JM, Dehom S, Modeste N, Sealy DA, Tarleton HP (2020) Heart rate variability and cardiovascular adaptations among cancer-survivors following a 26-week exercise intervention, Integr. Cancer Ther 19:1–11. https://doi.org/10.1177/1534735420969816

24. McKune AJ, Peters B, Ramklass SS, van Heerden J, Roberts C, Krejčí J, Botek M (2017) Autonomic cardiac regulation, blood pressure and cardiorespiratory fitness responses to different training doses over a 12 week group program in the elderly. Arch Gerontol Geriatr 70:130–135. https://doi.org/10.1016/j.archger.2017.01.012
25. Picard M, Tauveron I, Magdasy S, Benichou T, Bagheri R, Ugbolue UC, Navel V, Dutheil F (2021) Effect of exercise training on heart rate variability in type 2 diabetes mellitus patients: a systematic review and meta-analysis. PLoS ONE 16:1–26. https://doi.org/10.1371/journal.pone.0251863
26. Borresen J, Lambert MI (2008) Autonomic control of heart rate during and after exercise: Measurements and implications for monitoring training status. Sport Med 38:633–646. https://doi.org/10.2165/00007256-200838080-00002
27. Arai Y, Saul JP, Albrecht P, Hartley LH, Lilly LS, Cohen RJ, Colucci WS (1989) Modulation of cardiac autonomic activity during and immediately after exercise. Am J Physiol 256:H132–H141. https://doi.org/10.1152/ajpheart.1989.256.1.H132
28. Botek M, Stejskal P, Krejci J, Jakubec A, Gaba A (2010) Vagal threshold determination. Effect of age and gender, Int J Sports Med 31:768–772 (2010). https://doi.org/10.1055/s-0030-1263141
29. Perini R, Orizio C, Baselli G, Cerutti S, Veicsteinas A (1990) The influence of exercise intensity on the power spectrum of heart rate variability. Eur J Appl Physiol Occup Physiol 61:143–148. https://doi.org/10.1007/BF00236709
30. Buchheit M, Papelier Y, Laursen PB, Ahmaidi S (2007) Noninvasive assessment of cardiac parasympathetic function: postexercise heart rate recovery or heart rate variability? Am J Physiol Heart Circ Physiol 293:H8–H10. https://doi.org/10.1152/ajpheart.00335.2007
31. Stejskal P, Rechbergová J, Salinger J, Šlachta R, Elfmark M, Kalina M, Jurča R, Řehová I (2001) Power spectrum of heart rate variability in exercising humans: the effect of exercise intensity. Sport Med Train Rehabil 10:39–57. https://doi.org/10.1080/15438620109512095
32. Hautala A, Tulppo MP, Mäkikallio TH, Laukkanen R, Nissilä S, Huikuri HV (2001) Changes in cardiac autonomic regulation after prolonged maximal exercise. Clin Physiol 21:238–245. https://doi.org/10.1046/j.1365-2281.2001.00309.x
33. Seiler S, Haugen O, Kuffel E (2007) Autonomic recovery after exercise in trained athletes: intensity and duration effects. Med Sci Sports Exerc 39:1366–1373. https://doi.org/10.1249/mss.0b013e318060f17d
34. Daanen HAM, Lamberts RP, Kallen VL, Jin A, Van Meeteren NLU (2012) A systematic review on heart-rate recovery to monitor changes in training status in athletes. Int J Sports Physiol Perform 7:251–260. https://doi.org/10.1123/ijspp.7.3.251
35. Ackland GL, Abbott TEF, Cain D, Edwards MR, Sultan P, Karmali SN, Fowler AJ, Whittle JR, MacDonald NJ, Reyes A, Paredes LG, Stephens RCM, Del Arroyo AG, Woldman S, Archbold RA, Wragg A, Kam E, Ahmad T, Khan AW, Niebrzegowska E, Pearse RM (2019) Preoperative systemic inflammation and perioperative myocardial injury: prospective observational multicentre cohort study of patients undergoing non-cardiac surgery. Br J Anaesth 122:180–187. https://doi.org/10.1016/j.bja.2018.09.002
36. Jae SY, Heffernan KS, Yoon ES, Lee MK, Fernhall B, Park WH (2009) The inverse association between cardiorespiratory fitness and C-reactive protein is mediated by autonomic function: a possible role of the cholinergic antiinflammatory pathway. Mol Med 15:291–296. https://doi.org/10.2119/molmed.2009.00057
37. Weber CS, Thayer JF, Rudat M, Wirtz PH, Zimmermann-Viehoff F, Thomas A, Perschel FH, Arck PC, Deter HC (2010) Low vagal tone is associated with impaired post stress recovery of cardiovascular, endocrine, and immune markers. Eur J Appl Physiol 109:201–211. https://doi.org/10.1007/s00421-009-1341-x
38. Brooks GA, Arevalo JA, Osmond AD, Leija RG, Curl CC, Tovar AP (2022) Lactate in contemporary biology: a phoenix risen. J Physiol 600:1229–1251. https://doi.org/10.1113/JP280955
39. Nicolson GL, de Mattos GF, Settineri R, Costa C, Ellithorpe R, Rosenblatt S, La Valle J, Jimenez A, Ohta S (2016) Clinical effects of hydrogen administration: from animal and human diseases to exercise medicine. Int J Clin Med 7:32–76. https://doi.org/10.4236/ijcm.2016.71005

40. Calbet JAL, Martín-Rodríguez S, Martin-Rincon M, Morales-Alamo D (2020) An integrative approach to the regulation of mitochondrial respiration during exercise: focus on high-intensity exercise. Redox Biol 35:1–10. https://doi.org/10.1016/j.redox.2020.101478
41. Carrì MT, Valle C, Bozzo F, Cozzolino M (2015) Oxidative stress and mitochondrial damage: Importance in non-SOD1 ALS. Front Cell Neurosci 9:1–6. https://doi.org/10.3389/fncel.2015.00041
42. Westerblad H, Allen DG (2011) Emerging roles of ROS/RNS in muscle function and fatigue. Antioxid Redox Signal 15:2487–2499. https://doi.org/10.1089/ars.2011.3909
43. Harrison DG (2014) Basic science: pathophysiology: oxidative stress. J Am Soc Hypertens 8:601–603. https://doi.org/10.1016/j.jash.2014.07.002
44. Lumini JA, Magalhães J, Oliveira PJ, Ascensão A (2008) Beneficial effects of exercise on muscle mitochondrial function in diabetes mellitus. Sport Med 38:735–750. https://doi.org/10.2165/00007256-200838090-00003
45. Ohsawa I, Ishikawa M, Takahashi K, Watanabe M, Nishimaki K, Yamagata K, Katsura KI, Katayama Y, Asoh S, Ohta S (2007) Hydrogen acts as a therapeutic antioxidant by selectively reducing cytotoxic oxygen radicals. Nat Med 13:688–694. https://doi.org/10.1038/nm1577
46. Ohta S (2014) Molecular hydrogen as a preventive and therapeutic medical gas: Initiation, development and potential of hydrogen medicine. Pharmacol Ther 144:1–11. https://doi.org/10.1016/j.pharmthera.2014.04.006
47. Qiu P, Liu Y, Zhang J (2019) Recent advances in studies of molecular hydrogen against sepsis. Int J Biol Sci 15:1261–1275. https://doi.org/10.7150/ijbs.30741
48. Slezak J, Kura B, LeBaron TW, Singal PK, Buday J, Barancik M (2021) Oxidative stress and pathways of molecular hydrogen effects in medicine. Curr Pharm Des 27:610–625. https://doi.org/10.2174/1381612826666200821114016
49. Botek M, Krejčí J, McKune A, Valenta M, Sládečková B (2021) Hydrogen rich water consumption positively affects muscle performance, lactate response, and alleviates delayed onset of muscle soreness after resistance rraining. J Strength Cond Res 1–8. https://doi.org/10.1519/JSC.0000000000003979
50. Botek M, Krejčí J, McKune AJ, Sládečková B, Naumovski N (2019) Hydrogen rich water improved ventilatory, perceptual and lactate responses to exercise. Int J Sports Med 40:879–885. https://doi.org/10.1055/a-0991-0268
51. Ostojić SM, Stojanović MD, Calleja-Gonzalez J, Obrenović MD, Veljović D, Medjedović B, Kanostrevac K, Stojanović M, Vukomanović B (2011) Drinks with alkaline negative oxidative reduction potential improve exercise performance in physically active men and women: double-blind, randomized, placebo-controlled, corss-over trial of efficacy and safety. Serbian J Sport Sci 5:83–89
52. Gvozdjáková A, Kucharská J, Kura B, Vančová O, Rausová Z, Sumbalová Z, Uličná O, Slezák J (2020) A new insight into the molecular hydrogen effect on coenzyme Q and mitochondrial function of rats. Can J Physiol Pharmacol 98:29–34. https://doi.org/10.1139/cjpp-2019-0281
53. Murakami Y, Ito M, Ohsawa I (2017) Molecular hydrogen protects against oxidative stress-induced SH-SY5Y neuroblastoma cell death through the process of mitohormesis. PLoS ONE 12:1–14. https://doi.org/10.1371/journal.pone.0176992
54. Ara J, Fadriquela A, Ahmed F, Bajgai J, Sajo MEJ, Lee SP, Kim TS, Jung JY, Kim CS, Kim SK, Shim KY, Lee KJ (2018) Hydrogen water drinking exerts antifatigue effects in chronic forced swimming mice via antioxidative and anti-inflammatory activities. Biomed Res Int 2018:1–9. https://doi.org/10.1155/2018/2571269
55. Botek M, Krejčí J, McKune AJ, Sládečková B (2020) Hydrogen-rich water supplementation and up-hill running performance: effect of athlete performance level. Int J Sports Physiol Perform 15:1193–1196. https://doi.org/10.1123/ijspp.2019-0507
56. Javorac D, Stajer V, Ratgeber L, Betlehem J, Ostojic S (2019) Short-term H2 inhalation improves running performance and torso strength in healthy adults. Biol Sport 36:333–339. https://doi.org/10.5114/biolsport.2019.88756

57. Timón R, Olcina G, González- CustodioA, Camacho-Cardenosa M, Camacho-Cardenosa A, Martínez Guardado I (2021) Effects of 7-day intake of hydrogen-rich water on physical performance of trained and untrained subjects. Biol Sport 38:269–275. https://doi.org/10.5114/biolsport.2020.98625
58. Aoki K, Nakao A, Adachi T, Matsui Y, Miyakawa S (2012) Pilot study: effects of drinking hydrogen-rich water on muscle fatigue caused by acute exercise in elite athletes. Med Gas Res 2:1–6. https://doi.org/10.1186/2045-9912-2-12
59. Botek M, Khanna D, Krejčí J, Valenta M, McKune A, Sládečková B, Klimešová I (2022) Molecular hydrogen mitigates performance decrement during repeated sprints in professional soccer players. Nutrients 14:1–10. https://doi.org/10.3390/nu14030508
60. Da Ponte A, Giovanelli N, Nigris D, Lazzer S (2018) Effects of hydrogen rich water on prolonged intermittent exercise. J Sports Med Phys Fitness 58:612–621. https://doi.org/10.23736/S0022-4707.17.06883-9.
61. Zhou K, Liu M, Wang Y, Liu H, Manor B, Bao D, Zhang L, Zhou J (2023) Effects of molecular hydrogen supplementation on fatigue and aerobic capacity in healthy adults: a systematic review and meta-analysis. Front Nutr 10:1–18. https://doi.org/10.3389/fnut.2023.1094767
62. Ooi CH, Ng SK, Omar EA (2020) Acute ingestion of hydrogen-rich water does not improve incremental treadmill running performance in endurance-trained athletes. Appl Physiol Nutr Metab 45:513–519. https://doi.org/10.1139/apnm-2019-0553
63. Nogueira JE, Amorim MR, Pinto AP, da Rocha AL, da Silva ASR, Branco LGS (2021) Molecular hydrogen downregulates acute exhaustive exercise-induced skeletal muscle damage. Can J Physiol Pharmacol 99:812–820. https://doi.org/10.1139/cjpp-2020-0297
64. Valenta M, Botek M, Krejčí J, McKune A, Sládečková B, Neuls F, Bajgar R, Klimešová I (2022) Acute pre-exercise hydrogen rich water intake does not improve running performance at maximal aerobic speed in trained track and field runners: a randomized, double-blind, placebo-controlled crossover study. PLoS ONE 17:1–14. https://doi.org/10.1371/journal.pone.0279307
65. Guan WJ, Wei CH, Chen AL, Sun XC, Guo GY, Zou X, Shi JD, Lai PZ, Zheng ZG, Zhong NS (2020) Hydrogen/oxygen mixed gas inhalation improves disease severity and dyspnea in patients with Coronavirus disease 2019 in a recent multicenter, open-label clinical trial. J Thorac Dis 12:3448–3452. https://doi.org/10.21037/jtd-2020-057
66. Botek M, Krejčí J, Valenta M, McKune A, Sládečková B, Konečný P, Klimešová I, Pastucha D (2022) Molecular hydrogen positively affects physical and respiratory function in acute post-COVID-19 patients: a new perspective in rehabilitation. Int J Environ Res Public Health 19:1–14. https://doi.org/10.3390/ijerph19041992
67. Ichihara M, Sobue S, Ito M, Ito M, Hirayama M, Ohno K (2015) Beneficial biological effects and the underlying mechanisms of molecular hydrogen—comprehensive review of 321 original articles. Med Gas Res 5:1–21. https://doi.org/10.1186/s13618-015-0035-1
68. He F, Li J, Liu Z, Chuang CC, Yang W, Zuo L (2016) Redox mechanism of reactive oxygen species in exercise. Front Phys 7:1–10. https://doi.org/10.3389/fphys.2016.00486
69. Rosenwinkel ET, Bloomfield DM, Arwady MA, Goldsmith RL (2001) Exercise and autonomic function in health and cardiovascular disease. Cardiol Clin 19:369–387. https://doi.org/10.1016/s0733-8651(05)70223-x
70. Mäkikallio TH, Huikuri HV, Mäkikallio A, Sourander LB, Mitrani RD, Castellanos A, Myerburg RJ (2001) Prediction of sudden cardiac death by fractal analysis of heart rate variability in elderly subjects. J Am Coll Cardiol 37:1395–1402. https://doi.org/10.1016/s0735-1097(01)01171-8
71. Botek M, Sládečková B, Krejčí J, Pluháček F, Najmanová E (2021) Acute hydrogen-rich water ingestion stimulates cardiac autonomic activity in healthy females. Acta Gymnica. 51:1–6. https://doi.org/10.5507/ag.2021.009
72. Mizuno K, Sasaki AT, Ebisu K, Tajima K, Kajimoto O, Nojima J, Kuratsune H, Hori H, Watanabe Y (2018) Hydrogen-rich water for improvements of mood, anxiety, and autonomic nerve function in daily life. Med Gas Res 7:247–255. https://doi.org/10.4103/2045-9912.222448

73. Botek M, Krejčí J, Neuls F, Novotný J (2013) Effect of modified method of autonomic nervous system activity assessment on results of heart rate variability analysis. Acta Univ Palacki Olomuc Gymnica 43:39–46. https://doi.org/10.5507/ag.2013.011
74. Yamamoto Y, Hughson RL (1991) Coarse-graining spectral analysis: new method for studying heart rate variability. J Appl Physiol 71:1143–1150. https://doi.org/10.1152/jappl.1991.71.3.1143
75. Wang HM, Huang SC (2012) SDNN/RMSSD as a surrogate for LF/HF: a revised investigation. Model Simul Eng 2012:1–8. https://doi.org/10.1155/2012/931943
76. Malik M (1997) Time-domain measurement of heart rate variability. Card Electrophysiol Rev 1:329–334. https://doi.org/10.1023/A:1009912905325
77. Holm S (1979) A simple sequentially rejective multiple test procedure. Scand J Stat 6:65–70
78. Cohen J (1988) Statistical power analysis for the behavioral sciences, 2nd edn. Lawrence Erlbaum Associates, Hillsdale, NJ
79. Bernardi L, Keller F, Sanders M, Reddy PS, Griffith B, Meno F, Pinsky MR (1989) Respiratory sinus arrhythmia in the denervated human heart. J Appl Physiol 67:1447–1455. https://doi.org/10.1152/jappl.1989.67.4.1447
80. Malik M, Camm AJ (eds) (1995) Heart rate variability. Futura, New York, NY
81. Pomeranz B, Macaulay RJ, Caudill MA, Kutz I, Adam D, Gordon D, Kilborn KM, Barger AC, Shannon DC, Cohen RJ, Benson H (1985) Assessment of autonomic function in humans by heart rate spectral analysis. Am J Physiol 248:H151–H153. https://doi.org/10.1152/ajpheart.1985.248.1.H151
82. Furlan R, Porta A, Costa F, Tank J, Baker L, Schiavi R, Robertson D, Malliani A, Mosqueda-Garcia R (2000) Oscillatory patterns in sympathetic neural discharge and cardiovascular variables during orthostatic stimulus. Circulation 101:886–892. https://doi.org/10.1161/01.cir.101.8.886
83. Pagani M, Lombardi F, Guzzetti S, Rimoldi O, Furlan R, Pizzinelli P, Sandrone G, Malfatto G, Dell'Orto S, Piccaluga E (1986) Power spectral analysis of heart rate and arterial pressure variabilities as a marker of sympatho-vagal interaction in man and conscious dog. Circ Res 59:178–193. https://doi.org/10.1161/01.res.59.2.178
84. Matsumoto A, Yamafuji M, Tachibana T, Nakabeppu Y, Noda M, Nakaya H (2013) Oral "hydrogen water" induces neuroprotective ghrelin secretion in mice. Sci Rep 3:1–5. https://doi.org/10.1038/srep03273
85. Nagaya N, Kojima M, Uematsu M, Yamagishi M, Hosoda H, Oya H, Hayashi Y, Kangawa K (2001) Hemodynamic and hormonal effects of human ghrelin in healthy volunteers. Am J Physiol Regul Integr Comp Physiol 280:R1483–R1487. https://doi.org/10.1152/ajpregu.2001.280.5.R1483
86. Billman GE (2013) The LF/HF ratio does not accurately measure cardiac sympatho-vagal balance. Front Physiol 4:1–5. https://doi.org/10.3389/fphys.2013.00026
87. Giles D, Draper N, Neil W (2016) Validity of the Polar V800 heart rate monitor to measure RR intervals at rest. Eur J Appl Physiol 116:563–571. https://doi.org/10.1007/s00421-015-3303-9
88. Imai K, Sato H, Hori M, Kusuoka H, Ozaki H, Yokoyama H, Takeda H, Inoue M, Kamada T (1994) Vagally mediated heart rate recovery after exercise is accelerated in athletes but blunted in patients with chronic heart failure. J Am Coll Cardiol 24:1529–1535. https://doi.org/10.1016/0735-1097(94)90150-3
89. Pierpont GL, Stolpman DR, Gornick CC (2000) Heart rate recovery post-exercise as an index of parasympathetic activity. J Auton Nerv Syst 80:169–174. https://doi.org/10.1016/s0165-1838(00)00090-4
90. Rodríguez-Fernández A, Sanchez-Sanchez J, Ramirez-Campillo R, Nakamura FY, Rodríguez-Marroyo JA, Villa-Vicente JG (2019) Relationship between repeated sprint ability, aerobic capacity, intermittent endurance, and heart rate recovery in youth soccer players. J Strength Cond Res 33:3406–3413. https://doi.org/10.1519/JSC.0000000000002193
91. Dong G, Fu J, Bao D, Zhou J (2022) Short-term consumption of hydrogen-rich water enhances power performance and heart rate recovery in dragon boat athletes: Evidence from a pilot study. Int J Environ Res Public Health 19:1–9. https://doi.org/10.3390/ijerph19095413

92. Drid P, Trivic T, Casals C, Trivic S, Stojanovic M, Ostojic SM (2016) Is molecular hydrogen beneficial to enhance post-exercise recovery in female athletes? Sci Sports 31:207–213. https://doi.org/10.1016/j.scispo.2016.04.010
93. Ohta S (2023) Molecular hydrogen may activate the transcription factor Nrf2 to alleviate oxidative stress through the hydrogen-targeted porphyrin. Aging Pathobiol Ther 5:25–32 (2023). https://doi.org/10.31491/APT.2023.03.104
94. Shin DS, Jung SH, Hong EY, Shin YH, Park JY, Chung MH, Ro JY (2018) Remval effect of hydrogen water drinking on exercise-induced production of reactive oxygen species in adult men and women. Exerc Sci 27:289–295. https://doi.org/10.15857/ksep.2018.27.4.289
95. Bessman SP, Geiger PJ (1981) Transport of energy in muscle: the phosphorylcreatine shuttle. Science (80-.). 211:448–452. https://doi.org/10.1126/science.6450446
96. Nakamura FY, Soares-Caldeira LF, Laursen PB, Polito MD, Leme LC, Buchheit M (2009) Cardiac autonomic responses to repeated shuttle sprints. Int J Sports Med 30:808–813. https://doi.org/10.1055/s-0029-1234055
97. Campese VM, Ye S, Zhong H, Yanamadala V, Ye Z, Chiu J (2004) Reactive oxygen species stimulate central and peripheral sympathetic nervous system activity. Am J Physiol Heart Circ Physiol 287:H695–H703. https://doi.org/10.1152/ajpheart.00619.2003
98. Yu Y, Zhong MK, Li J, Sun XL, Xie GQ, Wang W, Zhu GQ (2007) Endogenous hydrogen peroxide in paraventricular nucleus mediating cardiac sympathetic afferent reflex and regulating sympathetic activity. Pflugers Arch Eur J Physiol 454:551–557. https://doi.org/10.1007/s00424-007-0256-9

Chapter 6
The Clinical Use of Hydrogen as a Medical Treatment

Yunbo Xie and Guohua Song

Abstract It has been demonstrated that hydrogen molecules possess biological effects, and furthermore, they are colorless, non-toxic, and have a small molecular weight, enabling them to traverse the blood–brain barrier. This review synthesizes more than 100 publications on the use of hydrogen therapy in clinical ailments and categorizes the applications of hydrogen medicine into nine major systemic diseases based on the International Classification of Diseases (ICD)-11. The efficacy of hydrogen therapy is influenced by the in vivo metabolic kinetics associated with different administration routes. In this review, we examine the utilization of hydrogen molecules via various delivery methods and their impact on the treatment of clinical diseases, along with the mechanisms underlying their biological effects.

Keywords Molecular hydrogen · Clinical use · Medical treatment

Introduction

Molecular hydrogen (H_2) is colorless, odorless, non-toxic [1], and safe [2, 3] preserving cells against common pathologies by antioxidation, anti-inflammation, and regulating apoptotic and/or repair processes [4, 5] as a new "biological gas" [6–8]. One of the earliest reports that H_2 has medicinal properties dates back to 1798, which discussed the therapeutic use of H_2 in respiration systems [9]. The possible medical benefits of H_2 were again reported in 1975 when Dole et al. discovered significant favorable biological effects of hyperbaric hydrogen on skin cancer in mice, and the result was published in *Science* [10]. However, it did not attract enough attention from the scientific community. In 2007, Ohsawa et al. demonstrated that H_2 could develop neuroprotective effects following middle cerebral artery occlusion [11], which was a milestone achievement published in *Nature Medicine*. From then on, hydrogen medicine has gained momentum, and most recently, it has been

Y. Xie · G. Song (✉)
School of Basic Medical Sciences, Shandong First Medical University and Shandong Academy of Medical Science, Jinan 250117, China
e-mail: ghsong@sdfmu.edu.cn

J. Slezak and B. Kura (eds.), *Molecular Hydrogen in Health and Disease*, Advances in Biochemistry in Health and Disease 27, https://doi.org/10.1007/978-3-031-47375-3_6

enclosed within the protocol for the management of patients with COVID-19 in China. (Chinese Clinical Guidance (7th edition) for COVID-19 respiratory disease identification and Treatment, issued by China National Health Commission) [12]. As of 2022, over 2000 articles on hydrogen medicine have been published, including over 100 preliminary articles on 9 major human systems [13] (Fig. 6.1).

Different routes of hydrogen administration can improve clinical trial endpoints or some of these disease markers [14]. Inhaling 4% hydrogen gas (4% H_2) [15], ingesting hydrogen-enriched water (HRW) [16], injecting hydrogen-rich saline (HRS) [17], taking hydrogen-water baths [18], ingesting solid hydrogen carriers, and consuming precursors that produce hydrogen (such as lactulose and L-arabinose) [19–22] are all common ways to administer hydrogen. Additionally, several hydrogen-targeted nanomaterials [2, 23] have emerged with the development of science and technology. Our previous studies on animals suggested that 4% H_2 in mice is quickly absorbed and short-acting, whereas HRW is long-lasting in mice but needs to be consumed numerous times [24]. According to current studies, the hydrogen effect of inhaling H_2 and drinking HRW is transient yet significant. Therefore, controlled and targeted release H_2 loading is crucial to increasing its therapeutic impact.

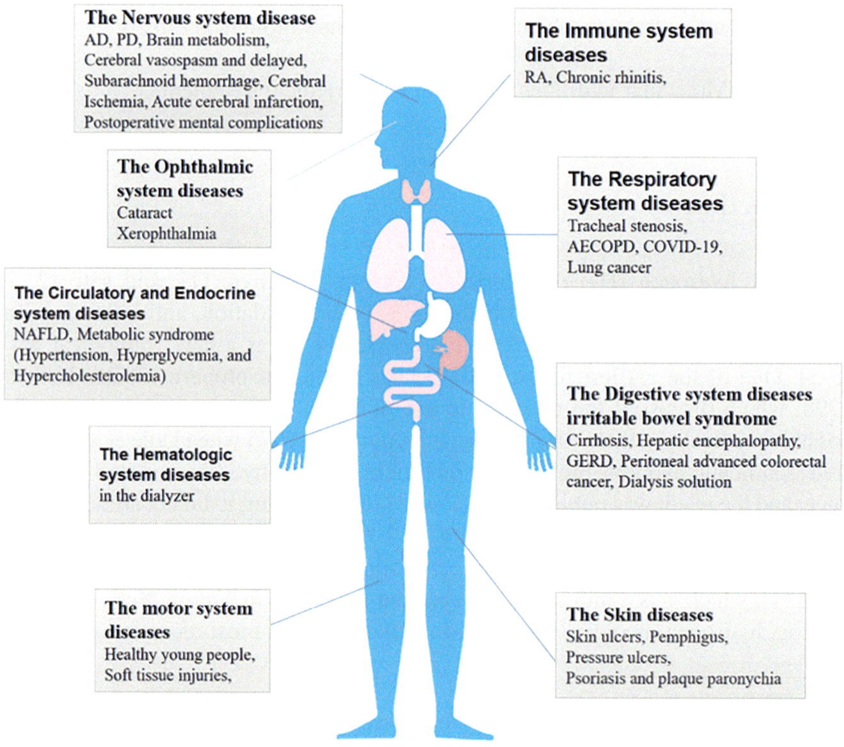

Fig. 6.1 The use of hydrogen medicine in clinical diseases

Clinical experimental data serves as evidence and a foundation for clinical application. Even though the research scale of hydrogen medications has risen quickly in the last 16 years [6], they have been used for a variety of systemic disorders [13], and over 100 clinical articles on hydrogen medicine and the potential mechanisms by which these effects can be exerted have been summarized. However, no large-scale double-blind controlled clinical investigations have been conducted so far, thereby creating an urgent demand for large-scale, long-term multicenter clinical research to confirm its efficacy. This summary lays the theoretical groundwork for larger-scale clinical research.

The Clinical Use of Hydrogen as a Medical Treatment

The Respiratory System Diseases

The respiratory system's structure is identified by bone or cartilage that serves as a support structure to prevent the airway walls from collapsing and maintain smooth airflow in and out. Inspiratory dyspnea, which is a sensation of difficulty breathing in, can have acute or chronic onset based on restrictions in airflow caused by various factors [25]. Tracheal stenosis and Acute exacerbation of chronic obstructive pulmonary disease (AECOPD) are frequent causes of acute dyspnea and require immediate attention [26, 27]. The current treatment measures for acute dyspnea include oxygen therapy [28, 29], non-invasive mechanical ventilation, and gaseous helium (He) inhalation. However, these treatments have disadvantages such as invasiveness, high cost, difficulty in storage and transportation, and low effectiveness [30–33]. Therefore, it is crucial to find a satisfactory method to manage the emergency care of acute dyspnea.

A prospective study employing self-control demonstrated that breathing H_2-O_2 (H_2:O_2 = 66%:33%, 6 L/min for 15 min) decreased inspiratory resistance, reduced diaphragmatic electromyography (EMGdi), and alleviated acute inspiratory dyspnea in 35 patients with severe acute tracheal stenosis. Increased IOS resistance measurements and Borg score; vital indicators were unaffected [34]. Importantly, no adverse reactions occurred in patients during the experiment [34]. In addition, this finding was also verified in the article on AECOPD treatment, which showed that inhalation of the hydrogen/oxygen mixture resulted in a more significant improvement in AECOPD symptoms, including dyspnea, cough, and sputum, and that the safety and tolerability of this treatment regimen were acceptable [35]. According to a recent study, administering inhaled hydrogen to patients with concurrent COVID-19-induced acute respiratory distress significantly improved the patients' physical health (improved 6-min walk test (6 MWT) distance and respiratory function forced vital capacity (FVC), second first–second expiratory volume (FEV1), and also improved oxygen saturation in hypoxic patients [36, 37]. In addition to this, Xu et al. found that hydrogen inhalation can be used to restore some immune system functions before lung cancer patients receive formal treatment [38] and that adjunctive hydrogen inhalation during treatment reduces most drug-related adverse effects, and further studies have found that H_2 can independently inhibit the growth of lung cancer cells [39]. However, the above experimental results are all small sample experiments and no large-scale population experiments have been conducted, so a rational view of hydrogen therapy is still needed.

H_2 is believed to have a variety of beneficial effects of its small molecular weight, which allows it to easily permeate tissues and organs. Additionally, H_2 possesses physical properties similar to those of He, which may reduce airway resistance in narrow tracheal passages. Moreover, H_2 selectively inhibits the excessive production of cytotoxic ROS, thereby countering oxidative stress (OS), without affecting non-cytotoxic ROS [37]. H_2 also inhibits the production of pro-inflammatory cytokines like interleukin-6 (IL-6), which play a key role in the development of inflammation [40–42], In addition, H_2 induces the synthesis of heat shock proteins, which are instrumental in protecting cells from damage. Furthermore, H_2 improves mitochondrial bioenergy, and high concentrations of H_2 do not cause any toxic reactions [43]. Importantly, H_2 can be produced easily and inexpensively by electrolyzing water, making it a convenient and readily available resource even in emergency settings such as ambulances [34].

The Digestive System Diseases

Human endogenous hydrogen is prevalent in the normal environment of human cells and is created by bacteria in the large intestine, accounting for around 74% of the total gas. Small intestinal bacterial overgrowth (SIBO) is defined as aberrant bacterial overgrowth in the small intestine, which is associated with mucosal inflammation [44], irritable bowel syndrome (IBS) [45], cirrhosis [45], and hepatic encephalopathy (HE) [46]. The small intestine does not produce H_2, but oral administration of a glucose substrate causes bacteria in the small intestinal tract to produce metabolites such as H_2, and the presence of gastrointestinal disease is determined by observing changes in the level of hydrogen in exhaled gas, a procedure known as the hydrogen breathing test [46].

In clinical applications, the hydrogen breath test is a noninvasive test for monitoring gastrointestinal function in humans. The first clinical study of hydrogen medicine in the digestive tract dates back to 1998, when Chen et al. also used the breath test to diagnose SIBO in patients with cirrhosis by assessing H_2 and methane levels [47]. More recently, the hydrogen breath test was found to be associated with liver function grading and clinical Child–Pugh classification in patients with HE, and rifaximin may be more effective in HE with H_2-SIBO [48]. In addition, a report showed that HRW consumption increased bifidobacteria levels, although no specific statistics were provided [49]. The same conclusion was found in a further study by Koyama et al. with detailed data, but they did not find differences in the diversity of the intestinal microbiota [50]. All of these studies discovered that H_2 can be used to assist in the diagnosis of gastrointestinal problems. However, the mechanism is unknown.

Osonoi's team found that acarbose, an α-glucosidase inhibitor, could reduce peripheral blood IL-1β mRNA levels in patients with type 2 diabetes mellitus (T2DM) by accelerating the production of hydrogen in the small intestine. The study included 16 Japanese patients with T2DM who received a test meal without (day 1) or with acarbose (day 2) at breakfast and then tested the level of glucose in the blood, H_2 in the breath, and mRNA levels of inflammatory factors (IL-1β and IL-8) in peripheral blood. It was found that H_2 production was negatively correlated with IL-1β mRNA expression in peripheral leukocytes after administration of a single dose of acarbose. Since the article is self-controlled research on T2DM patients, it is not yet clear whether the mRNA levels of these inflammatory cytokines in peripheral blood leukocytes of T2DM patients are higher than those of healthy individuals [51]. Another clinical study on gastroesophageal reflux disease (GERD) found that HRW drinking improved cellular redox status and enhanced the quality of life in GERD, this study consisted of 84 patients with GERD who received control treatment (PPI + tap water) or experimental treatment (PPI + HRW) for 3 months, and a GERD health-related quality of life questionnaire, as well as a derivative reactive oxygen metabolite (d-ROM) test, a biological antioxidant potential (BAP) test, superoxide anion, nitric oxide, and malondialdehyde measurements, were offered to the patients, the results showed changes in the values of the d-ROMs test and the BAP test [52].

Translational research of peritoneal dialysis solution with dissolved H_2 found that there were significant alterations in surrogate markers, such as elevated mesothelin and CA125, in the effluent in some instances, which suggested that H_2 promoted mesothelial regeneration, however, in this article, the effectiveness is limited as only 6 patients participated and the experimental period was 14 days, so further research needs more patients to take part in a long experimental periodicity [53].

In patients with cancer, immune status plays an important effect, decreasing in the peroxisome proliferator-activated receptor γ coactivator 1α (PGC-1α) leads to loss of $CD8^+$ T cells and Programmed cell death 1 (PD-1, as a marker of exhausted T cells), causing insufficient cellular energy and poor prognosis of cancer patients [54, 55]. Baba et al. found that inhaled H_2-O_2 (680,000 ppm H_2 and 320,000 ppm O_2, 3 months) restores exhausted CD8 + T cells possibly by activating mitochondria via PGC-1α in patients with advanced colorectal cancer to improve prognosis [56]. Importantly, the researchers also developed a new patient classification system to assist in predicting prognosis and therapeutic response [56]. A Further study of five elderly patients with complete metastasis after colorectal cancer found that after giving them Ag^+/H_2 water (1.5 ppm, starting 30 min before three meals, 520 ml three times a day), the results showed that carcinoembryonic antigen (CEA) and carbohydrate antigen (CA) 19–9 decreased to normal levels, and three of the subjects (liver (two cases) and lung (one case)) had complete disappearance of metastatic cancer foci [57]. The findings are not enough to be used as therapy recommendations due to the small number of patients in the current study, but if no other therapy is available for patients with terminal cancer, it may be a good option to perform such adjuvant or alternative therapies.

The Hematologic System Diseases

OS increases during hemodialysis (HD), exposing patients to the risk of major cardiovascular disease and shortened life expectancy. Antioxidant therapy is essential for hemodialysis patients for this reason [58]. Hydrogen has attracted attention because of its antioxidant properties [59].

Nakayama et al. found that OS could be reduced "in the dialyzer" using H_2-rich dialysis solutions. In a clinical trial, 8 HD patients received both standard and hydrogen-rich solutions (mean 50 p.p.b. H2), and the results of the study showed that the use of hydrogen-rich dialysis solutions reduced serum levels of total and reduced glutathione, increased levels of hydrogen peroxide and the average ratio of oxidized albumin [58].

The Skin Diseases

Skin is an effective barrier against bacterial infections. Many bacteria come in contact with or stay on the skin, but they rarely cause infections. Bacterial skin infections can affect anything from a single spot to the entire surface of the skin. They also vary in severity, from inconsequential to life-threatening. Research shows that H_2 has been successfully utilized in the treatment of skin ulcers [60], pemphigus [61], pressure ulcers [62], psoriasis, and plaque paronychia [18].

For topical wet dressing therapy, two patients with papulosquamous dermatosis received HRW (1.6 ppm twice a day for one hour), and the patients' wounds were examined after 4 weeks, amazing neither a bacterial nor a fungal was discovered, and the wound was fully recovered after 10 weeks [61]. This clinical instance merely demonstrates the ability of HRW to prevent the development of bacteria and fungus, but no in-depth investigation into the mechanism by which HRW prevents the development of flora has been made. Additional research has demonstrated that H_2 improves type I collagen rebuilding in dermal fibroblasts, enhances mitochondrial reducing capacity and decreases ROS in epidermal keratinocytes, all of which enhance quicker skin wound healing [62]. ROS has been found to play a role in the etiology of chronic inflammatory diseases [41, 63]. Given that psoriasis and plaque parapsoriasis are chronic inflammatory skin conditions, we looked at 41 patients with psoriasis and plaque parapsoriasis' vital signs, concurrent symptoms, side effects, and assessments of their psoriasis activity (PASI scores and images). We discovered that dermatological symptoms significantly improved after 4 weeks of treatment. This demonstrates that treating chronic inflammatory dermatoses with HRW or hydrogen water baths (10–15 min/twice weekly) is a possibility [18]. However, more large-scale randomized placebo-controlled trials are needed to support and extend these findings [18].

The Ophthalmic System Diseases

Elderly cataract patients are more common as the population ages and xerophthalmia patients are more common as society develops, electronic devices multiply, and lifestyles change. Due to its ability to have anti-inflammatory and antioxidant characteristics, perhaps H_2 can treat both ailments at the same time.

A prospective, randomized, placebo-controlled, crossover study with SUPER H_2 (30 min) in participants ($n = 10$) discovered that administering a sustained H_2-generating supplement increased human exhaled H_2 concentration and greatly improved tear stability and dry eye symptoms [64]. Another quick, straightforward, randomized, double-blind clinical trial with 32 clinical phacoemulsification patients discovered that H_2 dissolved in irrigation solution minimized corneal endothelium damage during phacoemulsification [65]. This may be due to the ability of hydrogen to alleviate corneal injury caused by OS during cataract ultrasound emulsification [65].

The Nervous System Disease

Neurodegenerative disorders such as Alzheimer's disease (AD) and Parkinson's disease (PD) are becoming increasingly common as average life expectancy rises. OS is one of the causes of these disorders [66]. Hydrogen has pleiotropic biological activity and easily crosses the blood–brain barrier [67], so diverse ingestion pathways promote neurodegenerative disorders and brain metabolism.

The Hattori team conducted a thorough investigation into hydrogen medicine use in PD. The Total Unified Parkinson's Disease Rating Scale (UPDRS) improved when levodopa medication was taken along with consuming HRW (1000 mL/day for 1 year), according to the findings of the first randomized double-blind, placebo-controlled, parallel trial in PD patients [68]. However, this study's small sample size and brief trial period were a drawback. Accordingly, the team carried out a second double-blind multicenter trial and evaluated the patients' overall UPDRS scores at the eighth, twenty-four, forty-eight, and seventy-two-week points [69]. But this multicenter trial has not been mentioned to this point in the pertinent outcome literature. It was discovered through the aforementioned research that drinking HRW is successful in treating PD. Will taking levodopa while inhaling H_2 be beneficial as well? H_2 was inhaled twice daily for one-hour sessions in a subsequent 16-week randomized, double-blind, placebo-controlled research to treat PD, however, the UPDRS scores did not improve as a result [70]. Next, the team carried out a larger trial in PD patients who did not use levodopa, and the results showed that drinking HRW (0.16 ppm; $n = 30$) was safe but that there was no positive advantage from doing so [71]. According to the aforementioned experimental results, eating HRW can improve levodopa's efficiency and reduce PD symptoms, however, breathing in H_2 is unable to offer the same advantages. More research is required on the underlying molecular pathways. Photobiological modulation (PBM) can improve mitochondrial function, stimulate adenosine triphosphate generation, and so decrease PD symptoms; however, this process is accompanied by an increase in the creation of reactive ROS [72–74]. H_2 is a powerful and selective antioxidant that mitigates the effects of ROS [67]. 18 PD patients were given daily PBM + H_2 therapy for 2 weeks, according to a small-scale, open-label, single-center phase I/IIa clinical trial [74]. Both the negative event and the UPDRS scores were noted, and the result showed that the UPDRS score dramatically improved after the first week and remained better until the end of the course of treatment, furthermore, there were no records of any negative incidents [74]. The UPDRS score marginally increased after a week without medication, but the improvement remained substantial compared to the baseline [74].

Although H_2 inhalation did not help PD patients' symptoms, studies have shown that it does help AD patients' symptoms, and the relief lasts for up to 6 months even in the absence of H_2 medication [75]. In this study, 8 patients with AD inhaled H_2 (3%, 1 h/twice daily) for 6 months, then stopped using H_2 and followed up for 1 year. The Alzheimer's Disease Cognitive Subscale Disease Assessment Scale (ADAS-cog) was used to evaluate the efficacy of H_2 after use. In addition, diffusion tensor imaging (DTI) and advanced magnetic resonance imaging (MRI) were used to assess hippocampal neuron integrity in the brain, and the results of this study showed that ADAS-cog was decreased, and hippocampal neurons became more intact [75]. Interestingly, drinking HRW only improved ADSA-cog scores in patients with the apolipoprotein E4 (APOE-4) genotype who had AD [76]. Although research has shown that different methods of administering hydrogen have different therapeutic benefits on neurodegenerative disorders, altogether, hydrogen therapy would be a significant intervention to help treat neurodegenerative diseases.

As we age, the ability to steadily supply the brain with the energy needed to sustain life functions becomes more critical for maintaining cognition and brain function [77]. A superficial study has shown that HRW consumption can improve quality of life by enhancing mood and stress resistance [78]. This might be because HRW enhances brain metabolism, raises the choline-to-creatine ratio in the frontal and paracentral brain, and boosts attention [79]. Because the participants in this randomized controlled crossover design were healthy young adults, no experiments were conducted on older adults or clinical populations with brain disease, so it is critical to explore the effects of HRW on this population. Although both coffee and HRW can have an alerting effect, HRW does not have any side effects [80] and they work through different metabolic pathways in the brain. HRW may become a new therapy for neurological disorders [81].

Takeuchi et al. administered HRW (1.6 ppm, 200 ml/h, 14 days) intravenously to the patient, with a trial period of 14 days [80, 81]. Studies have found that immediate intravenous infusion of hydrogen-rich fluid combined with an intra-cisternal infusion of magnesium sulfate reduced serum malondialdehyde neuron-specific enolase and improved Barthel's index reduced the incidence of cerebral vasospasm and delayed cerebral ischemia and improved the clinical prognosis of patients with subarachnoid hemorrhage, suggesting that H_2 can be used as an adjunctive treatment with no toxic side effects [80]. The latest research shows that hydrogen therapy reduces cerebrospinal fluid lactate levels in patients with subarachnoid hemorrhage [82]. The neuroprotective effect of H_2 has also been observed in patients with acute cerebral infarction [83]. According to Takanami et al., MRI data revealed that the H_2 group had less severe pathological and recovered more quickly at the site of cerebral infarction after patients received 3% H_2 for an hour twice daily for seven days. The results of the Barthel Index (BI) and National Institute of Health Stroke Scale (NIHSS) were also superior to those of the control group [83]. Furthermore, research indicates that breathing hydrogen decreases inflammatory reactions, helping seniors without heart disease in avoiding postoperative mental complications. This study focused on assessing the incidence of postoperative psychiatric problems in elderly patients who had received prophylactic hydrogen inhalation and the result showed that the incidence rate (12%) of the preventative hydrogen group was only half as high as the incidence rate (24%) of the control group. Additionally, the postoperative hydrogen group had a significantly lower level of C-reactive protein than the control group [84].

The Immune System Diseases

The immune system of the body is a sophisticated and reliable mechanism that quickly detects and eliminates invasive microorganisms, shielding the body from sickness and infection. However, overprotection can also result in illness. Rheumatoid arthritis (RA) is a chronic systemic autoimmune disease. Chronic inflammation associated with RA often damages the skin, subcutaneous tissue, and lungs [85, 86]. HRW is of high concern because of its anti-inflammatory properties. Nagao el at. found that dropped infused intravenously (DIV) 1 ppm H_2-dissolved saline (H_2-saline) for 5 days in RA patients, the disease activity score in 28 joints (DAS28), tumor necrosis factor-α (TNF-α), interleukin-6 (IL-6), matrix metalloproteinase-3 (MMP-3), and urinary 8-hydroxydeoxyguanosine (8-OHdG) was measured, and the result showed that IL-6 and 8-OHdG were significantly decreased. In conclusion, a DIV of H_2-saline can treat RA and other chronic age-related illnesses while having little to no negative effects on one's health [87]. A study was ordered to find that hydrogen therapy can treat chronic rhinitis, especially allergic rhinitis, which has allergic inflammation as its pathophysiology. The procedure consists of rinsing the nasal passages with hydrogen saline. According to the study, CR patients using HRS nasal rinses had better nasal symptoms and lower levels of ECP in nasal secretions. This was particularly evident in AR patients, who had successful outcomes [88]. Hydrogen therapy has a good safety profile and plays an active therapeutic and complementary function in the treatment of chronic inflammatory diseases associated with the immune system. Therefore, it can be applied in clinical situations.

The Circulatory and Endocrine System Diseases

Metabolic Syndrome (hypertension, hyperglycemia, and hypercholesterolemia) and Non-Alcoholic Fatty Liver Disease (NAFLD) are both primarily influenced by obesity [89]. OS plays a crucial role in this type of disease [90]. H_2 increases brain metabolism, as was already mentioned. So how does it impact the body's overall metabolism? Ostojic et al. discovered that when overweight women consumed hydrogen-producing minerals (6 ppm/day), after 4 weeks, their blood triglyceride (TC), fasting serum insulin levels, body fat, and arm fat index all drastically decreased [91]. Additional research has demonstrated that H_2 therapy can also boost mitochondrial function, primarily through altered of lactate, pyruvate, and coenzyme Q10 [91]. This could be the theoretical underpinning for hydrogen therapy to treat circulatory metabolic disorders. Thirty T2DM patients participated in a study by Yoshikawa et al. which was a crossover, randomized, double-blinded, and placebo-controlled, and the volunteers showed significant reductions in low-density lipoprotein, low-density lipoprotein (VLDL), urinary 8-isopropyl protein, oxidized low-density lipoprotein (ox-LDL) and free fatty acids (FFA) after 16 weeks of drinking 900 ml of HRW daily, and their plasma levels of adiponectin and extracellular superoxide dismutase were also significantly increased, and these may suggest that H_2 improves lipid and glucose metabolism in patients with type 2 diabetes or impaired glucose tolerance by reducing OS [92].

Inhaling a low-dose H_2-O_2 (66% H_2/33% O_2 treated 4 h per day for 2 weeks) mixture has been shown to lower blood pressure, as well as the hormones (Angiotensin II, Aldosterone, Aldosterone-to-renin ratio, Cortisol) and stress responses in the renin–angiotensin–aldosterone system in middle-aged and elderly hypertensive patients, according to our prior research [93]. In this experiment, H_2-O_2 is inhaled by the patient without the need for extra liquid input, but it can also lower the patient's TC [93]. As a result, it also offers high-cholesterol patients who are unable to drink a lot of liquids a new auxiliary lipid-lowering option. Our previous research showed that HRW has an effective lipid-lowering effect and enhances the anti-atherosclerotic effect of HDL in patients with hypercholesterolemia [90, 94, 95]. Patients with hypercholesterolemia received HRW (0.9 L/day) to examine the quantity, content, and biological activity of plasma lipoprotein in a double-blind, randomized, and placebo-controlled trial. Reduced TC, LDL, apolipoprotein B100, and IL-6, and increased apolipoprotein M while promoting the outflow of ATP-binding cassette transporter A1 [94]. According to the findings, hydrogen therapy has a critical role in the treatment of adjuvant metabolic syndrome [95]. This is consistent with Slezak's team's findings [96]. 60 individuals (30 men and 30 women) with metabolic syndrome participated in a randomized, double-blind, placebo-controlled experiment. For 24 weeks, patients received high HRW doses (>5.5 mmol H_2 per day). The findings demonstrated that supplementation with high doses of HRW significantly decreased blood cholesterol and glucose levels compared to the control group, attenuated serum hemoglobin A1c. Additionally, H_2 can also reduce waist-hip ratio and body mass index, overall, H_2 may potentially be beneficial in reducing metabolic syndrome risk factors [96].

The liver's excessive fat deposition is a significant contributor to NAFLD. There are now two articles on the use of hydrogen medicine in NAFLD patients, and in both, it was discovered that, at various doses, H2 dramatically reduced liver fat deposition and enhanced liver enzymes. The first experiment [97] employed HRW (1 L/day, 3 mM of H_2) for 28 days, and the second [98] used H_2–O_2 (66% hydrogen and 33% oxygen, 3 L/min) for 1 h every day. In a word, H_2 decreased serum levels of LDL, AST, ALT, MDA, SOD, TNF, and IL-6 [97, 98].

Post-cardiac arrest syndrome (PCAS) is the term used to describe the occurrence of reperfusion injury following the restoration of autonomic circulation following cardiac arrest, which may result in organ dysfunction in one or more instances. According to Tamura et al.'s research, treating PCAS patients with a combination of hydrogen inhalation (HI) and target temperature management (TTM) may be safe [99], The researchers undertook a prospective randomized controlled experiment to evaluate the effects of HI in PCAS patients, but no article has been published yet [100]. According to another study in patients with ST-segment elevation myocardial infarction, a significant improvement in LV beat volume index and LV ejection fraction was found in the HI (1.3% H_2, 26% O_2) group after 6 months of follow-up, indicating that HI during PCI is feasible, safe, and promotes reverse LV remodeling at 6 months after STEMI [101].

The Motor System Diseases

Skeletal muscles generally produce ROS to complement their daily energy needs, but long-term or intense exercise may increase ROS production. High quantities of reactive oxygen species, on the other hand, lead to contractile dysfunction, and eventually results in muscular weakening and fatigue [102]. Given its anti-inflammatory and antioxidant qualities, hydrogen was chosen for inclusion in the exercise system [103]. Furthermore, HRW can alkalinize blood pH, which aids in the relief of exercise-induced acidosis [104]. This section will discuss the impact of HRW ingestion on the exercise system.

Researchers experimented with 12 triathletes to test the effect of HRW on endurance performance, starting with volunteers pedaling at 65% of their maximum oxygen consumption for 60 min, then resting for 5 min before resuming training until they were completely tired. During the pedal test, blood parameters, tissue temperature, respiratory variables, and time to exhaustion (TTE) were measured, and the results revealed that consuming HRW during endurance exercise in a hot environment lowered energy expenditure but had no effect on TTE [105]. Michal et al.'s results also support this viewpoint, they began a randomized, double-blind, placebo-controlled crossover trial test, which observed the impact of acute, pre-exercise HRW intakes on running time to exhaustion at maximal aerobic speed in trained track and field runners, a total of 1260 ml of HRW was administered to 24 young persons in four doses, 120 min (420 ml), 60 min (420 ml), 30 min (210 ml), and 10 min (210 ml) before exercise, measured TTE, cardiovascular variables, and lactate concentration in the blood after exercise, the results showed no significant changes in the HRW group compared to the control group [106]. Another study also found that drinking HRW before running was not adequately ergogenic in endurance-trained athletes, in this research, volunteers drank HRW (290 mL) before exercise, and blood lactate, glucose, $HCO3^-$, pH, and lactate concentration were examined [107].

According to the aforementioned findings, supplied HRW before exercise does not provide energy and enhance endurance for athletes. Perhaps it has something associated with their physical condition. In a randomized controlled double-blind trial, 16 men ran two 4.2 km hill climb races in one week, drinking either hydrogen-rich water or a placebo before the race. The results showed that drinking HRW lowered the race time for the slowest runners, but it had no influence on exercise speed for the quicker runners [108]. However, ingestion of HRW has been demonstrated to enhance PCO_2 and bicarbonate concentration in participants at rest during high-intensity intermittent training (HIIT), as well as peak power in the first round [109]. Another study on HRW for sustained intermittent exercise found that drinking 2L of HRW (pH 9.8, ORP −180 mV, free Hydrogen 450 ppb) per day may aid in maintaining peak power output (PPO) over 30-min sprints to exhaustion [110].

There are many articles supporting that drinking HRW may be an effective and specific innovative treatment for exercise-induced oxidative stress and muscle fiber injury, as well as potential energy assistance for performance [107]. A double-blind, crossover and repeated measurement design experiment showed that after administering H_2–O_2 (68.0%:32.0%) to volunteers during the recovery period after exercise, it was found that it reduced fatigue after exercise by reducing oxidative damage (8-OHdG), however, there were no significant changes in the detection of jumping ability, pedal strength output, muscle strength evaluation, serum dimer reactive oxygen species, biological antioxidant capacity, serum creatine kinase, lactate dehydrogenase activity, and white blood cell count [111]. And more experts think that hydrogen therapy may also alleviate fatigue after exercise, as it can reduce lactate levels and increase exercise endurance [112–115]. Michal et al. found a beneficial effect of HRW consumption on lactate response, muscle performance, and the delayed onset of discomfort after resistance training, they conducted a randomized, double-blind placebo-controlled crossover trial where subjects were given the same exercise training and then assessed each group for a time, lactate, perceived fatigue score, creatine kinase, muscle soreness visual analog scale score, counter-exercise jump, and heart rate variability, showed that acute intermittent HRW rehydration improved muscle function reduced the lactate response and also alleviated the symptoms of delayed onset of muscle soreness [112]. In 2019, the findings of two distinct experimental teams were consistent. The Ohta team was one, and they also came to the same conclusion that drinking hydrogen water improved endurance and decreased mental weariness. They carried out a randomized, double-blind, placebo-controlled research in which subjects received 500 mL of HRW (0.8–1 ppm) to test this notion. Based on maximal oxygen uptake and the Borg scale, the H_2 group demonstrated considerable improvements in both endurance and fatigue [113]. Acute pre-exercise HRW supplementation was found to reduce serum levels of lactate at higher exercise intensities, improved exercise-induced perception of effort, and improved ventilatory efficiency, according to Naumovski's team, who also evaluated last-minute scores for cardiorespiratory variables, lactate, and perceived fatigue (RPE) at each step [114].

Soft tissue injuries are characterized by an acute inflammatory response with clinical manifestations of pain, redness, swelling, bruising, and functional impairment. Hydrogen-rich water may have a positive adjunctive effect on the treatment of soft tissue injuries due to its anti-inflammatory properties [116, 117]. First, within the first 24 h after Acute ankle sprain (AAS) treatment in professional athletes, HRW was found to be just as effective in reducing joint swelling and pain while restoring range of motion and balance compared to traditional methods for soft tissue injuries as the RICE (rest, ice, compression, elevation) protocol [116]. Further research on 36 professional athletes given oral hydrogen tablets (2 g per day), and oral hydrogen tablets (2 g per day) + topical hydrogen enriched pack (6 times per day for 20 min) during the first 24 h after injury found that oral + topical hydrogen interventions reduction in plasma viscosity compared to controls, and then helped injury limb flexion and extension to return to normal joint range of motion more quickly [117]. In both studies, it is suggested that HRW is beneficial in the treatment of AAS and that the combination of the two delivery routes has better results.

Remaining Issues and Challenges

Recent research has demonstrated the promising efficacy of hydrogen medicine, suggesting it could become a complementary medical gas. However, further investigation is needed regarding the relationship between dosage and response time. Additionally, there appears to be a lack of targeting in the hydrogen administration process, and more research and development of targeted hydrogen release and control drugs is necessary. We also found that in vivo hydrogen content has not been accurately measured, and further investigation is needed to fully comprehend the mechanisms by which hydrogen therapy works in various diseases.

Conclusion and Perspectives

Overall, the development of hydrogen medicine has made remarkable advancements. This paper classifies diseases into 9 major systems based on human physiological functions. Each system is then summarized in terms of the application environment for different hydrogen donors. For example, patients with respiratory diseases, circulatory system issues, or cancer who cannot consume more fluids can benefit from hydrogen inhalation therapy. Exercise system diseases, like fatigue relief and post-exercise lactic acid accumulation, and metabolic diseases should incorporate drinking hydrogen-rich water into their treatment plans. Meanwhile, hydrogen-rich saline is primarily used in cases of skin and eye diseases. As technology advances, more hydrogen donors will likely emerge, and we anticipate that hydrogen therapy will become an increasingly popular complementary treatment for a wider range of diseases.

Sources of Funding This work was supported by the National Natural Science Foundation of China (no. 81873517, 81,670,422), and Young Taishan Scholars Program of Shandong Province (tsqn20161045).

References

1. Fontanari P, Badier M, Guillot C et al (2000) Changes in the maximal performance of inspiratory and skeletal muscles during and after the 7.1-MPa Hydra 10 record human dive. Eur J Appl Physiol 81(4):325–328
2. Zhou G, Goshi E, He Q (2019) Micro/nanomaterials-augmented hydrogen therapy. Adv Healthc Mater 8(16):e1900463
3. Cole AR, Sperotto F, Dinardo JA et al (2021) Safety of prolonged inhalation of hydrogen gas in air in healthy adults. Crit Care Explor 3(10):e543
4. Tian Y, Zhang Y, Wang Y et al (2021) Hydrogen, a novel therapeutic molecule, regulates oxidative stress, inflammation, and apoptosis. Front Physiol 12:789507
5. Ohta S (2014) Molecular hydrogen as a preventive and therapeutic medical gas: initiation, development and potential of hydrogen medicine. Pharmacol Ther 144(1):1–11
6. Zafonte RD, Wang L, Arbelaez CA et al (2022) Medical gas therapy for tissue, organ, and CNS protection: a systematic review of effects, mechanisms, and challenges. Adv Sci (Weinh) 9(13):e2104136
7. Hirano SI, Ichikawa Y, Kurokawa R et al (2020) A "philosophical molecule," hydrogen may overcome senescence and intractable diseases. Med Gas Res 10(1):47–49
8. Du D, Zhao L, Shen M et al (2020) Hydrogen medicine: a rising star in gas medicine. Tradit Med Mod Med 3(03):153–161
9. Cavallo T (1798) An essay on the medicinal properties of factitious airs: with an appendix, on the nature of blood. Author, and sold
10. Dole M, Wilson FR, Fife WP (1975) Hyperbaric hydrogen therapy: a possible treatment for cancer. Science 190(4210):152–154
11. Ohsawa I, Ishikawa M, Takahashi K et al (2007) Hydrogen acts as a therapeutic antioxidant by selectively reducing cytotoxic oxygen radicals. Nat Med 13(6):688–694
12. Artamonov MY, Martusevich AK, Pyatakovich FA et al (2023) Molecular hydrogen: from molecular effects to stem cells management and tissue regeneration. Antioxidants (Basel), 12(3)
13. Lebaron TW, Ohno K, Hancock JT (2023) The on/off history of hydrogen in medicine: will the interest persist this time around? Oxygen 3(1):143–162
14. Ostojic SM (2015) Molecular hydrogen: an inert gas turns clinically effective. Ann Med 47(4):301–304
15. Ono H, Nishijima Y, Adachi N et al (2012) A basic study on molecular hydrogen (H_2) inhalation in acute cerebral ischemia patients for safety check with physiological parameters and measurement of blood H_2 level. Med Gas Res 2(1):21
16. Shimouchi A, Nose K, Shirai M et al (2012) Estimation of molecular hydrogen consumption in the human whole body after the ingestion of hydrogen-rich water. Adv Exp Med Biol 737:245–250

17. Ono H, Nishijima Y, Adachi N et al (2012) Hydrogen (H_2) treatment for acute erythymatous skin diseases. A report of 4 patients with safety data and a non-controlled feasibility study with H_2 concentration measurement on two volunteers. Med Gas Res 2(1):14
18. Zhu Q, Wu Y, Li Y et al (2018) Positive effects of hydrogen-water bathing in patients of psoriasis and parapsoriasis en plaques. Sci Rep 8(1):8051
19. Liu BY, Qin SC (2019) Different types of molecular hydrogen donors and their pharmacokinetics in vivo. Sheng Li Xue Bao 71(2):371–377
20. Ji X, Zhang Q, Zheng W et al (2019) Morphological and molecular response of small intestine to lactulose and hydrogen-rich water in female piglets fed Fusarium mycotoxins contaminated diet. J Anim Sci Biotechnol 10:9
21. Zhao L, Wang Y, Zhang G et al (2019) L-arabinose elicits gut-derived hydrogen production and ameliorates metabolic syndrome in C57BL/6J Mice on High-Fat-Diet. Nutrients, 11(12)
22. Chen W, Zhang HT, Qin SC (2021) Neuroprotective effects of molecular hydrogen: a critical review. Neurosci Bull 37(3):389–404
23. Zhang L, Yu H, Tu Q et al (2021) New approaches for hydrogen therapy of various diseases. Curr Pharm Des 27(5):636–649
24. Liu B, Xie Y, Chen J et al (2021) Protective effect of molecular hydrogen following different routes of administration on d-galactose-induced aging mice. J Inflamm Res 14:5541–5550
25. Berliner D, Schneider N, Welte T et al (2016) The differential diagnosis of dyspnea. Dtsch Arztebl Int 113(49):834
26. Rahman NA, Fruchter O, Shitrit D et al (2010) Flexible bronchoscopic management of benign tracheal stenosis: long term follow-up of 115 patients. J Cardiothorac Surg 5:2
27. Vargas F, Bui HN, Boyer A et al (2005) Intrapulmonary percussive ventilation in acute exacerbations of COPD patients with mild respiratory acidosis: a randomized controlled trial [ISRCTN17802078]. Crit Care 9(4):R382-389
28. Skrinskas GJ, Hyland RH, Hutcheon MA (1983) Using helium-oxygen mixtures in the management of acute upper airway obstruction. Can Med Assoc J 128(5):555–558
29. Brochard L (2000) Non-invasive ventilation for acute exacerbations of COPD: a new standard of care. Thorax 55(10):817–818
30. Mcgarvey JM, Pollack CV (2008) Heliox in airway management. Emerg Med Clin North Am 26(4):905–920
31. Jaber S, Carlucci A, Boussarsar M et al (2001) Helium-oxygen in the postextubation period decreases inspiratory effort. Am J Respir Crit Care Med 164(4):633–637
32. Kemper KJ, Ritz RH, Benson MS et al (1991) Helium-oxygen mixture in the treatment of postextubation stridor in pediatric trauma patients. Crit Care Med 19(3):356–359
33. Boorstein JM, Boorstein SM, Humphries GN et al (1989) Using helium-oxygen mixtures in the emergency management of acute upper airway obstruction. Ann Emerg Med 18(6):688–690
34. Zhou ZQ, Zhong CH, Su ZQ et al (2019) Breathing hydrogen-oxygen mixture decreases inspiratory effort in patients with tracheal stenosis. Respiration 97(1):42–51
35. Zheng ZG, Sun WZ, Hu JY et al (2021) Hydrogen/oxygen therapy for the treatment of an acute exacerbation of chronic obstructive pulmonary disease: results of a multicenter, randomized, double-blind, parallel-group controlled trial. Respir Res 22(1):149
36. Botek M, Krejčí J, Valenta M et al (2022) Molecular hydrogen positively affects physical and respiratory function in acute post-COVID-19 patients: a new perspective in rehabilitation. Int J Environ Res Public Health 19(4)
37. Zolotarenko AD, Zolotarenko AD, Veziroglu A et al (2022) The use of ultrapure molecular hydrogen enriched with atomic hydrogen in apparatuses of artificial lung ventilation in the fight against virus COVID-19. Int J Hydrogen Energy 47(11):7281–7288
38. Chen JB, Kong XF, Mu F et al (2020) Hydrogen therapy can be used to control tumor progression and alleviate the adverse events of medications in patients with advanced non-small cell lung cancer. Med Gas Res 10(2):75–80
39. Chen JB, Kong XF, Qian W et al (2020) Two weeks of hydrogen inhalation can significantly reverse adaptive and innate immune system senescence patients with advanced non-small cell lung cancer: a self-controlled study. Med Gas Res 10(4):149–154

40. Rubini A (2010) IL-6 increases airway resistance in the rat. Cytokine 51(3):266–273
41. Ichihara M, Sobue S, Ito M et al (2015) Beneficial biological effects and the underlying mechanisms of molecular hydrogen—comprehensive review of 321 original articles. Med Gas Res 5:12
42. Luo P, Ding Y, He Y et al (2022) Hydrogen-oxygen therapy alleviates clinical symptoms in twelve patients hospitalized with COVID-19: a retrospective study of medical records. Medicine (Baltimore) 101(9):e27759
43. Alwazeer D, Liu FF, Wu XY et al (2021) Combating oxidative stress and inflammation in COVID-19 by molecular hydrogen therapy: mechanisms and perspectives. Oxid Med Cell Longev 2021:5513868
44. Ghosh G, Jesudian AB (2019) Small intestinal bacterial overgrowth in patients with cirrhosis. J Clin Exp Hepatol 9(2):257–267
45. Choung RS, Ruff KC, Malhotra A et al (2011) Clinical predictors of small intestinal bacterial overgrowth by duodenal aspirate culture. Aliment Pharmacol Ther 33(9):1059–1067
46. Maslennikov R, Pavlov C, Ivashkin V (2018) Small intestinal bacterial overgrowth in cirrhosis: systematic review and meta-analysis. Hepatol Int 12(6):567–576
47. Yang CY, Chang CS, Chen GH (1998) Small-intestinal bacterial overgrowth in patients with liver cirrhosis, diagnosed with glucose H2 or CH4 breath tests. Scand J Gastroenterol 33(8):867–871
48. Yokoyama K, Sakamaki A, Takahashi K et al (2022) Hydrogen-producing small intestinal bacterial overgrowth is associated with hepatic encephalopathy and liver function. PLoS ONE 17(2):e0264459
49. Sha JB, Zhang SS, Lu YM et al (2018) Effects of the long-term consumption of hydrogen-rich water on the antioxidant activity and the gut flora in female juvenile soccer players from Suzhou. China. Med Gas Res 8(4):135–143
50. Tanaka Y, Kiuchi M, Higashimura Y et al (2021) The effects of ingestion of hydrogen-dissolved alkaline electrolyzed water on stool consistency and gut microbiota: a double-blind randomized trial. Med Gas Res 11(4):138–144
51. Tamasawa A, Mochizuki K, Hariya N et al (2015) Hydrogen gas production is associated with reduced interleukin-1β mRNA in peripheral blood after a single dose of acarbose in Japanese type 2 diabetic patients. Eur J Pharmacol 762:96–101
52. Franceschelli S, Gatta DMP, Pesce M et al (2018) Modulation of the oxidative plasmatic state in gastroesophageal reflux disease with the addition of rich water molecular hydrogen: a new biological vision. J Cell Mol Med 22(5):2750–2759
53. Nakayama M, Watanabe K, Hayashi Y et al (2018) Translational research of peritoneal dialysis solution with dissolved molecular hydrogen. Contrib Nephrol 196:162–170
54. Lu X, Yang L, Yao D et al (2017) Tumor antigen-specific CD8+ T cells are negatively regulated by PD-1 and Tim-3 in human gastric cancer. Cell Immunol 313:43–51
55. Scharping NE, Menk AV, Moreci RS et al (2016) The tumor microenvironment represses T cell mitochondrial biogenesis to drive intratumoral T cell metabolic insufficiency and dysfunction. Immunity 45(2):374–388
56. Akagi J, Baba H (2019) Hydrogen gas restores exhausted CD8+ T cells in patients with advanced colorectal cancer to improve prognosis. Oncol Rep 41(1):301–311
57. Kondo M, Hayama T, Matsuoka D The evaluation of preventive effects of cancer progression by molecular hydrogen plus silver ion (Ag+) in elder patients with recurrent colorectal carcinomas.-A Pilot Study.
58. Terawaki H, Zhu WJ, Matsuyama Y et al (2014) Effect of a hydrogen (H_2)-enriched solution on the albumin redox of hemodialysis patients. Hemodial Int 18(2):459–466
59. Iida A, Nosaka N, Yumoto T et al (2016) The clinical application of hydrogen as a medical treatment. Acta Med Okayama 70(5):331–337
60. Zhao P, Dang Z, Liu M et al (2023) Molecular hydrogen promotes wound healing by inducing early epidermal stem cell proliferation and extracellular matrix deposition. Inflamm Regen 43(1):22

61. Yang F, Chen Z, Chen SA et al (2019) Skin ulcers infected with conditional pathogenic strains treated with local hydrogen water packing in two pemphigus vulgaris patients: case reports with follow-up for 2 months. Dermatol Ther 32(5):e13027
62. Li Q, Kato S, Matsuoka D et al (2013) Hydrogen water intake via tube-feeding for patients with pressure ulcer and its reconstructive effects on normal human skin cells in vitro. Med Gas Res 3(1):20
63. Saitoh Y, Yoshimura Y, Nakano K et al (2009) Platinum nanocolloid-supplemented hydrogendissolved water inhibits growth of human tongue carcinoma cells preferentially over normal cells. Exp Oncol 31(3):156–162
64. Kubota M, Kawashima M, Inoue S et al (2021) Randomized, crossover clinical efficacy trial in humans and mice on tear secretion promotion and lacrimal gland protection by molecular hydrogen. Sci Rep 11(1):6434
65. Igarashi T, Ohsawa I, Kobayashi M et al (2019) Effects of hydrogen in prevention of corneal endothelial damage during phacoemulsification: a prospective randomized clinical trial. Am J Ophthalmol 207:10–17
66. Lin MT, Beal MF (2006) Mitochondrial dysfunction and oxidative stress in neurodegenerative diseases. Nature 443(7113):787–795
67. Ohta S (2012) Molecular hydrogen is a novel antioxidant to efficiently reduce oxidative stress with potential for the improvement of mitochondrial diseases. Biochim Biophys Acta 1820(5):586–594
68. Yoritaka A, Takanashi M, Hirayama M et al (2013) Pilot study of H_2 therapy in Parkinson's disease: a randomized double-blind placebo-controlled trial. Mov Disord 28(6):836–839
69. Yoritaka A, Abe T, Ohtsuka C et al (2016) A randomized double-blind multi-center trial of hydrogen water for Parkinson's disease: protocol and baseline characteristics. BMC Neurol 16:66
70. Yoritaka A, Kobayashi Y, Hayashi T et al (2021) Randomized double-blind placebo-controlled trial of hydrogen inhalation for Parkinson's disease: a pilot study. Neurol Sci 42(11):4767–4770
71. Yoritaka A, Ohtsuka C, Maeda T et al (2018) Randomized, double-blind, multicenter trial of hydrogen water for Parkinson's disease. Mov Disord 33(9):1505–1507
72. Shaw VE, Spana S, Ashkan K et al (2010) Neuroprotection of midbrain dopaminergic cells in MPTP-treated mice after near-infrared light treatment. J Comp Neurol 518(1):25–40
73. Ying R, Liang HL, Whelan HT et al (2008) Pretreatment with near-infrared light via light-emitting diode provides added benefit against rotenone- and MPP+-induced neurotoxicity. Brain Res 1243:167–173
74. Hong CT, Hu CJ, Lin HY et al (2021) Effects of concomitant use of hydrogen water and photobiomodulation on Parkinson disease: a pilot study. Med (Baltimore) 100(4):e24191
75. Ono H, Nishijima Y, Ohta S (2023) Therapeutic inhalation of hydrogen gas for Alzheimer's disease patients and subsequent long-term follow-up as a disease-modifying treatment: an open label pilot study. Pharmaceuticals 16(3):434
76. Nishimaki K, Asada T, Ohsawa I et al (2018) Effects of molecular hydrogen assessed by an animal model and a randomized clinical study on mild cognitive impairment. Curr Alzheimer Res 15(5):482–492
77. Cunnane SC, Trushina E, Morland C et al (2020) Brain energy rescue: an emerging therapeutic concept for neurodegenerative disorders of ageing. Nat Rev Drug Discov 19(9):609–633
78. Mizuno K, Sasaki AT, Ebisu K et al (2017) Hydrogen-rich water for improvements of mood, anxiety, and autonomic nerve function in daily life. Med Gas Res 7(4):247–255
79. Todorovic N, Zanini D, Stajer V et al (2021) Hydrogen-rich water and caffeine for alertness and brain metabolism in sleep-deprived habitual coffee drinkers. Food Sci Nutr 9(9):5139–5145
80. Takeuchi S, Kumagai K, Toyooka T et al (2021) Intravenous hydrogen therapy with Intracisternal magnesium sulfate infusion in severe aneurysmal subarachnoid hemorrhage. Stroke 52(1):20–27
81. Takeuchi S, Mori K, Arimoto H et al (2014) Effects of intravenous infusion of hydrogen-rich fluid combined with intra-cisternal infusion of magnesium sulfate in severe aneurysmal

subarachnoid hemorrhage: study protocol for a randomized controlled trial. BMC Neurol 14:176
82. Peng Z, Li X-J, Pang C et al (2023) Hydrogen inhalation therapy regulates lactic acid metabolism following subarachnoid hemorrhage through the HIF-1α pathway. Biochem Biophys Res Commun 663:192–201
83. Ono H, Nishijima Y, Ohta S et al (2017) Hydrogen gas inhalation treatment in acute cerebral infarction: a randomized controlled clinical study on safety and neuroprotection. J Stroke Cerebrovasc Dis 26(11):2587–2594
84. Lin H, Du J, Tian Z et al (2022) Hydrogen gas treatment improves postoperative delirium and cognitive dysfunction in elderly noncardiac patients. J Pers Med 13(1)
85. Mcinnes IB, Schett G (2011) The pathogenesis of rheumatoid arthritis. N Engl J Med 365(23):2205–2219
86. Smith HS, Smith AR, Seidner P (2011) Painful rheumatoid arthritis. Pain Physician 14(5):E427-458
87. Ishibashi T, Sato B, Shibata S et al (2014) Therapeutic efficacy of infused molecular hydrogen in saline on rheumatoid arthritis: a randomized, double-blind, placebo-controlled pilot study. Int Immunopharmacol 21(2):468–473
88. Jin L, Fan K, Tan S et al (2022) The Beneficial Effects of Hydrogen-Rich Saline Irrigation on Chronic Rhinitis: A Randomized. Double-Blind Clinical Trial. J Inflamm Res 15:3983–3995
89. Stefan N, Cusi K (2022) A global view of the interplay between non-alcoholic fatty liver disease and diabetes. Lancet Diabetes Endocrinol 10(4):284–296
90. Nakao A, Toyoda Y, Sharma P et al (2010) Effectiveness of hydrogen rich water on antioxidant status of subjects with potential metabolic syndrome-an open label pilot study. J Clin Biochem Nutr 46(2):140–149
91. Korovljev D, Trivic T, Drid P et al (2018) Molecular hydrogen affects body composition, metabolic profiles, and mitochondrial function in middle-aged overweight women. Ir J Med Sci 187(1):85–89
92. Kajiyama S, Hasegawa G, Asano M et al (2008) Supplementation of hydrogen-rich water improves lipid and glucose metabolism in patients with type 2 diabetes or impaired glucose tolerance. Nutr Res 28(3):137–143
93. Liu B, Jiang X, Xie Y et al (2022) The effect of a low dose hydrogen-oxygen mixture inhalation in midlife/older adults with hypertension: a randomized, placebo-controlled trial. Front Pharmacol 13:1025487
94. Song G, Lin Q, Zhao H et al (2015) Hydrogen activates ATP-binding cassette transporter A1-dependent efflux Ex vivo and improves high-density lipoprotein function in patients with hypercholesterolemia: a double-blinded, randomized, and placebo-controlled trial. J Clin Endocrinol Metab 100(7):2724–2733
95. Song G, Li M, Sang H et al (2013) Hydrogen-rich water decreases serum LDL-cholesterol levels and improves HDL function in patients with potential metabolic syndrome. J Lipid Res 54(7):1884–1893
96. Lebaron TW, Singh RB, Fatima G et al (2020) The effects of 24-week, high-concentration hydrogen-rich water on body composition, blood lipid profiles and inflammation biomarkers in men and women with metabolic syndrome: a randomized controlled trial. Diabetes Metab Syndr Obes 13:889–896
97. Korovljev D, Stajer V, Ostojic J et al (2019) Hydrogen-rich water reduces liver fat accumulation and improves liver enzyme profiles in patients with non-alcoholic fatty liver disease: a randomized controlled pilot trial. Clin Res Hepatol Gastroenterol 43(6):688–693
98. Tao G, Zhang G, Chen W et al (2022) A randomized, placebo-controlled clinical trial of hydrogen/oxygen inhalation for non-alcoholic fatty liver disease. J Cell Mol Med 26(14):4113–4123
99. Tamura T, Hayashida K, Sano M et al (2016) Feasibility and safety of hydrogen gas inhalation for post-cardiac arrest syndrome—first-in-human pilot study. Circ J 80(8):1870–1873
100. Tamura T, Hayashida K, Sano M et al (2017) Efficacy of inhaled hydrogen on neurological outcome following BRain ischemia during post-cardiac arrest care (HYBRID II trial): study protocol for a randomized controlled trial. Trials 18(1):488

101. Katsumata Y, Sano F, Abe T et al (2017) The effects of hydrogen gas inhalation on adverse left ventricular remodeling after percutaneous coronary intervention for ST-elevated myocardial infarction- first pilot study in humans. Circ J 81(7):940–947
102. Powers SK, Jackson MJ (2008) Exercise-induced oxidative stress: cellular mechanisms and impact on muscle force production. Physiol Rev 88(4):1243–1276
103. Ostojic SM (2015) Molecular hydrogen in sports medicine: new therapeutic perspectives. Int J Sports Med 36(4):273–279
104. Ostojic SM (2012) Serum alkalinization and hydrogen-rich water in healthy men. Mayo Clin Proc 87(5):501–502
105. Ito H, Kabayma S, Goto K (2020) Effects of electrolyzed hydrogen water ingestion during endurance exercise in a heated environment on body fluid balance and exercise performance. Temp (Austin) 7(3):290–299
106. Valenta M, Botek M, Krejčí J et al (2022) Acute pre-exercise hydrogen rich water intake does not improve running performance at maximal aerobic speed in trained track and field runners: a randomized, double-blind, placebo-controlled crossover study. PLoS ONE 17(12):e0279307
107. Ooi CH, Ng SK, Omar EA (2020) Acute ingestion of hydrogen-rich water does not improve incremental treadmill running performance in endurance-trained athletes. Appl Physiol Nutr Metab 45(5):513–519
108. Botek M, Krejčí J, Mckune AJ et al (2020) Hydrogen-rich water supplementation and up-hill running performance: effect of athlete performance Level. Int J Sports Physiol Perform 1–4
109. Alharbi AAD, Iwamoto N, Ebine N et al (2022) The acute effects of a single dose of molecular hydrogen supplements on responses to ergogenic adjustments during high-intensity intermittent exercise in humans. Nutrients 14(19)
110. Da Ponte A, Giovanelli N, Nigris D et al (2018) Effects of hydrogen rich water on prolonged intermittent exercise. J Sports Med Phys Fitness 58(5):612–621
111. Shibayama Y, Dobashi S, Arisawa T et al (2020) Impact of hydrogen-rich gas mixture inhalation through nasal cannula during post-exercise recovery period on subsequent oxidative stress, muscle damage, and exercise performances in men. Med Gas Res 10(4):155–162
112. Botek M, Krejčí J, Mckune A et al (2022) Hydrogen rich water consumption positively affects muscle performance, lactate response, and alleviates delayed onset of muscle soreness after resistance training. J Strength Cond Res 36(10):2792–2799
113. Mikami T, Tano K, Lee H et al (2019) Drinking hydrogen water enhances endurance and relieves psychometric fatigue: a randomized, double-blind, placebo-controlled study (1). Can J Physiol Pharmacol 97(9):857–862
114. Botek M, Krejčí J, Mckune AJ et al (2019) Hydrogen rich water improved ventilatory, perceptual and lactate responses to exercise. Int J Sports Med 40(14):879–885
115. Aoki K, Nakao A, Adachi T et al (2012) Pilot study: effects of drinking hydrogen-rich water on muscle fatigue caused by acute exercise in elite athletes. Med Gas Res 2:12
116. Ostojic SM, Vukomanovic B, Calleja-Gonzalez J et al (2014) Effectiveness of oral and topical hydrogen for sports-related soft tissue injuries. Postgrad Med 126(5):187–195
117. Javorac D, Stajer V, Ratgeber L et al (2021) Hydrotherapy with hydrogen-rich water compared with RICE protocol following acute ankle sprain in professional athletes: a randomized non-inferiority pilot trial. Res Sports Med 29(6):517–525

Chapter 7
Homeostatic and Endocrine Response Underlying Protective Effects by Molecular Hydrogen

Mami Noda and Eugene Iv. Nazarov

Abstract Molecular hydrogen (H_2) has multiple properties such as anti-apoptotic, anti-inflammatory and anti-oxidative properties, having wide variety of beneficial outcomes. Effects of H_2 are often slow to show the effects. The amount of H_2 in drinking water is far less than that of inhalation. H_2, especially in drinking water, can be used for therapy but rather for maintenance and improvement of the health condition or prevention of disorders. From mechanistic point of view, the problem is that pharmacokinetic of H_2 in drinking water is largely unknown, especially in the brain. This chapter focuses the long-term intake of H_2 and the mechanism is discussed from homeostatic and endocrinal point of view, trying to understand why H_2 has such a wide variety of effects. H_2 stimulates the production of gastrointestinal hormone, ghrelin, which stimulates growth hormone releasing, affects hypothalamic–pituitary–adrenal (HPA) axis, and also seems to affect hypothalamic–pituitary–gonadal (HPG) axis. Due to these endocrinological effects, H_2 has a corrective effect on the neuro-immuno-endocrine system and determines the functioning of the body's homeostatic system.

Keywords Homeostasis · Hormones · Medical gases · Molecular hydrogen · Endocrine system · Ghreline

M. Noda (✉)
Institute of Mitochondrial Biology and Medicine of Xi'an Jiaotong University School of Life Science and Technology, Xi'an, China
e-mail: maminoda39@gmail.com

E. Iv. Nazarov
Scientific Department of the International Association of Therapists Using Medical Gases, Odessa, Ukraine

J. Slezak and B. Kura (eds.), *Molecular Hydrogen in Health and Disease*, Advances in Biochemistry in Health and Disease 27, https://doi.org/10.1007/978-3-031-47375-3_7

Introduction

Among medical gases, molecular hydrogen (H_2) has multiple properties such as anti-apoptotic, anti-inflammatory and anti-oxidative properties [1]. Due to its beneficial effects, thousands of papers have been reported that H_2 has protective effects on any organs, tissue, and cells since the sensational report in the medical field [2]. Inhaled H_2 diffuses and permeates into every corner of the body due to its small molecular size, causing wide variety of effects. In addition to peripheral effects, efficacy of inhaled H_2 on neurological outcome, for example, following brain ischemia has been shown [3, 4], indicating that H_2 inhalation therapy saves lives and improves outcomes for patients. In periphery, especially in pulmonary function, it was a striking that H_2-O_2 inhalation particularly suitable for relieving dyspnea and other respiratory symptoms in patients with COVID-19 [5]. Later, more evidences in COVID-19 or post-COVID-19 patients have been shown [6, 7]. More recent studies showed that inhalation of H_2 improved functional state of red blood cells in chronic heart failure in both animal model [8] and clinical work in the early postoperative period [9], with H_2 concentration in the respiratory mixture of only 1.5 percent (at a H_2 flow rate of only 120 ml/min). Nevertheless, the results of such a low-level H_2 flux on myocardial safety during heart surgery turned out to be sufficient for effective protection of cardiomyocytes. It would be an important point that H_2 acts in very small doses. These results clearly indicate that H_2 therapy is useful and promising in emergency states.

On the other hand, beneficial effects of H_2 in drinking water has been proved in human as well. For example, improved biomarkers of inflammation and redox homeostasis with mild reduction in body mass index and waist-to-hip ratio after 24 weeks of H_2 intake has been reported [10].

The question is that why effects of drinking H_2-containing water, or sometimes inhaling H_2 gas as well, are so slow to show the effects. The amount of H_2 in drinking water is far less than that of inhalation. H_2 in drinking water can be used for therapy but rather for maintenance and improvement of the health condition or prevention of disorders. There have been many reports that drinking H_2 water is effective for neurological disorders such as Parkinson's disease in animal models [11, 12]. In clinical studies, only pilot study showed significant effects [13], not yet confirmed in randomized, double-blind, multicenter trial [14]. Controversial effects by inhalation of H_2 in Parkinson's disease patients were also reported; H_2 rather increased urinary 8-hydroxy-2-deoxyguanine and other reported stress [15]. These opposite effects of H_2 may indicate that its beneficial effects are partly or largely mediated by hormetic mechanisms proposed earlier [16].

On the other hand, relating to Alzheimer's disease, H_2-rich water is reported to attenuates amyloid β-induced cytotoxicity in cellular level [17], and in animal model, suppressing the decline of memory impairment and prolonging the mean lifespan by H_2-water group was reported [18]. In a randomized clinical study, only carriers of the apolipoprotein E4 (APOE4) genotype showed significant improvement in the scores after intake of H_2-group for 1 year [18], proposing that ApoE4 is a promising AD therapeutic target [19].

From mechanistic point of view, the problem is that pharmacokinetic of H_2 in drinking water is largely unknown, especially in the brain [20]. How can such a low concentration of H_2 work in neurological disorders? Mechanisms of H_2 on different organs and tissues might be complex, causing diverse effects, maintaining homeostasis under physiological conditions or preventing pathophysiological states, and consequently contributing healthful longevity, ageing prevention, and ageing-related disease therapy [21, 22]. There are many research trying to identify the direct H_2 target molecule, for example, the nuclear factor erythroid 2–related factor 2 (Nrf2) [23], Wnt/beta-catenin signaling [24], mitochondrial complex I [25], NADP/NADPH redox pathways [26], or Fe-porphyrin [27, 28], and the possible signaling mechanisms underlying the effects of H_2 on metabolic diseases has been proposed [29]. However, more attention should be paid to understand how H_2 works systemically, because H_2 works almost all organs, tissues, and cells at the same time and the consequence of all the responses become orchestrated. Moreover, intestinal flora-produced large amount of H_2 cannot mimic beneficial effects induced by small amount of exogenous H_2. On this point, it is indeed necessary to investigate the target of exogenous H_2, not endogenous H_2. Indeed, H_2 and gut microbiome have been intensively investigated recently [29, 30]. Human study has also been performed recently [31].

Therefore, in this chapter, long-term intake of H_2, not acute effects in the H_2-inhalation treatment mentioned above, is focused, and discussed from homeostatic and endocrine point of view, trying to understand why H_2 has such a wide variety of effects.

Concentration of H_2 in the Brain After Inhalation of H_2 or Drinking H_2-Containing Water

Using rats, it is unlikely that H_2 in drinking water reaches in the brain parenchyma, while H_2 easily reaches to the brain by inhalation (Fig. 7.1). Since H_2 in drinking water barely reaches to the brain, it is unlikely that H_2 directly affects brain tissue or cells. Nevertheless, drinking H_2-containing water for certain period resulted in distinct effects in neurodegenerative diseases not only Parkinson's disease and Alzheimer's disease, ischemia-induced neuronal damage as mentioned above, but also in stress-induced memory impairments [32] and more in animal models. To explain the effects on the neurological symptoms after long-term intake of H_2 water, there must be indirect effects, not direct effects as reported as a scavenger of •OH, the most toxic reactive oxygen species [2, 33].

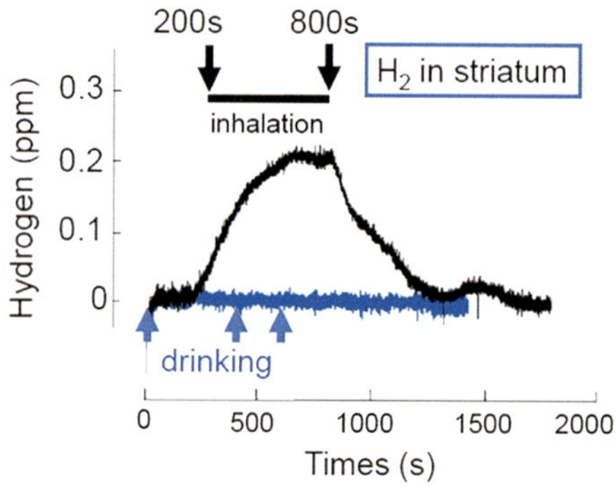

Fig. 7.1 Typical examples of H_2 concentration in striatum. **Black**: Anesthetized rat with H_2 electrode in the right striatum of the brain and current was recorded by voltammetry. The rat started to inhale H_2 gas at 200 s after the start of recording, then stopped at 800 s (modified from Supplementary Figure S2, [11]). **Blue**: The same rat without anesthesia drank H_2-saturated water in a glass bottle at the time of arrows while freely moving around

H_2 and Gastrointestinal Hormone

Ghrelin, a peptide of 28 amino acids and a growth hormone releasing acylated peptide from stomach, is a first hormone linking gastrointestinal-pituitary axis, being specific for growth hormone secretagogue receptor (GHS-R) [34]. Neuroprotective effects of ghrelin on dopaminergic neurons have been reported both in vitro [35, 36] and in vivo [37, 38]. GHS-R is highly expressed by dopaminergic neurons of the substantia nigra [39] and has been suggested that ghrelin protects nigrostriatal dopamine neurons via an uncoupling protein 2 (UCP2)-dependent mitochondrial mechanism [40]. Since drinking H_2 water induces ghrelin release [41], protective and preventive effect of drinking H_2 water can be connected to that of ghrelin at least on dopaminergic neurons. In an animal model of Parkinson's disease, protective effects of drinking H_2 water, for more than 4 days, was completely blocked by GHS-R inhibitor [41]. On the other hand, protective effects of drinking H_2 water was not cancelled in ghrelin-knockout mice [42], suggesting that ghrelin is not the only growth hormone releasing hormone induced by H_2. Other gastrointestinal hormones such as gastrin or secretin were not upregulated by H_2 water (not published). Therefore, other growth hormone-related hormones, instead of ghrelin, remain to be investigated. The mechanism how H_2 stimulate the production of ghrelin also remains unclear, except the evidence that it is independent on cyclic AMP [43].

If ghrelin is one of the major responsible hormones upregulated by drinking H_2 water, the question is whether other beneficial effects of H_2 are also explained by those of ghrelin. Acyl ghrelin ameliorates oxidative stress and enhances the mitochondrial function, adenosine triphosphate generation and biogenesis, therefore acyl ghrelin could be used for treatment in both Parkinson's and Alzheimer's disease [44]. In animal models of focal cerebral ischemia, ghrelin administration can also improve neuronal cell survival as well as rescue memory deficits [45]. Not only

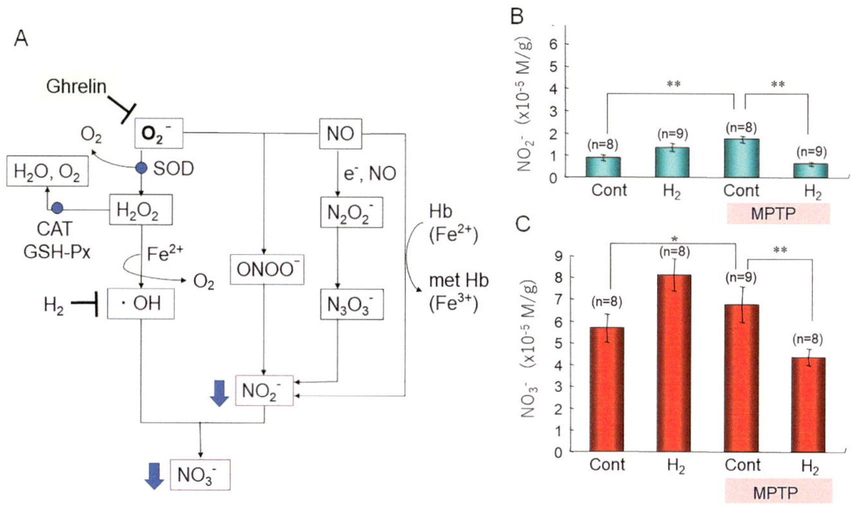

Fig. 7.2 Inhibition of reactive oxygen species (ROS) and nitric oxides by H_2 and ghrelin. **A** Metabolic pathway of reactive oxides (ROS), nitric oxides (NOx), and inhibition site by H_2 and ghrelin. Catalase, CAT; plasma glutathione peroxidase, GSH-Px; superoxide dismutase, SOD. **B, C** Formation of nitrite ion (NO_2^-) (B), and nitrate ion (NO_3^-) (C), in substantia nigra from MPTP-induced Parkinson's disease model mice (male C57/BL6J, 8–9-week-old), measured by HPLC-Griess system. n: number of animals. Cont: control, H_2: after oral intake of H_2 water for 7 days. *P < 0.05, **P < 0.01 are significantly inhibited by chronic intake of H_2 water (unpublished data)

in the nervous system, but also vascular protection by ghrelin has been reviewed [46]. Acyl ghrelin plays in maintaining normal endothelial function by maintaining the balance of vasodilator-vasoconstrictor factors, inhibiting inflammatory cytokine production and immune cell recruitment to sites of vascular injury and by promoting angiogenesis. Acyl ghrelin has been consistently shown to negatively regulate proinflammatory pathways and reduce oxidative and nitrosative stress (Fig. 7.2a). Since both acute and chronic inflammation and an excess of reactive oxygen and nitrogen species impair many functions, important inhibitory actions of H_2-induced ghrelin can be considered to act to maintain organ homeostasis and survival. Indeed, H_2 water-induced attenuation of nitric oxides in Parkinson's disease model mice was confirmed (Fig. 7.2b, c).

H_2 and Hypothalamic–Pituitary–Adrenal (HPA) Axis

Other medical gases such as ozone and xenon are known to affect HPA axis [1]. H_2 also affect HPA axis and is reported to increase resilience to stress in mice, attenuating stress-induced increase in corticosterone (CORT) and adrenocorticotropic hormone (ACTH) [47]. H_2 also inhibit stress-induced pro-inflammatory cytokines such as tumor necrosis factor-alpha (TNF-α) and interleukin 6 (IL6) [47]. In this study, H_2

was given via inhalation, not in drinking water. H_2-O_2 [67%:33% (V/V)] mixed gas was used for inhalation in a closed box, with actual concentration of H_2 gas was kept in level of approximately 65%. Mice were in the box for 1 h or 3 h daily for 14 consecutive days. It is suggested that H_2 affects different endocrine system with different way of administration.

Though H_2 water was not as effective as inhalation of H_2, significant decrease in visceral fat was observed after long term (6 months) of drinking H_2 water in rats [26]. It was due to H_2-induced alteration of liver metabolism which is accompanied by lipolysis by modulating NADP/NADPH redox pathways. With H_2 inhalation, serum level of epinephrine, a lipolytic stimulator, was also observed [26]. These results suggest that intake of H_2 may stimulate sympathetic nerve system.

Modulation of autonomic nerve function by H_2 water in daily life had been reported in human [48]. After 4 weeks of H_2 water, K6 score for mood and anxiety and the low-frequency component power for autonomic nerve function were significantly improved.

Since exercise may improve control over the HPA-axis response to stress [49], studies on H_2 supplementation and exercise may indicate the involvement of H_2 in modulation of HPA-axis as well. According to the meta-analysis, H_2 supplementation alleviates fatigue in healthy adults [50], helping to understand the beneficial effects of H_2 in modulating HPA-axis.

H_2 and Hypothalamic–Pituitary–Gonadal (HPG) Axis

Using a transgenic Alzheimer's disease mouse model, drinking H_2-water for 3 months significantly improved cognitive behavior in female mice without affecting amyloid beta (Aβ) clearance, and reversed the fall in brain estrogen level, estrogen receptor-β (ERβ), and brain-derived neurotrophic factor (BDNF) expressions [51]. These results imply that H_2 may affects HPG axis as well as HPA axis or through stimulating growth hormone, consequently preventing the drop of estrogen level. The reason why beneficial effects of H_2 water were more profound in the brains of female Alzheimer's disease mice than in those of males shall be explained by brain estrogen receptor β (ERβ)-BDNF signaling in Alzheimer's disease pathogenesis. In male rats, drinking H_2 water also alleviated the decrease of serum testosterone level induced by electromagnetic pulse which causes oxidative phosphorylation, TNF signaling pathways, and cytokine-receptor interactions [52]. Though the mechanism of H_2 was explained by decrease in the testicular apoptosis rate and apoptosis protein level, increased sperm motility, and improved antioxidative capacity, involvement of HPG axis can also contribute to the maintenance of testosterone production. Restoring the gonadal system consequently alleviate reproductive system damage in both male and female.

Homeostatic Action of H_2

The physiological activity of H_2 in various fields of biology and medicine is encountered through hormonal system reactions. Often it is considered that these reactions occur due to the influence on lipid peroxidation [53, 54]. The effect of H_2 on the body, as well as other medical gases such as ozone and xenon having a corrective effect on the neuro-immuno-endocrine system, determines the functioning of the body's homeostatic system [1]. The systemic effects of H_2 are a combination of gastrointestinal hormone, HPA axis, HPG axis, and autonomic nerve system (Fig. 7.3). When we focus on the stress response and sympathetic nervous system, it looks controversial but may be understandable. Though H_2 suppresses stress-induced excess release of corticosterone (CORT) or adrenocorticotropic hormone (ACTH), H_2 stimulates sympathetic nerve and increases plasma noradrenalin. One of the effects of noradrenalin is stimulating the release of ghrelin [55, 56]. Ghrelin, in turn, modulates sympathetic nerve system [57] as negative feedback system.

H_2 in drinking water goes through digestive organs, inducing the release of gastrointestinal hormone, ghrelin. Released ghrelin in the plasma can reach to the brain, protecting neurons especially in substantia nigra where ghrelin receptors are robustly expressed. At the same time, ghrelin stimulates the release of growth hormone releasing hormone (GHRH) and growth hormone. Stimulation of growth hormone release may affect may organs and induce various hormones including sex hormone such as estrogen and testosterone through hypothalamic–pituitary–gonadal (HPG) axis. Considering the hypothalamic–pituitary–adrenal (HPA) axis, ghrelin is reported to inhibit stress-induced release of corticosterone (CORT) and adrenocorticotropic hormone (ACTH). On the other hand, H_2 stimulate sympathetic nerves, increasing plasma noradrenalin (NA), stimulating the release of ghrelin. Crosstalk of endocrine system and autonomic nervous system contributes to the systemic effects of H_2.

Effects of medical gases are not all beneficial. It must be aware that medical gases other than H_2 also have side effects. Even with H_2, unconditional reflex should be considered. For example, the delayed anti-inflammatory effect of H_2 can be associated with the early local pro-inflammatory effect [15].

As for the mechanism on the H_2 action at the molecular level, it remains an open question. The modulation of gene expression can be involved in H_2-induced upregulation of the hormones, though how $H_{2,}$ the smallest molecule, works at the genetic level remains unknown. H_2 is also speculated to work in mitochondria, reducing the mitochondrial membrane potential [25]. Indeed, taking H_2 in drinking water for a couple of months increased the production of adenosine trisphosphate and coenzyme Q [58]. However, it may not be enough to explain the wide variety of beneficial effects of H_2. More precise mechanisms should be investigated to understand numerous effects of H_2.

As a conclusion, the corrective effects of H_2 on the neuro-immuno-endocrine system may be explained systematically and more analyses on neuro-endocrine-immune system of homeostasis stay awaited.

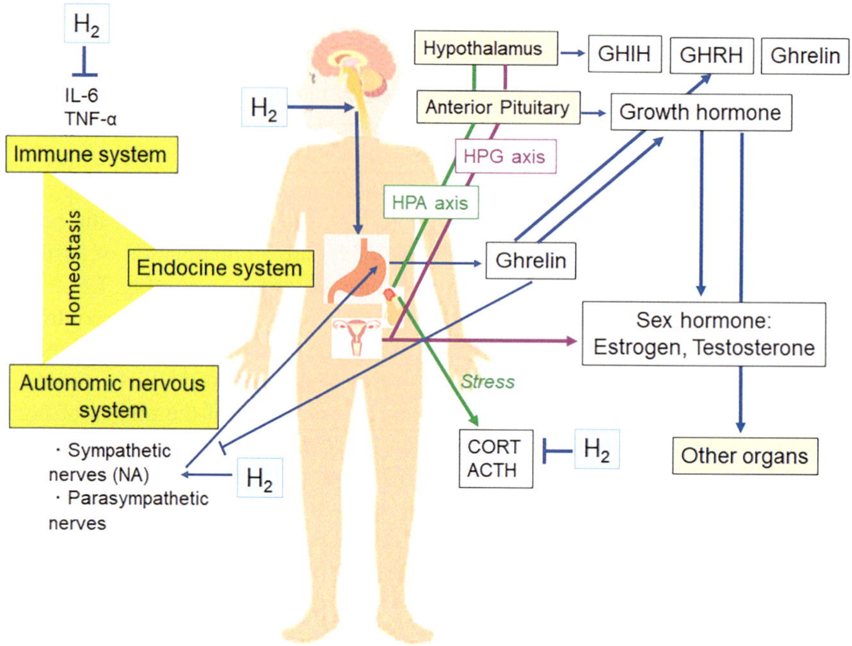

Fig. 7.3 H_2 in drinking water goes through digestive organs, inducing the release of gastrointestinal hormone, ghrelin. Released ghrelin in the plasma can reach to the brain, protecting neurons especially in substantia nigra where ghrelin receptors are robustly expressed. At the same time, ghrelin stimulates the release of growth hormone releasing hormone (GHRH) and growth hormone. Stimulation of growth hormone release may affect may organs and induce various hormones including sex hormone such as estrogen and testosterone through hypothalamic–pituitary–gonadal (HPG) axis. Considering the hypothalamic–pituitary–adrenal (HPA) axis, ghrelin is reported to inhibit stress-induced release of corticosterone (corticosterone (CORT)) and adrenocorticotropic hormone (adrenocorticotropic hormone (ACTH)). On the other hand, H_2 stimulate sympathetic nerves, increasing plasma noradrenalin (NA), stimulating the release of ghrelin. Crosstalk of endocrine system and autonomic nervous system contributes to the systemic effects of H_2

References

1. Nazarov EI, Khlusov IA, Noda M (2021) Homeostatic and endocrine responses as the basis for systemic therapy with medical gases: ozone, xenon and molecular hydrogen. Med Gas Res 11(4):174–186
2. Ohsawa I, Ishikawa M, Takahashi K et al (2007) Hydrogen acts as a therapeutic antioxidant by selectively reducing cytotoxic oxygen radicals. Nat Med 13(6):688–694
3. Ono H, Nishijima Y, Ohta S et al (2017) Hydrogen gas inhalation treatment in acute cerebral infarction: a randomized controlled clinical study on safety and neuroprotection. J Stroke Cerebrovasc Dis 26(11):2587–2594
4. Tamura T, Suzuki M, Homma K et al (2023) Efficacy of inhaled hydrogen on neurological outcome following brain ischaemia during post-cardiac arrest care (HYBRID II): a multi-centre, randomised, double-blind, placebo-controlled trial. EClinicalMedicine 58:101907

5. Guan WJ, Wei CH, Chen AL et al (2020) Hydrogen/oxygen mixed gas inhalation improves disease severity and dyspnea in patients with Coronavirus disease 2019 in a recent multicenter, open-label clinical trial. J Thorac Dis 12(6):3448–3452
6. Singh RB, Halabi G, Fatima G et al (2021) Molecular hydrogen as an adjuvant therapy may be associated with increased oxygen saturation and improved exercise tolerance in a COVID-19 patient. Clin Case Rep 9(11):e05039
7. Botek M, Krejci J, Valenta M et al (2022) Molecular hydrogen positively affects physical and respiratory function in acute post-COVID-19 patients: a new perspective in rehabilitation. Int J Environ Res Public Health 19(4)
8. Deryugina AV, Danilova DA, Pichugin VV et al (2023) The Effect of molecular hydrogen on functional states of erythrocytes in rats with simulated chronic heart Failure. Life (Basel) **13**(2)
9. Deryugina AV, Danilova DA, Brichkin YD et al (2023) Molecular hydrogen exposure improves functional state of red blood cells in the early postoperative period: a randomized clinical study. Med Gas Res 13(2):59–66
10. LeBaron TW, Singh RB, Fatima G et al (2020) The effects of 24-week, high-concentration hydrogen-rich water on body composition, blood lipid profiles and inflammation biomarkers in men and women with metabolic syndrome: a randomized controlled trial. Diabetes Metab Syndr Obes 13:889–896
11. Fujita K, Seike T, Yutsudo N et al (2009) Hydrogen in drinking water reduces dopaminergic neuronal loss in the 1-methyl-4-phenyl-1,2,3,6-tetrahydropyridine mouse model of Parkinson's disease. PLoS ONE 4(9):e7247
12. Fu Y, Ito M, Fujita Y et al (2009) Molecular hydrogen is protective against 6-hydroxydopamine-induced nigrostriatal degeneration in a rat model of Parkinson's disease. Neurosci Lett 453(2):81–85
13. Yoritaka A, Takanashi M, Hirayama M et al (2013) Pilot study of H(2) therapy in Parkinson's disease: a randomized double-blind placebo-controlled trial. Mov Disord 28(6):836–839
14. Yoritaka A, Ohtsuka C, Maeda T et al (2018) Randomized, double-blind, multicenter trial of hydrogen water for Parkinson's disease. Mov Disord 33(9):1505–1507
15. Hirayama M, Ito M, Minato T et al (2018) Inhalation of hydrogen gas elevates urinary 8-hydroxy-2'-deoxyguanine in Parkinson's disease. Med Gas Res 8(4):144–149
16. Murakami Y, Ito M, Ohsawa I (2017) Molecular hydrogen protects against oxidative stress-induced SH-SY5Y neuroblastoma cell death through the process of mitohormesis. PLoS ONE 12(5):e0176992
17. Lin CL, Huang WN, Li HH et al (2015) Hydrogen-rich water attenuates amyloid beta-induced cytotoxicity through upregulation of Sirt1-FoxO3a by stimulation of AMP-activated protein kinase in SK-N-MC cells. Chem Biol Interact 240:12–21
18. Nishimaki K, Asada T, Ohsawa I et al (2018) Effects of molecular hydrogen assessed by an animal model and a randomized clinical study on mild cognitive impairment. Curr Alzheimer Res 15(5):482–492
19. Safieh M, Korczyn AD, Michaelson DM (2019) ApoE4: an emerging therapeutic target for Alzheimer's disease. BMC Med 17(1):64
20. Ichihara G, Katsumata Y, Moriyama H et al (2021) Pharmacokinetics of hydrogen after ingesting a hydrogen-rich solution: a study in pigs. Heliyon 7(11):e08359
21. Fu Z, Zhang J, Zhang Y (2022) Role of molecular hydrogen in ageing and ageing-related diseases. Oxid Med Cell Longev 2022:2249749
22. Rahman MH, Jeong E-S, You H-S et al (2023) Redox-mechanisms of molecular hydrogen promote healthful longevity. Antioxidants 12:988
23. Kawamura T, Wakabayashi N, Shigemura N et al (2013) Hydrogen gas reduces hyperoxic lung injury via the Nrf2 pathway in vivo. American journal of physiology. Lung Cellular Molecular Physiology 304(10):L646–656
24. Lin Y, Ohkawara B, Ito M et al (2016) Molecular hydrogen suppresses activated Wnt/beta-catenin signaling. Sci Rep 6:31986
25. Ishihara G, Kawamoto K, Komori N et al (2020) Molecular hydrogen suppresses superoxide generation in the mitochondrial complex I and reduced mitochondrial membrane potential. Biochem Biophys Res Commun 522(4):965–970

26. Adzavon YM, Xie F, Yi Y et al (2022) Long-term and daily use of molecular hydrogen induces reprogramming of liver metabolism in rats by modulating NADP/NADPH redox pathways. Sci Rep 12(1):3904
27. Jin Z, Zhao P, Gong W et al (2023) Fe-porphyrin: a redox-related biosensor of hydrogen molecule. Nano Res 16:2020–2025
28. Sun X, Ohta S, Zhang JH (2023) Discovery of a hydrogen molecular target. Med Gas Res 13(2):41–42
29. Xie F, Song Y, Yi Y et al (2023) Therapeutic potential of molecular hydrogen in metabolic diseases from bench to bedside. Pharmaceuticals (Basel) 16(4)
30. Ivanova AY, Shirokov IV, Toshchakov SV et al (2023) Effects of coenzyme Q10 on the biomarkers (Hydrogen, Methane, SCFA and TMA) and composition of the gut microbiome in rats. Pharmaceuticals 16(5):686
31. Liang B, Shi L, Du D et al (2023) Hydrogen-rich water ameliorates metabolic disorder via modifying gut microbiota in impaired fasting glucose patients: a randomized controlled study. Antioxidants 12:1245
32. Nagata K, Nakashima-Kamimura N, Mikami T et al (2009) Consumption of molecular hydrogen prevents the stress-induced impairments in hippocampus-dependent learning tasks during chronic physical restraint in mice. Neuropsychopharmacology 34(2):501–508
33. Ohta S (2012) Molecular hydrogen is a novel antioxidant to efficiently reduce oxidative stress with potential for the improvement of mitochondrial diseases. Biochem Biophys Acta 1820(5):586–594
34. Kojima M, Hosoda H, Date Y et al (1999) Ghrelin is a growth-hormone-releasing acylated peptide from stomach. Nature 402(6762):656–660
35. Dong J, Song N, Xie J et al (2009) Ghrelin antagonized 1-methyl-4-phenylpyridinium (MPP(+))-induced apoptosis in MES23.5 cells. J Molecular Neurosc: MN 37(2):182–189
36. Liu L, Xu H, Jiang H et al (2010) Ghrelin prevents 1-methyl-4-phenylpyridinium ion-induced cytotoxicity through antioxidation and NF-kappaB modulation in MES23.5 cells. Exp Neurol 222(1):25–29
37. Jiang H, Li LJ, Wang J et al (2008) Ghrelin antagonizes MPTP-induced neurotoxicity to the dopaminergic neurons in mouse substantia nigra. Exp Neurol 212(2):532–537
38. Moon M, Kim HG, Hwang L et al (2009) Neuroprotective effect of ghrelin in the 1-methyl-4-phenyl-1,2,3,6-tetrahydropyridine mouse model of Parkinson's disease by blocking microglial activation. Neurotox Res 15(4):332–347
39. Zigman JM, Jones JE, Lee CE et al (2006) Expression of ghrelin receptor mRNA in the rat and the mouse brain. J Comp Neurol 494(3):528–548
40. Andrews ZB, Liu ZW, Walllingford N et al (2008) UCP2 mediates ghrelin's action on NPY/AgRP neurons by lowering free radicals. Nature 454(7206):846–851
41. Matsumoto A, Yamafuji M, Tachibana T et al (2013) Oral 'hydrogen water' induces neuroprotective ghrelin secretion in mice. Sci Rep 3:3273
42. Yoshii Y, Inoue T, Uemura Y et al (2017) Complexity of stomach-brain interaction induced by molecular hydrogen in Parkinson's disease model mice. Neurochem Res 42(9):2658–2665
43. Noda M, Uemura Y, Yoshii Y et al (2019) Circulating messenger for neuroprotection induced by molecular hydrogen. Can J Physiol Pharmacol 97(10):909–915
44. Reich N, Holscher C (2020) Acylated ghrelin as a multi-targeted therapy for Alzheimer's and Parkinson's disease. Front Neurosci 14:614828
45. Spencer SJ, Miller AA, Andrews ZB (2013) The role of ghrelin in neuroprotection after ischemic brain injury. Brain Sci 3(1):344–359
46. Pearson JT, Shirai M, Sukumaran V et al (2019) Ghrelin and vascular protection. Vasc Biol 1(1):H97–H102
47. Gao Q, Song H, Wang XT et al (2017) Molecular hydrogen increases resilience to stress in mice. Sci Rep 7(1):9625
48. Mizuno K, Sasaki AT, Ebisu K et al (2017) Hydrogen-rich water for improvements of mood, anxiety, and autonomic nerve function in daily life. Med Gas Res 7(4):247–255

49. Hare BD, Beierle JA, Toufexis DJ et al (2014) Exercise-associated changes in the corticosterone response to acute restraint stress: evidence for increased adrenal sensitivity and reduced corticosterone response duration. Neuropsychopharmacology 39(5):1262–1269
50. Zhou K, Liu M, Wang Y et al (2023) Effects of molecular hydrogen supplementation on fatigue and aerobic capacity in healthy adults: a systematic review and meta-analysis. Front Nutr 10:1094767
51. Hou C, Peng Y, Qin C et al (2018) Hydrogen-rich water improves cognitive impairment gender-dependently in APP/PS1 mice without affecting Abeta clearance. Free Radical Res 52(11–12):1311–1322
52. Ma L, Hao W, Feng WB et al (2022) Molecular hydrogen reduces electromagnetic pulse-induced male rat reproductive system damage in a rodent model. Oxid Med Cell Longev 2022:3469474
53. Iuchi K, Imoto A, Kamimura N et al (2016) Molecular hydrogen regulates gene expression by modifying the free radical chain reaction-dependent generation of oxidized phospholipid mediators. Sci Rep 6:18971
54. Iuchi K, Nishimaki K, Kamimura N et al (2019) Molecular hydrogen suppresses free-radical-induced cell death by mitigating fatty acid peroxidation and mitochondrial dysfunction. Can J Physiol Pharmacol 97(10):999–1005
55. Zhao TJ, Sakata I, Li RL et al (2010) Ghrelin secretion stimulated by β1-adrenergic receptors in cultured ghrelinoma cells and in fasted mice. Proc Natl Acad Sci USA 107(36):15868–15873
56. Gagnon J, Anini Y (2012) Insulin and norepinephrine regulate ghrelin secretion from a rat primary stomach cell culture. Endocrinology 153(8):3646–3656
57. Lambert E, Lambert G, Ika-Sari C et al (2011) Ghrelin modulates sympathetic nervous system activity and stress response in lean and overweight men. Hypertension 58(1):43–50
58. Gvozdjakova A, Kucharska J, Kura B et al (2020) A new insight into the molecular hydrogen effect on coenzyme Q and mitochondrial function of rats. Can J Physiol Pharmacol 98(1):29–34

Chapter 8
Radiation-Induced Heart Disease: Potential Role for Molecular Hydrogen

Branislav Kura, Patricia Pavelkova, Barbora Kalocayova, and Jan Slezak

Abstract Radiation-induced heart disease (RIHD) is a common complication of mediastinal radiotherapy. RIHD includes structural and functional abnormalities of the pericardium, coronary vessels, myocardium, valves, and conduction system. The underlying pathological mechanisms are complex, mainly related to endothelial cell injury, oxidative stress, and inflammation. Radiation can cause cardiomyocyte death, tissue fibrosis and ultimately may end up with heart failure. To overcome these complications, there is a need to look for specific therapeutic interventions, which are still missing. Molecular hydrogen (H_2) has been recognized as a molecule with antioxidant, anti-inflammatory, and anti-apoptotic protective effects in different disease settings. In vitro as well as in vivo studies demonstrated that H_2 exerted preventive or therapeutic effects on radiation-induced injury, including RIHD. H_2 could be effective to mitigate RIHD through various mechanisms, e.g. selective neutralization of hydroxyl radicals, protection against inflammatory and apoptotic damage, anti-fibrotic and anti-hypertrophic effects, etc. More research is needed to elucidate further mechanisms of H_2 action, and to verify the effectiveness of H_2 therapy in clinical trials.

Keywords Inflammation · Molecular hydrogen · Oxidative stress · Radiation-induced heart disease

---

B. Kura (✉) · P. Pavelkova · B. Kalocayova · J. Slezak
Centre of Experimental Medicine, Institute for Heart Research, Slovak Academy of Sciences, Dubravska cesta 9, 841 04 Bratislava, Slovakia
e-mail: branislav.kura@savba.sk

P. Pavelkova
e-mail: patricia.pavelkova@savba.sk

B. Kalocayova
e-mail: barbora.kalocayova@savba.sk

J. Slezak
e-mail: jan.slezak@savba.sk

P. Pavelkova
Department of Animal Physiology and Ethology, Faculty of Natural Sciences, Comenius University, Bratislava, Slovakia

J. Slezak and B. Kura (eds.), *Molecular Hydrogen in Health and Disease*, Advances in Biochemistry in Health and Disease 27, https://doi.org/10.1007/978-3-031-47375-3_8

Introduction

Radiation therapy is still widely used in the treatment of numerous oncological diseases. Radiotherapy aims to kill tumor cells; however, it may have deleterious side effects on the surrounding healthy tissues. If applied to the mediastinal area (e.g. to treat breast, lung, esophageal cancer, or Hodgkin's lymphoma), unwanted irradiation of the heart may occur as it is in close anatomical proximity. The syndrome of unwanted cardiovascular side effects of mediastinal irradiation is called radiation-induced heart disease (RIHD) [1]. A number of large clinical studies have confirmed that radiation therapy increases the risk of heart disease-related mortality [2, 3]. Although morbidity of RIHD has been reduced by the use of novel radiation techniques, studies have shown that modern technology does not fully eliminate the risk of RIHD [4].

The pathological spectrum of RIHD includes conduction abnormalities, valvular disease, coronary artery disease, pericardial disease, cardiomyopathy, and myocardial fibrosis [5]. These are a result of various mechanisms interacting with each other through multiple complex pathways. Radiation increases oxidative stress, with subsequent inflammatory response that includes activation of nuclear factor-kappa B (NF-κB) [6]. The inflammatory cascade is initiated by the vascular injury and endothelial dysfunction. Damaged cells induce the secretion of proinflammatory cytokines and growth factors, including transforming growth factor-beta (TGF-β), which is closely related to the development of tissue fibrosis. Myocardial fibrosis is considered the final stage of RIHD and is characterized by the excess collagen deposition in the damaged cardiac tissue, ultimately leading to heart failure [7].

As inflammatory response and oxidative stress play significant roles in developing RIHD, anti-inflammatory and antioxidant compounds can demonstrate radioprotective effects. Molecular hydrogen (H_2) acts as an antioxidant by selectively reducing hydroxyl radicals (·OH) and peroxynitrite ($ONOO^-$) as well as via modulation of transcription factors. Anti-inflammatory and cytoprotective effects have also been well described [8]. H_2 can mitigate different pathophysiological processes related to radiation damage, like fibrosis, hypertrophy, or hypoxia. Animal experiments and clinical trials have proved that H_2 exhibits radioprotective effects in different organs [9].

H_2 is a very small molecule, what allows it to easily pass through the cell membrane and diffuse into the cytoplasm. In a short time, H_2 is able to penetrate into subcellular organelles, like mitochondria and nucleus. H_2 can also easily cross the blood–brain barrier [10]. Other advantages include wide spectrum of H_2 administration methods, including inhalation of hydrogen gas, oral intake of hydrogen-rich water (HRW), and injection with hydrogen-rich saline (HRS) [11].

This article will review the pathophysiology of RIHD and potential use of molecular hydrogen to mitigate this negative side effect of radiotherapy. Focus of the manuscript will be oriented on the putative mechanisms of H_2 action with some insight into future perspectives.

Pathophysiology of RIHD

Understanding the pathogenesis of RIHD is very important to evaluate the feasible therapeutic targets and possible treatment options. Radiation therapy simultaneously causes damage to the macrovasculature, the microvasculature, and the myocardium [12]. A substantial body of evidence suggests that the radiation-induced immediate oxidative damage of macromolecules, including DNA, proteins, and lipids, is the initiating event in RIHD. The increased oxidative stress triggers other biological processes, including endothelial cell injury, acute inflammation, and the various forms of cell death [13]. Several compensatory mechanisms, including endothelial cell proliferation and cardiac hypertrophy, develop in the sublethally damaged surviving cells [14]. If these compensatory mechanisms are exhausted, chronic inflammatory processes and fibrosis play the central role in the disease progression, ending up with heart failure (Fig. 8.1) [15].

Oxidative Stress

Like other heart diseases, oxidative stress also plays an important role in RIHD [16]. Approximately 80% of tissues and cells forms water. Therefore, the initial radiation-induced cellular damage is mostly caused by the radiolysis of water generating reactive oxygen species (ROS) leading to indirect effects. The major reactive species produced in the radiolysis of water are hydroxyl radical (·OH), superoxide anion ($O_2^{\cdot-}$), and hydrogen peroxide (H_2O_2) [17]. Ionizing radiation may upregulate inducible nitric oxide synthase (iNOS), thereby generating a large amount of nitric oxide (·NO), which can react with $O_2^{\cdot-}$ to form peroxynitrite ($ONOO^-$) and secondarily other reactive nitrogen species (RNS) [18].

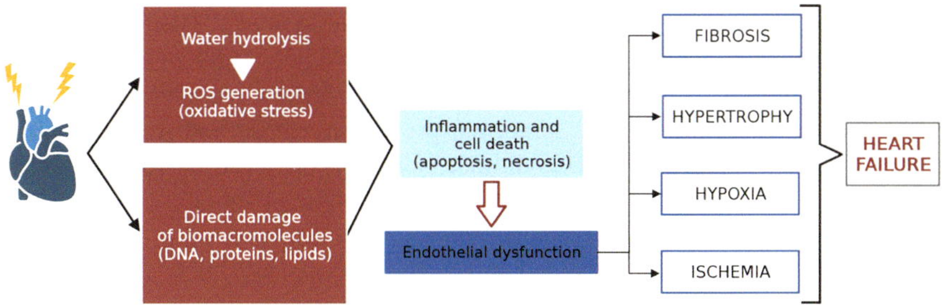

Fig. 8.1 Pathophysiological mechanisms of radiation-induced heart disease (RIHD). After irradiation, either DNA and other macromolecules are directly damaged in the heart, or they are damaged by the effect of ROS, which are formed by hydrolysis of water. Both mechanisms lead to inflammatory processes or cell death primarily damaging endothelium. Further, this causes many pathological conditions of the heart and blood vessels, leading to heart failure

Accumulated ROS/RNS may cause macromolecular damage. In addition to DNA damage, ROS/RNS can also lead to peroxidation of lipids and proteins and activate multiple signaling pathways [19]. Oxidative stress causes damage to both mitochondrial and nuclear DNA. 8-hydroxy-2′-deoxyguanosine (8-OHdG) is considered as the marker of oxidative DNA damage. As a result of lipid peroxidation, reactive lipid derivatives are formed, mainly trans-4-hydroxy-2-nonenal (4-HNE) or malondialdehyde (MDA), which are used as markers of the lipid peroxidation. Oxidative modification of proteins results mainly in the formation of protein carbonyl derivatives [19]. Oxidative stress markers have been found to be elevated in different animal models after irradiation [20, 21].

Radiation can directly disrupt the respiratory chain of mitochondria, leading to respiratory chain dysfunction and thus reducing ATP production, increasing ROS production, reducing antioxidant capacity, and inducing apoptosis [22]. The free radicals produced by radiation can upregulate several enzymes including nicotinamide adenine dinucleotide phosphate oxidase (NADPH oxidase), lipoxygenases (LOXs), and cyclooxygenases (COXs), which are other sources of on-going ROS generation [23]. ROS promote inflammation via their interaction with NF-κB. NF-κB upregulation was detected 4 to 500 weeks post-irradiation in small vessels of the neck in humans [24].

Oxidative and nitrative changes might continue to present months or years after the initial radiation exposure, presumably due to continuously increased generation of ROS/RNS via different mechanisms, e.g. mitochondrial damage, inflammatory and cell death processes, overexpression of ROS-generating enzymes in cardiac tissue, and insufficient antioxidant mechanisms [16].

Endothelial Cell Injury and Inflammation

Cardiac myocytes are relatively resistant to radiation damage. However, endothelial cells remain sensitive to radiation, and the pathophysiology of RIHD appears to be primary associated with damage to endothelial cells [6]. Excessive ROS production represents the initial cause of endothelial dysfunction. At first, endothelial activation is observed, characterized by an abnormal pro-inflammatory and pro-thrombotic action of the endothelial cells. This ultimately leads to reduced nitric oxide (NO) bioavailability, impairment of the vascular tone and other endothelial changes [25].

Damaged endothelial cells secrete adhesion molecules and growth factors prompting activation of the acute inflammatory response. NF-κB is a mediator of this process [26]. Adhesion molecules facilitate inflammatory cell recruitment to the damaged endothelium, predominantly representing neutrophil granulocytes in the acute phase. Inflammatory cells release pro-inflammatory and pro-fibrotic cytokines, including tumor necrosis factor-alpha (TNF-α), interleukin-6 (IL-6), IL-8, TGF-β, platelet-derived growth factor (PDGF), fibroblast growth factor (FGF), etc. [15].

In addition to inflammatory changes, changes in coagulation and platelet function caused by increased release of von Willebrand factor (vWF) in endothelial cells were observed in vessels after heart irradiation. Platelet adhesion as well as thrombus formation in microvessels and arteries increased [27]. Radiation decreases levels of the anticoagulant, thrombomodulin, via inhibition by TGF-β, what impairs endogenous fibrinolysis and promotes coagulation [28].

One of the key consequences of endothelial dysfunction is impairment of endothelium-dependent vasodilation due to reduced bioavailability of vasodilators, particularly NO [29]. After exposure of endothelial cells to ionizing radiation, NO is rapidly deactivated by superoxide radicals, resulting in the formation of peroxynitrites [30]. Irradiation-induced oxidative stress also causes endothelial NO synthase (eNOS) uncoupling, resulting in diminished release of NO [31]. Endothelium-dependent vasodilation seems to be compromised even after longer time periods, as detected in breast cancer radiotherapy patients more than 3 year after radiotherapy [29].

Irradiation to mediastinal structures may influence cardiac capillary endothelial cells causing their proliferation, injury, swelling and degeneration, and significantly reduce the number of capillaries. Although endothelial cells can regenerate, capillary network damage is irreversible [32]. The heart is the main oxygen consuming organ in the human body, and therefore decreased blood supply can give rise to myocardial hypoxia which will aggravate the myocardial injury. Myocardial ischemia and hypoxia, inflammatory responses, collagen deposition, and proliferation of endothelial cells and fibroblasts lead to tissue remodeling, cardiac fibrosis, and atherosclerosis [33].

Cell Death

Irradiation can induce cell death in different cell types of cardiac tissue, including cardiomyocytes, endothelial cells, fibroblasts, as well as cells of the conduction system [14]. According to the current knowledge, the critical aspects in cell death pathways in RIHD are irreversible damage of the structure of cellular compartments and molecules, mitochondrial dysfunction, and endoplasmic reticulum (ER) stress [34].

Ionizing radiation induces a wide range of DNA lesions, including base damage, single-strand breaks, and double-strand breaks directly, but also indirectly via formation of ROS. If the damaged DNA is not adequately repaired, cells can go into apoptosis [35].

Mitochondrial dysfunction is closely related to ER stress. After irradiation of cardiomyocytes, the stimulated ER releases calcium ions into the cytoplasm, leading to mitochondrial calcium overload and mitochondrial swelling [22]. It has been shown that exposure to radiation leads to increased expression and activation of pro-apoptotic protein Bax, what leads to its translocation and insertion into the mitochondrial outer membrane. This enhances the permeability of mitochondrial membrane and reduces the mitochondrial membrane potential, which is closely related to the opening of mitochondrial permeability transition pore (mPTP) [36]. Permeable mitochondrial membrane releases pro-apoptotic factors, such as cytochrome c, what initiates the intrinsic pathway of apoptosis [22].

The increased permeability of mitochondrial membrane also leads to a cascade reaction, which produces a large amount of ROS. ROS further promotes the release of calcium from the ER, leading to sarcomere hypercontractions and calcium overload in mitochondria, thus increasing the generation of ROS. This may lead to long-term toxicity induced by radiation, which eventually leads to cell cycle arrest [37].

Fibrosis and Hypertrophy

Radiation-induced myocardial fibrosis is considered the final stage of RIHD and is characterized by the excess collagen deposition in the damaged cardiac tissue [7]. It is the result of multiple converging pathways including inflammation, oxidative stress, and chronic changes in gene expression. The inflammatory pathway is likely the main mediator of fibrosis [38].

The initiation stage of fibrogenesis is driven by the radiation-induced primary vascular endothelial cell injury. TGF-β is considered a key factor triggering the fibrotic process after irradiation. Its expression is continuing in irradiated tissues [39]. TGF-β is known as a fibroblast mediator and can induce fibroblast differentiation into myofibroblasts, which intensify the secretion of collagen. In the irradiated hearts, there was a significant increase in total tissue collagen concentration compared to non-irradiated hearts. Both Type I and Type III collagen were increased; however, there was a disproportionate increase in the amount of Type I collagen [40].

Radiation-induced TGF-ß expression can be stimulated via different mechanisms. ROS up-regulate TGF-β, stimulate the production of Collagen I, and support radiation-induced fibrosis. DNA damage induced after irradiation can lead to cell death. Apoptosis and senescence are responsible to induce macrophages to release TGF-β [41]. Chronic hypoxia from microvessel damage leads to upregulation of hypoxia inducible factor α (HIF1-α), which is another stimulator of TGF-β [38].

Radiation leads to fibrosis of different layers of the heart, including the myocardium, the endocardium, and the epicardium [7]. Radiation-induced fibrotic damage has been also reported in valves. Fibrotic damage in the valves is unlikely related to microvascular damage as the heart valves are avascular. Post-mortem investigations in patients receiving at least 35 Gy to heart showed focal thickening of the valvular endocardium by elastic fibers [42].

The progressive and diffuse interstitial fibrosis leads to decreased tissue elasticity and contractility as well as chronic hypoxia by separating and replacing the cardiomyocytes [16]. Fibrosis affecting the heart's conduction system or cardiomyocyte conduction could interfere with the transmission of electrophysiological signals, causing arrhythmias [43].

After radiation-induced acute cell damage and death, a compensatory hypertrophy is initiated in the surviving cardiomyocytes to compensate for declined cardiac function due to the loss of cardiomyocytes. Hypertrophy of cardiomyocytes may also develop as a result of insufficient function of fibrotic valves. Chronic hypoxia and repetitive ischemia could play a role in the development of compensatory hypertrophy [44].

Beneficial Effects of H_2 Application to Prevent RIHD

Hydrogen is the lightest and most abundant chemical element. H_2 has a relatively small molecular mass, which helps it penetrate cell membranes and disperse to the cytoplasm, mitochondria, nucleus, and other organelles to exert the biological activities. Moreover, H_2 passes through the blood–brain barrier [45]. There is a wide range of H_2 delivery methods. H_2 is primarily administered via inhalation, drinking H_2-rich water, or injection of H_2 saline. In recent years, nanomaterial delivery systems have also been developed [8].

Many studies have explored protective or beneficial effects of H_2 in different disease settings, including ischemia/reperfusion (I/R) injury [46, 47], organ transplantation [48, 49], metabolic syndromes [50], inflammation [51, 52], as well as radiation injury [20, 53]. The radioprotective effects of H_2 are thought to be related to the scavenging of radiation-induced free radicals but are also suggested to be indirectly related to anti-inflammatory and anti-apoptotic properties, as well as the regulation of gene expression (Fig. 8.2) [9].

Antioxidant Activity

The antioxidant effects of H_2 lie in its specific ability to bind free radicals, especially the hydroxyl radical, which is produced in high quantities during irradiation. Ohsawa et al. [10] showed that H_2 is an antioxidant that selectively reduces highly reactive ROS and RNS, such as $\cdot OH$ and $ONOO^-$, but does not react with other ROS with physiological function such as superoxide anions (O_2^-) and hydrogen peroxide (H_2O_2). In an *in vitro* model, H_2 has been shown to protect hyaluronic acid molecule from degradation induced by hydroxyl radicals [54]. Terasaki et al. [55] demonstrated that H_2 can reduce cellular ROS ($\cdot OH$ and $ONOO^-$) produced by irradiation in living lung epithelial cells. Another study reported H_2 as an effective radioprotective agent on immune system by scavenging ROS [56].

Radiation-induced reactive species may cause oxidative damage to biomolecules. According to Chuai et al. [57], pre-treatment of germ cells with H_2 before irradiation significantly suppressed the reaction of hydroxyl radicals and the cellular macromolecules, thereby lipid peroxidation, protein carbonylation and oxidative DNA damage. The studies have evaluated different oxidative stress markers, mainly 8-OHdG as an indicator of DNA oxidation, and MDA as an indicator of lipid oxidation. Kura et al. [58] showed that the levels of MDA increased in the rat myocardium after

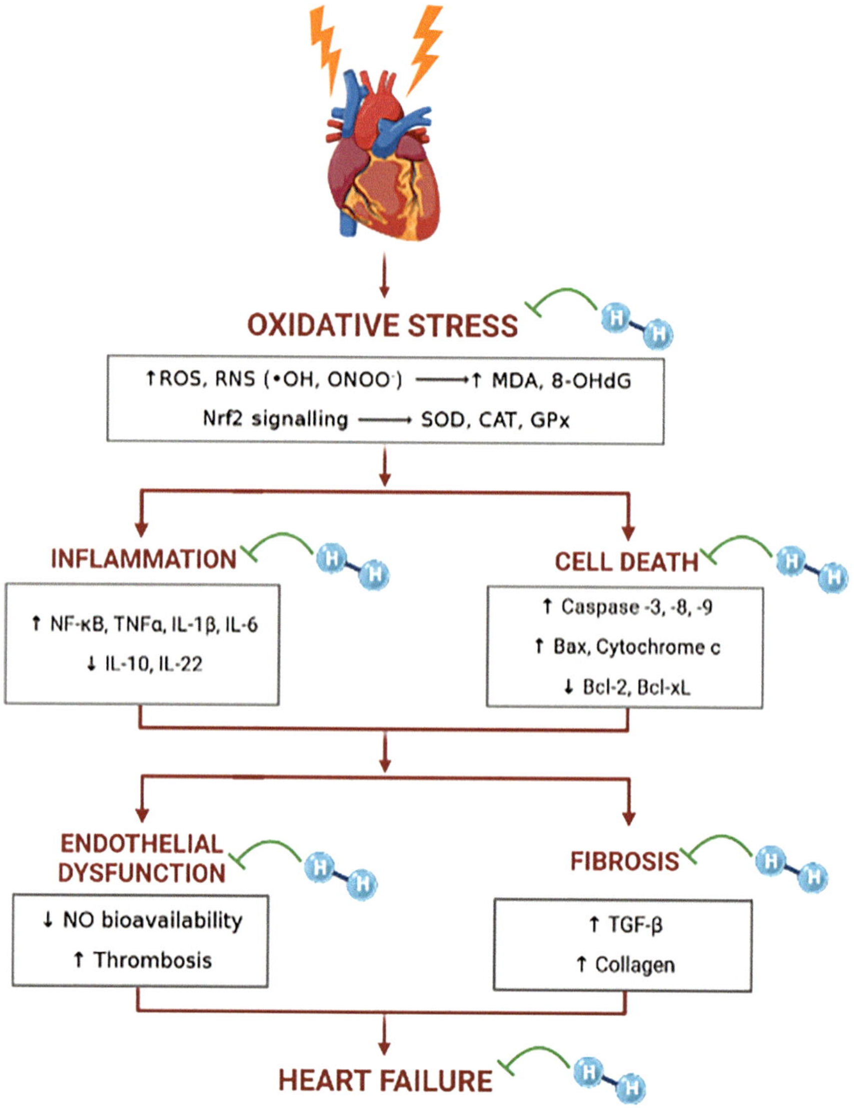

Fig. 8.2 Possible biological effects of H_2 in radiation-induced heart disease. The radioprotective effects of H_2 are related to multiple pathways via its free radicals scavenging, anti-inflammatory, anti-apoptotic, or anti-fibrotic action. ROS – reactive oxygen species, RNS – reactive nitrogen species, MDA – malondialdehyde, 8-OHdG – 8-hydroxy-2′-deoxyguanosine, Nrf2 – Nuclear factor erythroid 2-related factor 2, SOD – superoxide dismutase, CAT – catalase, GPx – glutathione peroxidase, NF-κB – nuclear factor kappa B, TNF-α – tumor necrosis factor alpha, IL-1β – interleukin 1 beta, Bax – Bcl-2-associated X protein, Bcl-2 – B-cell lymphoma-2, NO – nitric oxide, TGF-β – transforming growth factor beta

irradiation. Treatment with HRW reduced these values to the level of non-irradiated controls. In another study, administration of HRW lowered the levels of MDA and 8-OHdG in the brain of irradiated rats [53]. Similar protective results of H_2 have been described by other authors [20, 59].

H_2 is also able to provide antioxidant protection via stimulation of endogenous antioxidants. Many studies demonstrated that H_2 promotes the expression of nuclear factor erythroid 2–related factor 2 (Nrf2), which regulates the expression of innate antioxidant and cytoprotective enzymes [60, 61]. Mei et al. [62] examined radioprotective effects of H_2 using human keratinocyte HaCaT cells. H_2 protected cells from radiation injury among other things via increasing superoxide dismutase (SOD) and glutathione (GSH) activities. Gharib et al. [63] found that H_2 increases SOD activity and reduces MDA level in a model of liver inflammation. In a rat model, hydrogen inhalation alleviated brain injury via reduction of oxidative stress and stimulation of enzymatic activities of the endogenous antioxidants SOD and catalase (CAT) [64]. H_2 inhalation reduced ROS accumulation and up-regulated antioxidant enzyme activities [SOD, CAT, glutathione peroxidase (GPx)] in skin tissue after cutaneous ischemia/reperfusion injury [65].

Anti-inflammatory Effect

The first evidence of hydrogen's anti-inflammatory properties was demonstrated in a study by Gharib et al. [63]. In a mouse model of chronic liver inflammation, they found declined circulating TNF-α after H_2 inhalation for two weeks. The results of other studies also confirm that molecular hydrogen reduces the expression of various pro-inflammatory cytokines, including TNF-α, NF-κB and some interleukins [66, 67]. As inflammation is considered one of the main mediators of radiation-induced injury, anti-inflammatory action of H_2 constitutes one of its possible radioprotective mechanisms. In the work by Kura et al. [58], H_2 significantly reduced MDA and TNF-α levels in the myocardium damaged by irradiation. Zhou et al. [59] examined the radioprotective effects of H_2 on a rat skin model. They showed that H_2 significantly reduced MDA and IL-6 levels in the damaged skin. HRS protected rat skin tissue against UVB radiation injury by decreasing the production of pro-inflammatory cytokines TNF-α, IL-1β, and IL-6 [68].

Anti-inflammatory action of H_2 is also executed via increasing the levels of anti-inflammatory cytokines. Inhalation of H_2 was found to attenuate exercise-induced inflammation in the hippocampus by upregulating the anti-inflammatory cytokine IL-10 [69]. HRW treatment prevented ethanol-induced fatty liver in mice. Intake of HRW reduced levels of pro-inflammatory cytokines TNF-α and IL-6, but increased levels of anti-inflammatory molecules IL-10 and IL-22 [70].

Molecular hydrogen has been shown to be effective in restoring the function of endothelial cells, which can be damaged by irradiation. Treatment with HRS alleviates vascular dysfunction through reducing oxidative stress, suppressing inflammation, preserving mitochondrial function, and enhancing nitric oxide bioavailability

[71]. Sakai et al. [72] have shown that the daily consumption of water containing H_2 may protect the vasculature via neutralization of detrimental ROS to preserve NO bioavailability as well as suppression of the inflammatory reactions. Wang et al. [73] have reported that hydrogen may inhibit platelet activation and prevent thrombosis formation.

Cytoprotective Action

Radioprotective effects of H_2 are related not only to increased antioxidant and anti-inflammatory protection of cells, but also to the modulation of cell death pathways. Hong et al. [74] determined the anti-apoptotic effect of H_2 in rat brain cells. In their study, hydrogen increased the expression of the anti-apoptotic protein Bcl-2 and decreased the pro-apoptotic protein Bax and caspase 3. The activity of caspase-3 was also suppressed after H_2 treatment in a model of cardiac ischemia/reperfusion injury in rats, thereby reduced cell apoptosis [46]. In the experiment using intestinal crypt epithelial cell line IEC-6, the authors showed that H_2 inhibits mitochondrial depolarization, cytochrome c release, and the activities of caspase-3, and caspase-9. They further demonstrated increased expression of Bcl-xl and Bcl-2 and decreased expression of Bax protein after H_2 treatment [75]. HRS promoted the recovery of renal function after I/R injury via reducing the expression of pro-apoptotic proteins Bcl-2, caspase-3, caspase-9, caspase-8 and increasing the level of anti-apoptotic Bax [76].

Terasaki et al. [55] found that H_2 protects *in vitro* cultured lung epithelial cells A549 against damage from oxidative stress induced by irradiation, as evidenced by reducing the apoptotic response and cell viability. H_2 significantly reduced levels of Bax and active caspase 3 as markers of apoptosis in irradiated cells. In another study, treating cells with H_2 before irradiation significantly inhibited radiation-induced human lymphocyte cells (AHH-1) apoptosis [20].

Anti-fibrotic and Anti-hypertrophic Action

Oxidative stress and subsequent inflammatory process induced by radiation lead to the development of tissue fibrosis, which is characterized by increased collagen deposition. TGF-β serves as the primary mediator in this response [77]. Molecular hydrogen has been found to suppress the profibrotic cytokine TGF-β in different studies. H_2 treatment proved to be able to mitigate oxidative stress and subsequent longer term fibrotic damage of lungs in irradiated mice as evidenced by the decreased level of TGF-β and type III collagen [55]. In another model of lung injury, hydrogen inhalation reduced alveolar fibrosis, attenuated the upregulation of pro-inflammatory interleukins and suppressed the expression of TGF-β [78]. Gao et al. [79] demonstrated that hydrogen inhalation attenuates pulmonary fibrosis by inhibiting oxidative stress, TGF-β and suppressing collagen I generation. H_2 has been also found effective to suppress fibrosis in a model of kidney injury. Treatment with HRW alleviated renal fibrosis via regulation of TGF-β induced expression of its downstream molecule Sirtuin 1 [80]. In spontaneous hypertensive rats, HRS reduced oxidative stress and fibrosis by inhibiting the TGF-β signaling pathway [81].

Yu and Zheng [82] explored the effect of HRS treatment on left ventricular hypertrophy in spontaneously hypertensive rats. HRS treatment attenuated left ventricular hypertrophy through abating oxidative stress, suppressing inflammatory process, and preserving mitochondrial function. Zhang et al. [83] detected that H_2 inhibits isoproterenol-induced cardiac hypertrophy via antioxidative pathways. In another study, HRS was found to mitigate cardiac hypertrophy induced by pressure overload in rats and reduces atrial fibrosis and fibrillation, possibly via inhibition of the JAK-STAT signaling pathway [84].

Evidence for H_2 Action to Mitigate Radiation-Induced Injury

H_2 has been shown to display wide range of activities, including antioxidant, anti-inflammatory, cytoprotective, anti-fibrotic action, what might be beneficial to alleviate radiation-induced damage. The potential application of H_2 to mitigate radiation-induced injury has been studied in different organs including heart, lung, brain, intestine, and skin [9]. The studies are summarized in Table 8.1.

Qian et al. [20] tested cardioprotective effects of H_2 in irradiated mice. In this study, pretreatment with H_2 protected myocardium from radiation-induced damage, decreased levels of myocardial MDA, 8-OHdG, and increased myocardial endogenous antioxidants SOD and GSH. Mice treated with H_2 showed improved survival rate compared to control irradiated group. Slezak et al. [85] demonstrated, that in the hearts of rats injured by irradiation H_2 administration led to a significant decrease in MDA and TNF-α levels. In another study, Kura et al. [58] irradiated rats in the mediastinal area. Application of H_2 for 45 days alleviated oxidative and inflammatory damage of irradiated myocardium as shown by decreased levels of MDA and TNF-α. After H_2 treatment, authors also demonstrated normalization of selected miRNAs connected to fibrosis and hypertrophy (miRNA-1, -15b, -21) in irradiated rat myocardium.

Radioprotective effects of H_2 have been also demonstrated on the lungs of mice after whole thorax irradiation. H_2 treatment reduced oxidative stress and apoptosis within 1 week after irradiation. Levels of MDA, 8-OHdG, Bax, the profibrotic cytokine TGF-β1, and Bax mitochondrial translocation were significantly reduced in lung tissue from irradiated mice after H_2 treatment. Radioprotective effects of H_2 treatment persisted 5 months after exposure to irradiation, during the late phase of damage, as reflected by suppression of lung fibrosis [55].

Table 8.1 In vivo studies documenting radioprotective effects of molecular hydrogen. HRW – hydrogen-rich water, MDA – malondialdehyde, 8-OHdG – 8-hydroxy-2′-deoxyguanosine, SOD – superoxide dismutase, GSH – glutathione, TNF-α – tumor necrosis factor alpha, Bax – Bcl-2-associated X protein, TGF-β – transforming growth factor beta, IL-6 – interleukin 6, MPO – myeloperoxidase, EGF – endothelial growth factor

Organ	Species	H_2 administration method	Radioprotective effects	References
Heart	Mice	HRW	↓MDA, 8-OHdG ↑SOD, GSH	[20]
	Rat	HRW/inhalation/HRS	↓MDA, TNF-α	[85]
	Rat	HRW	↓MDA, TNF-α	[58]
Lung	Mice	HRW/inhalation	↓MDA, 8-OHdG ↓Bax, TGF-β1	[55]
Brain	Rat	HRS	↓MDA, 8-OHdG ↑SOD	[86]
	Rat	HRW	↓MDA, 8-OHdG ↑SOD, GSH	[53]
Intestine	Mice	HRS intraperitoneally	↓MDA, 8-OHdG ↑SOD, GSH	[87]
	Mice	HRS intraperitoneally	↓MDA, 8-OHdG ↑SOD ↓TNF-α, IL-6, MPO	[75]
Skin	Rat	Inhalation	↓MDA, 8-OHdG	[88]
	Rat	HRW	↓MDA, IL-6 ↑SOD, EGF	[59]

Several works investigated the protective effect of H_2 treatment on the brain damaged by irradiation. In a study by Huo et al. [86], H_2 was demonstrated to alleviate irradiation-induced acute brain injury in rats. H_2 lowered oxidative stress markers MDA and 8-OHdG and increased the content of endogenous antioxidant SOD as measured in the brain homogenate. They also found the damage degree in the nerve cells of hippocampus was less in the hydrogen group. Another study indicates that HRW has a protective effect on radiation-induced cognitive dysfunction, and that the possible mechanisms mainly involve antioxidant and anti-inflammatory action of H_2. The authors recorded enhancement of GSH concentration and SOD activity as well as decrease in MDA and 8-OHdG concentration in cerebral cortex in the irradiated group after treatment with HRW [53].

H_2 has been proven as effective to mitigate radiation-induced intestinal damage. In a work by Qian et al. [87], radiation damaged the intestine of mice, as evidenced by mucosal neutrophil infiltration, loss of villous epithelium, and shortening of the villi. Intraperitoneal administration of HRS improved all of these histological findings. Study also revealed decreased plasma MDA and intestinal 8-OHdG levels and increased plasma levels of endogenous antioxidants GSH and SOD. Qiu et al. [75] demonstrated that HRS significantly reduced radiation-induced intestinal mucosal damage, improved intestinal function, and increased the survival rate of irradiated mice. H_2 treatment mitigated radiation-induced oxidative stress via lowering of MDA, 8-OHdG and stimulating the activity of SOD. In addition, H_2 significantly reduced the plasma levels of inflammatory molecules TNF-α and IL-6 as well as the activity of myeloperoxidase, what indicates attenuation of radiation-induced inflammatory response.

In a radiation-induced skin injury model, Watanabe et al. [88] detected the protective effect of H_2 inhalation on radiation-induced dermatitis. The authors determined increase in the serum concentration of MDA and 8-OHdG in the non- and post-inhalation groups compared with the pre-inhalation group, but these were not statistically significant. Zhou et al. [59] investigated the healing effect of HRW on radiotherapy-induced skin injury. After H_2 treatment, they observed shortened healing time and increased healing rate of damaged skin. This might be connected to its antioxidant and anti-inflammatory effects, as H_2 reduced the levels of MDA, IL-6 and increased the activity of SOD. The endothelial growth factor (EGF) level was also significantly increased.

Future Perspectives

Up to now, several substances have been tested to protect normal tissues around tumors from radiation damage during radiotherapy. Among them, amifostine is the only radioprotective agent approved for clinical use. However, its use is accompanied by side effects such as hypotension, nausea, and vomiting [89].

H_2 is a molecule displaying preventive and therapeutic effects on a wide variety of diseases including radiation-induced disorders [54]. It has many advantages, like rapid diffusion through biological membranes and barriers, or the range of administration methods. Moreover, H_2 is generally considered to be a safe agent. Most clinical trials revealed no side effects after H_2 administration [90].

Studies demonstrated, that H_2 exerts its radioprotective effects without compromising anti-tumor effects. Kang et al. [91] reported that HRW improved the quality of life of liver cancer patients receiving radiotherapy. There was no difference in tumor response to radiotherapy between the treated and non-treated group. Hirano et al. [92] have shown that inhalation of H_2 alleviated radiotherapy-induced bone marrow damage without compromising the anti-tumor effects.

Based on this information, H_2 has good potential for the use in clinics to mitigate radiation-induced normal tissue injury. However, further research is warranted to confirm the outcomes in larger clinical studies.

Conclusion

Patients receiving radiotherapy in the mediastinal area are at an increased risk of radiation-induced heart disease (RIHD). Although modern radiation techniques have decreased the incidence of RIHD, it remains a major cause of morbidity and mortality among cancer survivors. There are several pathways involved in the development of RIHD, including microvascular damage, inflammation, and fibrosis, although other pathways may contribute. Understanding the exact molecular mechanisms in the progression of RIHD is necessary for developing preventive and therapeutic strategies without attenuating the effect of radiotherapy on cancer cells.

H_2 is an effective scavenger of reactive oxygen and nitrogen species. Beneficial effects are also manifested indirectly, via inhibition of inflammatory, cell death and other pathways. Management of inflammation and oxidative damage is one of the possible strategies for the amelioration of the toxicity caused by radiation. Therefore, H_2 may have potential clinical application as a radioprotective agent. The advantage of H_2 application lies in its ability to easily pass through the cell membranes to reach the site of action. Other benefit is the wide range of H_2 delivery methods. Based on the studies conducted so far, it seems that H_2 application has no or minor side effects with no negative effect on radiotherapy outcomes. Much more research is necessary to elucidate detailed mechanisms, and to verify the extent of effectiveness of H_2 treatment.

Competing Interest This work was supported by grant of Slovak research and development agency (APVV-19-317) and by grant of Slovak Academy of Sciences (VEGA 2/0092/22, VEGA 2/0148/22). Authors declare no conflict of interests.

References

1. Sárközy M, Varga Z, Gáspár R et al (2021) Pathomechanisms and therapeutic opportunities in radiation-induced heart disease: from bench to bedside. Clin Res Cardiol 110:507–531. https://doi.org/10.1007/s00392-021-01809-y
2. van Leeuwen FE, Ng AK (2016) Long-term risk of second malignancy and cardiovascular disease after Hodgkin lymphoma treatment. Hematology 2016:323–330. https://doi.org/10.1182/asheducation-2016.1.323
3. Little MP, Azizova TV, Richardson DB et al (2023) Ionising radiation and cardiovascular disease: systematic review and meta-analysis. BMJ 380:e072924. https://doi.org/10.1136/bmj-2022-072924
4. Koutroumpakis E, Palaskas NL, Lin SH et al (2020) Modern radiotherapy and risk of cardiotoxicity. Chemotherapy 65:65–76. https://doi.org/10.1159/000510573
5. Boerma M, Sridharan V, Mao X-W et al (2016) Effects of ionizing radiation on the heart. Mutat Res Mutat Res 770:319–327. https://doi.org/10.1016/j.mrrev.2016.07.003
6. Donnellan E, Phelan D, McCarthy CP et al (2016) Radiation-induced heart disease: a practical guide to diagnosis and management. Cleve Clin J Med 83:914–922. https://doi.org/10.3949/ccjm.83a.15104

7. Wang B, Wang H, Zhang M et al (2020) Radiation-induced myocardial fibrosis: mechanisms underlying its pathogenesis and therapeutic strategies. J Cell Mol Med 24:7717–7729. https://doi.org/10.1111/jcmm.15479
8. Tian Y, Zhang Y, Wang Y et al (2021) Hydrogen, a novel therapeutic molecule, regulates oxidative stress, inflammation, and apoptosis. Front Physiol 12. https://doi.org/10.3389/fphys.2021.789507
9. Hirano S, Ichikawa Y, Sato B et al (2021a) Molecular hydrogen as a potential clinically applicable radioprotective agent. Int J Mol Sci 22:4566. https://doi.org/10.3390/ijms22094566
10. Ohsawa I, Ishikawa M, Takahashi K et al (2007) Hydrogen acts as a therapeutic antioxidant by selectively reducing cytotoxic oxygen radicals. Nat Med 13:688–694. https://doi.org/10.1038/nm1577
11. Russell G, Nenov A, Kisher H, Hancock JT (2021) Molecular hydrogen as medicine: an assessment of administration methods. Hydrogen 2:444–460. https://doi.org/10.3390/hydrogen2040025
12. Andratschke N, Maurer J, Molls M, Trott KR (2011) Late radiation-induced heart disease after radiotherapy. Clinical importance, radiobiological mechanisms and strategies of prevention. Radiother Oncol 100:160–166. https://doi.org/10.1016/j.radonc.2010.08.010
13. Slezák J, Surový J, Buday J, Kura B (2017) Clinics in oncology molecular hydrogen as a novel therapeutic tool in situations of increased production of free radicals. Clin Oncol 2:1367
14. Tapio S (2016) Pathology and biology of radiation-induced cardiac disease. J Radiat Res 57:439–448. https://doi.org/10.1093/jrr/rrw064
15. Taunk NK, Haffty BG, Kostis JB, Goyal S (2015) Radiation-induced heart disease: pathologic abnormalities and putative mechanisms. Front Oncol 5:39. https://doi.org/10.3389/fonc.2015.00039
16. Cuomo JR, Sharma GK, Conger PD, Weintraub NL (2016) Novel concepts in radiation-induced cardiovascular disease. World J Cardiol 8:504–519. https://doi.org/10.4330/wjc.v8.i9.504
17. Azzam EI, Jay-Gerin J-P, Pain D (2012) Ionizing radiation-induced metabolic oxidative stress and prolonged cell injury. Cancer Lett 327:48–60. https://doi.org/10.1016/j.canlet.2011.12.012
18. Radi R (2018) Oxygen radicals, nitric oxide, and peroxynitrite: redox pathways in molecular medicine. Proc Natl Acad Sci 115:5839–5848. https://doi.org/10.1073/pnas.1804932115
19. Ping Z, Peng Y, Lang H et al (2020) Oxidative stress in radiation-induced cardiotoxicity. Oxid Med Cell Longev 2020:3579143. https://doi.org/10.1155/2020/3579143
20. Qian L, Cao F, Cui J et al (2010) The potential cardioprotective effects of hydrogen in irradiated mice. J Radiat Res 51:741–747. https://doi.org/10.1269/jrr.10093
21. Yilmaz H, Karakoc Y, Tumkaya L et al (2022) The protective effects of red ginseng and amifostine against renal damage caused by ionizing radiation. Hum Exp Toxicol 41:096032712211430. https://doi.org/10.1177/09603271221143029
22. Livingston K, Schlaak RA, Puckett LL, Bergom C (2020) The role of mitochondrial dysfunction in radiation-induced heart disease: from bench to bedside. Front Cardiovasc Med 7:20. https://doi.org/10.3389/fcvm.2020.00020
23. Wei J, Wang B, Wang H et al (2019) Radiation-induced normal tissue damage: oxidative stress and epigenetic mechanisms. Oxid Med Cell Longev 2019:1–11. https://doi.org/10.1155/2019/3010342
24. Halle M, Gabrielsen A, Paulsson-Berne G et al (2010) Sustained inflammation due to nuclear factor-kappa B activation in irradiated human arteries. J Am Coll Cardiol 55:1227–1236. https://doi.org/10.1016/j.jacc.2009.10.047
25. Incalza MA, D'Oria R, Natalicchio A et al (2018) Oxidative stress and reactive oxygen species in endothelial dysfunction associated with cardiovascular and metabolic diseases. Vascul Pharmacol 100:1–19. https://doi.org/10.1016/j.vph.2017.05.005
26. Halle M, Hall P, Tornvall P (2011) Cardiovascular disease associated with radiotherapy: activation of nuclear factor kappa-B. J Intern Med 269:469–477. https://doi.org/10.1111/j.1365-2796.2011.02353.x
27. Boerma M, Kruse JJCM, van Loenen M et al (2004) Increased deposition of von Willebrand factor in the rat heart after local ionizing irradiation. Strahlentherapie und Onkol 180:109–116. https://doi.org/10.1007/s00066-004-1138-0

28. Pathak R, Shao L, Ghosh SP et al (2015) Thrombomodulin contributes to gamma tocotrienol-mediated lethality protection and hematopoietic cell recovery in irradiated mice. PLoS ONE 10:e0122511. https://doi.org/10.1371/journal.pone.0122511
29. Beckman JA, Thakore A, Kalinowski BH et al (2001) Radiation therapy impairs endothelium-dependent vasodilation in humans. J Am Coll Cardiol 37:761–765. https://doi.org/10.1016/S0735-1097(00)01190-6
30. Hong C-W, Kim Y-M, Pyo H et al (2013) Involvement of inducible nitric oxide synthase in radiation-induced vascular endothelial damage. J Radiat Res 54:1036–1042. https://doi.org/10.1093/jrr/rrt066
31. Pathak R, Cheema A, Boca S et al (2015) Modulation of radiation response by the tetrahydrobiopterin pathway. Antioxidants 4:68–81. https://doi.org/10.3390/antiox4010068
32. Carr ZA, Land CE, Kleinerman RA et al (2005) Coronary heart disease after radiotherapy for peptic ulcer disease. Int J Radiat Oncol 61:842–850. https://doi.org/10.1016/j.ijrobp.2004.07.708
33. Seemann I, te Poele JAM, Hoving S, Stewart FA (2014) Mouse bone marrow-derived endothelial progenitor cells do not restore radiation-induced microvascular damage. ISRN Cardiol 2014:506348. https://doi.org/10.1155/2014/506348
34. Wang H, Wei J, Zheng Q et al (2019) Radiation-induced heart disease: a review of classification, mechanism and prevention. Int J Biol Sci 15:2128–2138. https://doi.org/10.7150/ijbs.35460
35. Baselet B, Rombouts C, Benotmane AM et al (2016) Cardiovascular diseases related to ionizing radiation: the risk of low-dose exposure (Review). Int J Mol Med 38:1623–1641. https://doi.org/10.3892/ijmm.2016.2777
36. Sridharan V, Aykin-Burns N, Tripathi P et al (2014) Radiation-induced alterations in mitochondria of the rat heart. Radiat Res 181:324. https://doi.org/10.1667/RR13452.1
37. Yamamori T, Yasui H, Yamazumi M et al (2012) Ionizing radiation induces mitochondrial reactive oxygen species production accompanied by upregulation of mitochondrial electron transport chain function and mitochondrial content under control of the cell cycle checkpoint. Free Radic Biol Med 53:260–270. https://doi.org/10.1016/j.freeradbiomed.2012.04.033
38. Yarnold J, Vozenin Brotons M-C (2010) Pathogenetic mechanisms in radiation fibrosis. Radiother Oncol 97:149–161. https://doi.org/10.1016/j.radonc.2010.09.002
39. Martin M, Lefaix J-L, Delanian S (2000) TGF-β1 and radiation fibrosis: a master switch and a specific therapeutic target? Int J Radiat Oncol 47:277–290. https://doi.org/10.1016/S0360-3016(00)00435-1
40. Chello M, Mastroroberto P, Romano R et al (1996) Changes in the proportion of types I and III collagen in the left ventricular wall of patients with post-irradiative pericarditis. Cardiovasc Surg 4:222–226. https://doi.org/10.1016/0967-2109(96)82320-9
41. Farhood B, Khodamoradi E, Hoseini-Ghahfarokhi M et al (2020) TGF-β in radiotherapy: mechanisms of tumor resistance and normal tissues injury. Pharmacol Res 155:104745. https://doi.org/10.1016/j.phrs.2020.104745
42. Brosius FC, Waller BF, Roberts WC (1981) Radiation heart disease. Am J Med 70:519–530. https://doi.org/10.1016/0002-9343(81)90574-X
43. Slama MS, Guludec D, Sebag C et al (1991) Complete atrioventricular block following mediastinal irradiation: a report of six cases. Pacing Clin Electrophysiol 14:1112–1118. https://doi.org/10.1111/j.1540-8159.1991.tb02842.x
44. O'Donnell L, O'Neill T, Toner M et al (1986) Myocardial hypertrophy, fibrosis and infarction following exposure of the heart to radiation for Hodgkin's disease. Postgrad Med J 62:1055–1058. https://doi.org/10.1136/pgmj.62.733.1055
45. Zhou Z-Q, Zhong C-H, Su Z-Q et al (2019) Breathing hydrogen-oxygen mixture decreases inspiratory effort in patients with tracheal stenosis. Respiration 97:42–51. https://doi.org/10.1159/000492031
46. Sun Q, Kang Z, Cai J et al (2009) Hydrogen-rich saline protects myocardium against ischemia/reperfusion injury in rats. Exp Biol Med 234:1212–1219. https://doi.org/10.3181/0812-rm-349
47. Malý O, Zajak J, Hyšpler R et al (2019) Inhalation of molecular hydrogen prevents ischemia-reperfusion liver damage during major liver resection. Ann Transl Med 7:774–774. https://doi.org/10.21037/atm.2019.11.43

48. Noda K, Tanaka Y, Shigemura N et al (2012) Hydrogen-supplemented drinking water protects cardiac allografts from inflammation-associated deterioration. Transpl Int 25:1213–1222. https://doi.org/10.1111/j.1432-2277.2012.01542.x
49. Saito M, Chen-Yoshikawa TF, Takahashi M et al (2020) Protective effects of a hydrogen-rich solution during cold ischemia in rat lung transplantation. J Thorac Cardiovasc Surg 159:2110–2118. https://doi.org/10.1016/j.jtcvs.2019.09.175
50. Zhang Y, Tan S, Xu J, Wang T (2018) Hydrogen therapy in cardiovascular and metabolic diseases: from bench to bedside. Cell Physiol Biochem 47:1–10. https://doi.org/10.1159/000489737
51. Tian R, Hou Z, Hao S et al (2016) Hydrogen-rich water attenuates brain damage and inflammation after traumatic brain injury in rats. Brain Res 1637:1–13. https://doi.org/10.1016/j.brainres.2016.01.029
52. Qiu P, Liu Y, Zhang J (2019) Recent advances in studies of molecular hydrogen against sepsis. Int J Biol Sci 15:1261–1275
53. Liu M, Yuan H, Yin J et al (2019) Effect of hydrogen-rich water on radiation-induced cognitive dysfunction in rats. Radiat Res 193:16. https://doi.org/10.1667/RR15464.1
54. Kura B, Bagchi AK, Singal PK et al (2019) Molecular hydrogen: potential in mitigating oxidative-stress-induced radiation injury. Can J Physiol Pharmacol 97:287–292. https://doi.org/10.1139/cjpp-2018-0604
55. Terasaki Y, Ohsawa I, Terasaki M et al (2011) Hydrogen therapy attenuates irradiation-induced lung damage by reducing oxidative stress. Am J Physiol Cell Mol Physiol 301:L415–L426. https://doi.org/10.1152/ajplung.00008.2011
56. Zhao S, Yang Y, Liu W et al (2014) Protective effect of hydrogen-rich saline against radiation-induced immune dysfunction. J Cell Mol Med 18:938–946. https://doi.org/10.1111/jcmm.12245
57. Chuai Y, Gao F, Li B et al (2012) Hydrogen-rich saline attenuates radiation-induced male germ cell loss in mice through reducing hydroxyl radicals. Biochem J 442:49–56. https://doi.org/10.1042/BJ20111786
58. Kura B, Kalocayova B, LeBaron TW et al (2019) Regulation of microRNAs by molecular hydrogen contributes to the prevention of radiation-induced damage in the rat myocardium. Mol Cell Biochem. https://doi.org/10.1007/s11010-019-03512-z
59. Zhou P, Lin B, Wang P et al (2019) The healing effect of hydrogen-rich water on acute radiation-induced skin injury in rats. J Radiat Res 60:17–22. https://doi.org/10.1093/jrr/rry074
60. Tamaki N, Orihuela-Campos RC, Fukui M, Ito H-O (2016) Hydrogen-rich water intake accelerates oral palatal wound healing via activation of the Nrf2/antioxidant defense pathways in a rat model. Oxid Med Cell Longev 2016:1–13. https://doi.org/10.1155/2016/5679040
61. Zhang B, Zhao Z, Meng X et al (2018) Hydrogen ameliorates oxidative stress via PI3K-Akt signaling pathway in UVB-induced HaCaT cells. Int J Mol Med 41:3653–3661. https://doi.org/10.3892/ijmm.2018.3550
62. Mei K, Zhao S, Qian L et al (2014) Hydrogen protects rats from dermatitis caused by local radiation. J Dermatolog Treat 25:182–188. https://doi.org/10.3109/09546634.2012.762639
63. Gharib B, Hanna S, Abdallahi OMS et al (2001) Anti-inflammatory properties of molecular hydrogen: investigation on parasite-induced liver inflammation. Comptes Rendus l'Académie des Sci - Ser III - Sci la Vie 324:719–724. https://doi.org/10.1016/S0764-4469(01)01350-6
64. Ji X, Liu W, Xie K et al (2010) Beneficial effects of hydrogen gas in a rat model of traumatic brain injury via reducing oxidative stress. Brain Res 1354:196–205. https://doi.org/10.1016/j.brainres.2010.07.038
65. Fang W, Wang G, Tang L et al (2018) Hydrogen gas inhalation protects against cutaneous ischaemia/reperfusion injury in a mouse model of pressure ulcer. J Cell Mol Med 22:4243–4252. https://doi.org/10.1111/jcmm.13704
66. Bin ZC, Tang YC, Xu XJ et al (2015) Hydrogen gas inhalation protects against liver ischemia/reperfusion injury by activating the NF-κB signaling pathway. Exp Ther Med 9:2114–2120. https://doi.org/10.3892/etm.2015.2385

67. Yang W-C, Li T, Wan Q et al (2022) Molecular Hydrogen mediates neurorestorative effects after stroke in diabetic rats: the TLR4/NF-κB inflammatory pathway. J Neuroimmune Pharmacol. https://doi.org/10.1007/s11481-022-10051-w
68. Guo Z, Zhou B, Li W et al (2012) Hydrogen-rich saline protects against ultraviolet B radiation injury in rats. J Biomed Res 26:365–371. https://doi.org/10.7555/JBR.26.20110037
69. Nogueira JE, de Deus JL, Amorim MR et al (2020) Inhaled molecular hydrogen attenuates intense acute exercise-induced hippocampal inflammation in sedentary rats. Neurosci Lett 715:134577. https://doi.org/10.1016/j.neulet.2019.134577
70. Lin C-P, Chuang W-C, Lu F-J, Chen C-Y (2017) Anti-oxidant and anti-inflammatory effects of hydrogen-rich water alleviate ethanol-induced fatty liver in mice. World J Gastroenterol 23:4920–4934. https://doi.org/10.3748/wjg.v23.i27.4920
71. Zheng H, Yu YS (2012) Chronic hydrogen-rich saline treatment attenuates vascular dysfunction in spontaneous hypertensive rats. Biochem Pharmacol 83:1269–1277. https://doi.org/10.1016/j.bcp.2012.01.031
72. Sakai T, Sato B, Hara K et al (2014) Consumption of water containing over 3.5 mg of dissolved hydrogen could improve vascular endothelial function. Vasc Health Risk Manag 10:591. https://doi.org/10.2147/VHRM.S68844
73. Wang Y, Wu Y, Han J et al (2019) Inhibitory effects of hydrogen on in vitro platelet activation and in vivo prevention of thrombosis formation. Life Sci 233:116700. https://doi.org/10.1016/j.lfs.2019.116700
74. Hong Y, Shao AW, Wang J et al (2014) Neuroprotective effect of hydrogen-rich saline against neurologic damage and apoptosis in early brain injury following subarachnoid hemorrhage: possible role of the Akt/GSK3β signaling pathway. PLoS One 9:e96212-undefined. https://doi.org/10.1371/journal.pone.0096212
75. Qiu X, Dong K, Guan J, He J (2020) Hydrogen attenuates radiation-induced intestinal damage by reducing oxidative stress and inflammatory response. Int Immunopharmacol 84:106517. https://doi.org/10.1016/j.intimp.2020.106517
76. Li J, Hong Z, Liu H et al (2016) Hydrogen-rich saline promotes the recovery of renal function after ischemia/reperfusion injury in rats via anti-apoptosis and anti-inflammation. Front Pharmacol 7:106-undefined. https://doi.org/10.3389/fphar.2016.00106
77. Straub JM, New J, Hamilton CD et al (2015) Radiation-induced fibrosis: mechanisms and implications for therapy. J Cancer Res Clin Oncol 141:1985–1994. https://doi.org/10.1007/s00432-015-1974-6
78. Aokage T, Seya M, Hirayama T et al (2021) The effects of inhaling hydrogen gas on macrophage polarization, fibrosis, and lung function in mice with bleomycin-induced lung injury. BMC Pulm Med 21:339. https://doi.org/10.1186/s12890-021-01712-2
79. Gao L, Jiang D, Geng J et al (2019) Hydrogen inhalation attenuated bleomycin-induced pulmonary fibrosis by inhibiting transforming growth factor-β1 and relevant oxidative stress and epithelial-to-mesenchymal transition. Exp Physiol 104:1942–1951. https://doi.org/10.1113/EP088028
80. Xing Z, Pan W, Zhang J et al (2017) Hydrogen rich water attenuates renal injury and fibrosis by regulation transforming growth factor-β induced Sirt1. Biol Pharm Bull 40:610–615. https://doi.org/10.1248/bpb.b16-00832
81. Liu L, Shi Q, Liu X et al (2022) Attenuation of myocardial fibrosis using molecular hydrogen by inhibiting the TGF-β signaling pathway in spontaneous hypertensive rats. Am J Hypertens 35:156–163. https://doi.org/10.1093/ajh/hpab159
82. Yu YS, Zheng H (2012) Chronic hydrogen-rich saline treatment reduces oxidative stress and attenuates left ventricular hypertrophy in spontaneous hypertensive rats. Mol Cell Biochem 365:233–242. https://doi.org/10.1007/s11010-012-1264-4
83. Zhang Y, Xu J, Long Z et al (2016) Hydrogen (H2) inhibits isoproterenol-induced cardiac hypertrophy via antioxidative pathways. Pathways Front Pharmacol 7:392. https://doi.org/10.3389/fphar.2016.00392
84. Wang C, Pan Z (2020) Hydrogen-rich saline mitigates pressure overload-induced cardiac hypertrophy and atrial fibrillation in rats via the JAK-STAT signalling pathway. J Int Med Res 48:030006052093641. https://doi.org/10.1177/0300060520936415

85. Slezak J, Surovy J, Buday J, Kura B (2017) Molecular hydrogen as a novel therapeutic tool in situations of increased production of free radicals. Clin Oncol 2:1367-undefined
86. Huo H, Yang S, Chen L et al (2012) Hydrogen-rich saline alleviation on the oxidative stress and early-phase radiation-induced brain injury in rats. Chinese J Radiol Med Prot 32:485–487
87. Qian L, Cao F, Cui J et al (2010) Radioprotective effect of hydrogen in cultured cells and mice. Free Radic Res 44:275–282. https://doi.org/10.3109/10715760903468758
88. Watanabe S, Fujita M, Ishihara M et al (2014) Protective effect of inhalation of hydrogen gas on radiation-induced dermatitis and skin injury in rats. J Radiat Res 55:1107–1113. https://doi.org/10.1093/jrr/rru067
89. King M, Joseph S, Albert A et al (2020) Use of amifostine for cytoprotection during radiation therapy: a review. Oncology 98:61–80. https://doi.org/10.1159/000502979
90. Ostojic S (2014) Molecular hydrogen in sports medicine: new therapeutic perspectives. Int J Sports Med 36:273–279. https://doi.org/10.1055/s-0034-1395509
91. Kang KM, Kang YN, Choi IB et al (2011) Effects of drinking hydrogen-rich water on the quality of life of patients treated with radiotherapy for liver tumors. Med Gas Res 1:11. https://doi.org/10.1186/2045-9912-1-11
92. Hirano SI, Aoki Y, Li XK, Ichimaru N, Takahara S, Takefuji Y (2021b) Protective effects of hydrogen gas inhalation on radiation-induced bone marrow damage in cancer patients: a retrospective observational study. Med Gas Res 11(3):104–109. https://doi.org/10.4103/2045-9912.314329; https://pubmed.ncbi.nlm.nih.gov/33942780/

Chapter 9
Short-Lasting Supplementation with Molecular Hydrogen and Vitamin E Upregulates Myocardial Connexin-43 in Irradiated and Non-irradiated Rat Heart

Barbara Szeiffova Bacova, Katarina Andelova, Matus Sykora, Branislav Kura, Barbora Kalocayova, Jan Slezak, and Narcis Tribulova

Abstract In this study we aimed to explore whether supplementation with antioxidants, molecular H_2 and vitamin E, may impact gap junction channel protein, connexin-43, in left ventricular tissue of irradiated rat hearts. Experiments were performed using 3-month-old male Wistar rats that were randomized into 6 groups. (1) Intact rats; (2) Intact rats treated with molecular H_2 (4% H_2 in inhalation chamber, 3×30 min per day) (3) intact rats supplemented with vitamin E (30 mg/kg body weight); (4) Irradiated rats after exposure to single dose of 10 Gy; (5) Irradiated rats treated with molecular H_2; (6) Irradiated rats treated with vitamin E. Left heart myocardium was used for analysis, at 9 days after irradiation. Findings indicate that

B. Szeiffova Bacova · K. Andelova · M. Sykora · B. Kura · B. Kalocayova · J. Slezak · N. Tribulova (✉)
Centre of Experimental Medicine SAS, Institute for Heart Research, Dúbravská cesta 9, 841 04 Bratislava, Slovak Republic
e-mail: narcis.tribulova@savba.sk

B. Szeiffova Bacova
e-mail: barbara.bacova@savba.sk

K. Andelova
e-mail: katarina.andelova@savba.sk

M. Sykora
e-mail: matus.sykora@savba.sk

B. Kura
e-mail: branislav.kura@savba.sk

B. Kalocayova
e-mail: barbora.kalocayova@savba.sk

J. Slezak
e-mail: jan.slezak@savba.sk

J. Slezak and B. Kura (eds.), *Molecular Hydrogen in Health and Disease*, Advances in Biochemistry in Health and Disease 27, https://doi.org/10.1007/978-3-031-47375-3_9

treatment with molecular H_2 and vitamin E up-regulates connexin-43 and PKCε not only in irradiated but also in intact rat hearts. It is challenging for further research to elucidate mechanisms.

Keywords Rat heart · Irradiation · Molecular H_2 · Vitamin E · Connexin-43 · PKCε

Introduction

It is generally accepted that oxidative stress and inflammation are crucial factors jeopardizing cardiovascular health [1–3] and can increase the heart's susceptibility to malignant arrhythmias [4, 5]. Redox disorders and increased level of reactive oxygen species (ROS) alter function of myocardial ion channels and gap junction connexin channels as well as disturb Ca^{2+} handling. Consequently, the heart is less electrically stable and prone to develop malignant arrhythmias mainly in the presence of arrhythmogenic substrate due to structural remodelling.

On the other hand, it should be emphasized that numerous cardioprotective compounds which exhibit antiarrhythmic properties are potent anti-oxidative and/ or anti-inflammatory agents [6]. It fits with the concept of effective mitochondria-targeted antioxidant therapy to fight cardiac arrhythmias [7]. Thus, the identification of targets downstream from ROS that impact arrhythmia, such as connexin-43 (Cx43), may yield effective antiarrhythmic therapies [5–7].

Dominant cardiac gap junction Cx43 channels ensure coupling among cardiac myocytes for rapid propagation of electrical and molecular signals that underlie synchronised cardiac function [5, 8]. While down-regulation of Cx43 or its abnormal myocardial topology in various pathologies (mostly accompanied by oxidative stress and inflammation) deteriorated coupling and increased susceptibility of the heart to life-threatening arrhythmias [5, 6, 9, 10].

One of the promising candidate to reduce oxidative stress and inflammation in cardiovascular disorders or after radiation injury, is molecular hydrogen (H_2) [11–13]. While α-tocopherol (vitamin E) regulates signalling pathways in inflammation [14, 15], attenuates the lipid peroxidation in cell membranes and provides myocardial protection against ischemia–reperfusion injury [16]. However, some interventional clinical trials with vitamin supplements provided inconclusive results, mostly due to inappropriate dosages and missing pharmacokinetics to assess vitamin quantity.

In this pilot study we aimed to explore whether supplementation with molecular H_2 or vitamin E may impact the protein levels and/or topology of Cx43 in left ventricular tissue of post-irradiated rat hearts. Besides, we examined venous and arterial capillary network in response to irradiation and treatment.

Material and Methods

Experiments performed on 30 males, 3-month-old Wistar rats were approved by the Animal Care and Use Committee of the Institute of Experimental Medicine, SAS, in agreement with guidelines and practices established by the Directive 2010/63/EU of the European Parliament on the Protection of Animals Used for Scientific Purposes. Animals were housed under standard conditions, at 22 ± 1 °C, 12-h light/dark cycles, with *ad libitum* access of rat chows and tap water.

Intact rats ($n = 15$) and irradiated rats ($n = 15$) with single dose of 10 Gy at the mediastinum area were used. Next day after irradiation, intact ($n = 10$) and irradiated ($n = 10$) animals were treated for 9 days with molecular H_2 (4% H_2 in inhalation chamber, 3 × 30 min per day) or vitamin E (Jamieson, on vehiculum biscuit in dose 30 mg/kg body weight once a day) and compared with non-treated controls ($n = 10$). After euthanasia (Thiopental VUAB 1.0 g, 60 mg/kg i.p.), the rat chest was open, and the heart was excised into ice-cold saline. It was followed by heart weight registration and cardiac tissue sampling. All samples were stored in a freezer at −80 °C until analysis.

Left ventricular frozen tissue was used for preparing 10 μm thick cryostat sections. Light microscopy examination of hematoxilin-eosin (HE) stained myocardial tissue sections was performed to assess the structure.

Besides, i*n situ* enzyme histochemistry was performed on cryostat sections to assess the myocardial capillary function and density. Activity of alkaline phosphatase (AlP) that is specific for arterial part of capillary network [17] and the activity of dipeptidyl petidase-4 (DPP4) that is specific for venous part of capillary network [18] were examined. In parallel, AlP or DPP4 positive capillaries point out myocardial capillary density. Quantitative microscopic images analysis was performed, as previously reported [6] to evaluate differences among experimental groups.

Immunofluorescence labelling of myocardial Cx43 (1:500, MAB3068, CHEMICON International, Inc., Temecula, CA, USA) was performed using cryostat sections to demonstrate topology of Cx43, while western blot analysis was performed to quantify both total Cx43 protein levels (1:5000, C6219, Sigma-Aldrich, Missouri, USA) and its serine 368 phosphorylated form (1:1000, sc-101660, Santa Cruz Biotechnology, Texas, USA). In parallel, protein levels of PKCε (1:1000, sc-214, Santa Cruz Biotechnology, Texas, USA), which phosphorylates Cx43 on serine 368 and modulates Cx43 channels function was determined as well. All these methods were performed as previously reported [19].

Statistical Evaluation

Differences between groups were evaluated using one-way analysis of variance (ANOVA) and Bonferroni's multiple comparison test. The Kolmogorov–Smirnov normality test was used to examine whether variables are normally distributed. Data were expressed as means ± standard deviations (SD); $p < 0.05$ was considered to be significant.

Results

As shown in Fig. 9.1, light microscopy examination did not reveal apparent structural injury of cardiomyocytes or enhanced infiltration with polymorphonuclears pointing out inflammation, in left ventricular myocardium 9 days after irradiation.

However, post-irradiated rat heart exhibited significant reduction of alkaline phosphatase activity as well as density of arterial capillary network, as demonstrated in Fig. 9.2. Treatment of irradiated rats with molecular H_2, while not with vitamin E, enhanced capillary density with preserved alkaline phosphatase activity.

Moreover, as shown in Fig. 9.3, there was a significant decrease of dipeptidyl peptidase-4 activity in venous part of capillary network in response to irradiation. Treatment of both intact and irradiated rats with either molecular H_2 or vitamin E did not affect activity of dipeptidyl peptidase-4 activity.

Representative images of Cx43 topology are demonstrated in Fig. 9.4. There is obvious prevalent polar distribution of Cx43 at the intercalated disc and less frequent distribution of Cx43 on lateral sides of the cardiomyocytes, regardless experimental

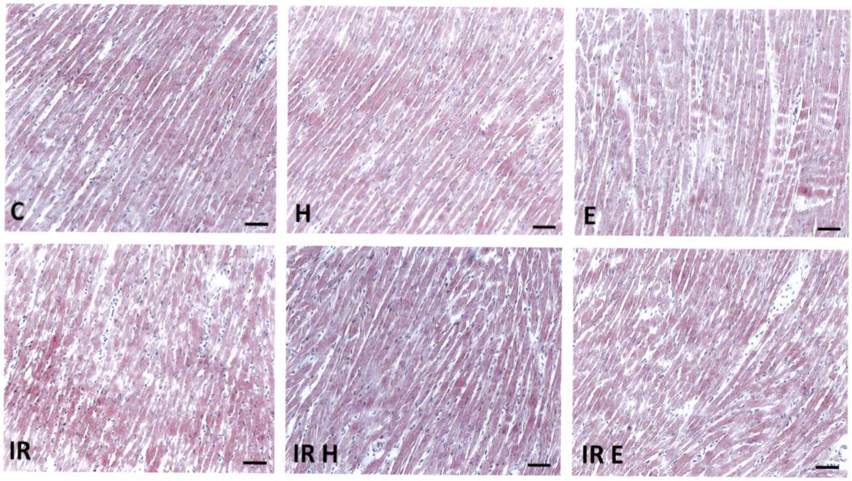

Fig. 9.1 Representative histological images of left heart ventricles of experimental rats after HE staining. Comparing to intact control rat heart there are no remarkable alterations in structure of the cardiomyocytes or extracellular space either in response to irradiation or treatment. C – intact controls, H – molecular H_2 treated group, E – vitamin E treated group, IR – irradiated group, IR H – irradiated group with H_2 treatment, IR E – irradiated group with vitamin E treatment. Scale bar represents 100 μm

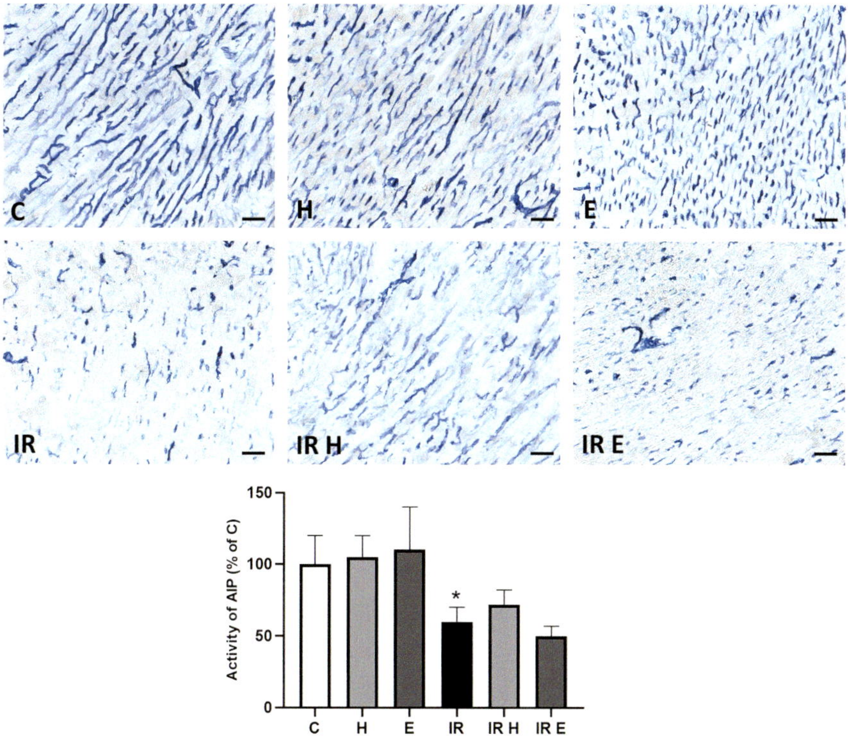

Fig. 9.2 Representative microscopic images demonstrating alkaline phosphatase activity in arterial capillaries of left heart ventricles of experimental rats. Comparing to intact control rat heart there is remarkable reduced activity in response to irradiation while tendency to increase after treatment with molecular H_2. C – intact controls, H – molecular H_2 treated group, E – vitamin E treated group, IR – irradiated group, IR H – irradiated group with H_2 treatment, IR E – irradiated group with vitamin E treatment. Scale bar represents 100 μm

groups. However, it seems that Cx43 positive labelling is enhanced in response to irradiation and treatment with molecular H_2.

Quantitative western blot analyses of total Cx43 protein and its functional phosphorylated form in left ventricular tissue of experimental rats is demonstrated on Fig. 9.5. There are no significant changes in Cx43 protein levels at 9 days post-irradiation. Surprisingly, treatment with either molecular H_2 or vitamin E significantly enhanced Cx43 protein levels in both irradiated and non-irradiated rat heart.

Myocardial protein levels of PKCε, which phosphorylates Cx43 on serine 368, did not change at 9 days post-irradiation but it was significantly increased in non-irradiated rat heart upon treatment with either molecular H_2 or vitamin E (Fig. 9.6).

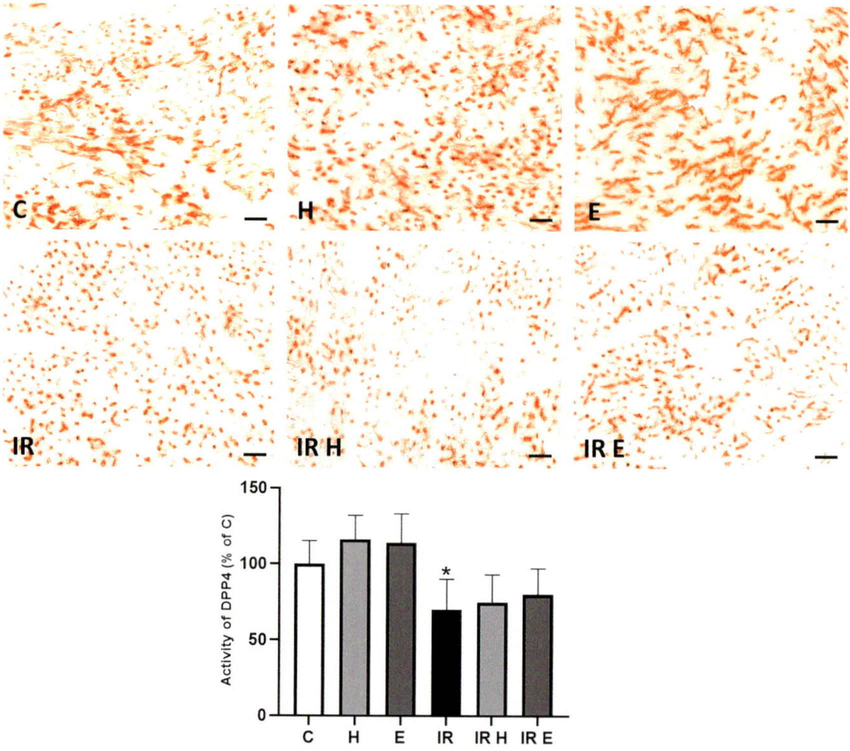

Fig. 9.3 Representative microscopic images demonstrating activity of dipeptidyl peptidase-4 in venous capillaries of left heart ventricles of experimental rats. Comparing to intact control rat heart there is significant decrease of enzyme activity in response to irradiation that was not affected by treatment. C – intact controls, H – molecular H_2 treated group, E – vitamin E treated group, IR – irradiated group, IR H – irradiated group with H_2 treatment, IR E – irradiated group with vitamin E treatment. Scale bar represents 100 μm

Discussion

This pilot study investigated whether treatment with potent cardio-protective agents, such as molecular H_2 and vitamin E, affect irradiated rat heart, 9-days after exposure to single dose of 10 Gy.

Microscopic examination did not reveal apparent structural changes of cardiomyocytes or extracellular space in left heart ventricle of post-irradiated rats. However, there was significant decrease of alkaline phosphatase activity in arterial capillaries of post-irradiated rats. The density of enzyme positive capillaries was reduced as well. These findings point out reduced myocardial arterial capillarization of irradiated rat heart. In contrast, treatment with molecular H_2 attenuated deleterious effect of irradiation and enhanced arterial capillary network. The latter was reported in aged hypertensive rats treated with omega-3 fatty acids [20] or in rats in response

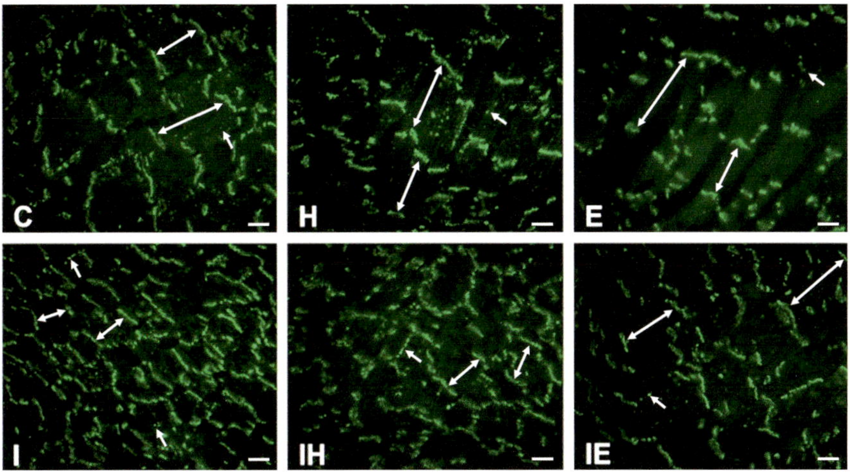

Fig. 9.4 Representative immunofluorescence images demonstrating topology of Cx43 in left heart ventricles of experimental rats. Note prevalent distribution of Cx43 at the intercalated disc (long arrows) while less frequent on lateral sides (short arrows) of the cardiomyocytes that is obvious in intact rats. Immunolabeling of Cx43 is enhanced in irradiated rat heart but to lesser extent in treated rats. C – intact controls, H – molecular H_2 treated group, E – vitamin E treated group, I – irradiated group, IH – irradiated group with H_2 treatment, IE – irradiated group with vitamin E treatment. Scale bar represents 100 μm

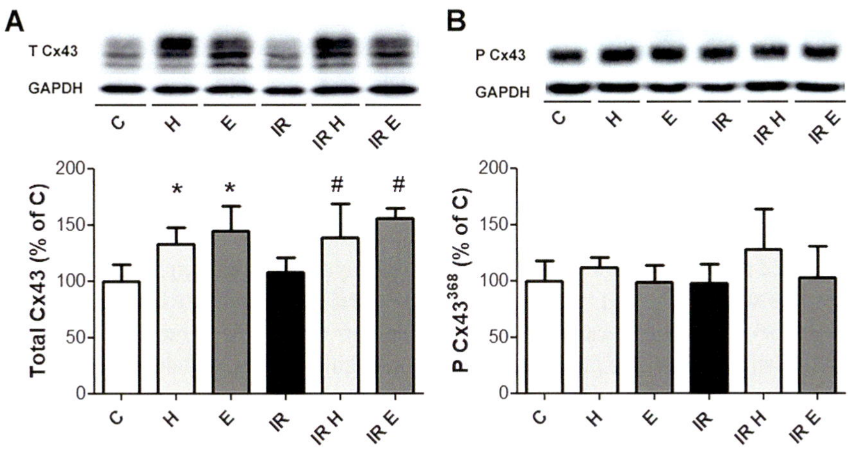

Fig. 9.5 Representative western blots of total Cx43 protein (A) and its phosphorylated form on serine 368 (B) in left heart ventricles of experimental rats. Columns pointing out quantitative evaluation revealed significant increase of total Cx43 protein levels after treatment of irradiated as well as intact rat heart. C – intact controls, H – molecular H_2 treated group, E – vitamin E treated group, IR – irradiated group, IR H – irradiated group with H_2 treatment, IR E – irradiated group with vitamin E treatment

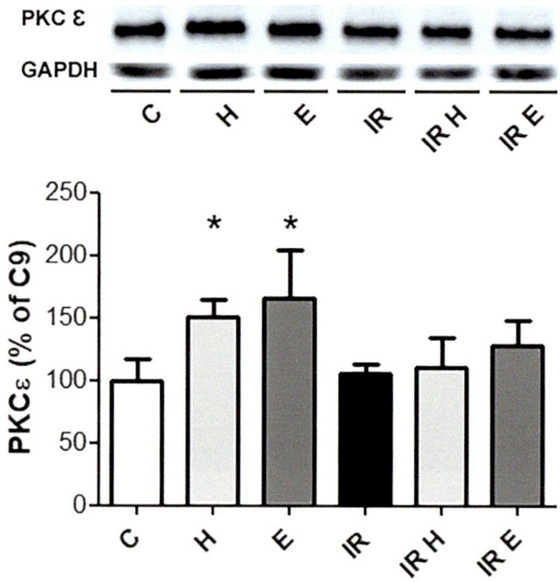

Fig. 9.6 Representative western blots of PKCε in left heart ventricles of experimental rats. There is significant increase of PKCε protein levels after treatment of intact rat heart. C – intact controls, H – molecular H_2 treated group, E – vitamin E treated group, IR – irradiated group, IR H – irradiated group with H_2 treatment, IR E – irradiated group with vitamin E treatment

to cold acclimation [21]. On the other hand, treatment with either molecular H_2 or vitamin E did not affect reduced activity of dipeptidyl peptidase-4 in venous capillary network of irradiated rat hearts. This enzyme participates in collagen metabolism and synthesis of endothelial capillary coat as well as in degradation of vasoactive peptides [18].

Immunolabeling of Cx43 did not show any abnormal cardiomyocyte distribution regardless experimental groups. There was prevalent immunopositivity of Cx43 at the gap junctions in intercalated discs and less frequent on lateral sides of the cardiomyocytes. This pattern was not altered either by irradiation or by treatment. There were no changes in Cx43 protein levels and its phosphorylated status in post-irradiated rat hearts examined at early period (9 days). These findings differ from our previous studies [22, 23], which showed increased Cx43 protein expression and its phosphorylation in rat heart that were examined 6 weeks after irradiation. It points out that single noxious signal might induce up-regulation of Cx43 during adaptation period.

Noteworthy, treatment with either molecular H_2 or vitamin E enhanced Cx43 protein levels in post-irradiated as well as intact rat hearts. This finding is challenging to investigate underlying mechanisms since antioxidants may affect myocardial Cx43 expression [6]. In this context it is interesting the fact that molecular H_2 as well as vitamin E significantly enhanced protein levels of PKCε in intact rat heart and moderately in post-irradiated rats. Cx43 is downstream target for PKCε phosphorylation that is associated with modulation Cx43 channels function and cytoprotection [24]. Despite demonstrated up-regulation of PKCε 6 weeks after irradiation we did not detect any changes 9 days after irradiation [22, 23].

In conclusion, findings of this pilot study stimulate further research to elucidate mechanisms of molecular H_2 and vitamin E involved in Cx43 and PKCε upregulation.

Funding This research was supported by VEGA grants 2/0002/20, 2/0092/22, and 2/0006/23, and the Slovak Research and Development Agency under contract No. 19-0317, 21-0410.

References

1. Tretter V, Hochreiter B, Zach ML, Krenn K, Klein KU (2022) Understanding cellular redox homeostasis : a challenge for precision medicine
2. Lawler PR, Bhatt DL, Godoy LC, Lüscher TF, Bonow RO, Verma S, Ridker PM (2021) Targeting cardiovascular inflammation: next steps in clinical translation. Eur Heart J 42:113–131. https://doi.org/10.1093/eurheartj/ehaa099
3. Liberale L, Montecucco F, Schwarz L, Lüscher TF, Camici GG (2021) Inflammation and cardiovascular diseases: lessons from seminal clinical trials. Cardiovasc Res 117:411–422. https://doi.org/10.1093/cvr/cvaa211
4. Adameova A, Shah AK, Dhalla NS (2020) Role of oxidative stress in the genesis of ventricular arrhythmias. Int J Mol Sci 21:1–16. https://doi.org/10.3390/ijms21124200
5. Andelova K, Benova TE, Bacova BS, Sykora M, Prado NJ, Diez ER, Hlivak P, Tribulova N (2021) Cardiac connexin-43 hemichannels and pannexin1 channels: provocative antiarrhythmic targets. Int J Mol Sci 22. https://doi.org/10.3390/ijms22010260
6. Andelova K, Bacova BS, Sykora M, Hlivak P, Barancik M, Tribulova N (2022) Mechanisms underlying antiarrhythmic properties of cardioprotective agents impacting inflammation and oxidative stress. Int J Mol Sci 23. https://doi.org/10.3390/ijms23031416
7. Sovari AA, Bonini MG, Dudley SC (2011) Effective antioxidant therapy for the management of arrhythmia. Expert Rev Cardiovasc Ther 9:797–800. https://doi.org/10.1586/erc.11.85
8. Tribulova N, Szeiffova Bacova B, Benova T, Viczenczova C (2015) Can we protect from malignant arrhythmias by modulation of cardiac cell-to-cell coupling? J Electrocardiol 48:434–440. https://doi.org/10.1016/j.jelectrocard.2015.02.006
9. Danik SB, Rosner G, Lader J, Gutstein DE, Fishman GI, Morley GE (2008) Electrical remodeling contributes to complex tachyarrhythmias in connexin43-deficient mouse hearts. FASEB J 22:1204–1212. https://doi.org/10.1096/fj.07-8974com
10. Dhein S, Salameh A (2021) Remodeling of cardiac gap junctional cell–cell coupling. Cells 10. https://doi.org/10.3390/cells10092422
11. Slezák J, Kura B, Frimmel K, Zálešák M, Ravingerová T, Viczenczová C, Okruhlicová L, Tribulová N (2016) Preventive and therapeutic application of molecular hydrogen in situations with excessive production of free radicals. Physiol Res 65:S11–S28. https://doi.org/10.33549/physiolres.933414
12. LeBaron TW, Kura B, Kalocayova B, Tribulova N, Slezak J (2019) A new approach for the prevention and treatment of cardiovascular disorders. Molecular hydrogen significantly reduces the effects of oxidative stress. Molecules 24. https://doi.org/10.3390/molecules24112076
13. Kura B, Bagchi AK, Singal PK, Barancik M, Lebaron TW, Valachova K, Šoltés L, Slezák J (2019) Molecular hydrogen: potential in mitigating oxidative-stress-induced radiation injury. Can J Physiol Pharmacol 97:287–292. https://doi.org/10.1139/cjpp-2018-0604
14. Sozen E, Demirel T, Ozer NK (2019) Vitamin E: regulatory role in the cardiovascular system. IUBMB Life 71:507–515. https://doi.org/10.1002/iub.2020
15. Violi F, Nocella C, Loffredo L, Carnevale R, Pignatelli P (2022) Interventional study with vitamin E in cardiovascular disease and meta-analysis. Free Radic Biol Med 178:26–41. https://doi.org/10.1016/j.freeradbiomed.2021.11.027

16. Álvarez-Ayuso L, Gómez-Heras SG, Jorge E, Guardiola JM, Torralba A, Granado F, Millán I, Roda JR, Calero P, Fernández-García H et al (2010) Vitamin E action on oxidative state, endothelial function and morphology in long-term myocardial preservation. Histol Histopathol 25:577–587. https://doi.org/10.14670/HH-25.577
17. Lojda Z, Gossrau R, Schiebler TH (1976) Enzymhistochemische Methoden
18. Lojda Z (1979) Studies on dipeptidyl(amino)peptidase IV (glycyl-proline naphthylamidase). Histochemistry 59:153–166. https://doi.org/10.1007/BF00495663
19. Bacova BS, Viczenczova C, Andelova K, Sykora M, Chaudagar K, Barancik M, Adamcova M, Knezl V, Benova TE, Weismann P et al (2020) Antiarrhythmic effects of melatonin and omega-3 are linked with protection of myocardial cx43 topology and suppression of fibrosis in catecholamine stressed normotensive and hypertensive rats. Antioxidants 9:1–19. https://doi.org/10.3390/antiox9060546
20. Mitašíková M, Šmidová S, Macsaliová A, Knezl V, Dlugošová K, Okruhlicová L, Weismann P, Tribulová N (2008) Aged male and female spontaneously hypertensive rats benefit from n-3 polyunsaturated fatty acids supplementation. Physiol Res 57:39–48. https://doi.org/10.33549/physiolres.931550
21. Andelova K, Szeiffova Bacova B, Sykora M, Pavelka S, Rauchova H, Tribulova N (2022) Cardiac Cx43 signaling is enhanced and TGF-β1/SMAD2/3 suppressed in response to cold acclimation and modulated by thyroid status in hairless SHRM. Biomedicines 10. https://doi.org/10.3390/biomedicines10071707
22. Viczenczova C, Kura B, Benova TE, Yin C, Kukreja RC, Slezak J, Tribulova N, Bacova BS (2018) Irradiation-induced cardiac connexin-43 and miR-21 responses are hampered by treatment with atorvastatin and aspirin. Int J Mol Sci 19:1–10. https://doi.org/10.3390/ijms19041128
23. Viczenczova C, Bacova BS, Benova TE, Kura B, Yin C, Weismann P, Kukreja R, Slezak J, Tribulova N (2016) Myocardial connexin-43 and PKC signalling are involved in adaptation of the heart to irradiation-induced injury: Implication of miR-1 and miR-21. Gen Physiol Biophys 35:215–222. https://doi.org/10.4149/gpb_2015038
24. Jeyaraman MM, Srisakuldee W, Nickel BE, Kardami E (2012) Connexin43 phosphorylation and cytoprotection in the heart. Biochim Biophys Acta—Biomembr 1818:2009–2013. https://doi.org/10.1016/j.bbamem.2011.06.023

Chapter 10
Molecular Hydrogen: A New Protective Tool Against Radiation-Induced Toxicity

Jana Vlkovicova, Branislav Kura, Patricia Pavelkova, and Barbora Kalocayova

Abstract Radiation damage is defined as exposure to ionizing radiation causing a series of organ injuries over a specified period of time. In this study, we investigated the effect of the mediastinum irradiation of Wistar male rats in a single dose of 10 Gy and subsequent 9-day treatment with selected antioxidants (molecular hydrogen—H_2, vitamin E, *Laminaria digitata*). Blood plasma biomarkers were analyzed to determine the extent of the irradiation effect on the kidney (creatinine, uric acid), liver (alanine aminotransferase, aspartate aminotransferase), metabolism (total cholesterol, triglycerides, glucose), and cytotoxicity (lactate dehydrogenase). Increased levels of markers in the plasma of rats indicate a deterioration of the general condition of the organism 9 days after irradiation of the mediastinum. Administration of selected compounds reduced levels of selected parameters in plasma, which indicates their protective effect. Our results show that among the selected antioxidants, H_2 appears to be the most effective radioprotectant in the observed period. We assume that H_2 could represent a novel and clinically applicable agent against radiation-induced toxicity.

Keywords Mediastinal irradiation · Antioxidant therapy · Molecular hydrogen · *Laminaria digitata* · Vitamin E

Introduction

Radiation injury is defined as an ionizing radiation exposure inducing a series of organ injuries within a specified time. The severity of organ injury depends on the radiation dose and the duration of radiation exposure. Acute radiation syndrome

J. Vlkovicova · B. Kura (✉) · P. Pavelkova · B. Kalocayova
Centre of Experimental Medicine, Institute for Heart Research, Slovak Academy of Sciences, Dubravska cesta 9, 841 04 Bratislava, Slovak Republic
e-mail: branislav.kura@savba.sk

P. Pavelkova
Department of Animal Physiology and Ethology, Faculty of Natural Sciences, Comenius University, Bratislava, Slovakia

J. Slezak and B. Kura (eds.), *Molecular Hydrogen in Health and Disease*, Advances in Biochemistry in Health and Disease 27, https://doi.org/10.1007/978-3-031-47375-3_10

generally includes DNA double-strand breaks, hematopoietic syndrome, cutaneous injury, brain hemorrhage, and splenomegaly within 30 days after radiation exposure [1]. Radiation therapy in the mediastinal area effectively treats several malignancies, such as Hodgkin's lymphoma and lung or breast cancer. It is known that radiation therapy induces vascular endothelium damage, promoting inflammation, and accelerating atherosclerosis. Moreover, the increased use of mediastinal radiotherapy to treat various thoracic malignancies has been associated with a wide variety of cardiovascular diseases, which may develop years after radiation therapy [2]. The health risks associated with direct radiation exposure have been extensively studied however, indirect, and delayed effects of radiation risk are still under consideration and require more evidence for comprehensive knowledge. Radiation has a wide range of biological effects that are not only limited to its direct exposure but also the number of indirect and delayed effects [3, 4]. It was documented that irradiated cell-conditioned medium exposure dominantly reduces cell survival and increases apoptosis [5]. Conversely, low-dose radiation exposure (less than 0.5 Gy) can result in the suppressed innate immune system, while establishing pro-inflammatory environments for adaptive immune cells. In peripheral organs and the brain, low-dose radiation commonly induces DNA damage and oxidative stresses, leading to systemic aberrations [6].

Radiation-induced "bystander effect" is defined as biological effect expressed after irradiation by cells whose nuclei have not been directly irradiated. These effects include DNA damage, chromosomal instability, mutation, and apoptosis. There is considerable evidence that ionizing radiation affects cells located near the site of irradiation, which respond individually and collectively as part of a large, interconnected web. The upregulation of oxidative metabolism in bystander cells suggests that the biological effects in these cells may be a consequence of oxidative stress. In addition to direct cellular damage via reactive oxygen and nitrogen species induced by radiation therapy, irradiated cells could also induce changes in distant non-irradiated cells through cell signaling molecules [7]. Initial studies demonstrated that cell culture media taken from irradiated cultures could be transferred to non-irradiated cultures and induce DNA damage [7]. In addition to bystander effects, the abscopal effect is a phenomenon seen when irradiation at a distinct anatomic site induces a systemic response throughout the body [8, 9]. The systemic oxidative stress induced by radiotherapy is strongly implicated in the processes of cytotoxicity and carcinogenesis in normal tissues, which can be counteracted by the free-radical scavenging properties of antioxidants [10]. As a potential health risk to human, ionizing radiation have been paid much attention and studies on anti-radiation drugs are becoming increasingly important [11]. It was documented that multiple antioxidants (*N*-acetyl-L-cysteine, sulforaphane, and resveratrol) eliminate excessive ROS levels and could alleviate the damage of radiation-induced bystander effects [12].

Molecular hydrogen (H_2) has been proposed as a new class of radioprotective agents [13, 14]. H_2 is an efficient antioxidant that quickly diffuses across cell membranes, reduces ROS such as $\cdot OH$ and $ONOO^-$, and suppresses damage caused by oxidative stress in various organs. Moreover, the beneficial effects of molecular hydrogen may be through the activation of the nuclear factor erythroid 2-related factor 2 pathway that promotes innate antioxidants and reduction of apoptosis, as well as inflammation [15]. Another mechanism of H_2 protective effects on irradiation-induced heart damage may be mediated by regulating miRNA-1, -15b, and -21 [16]. H_2-treatment of acute and chronic radiation-induced lung damage in mice, either by inhaled H_2 gas (for acute damage) or by imbibed H_2-enriched water (for chronic), has shown anti-inflammatory and anti-apoptotic effects in cultured lung epithelial cells [17]. Vitamin E is a well-known antioxidant, effective in scavenging free radicals generated by radiation exposure. Vitamin E analogs, collectively known as tocols, have been subject to active investigation for a long time as radioprotectors in patients undergoing radiotherapy and in the context of possible radiation accidents or terrorism scenarios [18]. Vitamin E is protective against the damaging effects of radiation, with less toxicity and side effects. Therefore, we propose the implementation of antioxidant protective agents as prevention so that the therapeutic index of radiation oncology treatments can be improved in the future [19]. The use of vitamins in natural form or supplementation can be useful to reduce the radiation effect on the body, organs and/or cells. Only four of the thirteen vitamins (A, C, D and E) have been detected with radioprotective properties, mainly vitamin E, followed by vitamins C, A and D [18, 20]. Similarly, various algae have demonstrated a variety of pharmacological properties and biological functions, including antioxidant and radioprotective effects [11, 21]. Algae, including *Laminaria digitata*, contain substances such as fucoidan, fucoxanthin or bioactive polysaccharides, which have been proven to have antioxidant and radiation-protective properties [22, 23]. Furthermore, fucoidan has been shown to protect against radiation-induced damage, both *in vitro* and *in vivo* [22].

The aim of this study was to figure out the damaging effect of mediastinal irradiation on selected biochemical parameters in blood plasma as an indicators of the health condition of rats 9 days after irradiation and the possibilities of ameliorating it by using different antioxidants.

Materials and Methods

Experimental Animals

All experiments with animals in this study were in compliance with the Ethics Committee of the Institute for Heart Research, Slovak Academy of Sciences, and protocols were approved by the State Veterinary and Food Administration of the Slovak Republic. Animals were treated in accordance with the Guide for the Care and Use of Laboratory Animals (8th edition, National Academies Press). Adult male Wistar rats (12 weeks old, 200–220 g) were obtained from Velaz Ltd., Lysolaje, Czech Republic. Animals were housed and bred under standard environmental conditions (12 h light/dark cycle, ambient temperature 22–24 °C, 45–65% humidity). Food and water were available during the whole experiment *ad libitum* and its consumption was monitored daily. At the end of the experiment, rats were euthanized by an overdose of anesthesia (Thiopental, 60 mg/kg) and by heart excision. All collected samples were stored at − 80 °C until analyses.

Experimental Model

Irradiation of experimental animals was performed at the Bory Hospital in Bratislava, Department of Radiology, Slovak Republic. A single dose of 10 Gy irradiation (4–5 Gy/min) was applied to the rat's mediastinum area (wide 19 mm) with a pulsed linear electron accelerator Electra Harmony Pro after intraperitoneal application of anesthesia (Zoletil 20 mg/kg, Domitor 0.4 mg/kg).

Treatment

In the study, rats were randomly divided into 5 groups: non-irradiated group: control rats (C), irradiated untreated rats (I), irradiated rats treated with H_2 (IH, 4% H_2 in inhalation chamber, 3 × 30 min per day), irradiated rats treated with *Laminaria digitata* administration (IL, *Laminaria digitata* powder from Thorverk—Island, incorporated into chow in concentration 3% *ad libitum*), irradiated rats treated with vitamin E (IE, obtained from Jamieson, on vehiculum biscquit in dose 100 mg/kg body weight, once a day), respectively groups received only vehiculum in the same frequency). Rats were treated daily after irradiation for 9 days when the experiment was terminated (Fig. 10.1). This time represents the sub-chronic phase, a period associated with adaptation mechanisms.

Materials

Plasma samples were analysed by using a DRI-CHEM (FUJIFILM) a dry chemistry analyzer which can perform multiple test parameters of Clinical Chemistry. We used following FUJI DRI-CHEM reagent slides: glucose (GLU—F15809528), creatinine (CRE—F15809475), lactatedehydrogenaze (LDH—F15809607), total cholesterol (TCHO—F15809669), triglycerides (TG—F15809671), uric acid (UA—F15809700), aspartate aminotransferase (AST—F15809542), alanine aminotransferase (ALT—F15809554). Data were expressed as means + S.E.M., the statistical significance of differences between the groups was analyzed by Student T-test. Differences were considered significant at $p < 0.05$.

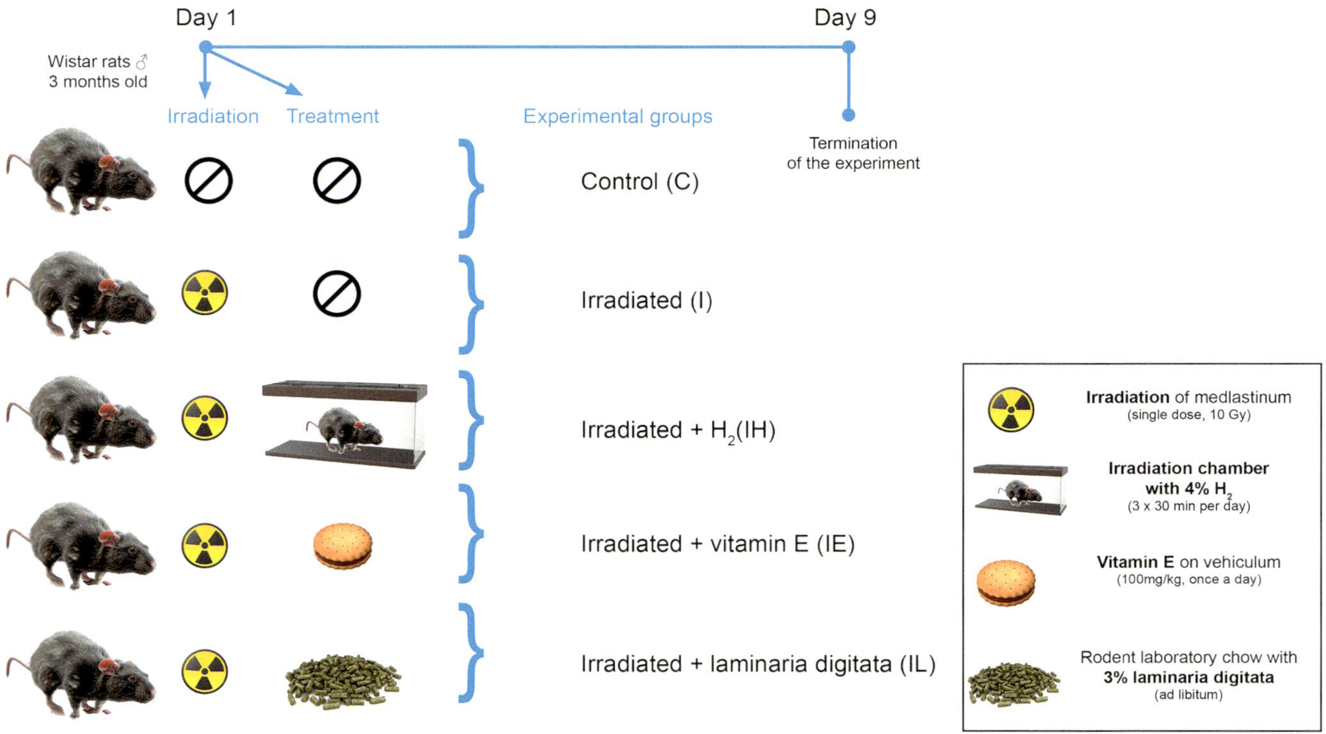

Fig. 10.1 Scheme of the biological model. Male Wistar rats were divided into 5 groups (n = 8 in each group): non-irradiated group: control group (C), irradiated (10 Gy, mediastinum) untreated group (I), irradiated and treated with H_2 administration (IH, 4% H_2 in inhalation chamber, 3 × 30 min per day), irradiated and treated with *Laminaria digitata* administration (IL, *Laminaria digitata* powder, incorporated into chow in concentration 3%, *ad libitum*), irradiated and treated with vitamin E (IE, 100 mg/kg body weight, once a day)

Results

Markers of hepatocellular injury:

In the 9 days following irradiation, ALT and AST levels were increased in the blood plasma compared to the control group. Treatment of irradiated rats with H_2, *Laminaria digitata*, and vitamin E resulted in a reduction of these parameters towards control levels (Fig. 10.2).

Markers of renal injury:

There was a statistically insignificant increase in CRE and UA levels in the irradiated untreated group compared to controls, whereas H_2 reduced this parameters most obviously when compare to *Laminaria digitata* or vitamin E (Fig. 10.3).

Markers of metabolism:

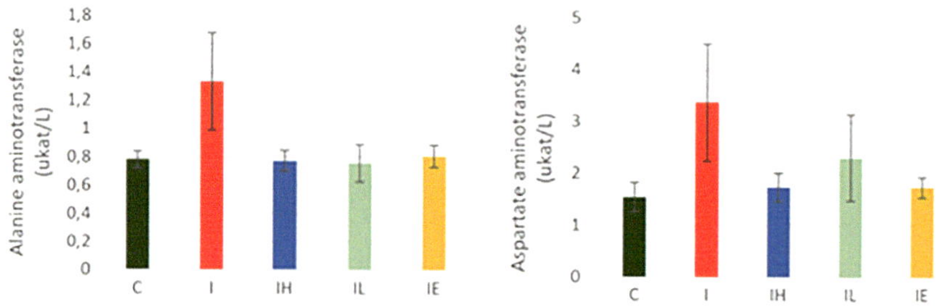

Fig. 10.2 Effect of mediastinal irradiation (10 Gy) and 9 days lasting treatment of H_2, vitamin E and *Laminaria digitata* on liver function indicators as plasma alanine aminotransferase (left) and aspartate aminotransferase (right). Experimental groups: un-irradiated control group (C), irradiated untreated group (I), irradiated, and treated with H_2 (IH), irradiated and treated with *Laminaria digitata* (IL), irradiated and treated with vitamin E (IE). Results are expressed as means ± SEM. $p < 0.05$ was considered significant

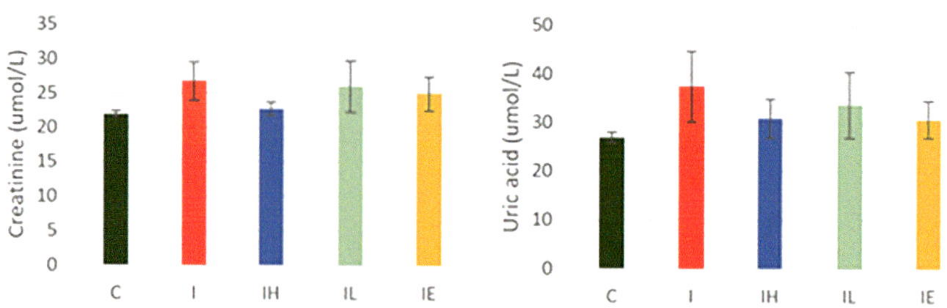

Fig. 10.3 Effect of mediastinal irradiation (10 Gy) and 9 days lasting treatment of H_2, vitamin E and *Laminaria digitata* on kidney function indicators as plasma uric acid (left), creatinine (right). Experimental groups: un-irradiated control group (C), irradiated untreated group (I), irradiated, and treated with H_2 (IH), irradiated, and treated with *Laminaria digitata* (IL), irradiated, and treated with vitamin E (IE). Results are expressed as means ± SEM. $p < 0.05$ was considered significant

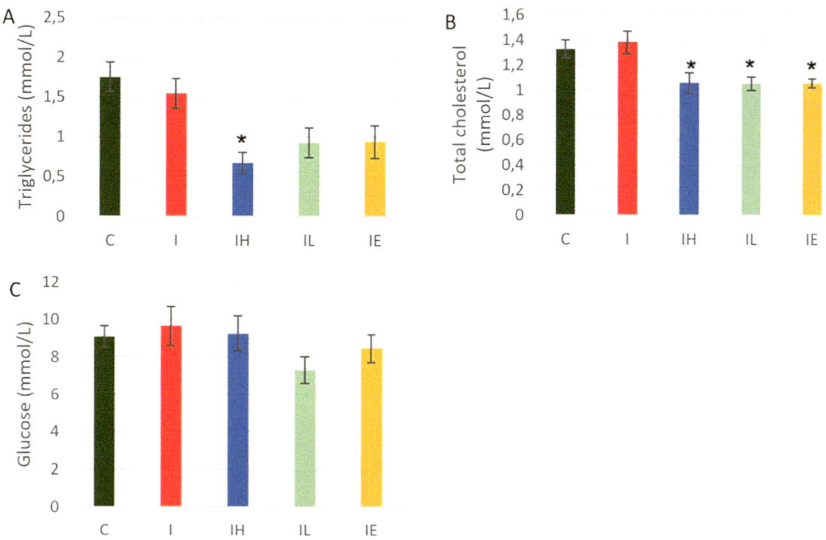

Fig. 10.4 Effect of mediastinal irradiation (10 Gy) and 9 days lasting treatment of H_2, vitamin E and *Laminaria digitata* on metabolism indicators as plasma levels of triglycerides (A), total cholesterol (B), and glucose (C). Experimental groups: un-irradiated control group (C), irradiated untreated group (I), irradiated, and treated with H_2 (IH), irradiated, and treated with *Laminaria digitata* (IL), irradiated, and treated with vitamin E (IE). Results are expressed as means ± SEM. $p < 0.05$ was considered significant

Both, TG and TCHO levels did not significantly change 9 days after mediastinal irradiation. However, administration of all three antioxidants reduced the levels of these parameters, H_2 to the greatest extent. Under these experimental conditions, the blood glucose levels were not significantly changed by either radiation or treatment (Fig. 10.4).

Markers of cytotoxicity:

Lactate dehydrogenase activity increased significantly nine days after irradiation. The administration of antioxidants changed its activity towards control levels, especially H_2 and vitamin E (Fig. 10.5).

Discussion

Ionizing radiation has for long found many applications in the medical field such as radiotherapy, radiology, and nuclear medicine. Radiation, on passing through living tissues, generates reactive free radicals, which interact with critical macromolecules such as DNA and proteins and can bring about cell damage and lead to cell dysfunction and death. Ionizing radiation has many side effects, and it is important to protect organisms from radiation-induced toxicity [24]. Mediastinal irradiation

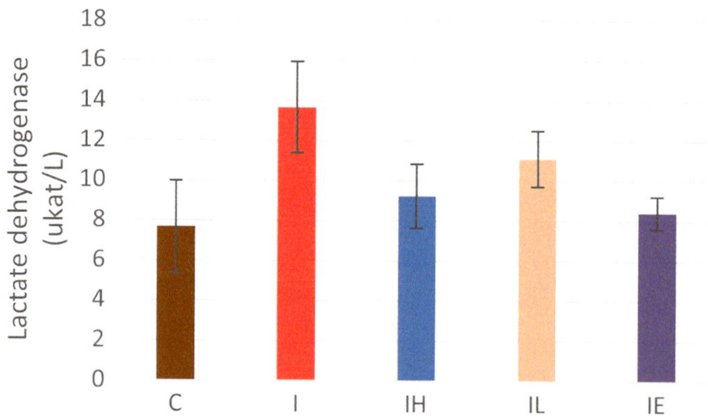

Fig. 10.5 Effect of mediastinal irradiation (10 Gy) and 9 days lasting treatment of H_2, vitamin E and *Laminaria digitata* on cytotoxicity marker lactate dehydrogenase. Experimental groups: unirradiated control group (C), irradiated untreated group (I), irradiated, and treated with H_2 (IH), irradiated, and treated with *Laminaria digitata* (IL), irradiated, and treated with vitamin E (IE). Results are expressed as means ± SEM. $p < 0.05$ was considered significant

is commonly used to treat various malignancies and can cause radiation injury to mediastinal structures, most importantly the cardiovascular system which can lead to radiation-induced heart disease [25]. It was documented that local heart irradiation in rat models induces alterations in mitochondrial morphology and function that last for several months after radiation exposure [26–29]. Although the molecular mechanisms of radiation-induced changes on not directly irradiated organs, so-called "abscopal effect", are still not clear, longstanding evidence suggests that radiation-induced inflammatory cytokines are a kind of important signaling factor. In general, cytokine activation is associated with both acute and delayed abscopal effects of irradiation and is accompanied by an increased activation of macrophages and subsequent "storm of cytokines" maintained at high levels up to 16 weeks after irradiation [30]. Inflammation factors could trigger oxidative stress and lead to an increase of ROS and NO [31, 32]. Radioprotective agents have been shown to have two main functions: radical scavenging and immunostimulating [22]. Radioprotectors are agents required to protect biological systems exposed to radiation, either naturally or through radiation leakage, and they protect normal cells from radiation injury in cancer patients undergoing radiotherapy [33]. Among these compounds with protective effects are antioxidants that prevent the propagation of chain reactions initiated by free radicals [34]. In this regard, many naturally occurring antioxidants exhibit a long-term protection, including post-irradiation protection against lethality and mutagenesis [34]. To minimalize radiation-induced toxicity and improve patient survival, it is necessary to understand the mechanisms of radiation-induced organ disease and develop new therapeutic agents [35]. Most of the available experimental studies are oriented to whole-body irradiation or examined exclusively the directly irradiated organ, only a few works dealt with the abscopal effect. In our study, we monitored the impact of a single dose of mediastinal irradiation in a dose of 10 Gy on selected plasma parameters as indicators of the overall condition/state of the organism. In addition, we have proved protective effects of selected antioxidants as H_2, vitamin E, and *Laminaria digitata* against radiation-induced toxicity. Our results point to an abscopal effect of mediastinal irradiation on the liver and kidney, as evidenced by the results from plasma analyses. Hepatocyte health biomarkers ALT and AST were increased after irradiation suggesting liver damage 9 days after exposure. Similarly, ALT, AST, and also ALP levels were increased in another study 3 days after abdominal irradiation [36]. In the acute period after irradiation, the liver seems to be more sensitive to the production of ROS, as another study showed that the liver was not affected at 120 days after 10 Gy of total body irradiation on male WAG/RijCmcr rats [37]. Moreover in the study with the 10 Gy of total body irradiation, the serum ALT activity was decreased by 21% at 100–120 days after irradiation, while AST activity was decreased by 15% at 40–60 days after total body irradiation [38]. Treatment with antioxidants could be protective against irradiation as was documented. Intraperitoneal application of vitamin E (30 mg) to rats has reversed irradiation-caused increase of ALT, AST, and ALP after 3 days of irradiation in a dose 1000-cGy [36]. The liver is responsible for cholesterol and also triacylglycerides synthesis and their increased levels are known risk factors for cardiac disease. Cholesterol and triacylglycerides can be obtained from the diet or it can be synthesized de novo. However, in our experiment, 9 days after the irradiation, the levels of the cholesterol and triacylglycerides were not significantly changed. These results are in agreement with previous study, where neither total nor partial body irradiation was not followed by changes in plasma total cholesterol and triacylglycerides within 40 days [37]. It is interesting that irradiated rats treated with H_2, vitamin E, and *Laminaria digitata* had levels of these parameters decreased, more intensively triacylglycerides than total cholesterol. Although there are no studies on 10 Gy irradiation of the mediastinum and treatment with these compounds for 9 days, in general, the effects of H_2 [39–41], vitamin E [42], and *Laminaria digitata* [43] on lowering the lipid profile have been confirmed. As the level of cholesterol and triacylglycerides in the diet of rats was constant, the lower concentration of the mentioned parameters in the blood after the irradiation with treatment is probably the result of their decreased uptake from the gut, decreased synthesis or increased clearance of cholesterol and triacylglycerides from the circulation. Using these substances seems to be cardioprotective, however, cardiotoxicity after radiotherapy includes myocardial ischemia and also hypertension, vascular events, valvular disease cardiomyopathy, myocarditis, pericardial disease, and arrhythmias [44]. Furthermore, our results do not indicate significant post-irradiation changes in blood glucose levels, or after treatment with selected compounds. Similarly, the peroxidase-glucose oxidase test performed one day after low-power laser irradiation of the salivary gland area did not detect any change in glucose concentration in the blood of control irradiated rats, however, glucose levels dropped significantly in diabetic rats contrary to controls [45]. This parameter seems to change over a longer period, as was reported in a study of where patients who received radiotherapy for gastroduodenal indolent lymphoma had increased levels of HbA1c glycated hemoglobin, which reflects the average blood glucose level over the previous 12 weeks. They concluded that patients who received radiotherapy had an increased risk of diabetes compared to those who did not [46].

As was mentioned before, mediastinal irradiation may lead to an abscopal effect and cause damage to other organs. According to a study [47], the abscopal effect on the brain and kidney is comparable to the effect of direct irradiation. It was documented that irradiation of a single dose (25 Gy) in the mediastinal area of rats induced remote deteriorating effects in other parts of the body despite their protection by lead shield [9]. In our samples, we were able to detect the increase in creatinine and uric acid levels in the irradiated untreated group compared to controls indicating kidney damage. Previously only a slight increase of creatine was observed 6 weeks after 25 Gy mediastinal irradiation [9]. The kidney is a radiosensitive abdominal organ susceptible to the development of nephropathy with hypertension, proteinuria and azotemia after irradiation [37, 48]. Moreover, it was documented that kidneys are sensitive organs to abscopal effect, as was described by the study oriented to alterations in the protein secondary structure to investigate the protein deformation in the brain, lung, and kidney of rats that were irradiated whole-body, cranially, or in the lower-limb (at a dose of 2 Gy γ—0.5 Gy/min). The results showed a decrease in α-helix spectral mass associated with an equivalent increase in the β-sheets, and random coils-turns/loops-turns in brains and kidneys of all of the irradiated groups [47]. Also, in a case report of a patient with bilateral renal involvement with diffuse histiocytic lymphoma, radiation therapy was given to only one kidney, but both kidneys were shown to respond to radiotherapy [49]. Furthermore, in our study we observed increased levels of creatinine and uric acid in the blood plasma just 9 days after mediastinum irradiation, indicating kidney damage as non-target effect caused by released of signaling factors from irradiated cells. Among the substances we tested, we recorded the best results with H_2, which is known to protect the kidneys in several animal studies with kidney disease, such as renal calculi, renal fibrosis, and drug-induced nephrotoxicity [50].

Lactate dehydrogenase performs a very important role in overall body metabolism. Different toxicological and pathological complications damage various organs, which ultimately results in leakage of this enzyme in serum. Hence, unusual lactate dehydrogenase level in serum serves as a significant biomarker of different diseases [51]. It is known that radiation by itself causes significant elevations of lactate dehydrogenase in serum [52, 53]. Similarly, in our study, we observed elevated activity of this enzyme in the plasma of rats 9 days after mediastinal irradiation. It seems that treatment with our selected compounds protected the body against radiocytotoxicity. Based on our results, we assume that inhalation of H_2 and vitamin E were the most effective to decrease the activity of lactate dehydrogenase in our experimental conditions. In addition, another study also observed the protective effects of vitamin E in the radiation model. The *i.p.* application of vitamin E (30 mg DL-α-tocopherol acetate) was followed by decreased activity of lactate dehydrogenase in the serum of rats 3 days after abdominal γ-irradiation (1000-cGy) [36]. The number of *in vivo* studies oriented to lactate dehydrogenase, irradiation and H_2 is limited. On the other hand, by *in vitro* study with intestinal crypt HIEC cells treated with H_2-enriched PBS solution and then applied γ-ray irradiation proved the radioprotective effect of H_2.

It was documented that the pre-treatment of HIEC cells with 0.1–0.4 mmol/L H_2 before γ-ray irradiation considerably boosted cell survival as compared to the cells only treated with irradiation at tested doses up to 8 Gy. Simultaneously, pretreatment with H_2 significantly reduced the lactate dehydrogenase leakage of HIEC cells [54]. Also *in vitro* model of ischemia/reperfusion injury and administration of H_2-rich medium to H9C2 cell culture was accompanied by improved cell viability and LDH release following hypoxia/reoxygenation in myocardial cells [55]. Based on our results we suggest that H_2 is more effective than vitamin E or *Laminaria digitata* in reversing the detrimental effect of irradiation. The potential mechanism of H_2, the smallest molecule, could be its antioxidant activity as was documented in many studies before [56–58] but also its ability to modify Keap1-Nrf2 signaling pathway, which plays a central role in protecting cells against oxidative and xenobiotic stresses [59]. Moreover, the radioprotective effects of H_2 and discusses the mechanisms of H_2, include also in intracellular responses including anti-inflammation, anti-apoptosis, and the regulation of gene expression [60]. It could be hypothesized that 9 days of application of H_2 after mediastinal irradiation scavenge excessive free oxygen radicals and thus eliminate oxidative stress in the organism.

Conclusion

Increased values of markers in the plasma of rats indicate a deterioration of the general condition of the organism 9 days after irradiation of the mediastinum with a single dose of 10 Gy. Administration of H_2, vitamin E and *Laminaria digitata* reduced selected parameters in plasma, which indicates their protective effect against radiation-induced toxicity. Our results show that among the selected antioxidants, H_2 appears to be the most effective compound, which most significantly suppressed the changes in plasma caused by radiation. We assume that H_2 could represent a novel and clinically applicable protective agent against radiation-induced toxicity.

Funding This research was supported by VEGA grants 2/0148/22 and 2/0092/22, and the Slovak Research and Development Agency under contract No. 19-0317.

References

1. Kiang JG, Olabisi AO (2019) Radiation: a poly-traumatic hit leading to multi-organ injury. Cell Biosci 9:25. https://doi.org/10.1186/s13578-019-0286-y
2. Raghunathan D, Khilji MI, Hassan SA, Yusuf SW (2017) Radiation-induced cardiovascular disease. Curr Atheroscler Rep 19:22. https://doi.org/10.1007/s11883-017-0658-x
3. Alexandrou AT, Li JJ (2014) Cell cycle regulators guide mitochondrial activity in radiation-induced adaptive response. Antioxid Redox Signal 20:1463–1480. https://doi.org/10.1089/ars.2013.5684

4. Hall EJ, Hei TK (2003) Genomic instability and bystander effects induced by high-LET radiation. Oncogene 22:7034–7042. https://doi.org/10.1038/sj.onc.1206900
5. Saeed Y, Xie B, Xu J et al (2014) Indirect effects of radiation induce apoptosis and neuroinflammation in neuronal SH-SY5Y cells. Neurochem Res 39:2334–2342. https://doi.org/10.1007/s11064-014-1432-x
6. Shin E, Lee S, Kang H et al (2020) Organ-specific effects of low dose radiation exposure: a comprehensive review. Front Genet 11. https://doi.org/10.3389/fgene.2020.566244
7. Prise KM, O'Sullivan JM (2009) Radiation-induced bystander signalling in cancer therapy. Nat Rev Cancer 9:351–360. https://doi.org/10.1038/nrc2603
8. Craig DJ, Nanavaty NS, Devanaboyina M et al (2021) The abscopal effect of radiation therapy. Futur Oncol 17:1683–1694. https://doi.org/10.2217/fon-2020-0994
9. Kaločayová B, Kovačičová I, Radošinská J et al (2019) Alteration of renal Na,K-ATPase in rats following the mediastinal γ-irradiation. Physiol Rep 7:1–11. https://doi.org/10.14814/phy2.13969
10. Martin OA, Yin X, Forrester HB et al (2016) Potential strategies to ameliorate risk of radiotherapy-induced second malignant neoplasms. Semin Cancer Biol 37–38:65–76. https://doi.org/10.1016/j.semcancer.2015.12.003
11. Qiong L, Jun L, Jun Y et al (2011) The effect of Laminaria japonica polysaccharides on the recovery of the male rat reproductive system and mating function damaged by multiple mini-doses of ionizing radiations. Environ Toxicol Pharmacol 31:286–294. https://doi.org/10.1016/j.etap.2010.11.006
12. Hu L, Yin X, Zhang Y et al (2021) Radiation-induced bystander effects impair transplanted human hematopoietic stem cells via oxidative DNA damage. Blood 137:3339–3350. https://doi.org/10.1182/blood.2020007362
13. Qian L, Shen J, Chuai Y, Cai J (2013) Hydrogen as a new class of radioprotective agent. Int J Biol Sci 9:887–894. https://doi.org/10.7150/ijbs.7220
14. Chuai Y, Qian L, Sun X, Cai J (2012) Molecular hydrogen and radiation protection. Free Radic Res 46:1061–1067. https://doi.org/10.3109/10715762.2012.689429
15. Kura B, Bagchi AK, Singal PK et al (2019) Molecular hydrogen: potential in mitigating oxidative-stress-induced radiation injury. Can J Physiol Pharmacol 97:287–292. https://doi.org/10.1139/cjpp-2018-0604
16. Kura B, Kalocayova B, LeBaron TW et al (2019) Regulation of microRNAs by molecular hydrogen contributes to the prevention of radiation-induced damage in the rat myocardium. Mol Cell Biochem 0:1–12. https://doi.org/10.1007/s11010-019-03512-z
17. Terasaki Y, Terasaki M, Shimizu A (2021) Protective effects of hydrogen against irradiation. Curr Pharm Des 27:679–686. https://doi.org/10.2174/1381612827666210119103545
18. Singh VK, Beattie LA, Seed TM (2013) Vitamin E: tocopherols and tocotrienols as potential radiation countermeasures. J Radiat Res 54:973–988. https://doi.org/10.1093/jrr/rrt048
19. Shirazi A, Mihandoost E, Mahdavi SR, Mohseni M (2012) Radio-protective role of antioxidant agents. Oncol Rev 6:16. https://doi.org/10.4081/oncol.2012.e16
20. Lledó I, Ibáñez B, Melero A et al (2023) Vitamins and radioprotective effect: a review. Antioxidants 12:611. https://doi.org/10.3390/antiox12030611
21. Abraham RE, Alghazwi M, Liang Q, Zhang W (2021) Advances on marine-derived natural radioprotection compounds: historic development and future perspective. Mar Life Sci Technol 3:474–487. https://doi.org/10.1007/s42995-021-00095-x
22. Rhee KH, Lee KH (2011) Protective effects of fucoidan against γ-radiation-induced damage of blood cells. Arch Pharm Res 34:645–651. https://doi.org/10.1007/s12272-011-0415-6
23. El Bakary NM, Thabet NM, El Fatih NM et al (2021) Fucoxanthin alters the apelin-13/APJ pathway in certain organs of γ-irradiated mice. J Radiat Res 62:600–617. https://doi.org/10.1093/jrr/rraa141
24. Hosseinimehr SJ (2007) Trends in the development of radioprotective agents. Drug Discov Today 12:794–805. https://doi.org/10.1016/j.drudis.2007.07.017
25. Tang D, Peng EW, Giri D et al (2009) Mediastinal irradiation and its effect on the cardiovascular system. Br J Hosp Med 70:222–224. https://doi.org/10.12968/hmed.2009.70.4.41626

26. Sridharan V, Aykin-Burns N, Tripathi P et al (2014) Radiation-induced alterations in mitochondria of the rat heart. Radiat Res 181:324. https://doi.org/10.1667/RR13452.1
27. Slezak J, Kura B, Ravingerová T et al (2015) Mechanisms of cardiac radiation injury and potential preventive approaches. Can J Physiol Pharmacol 93:737–753. https://doi.org/10.1139/cjpp-2015-0006
28. Viczenczova C, Bacova BS, Benova TE et al (2016) Myocardial connexin-43 and PKC signalling are involved in adaptation of the heart to irradiation-induced injury: implication of miR-1 and miR-21. Gen Physiol Biophys 35:215–222. https://doi.org/10.4149/gpb_2015038
29. Kura B, Yin C, Frimmel K et al (2016) Changes of microRNA-1, -15b and -21 levels in irradiated rat hearts after treatment with potentially radioprotective drugs. Physiol Res 65(Suppl 1):S129–S137
30. Calveley VL, Khan MA, Yeung IWT et al (2005) Partial volume rat lung irradiation: temporal fluctuations of in-field and out-of-field DNA damage and inflammatory cytokines following irradiation. Int J Radiat Biol 81:887–899. https://doi.org/10.1080/09553000600568002
31. Hei TK, Zhou H, Ivanov VN et al (2010) Mechanism of radiation-induced bystander effects: a unifying model. J Pharm Pharmacol 60:943–950. https://doi.org/10.1211/jpp.60.8.0001
32. Hu S, Shao C (2020) Research progress of radiation induced bystander and abscopal effects in normal tissue. Radiat Med Prot 1:69–74. https://doi.org/10.1016/j.radmp.2020.04.001
33. Kamran MZ, Ranjan A, Kaur N et al (2016) Radioprotective agents: strategies and translational advances. Med Res Rev 36:461–493. https://doi.org/10.1002/med.21386
34. Weiss JF, Landauer MR (2003) Protection against ionizing radiation by antioxidant nutrients and phytochemicals. Toxicology 189:1–20. https://doi.org/10.1016/S0300-483X(03)00149-5
35. Kim J, Jung Y (2017) Radiation-induced liver disease: current understanding and future perspectives. Exp Mol Med 49:e359–e359. https://doi.org/10.1038/emm.2017.85
36. Yanardag R, Bolkent S, Kizir A (2001) Protective effects of DL-α-tocopherol acetate and sodium selenate on the liver of rats exposed to gamma radiation. Biol Trace Elem Res 83:263–273. https://doi.org/10.1385/BTER:83:3:263
37. Lenarczyk M, Lam V, Jensen E et al (2013) Cardiac injury after 10 Gy total body irradiation: indirect role of effects on abdominal organs. Radiat Res 180:247–258. https://doi.org/10.1667/RR3292.1
38. Baker JE, Fish BL, Su J et al (2009) 10 Gy total body irradiation increases risk of coronary sclerosis, degeneration of heart structure and function in a rat model. Int J Radiat Biol 85:1089–1100. https://doi.org/10.3109/09553000903264473
39. Todorovic N, Fernández-Landa J, Santibañez A et al (2023) The effects of hydrogen-rich water on blood lipid profiles in clinical populations: a systematic review and meta-analysis. Pharmaceuticals 16:142. https://doi.org/10.3390/ph16020142
40. Zhao Z, Ji H, Zhao Y et al (2023) Effectiveness and safety of hydrogen inhalation as an adjunct treatment in Chinese type 2 diabetes patients: a retrospective, observational, double-arm, real-life clinical study. Front Endocrinol (Lausanne) 13. https://doi.org/10.3389/fendo.2022.1114221
41. Nakao A, Toyoda Y, Sharma P et al (2010) Effectiveness of hydrogen rich water on antioxidant status of subjects with potential metabolic syndrome—an open label pilot study. J Clin Biochem Nutr 46:140–149. https://doi.org/10.3164/jcbn.09-100
42. Leonard SW, Joss JD, Mustacich DJ et al (2007) Effects of vitamin E on cholesterol levels of hypercholesterolemic patients receiving statins. Am J Heal Pharm 64:2257–2266. https://doi.org/10.2146/ajhp070041
43. Ribeiro D, Alfaia C, Pestana J et al (2022) Influence of feeding weaned piglets with Laminaria digitata on the quality and nutritional value of meat. Foods 11:1024. https://doi.org/10.3390/foods11071024
44. Dreyfuss AD, Velalopoulou A, Avgousti H et al (2022) Preclinical models of radiation-induced cardiac toxicity: potential mechanisms and biomarkers. Front Oncol 12. https://doi.org/10.3389/fonc.2022.920867
45. Fukuoka CY, Torres Schröter G, Nicolau J, Simões A (2016) Low-power laser irradiation in salivary glands reduces glycemia in streptozotocin-induced diabetic female rats. J Biophotonics 9:1246–1254. https://doi.org/10.1002/jbio.201600175

46. Baek JY, Lim DH, Oh D et al (2022) Increased risk of diabetes after definitive radiotherapy in patients with indolent gastroduodenal lymphoma. Cancer Res Treat 54:294–300. https://doi.org/10.4143/crt.2021.073
47. Abouelsayed A, Hezma AM, El-Bahy GS, Abdelrazzak AB (2023) Modification of protein secondary structure as an indicator of radiation-induced abscopal effect: a spectroscopic investigation. Spectrochim Acta Part A Mol Biomol Spectrosc 287:122093. https://doi.org/10.1016/j.saa.2022.122093
48. Lenarczyk M, Laiakis EC, Mattson DL et al (2020) Irradiation of the kidneys causes pathologic remodeling in the nontargeted heart: a role for the immune system. FASEB BioAdvances 2:705–719. https://doi.org/10.1096/fba.2020-00071
49. Robins HI, Buchon JA, Varanasi VR, Weinstein AB (1981) The abscopal effect: demonstration in lymphomatous involvement of kidneys. Med Pediatr Oncol 9:473–476. https://doi.org/10.1002/mpo.2950090510
50. Wang B, Li Z, Mao L et al (2022) Hydrogen: a novel treatment strategy in kidney disease. Kidney Dis 8:126–136. https://doi.org/10.1159/000520981
51. Khan AA, Allemailem KS, Alhumaydhi FA et al (2020) The biochemical and clinical perspectives of lactate dehydrogenase: an enzyme of active metabolism. Endocrine, Metab Immune Disord—Drug Targets 20:855–868. https://doi.org/10.2174/1871530320666191230141110
52. Morgan RM, Kundomal YR, Hupp EW (1983) Serum lactate dehydrogenase (LDH) activity following exposures to cadmium and/or 60 Co gamma irradiation. J Environ Sci Heal Part A Environ Sci Eng 18:483–492. https://doi.org/10.1080/10934528309375117
53. Rave-Fränk M, Malik IA, Christiansen H et al (2013) Rat model of fractionated (2 Gy/day) 60 Gy irradiation of the liver: long-term effects. Radiat Environ Biophys 52:321–338. https://doi.org/10.1007/s00411-013-0468-7
54. Hu Q, Zhou Y, Wu S et al (2020) Molecular hydrogen: a potential radioprotective agent. Biomed Pharmacother 130:110589. https://doi.org/10.1016/j.biopha.2020.110589
55. Yao L, Chen H, Wu Q, Xie K (2019) Hydrogen-rich saline alleviates inflammation and apoptosis in myocardial I/R injury via PINK-mediated autophagy. Int J Mol Med. https://doi.org/10.3892/ijmm.2019.4264
56. Slezak J, Kura B, LeBaron TW et al (2021) Oxidative stress and pathways of molecular hydrogen effects in medicine. Curr Pharm Des 27:610–625. https://doi.org/10.2174/1381612826666200821114016
57. LeBaron TW, Kura B, Kalocayova B et al (2019) A new approach for the prevention and treatment of cardiovascular disorders. molecular hydrogen significantly reduces the effects of oxidative stress. Molecules 24:2076. https://doi.org/10.3390/molecules24112076
58. Huang L (2016) Molecular hydrogen: a therapeutic antioxidant and beyond. Med Gas Res 6:219–222. https://doi.org/10.4103/2045-9912.196904
59. Deshmukh P, Unni S, Krishnappa G, Padmanabhan B (2017) The Keap1–Nrf2 pathway: promising therapeutic target to counteract ROS-mediated damage in cancers and neurodegenerative diseases. Biophys Rev 9:41–56. https://doi.org/10.1007/s12551-016-0244-4
60. Hirano S, Ichikawa Y, Sato B et al (2021) Molecular hydrogen as a potential clinically applicable radioprotective agent. Int J Mol Sci 22:4566. https://doi.org/10.3390/ijms22094566

Chapter 11
Role of Matrix Metalloproteinases in Effects of Molecular Hydrogen

Barbora Bot'anská, Viktória Pecníková, Branislav Kura, Ján Slezák, and Miroslav Barančík

Abstract Molecular hydrogen plays a role in modulating several cellular functions and in several diseases were documented its pleiotropic therapeutic effects. Several lines of evidence indicates that an important role in mechanisms involved in molecular hydrogen effects play modulation of cellular antioxidant defenses, including intracellular and extracellular redox signaling. Matrix metalloproteinases (MMPs) are zinc-dependent endopeptidases that are involved in regulation of several cellular functions, in particular the modulation of extracellular matrix turnover. Moreover, oxidative stress-induced dysregulation of MMPs activities plays a crucial role in the development of pathological changes. Since it was described regulatory role of molecular hydrogen in modulation of several MMPs these enzymes represent potential targets of molecular hydrogen action. Our data demonstrated potential negative role of MMP-2 and MMP-28 in development of pathological changes induced by mediastinal irradiation (MI) of rats. Molecular hydrogen possessed beneficial effects against MI-induced changes in both matrix metalloproteinases. To the potential role of MMPs in protective effects of molecular hydrogen point also findings that inhalation of hydrogen-rich air was associated with protection against post-transplant complications and led to a partial reversal of MMP-9 activation. In conclusion, modulation of matrix metalloproteinases using molecular hydrogen could be promising strategy for treatment of several diseases, in which oxidative stress and tissue remodeling is an important partner.

B. Bot'anská · V. Pecníková · B. Kura (✉) · J. Slezák · M. Barančík
Centre of Experimental Medicine, Institute for Heart Research, Slovak Academy of Sciences, Dúbravská cesta 9, Bratislava 841 04, Slovakia
e-mail: branislav.kura@savba.sk

B. Bot'anská
e-mail: barbora.botanska@savba.sk

V. Pecníková
e-mail: viktoria.pecnikova@savba.sk

J. Slezák
e-mail: jan.slezak@savba.sk

M. Barančík
e-mail: miroslav.barancik@savba.sk

J. Slezak and B. Kura (eds.), *Molecular Hydrogen in Health and Disease*, Advances in Biochemistry in Health and Disease 27, https://doi.org/10.1007/978-3-031-47375-3_11

Keywords Molecular hydrogen · Matrix metalloproteinases · Oxidative stress

Introduction

Matrix metalloproteinases (MMPs) are enzymes with important role in remodeling extracellular matrix (ECM) and regulation of cellular functions such as proliferation, migration, differentiation, angiogenesis, apoptosis, and cell–cell interactions. Moreover, dysregulation of MMPs activities plays a crucial role in pathophysiology of several diseases such as cardiovascular, neurodegenerative, and metabolic diseases and cancer. Reactive oxygen species (ROS) and reactive nitrogen species (RNS), especially superoxide and peroxynitrite, are essential in regulating MMPs activities/ expression. Therefore, the fact that molecular hydrogen selectively reduces not only hydroxyl radicals but also peroxynitrite points to its regulatory role in the modulation of MMPs.

This book chapter focuses on the molecular hydrogen and matrix metalloproteinases and summarises knowledge about the involvement of matrix metalloproteinases in effects of molecular hydrogen at pathological stress conditions. Presented are also our experimental data documenting a link between positive therapeutic efffects of molecular hydrogen and modulation of matrix metalloproteinases at stress conditions induced by mediastinal irradiation of rats.

Molecular Hydrogen as Modulator of Cellular Functions

The physico-chemical properties of hydrogen molecule (low molecular weight, electrically neutral and nonpolar character) allow its easy entrance into cells and rapid diffusion across all biological cell membranes. Due to these unique physical and chemical properties can molecular hydrogen (H_2) reach the subcellular compartments, such as mitochondria and sarcoplasmic reticulum, which are the primary sites of reactive oxygen species (ROS). Here can be molecular hydrogen involved in selective reduction of highly toxic hydroxyl radicals and peroxynitrite levels [1]. This is also why molecular hydrogen is vital in modulating several cellular functions. Documented were its anti-oxidative [2], anti-inflammatory [3], anti-apoptotic effects, and autophagy modulation [4]. Due to these effects, H_2 is also a promissing therapeutic molecule. This is supported by the fact that pleiotropic therapeutic effects of H_2 in several animal disease models and many human diseases were demonstrated. It has been described positive protective effects of supplemental molecular hydrogen application in cardiovascular diseases [5], neurodegenerative diseases [6], inflammatory diseases [7], metabolic syndrome [8], diabetes [9], kidney disorders [10], and cancer [11]. The detailed and precise molecular mechanisms involved in molecular hydrogen effects are not known but current information indicates that an important role can have play modulation of cellular antioxidant defences, including intracellular and extracellular redox signaling. This is supported by findings documenting role of molecular hydrogen in the modulation of cellular responses to stress conditions, buffering of oxidative stress, reduction of endoplasmic reticulum stress, apoptosis inhibition, and autophagy machinery regulation [4, 6]. Molecular hydrogen can also influence gene expression, which would most likely be a downstream effect of an earlier action. From the point of modulation of oxidative stress, a possible mechanism underlying the cellular protection afforded by molecular hydrogen may be increased expression of anti-oxidant enzymes such as catalase, superoxide dismutase or heme oxygenase-1 [9, 12, 13].

Physiological Functions of Matrix Metalloproteinases

Matrix metalloproteinases (MMPs) represent a large family of zinc-dependent endopeptidases involved in regulating and remodelation of extracellular matrix (ECM). MMPs also play an essential role in cell behaviors such as proliferation, migration, differentiation, angiogenesis, and apoptosis [14–16]. Moreover, their role in regulating cell–cell interactions through affecting gap junctions and tight junctions proteins has also been documented [17, 18]. It has been shown that junction proteins, such as occludin, connexin-32, and connexin-43 are substrates of MMP-2 and MMP-9. Activities of MMPs can be regulated through post-translational modifications that include either proteolytic removal of the propeptide domain or zymogen modification by oxidative stress. In the latter mechanism, activation occurs through conformational changes induced by oxidative stress and powerful endogenous oxidants, such as superoxide and peroxynitrite [19]. The activities of MMPs are also tightly regulated by the endogenous tissue inhibitors of matrix metalloproteinases (TIMP-1, TIMP-2, TIMP-3, and TIMP-4) [20]. Several factors regulate MMPs protein levels and activity, such as ROS, growth factors, cytokines, and hormones. Increased production of ROS and following MMPs modulation (activation) involves also activating specific signaling pathways such as mitogen-activated protein kinases (MAPK) [16].

Role of Matrix Metalloproteinases in Responses to Pathological Stress Conditions

Dysregulation of MMPs activities and/or TIMPs plays a crucial role in the development of pathological changes. Uncontrolled MMPs activation in responses to pathological stress conditions is associated with remodeling and disturbed ECM turnover following functional alterations and tissue damage. The negative impact of oxidative stress- induced MMPs activation on cell–cell interactions and that through cleavage of junction protein has been reported [21].

MMPs play an essential role in the pathophysiology of several diseases such as cardiovascular, neurodegenerative, and metabolic diseases and cancer. As regards to cardiovascular diseases, increased MMPs, especially MMP-2 and MMP-9, were found to be responsible for structural disorganizations of the cardiac extracellular space associated with disruption of cardiac function, development of cardiac fibrosis, and progresssion of cardiovascular diseases (cardiomyopathies, heart failure) [22, 23]. Increased plasma MMPs activities can serve as potential biomarkers for estimation of cardiac injury. It has been found that increased circulating levels of MMP-2 and MMP-9 are associated with worse prognosis in patients with cardiovascular disease and were suggested to predict mortality [24]. Moreover, elevated MMP-2 and MMP-9 in circulation were found to be closely associated with impairment of both cardiac and vascular function and heart failure development [25]. Increased circulating and tissue MMP-2 and MMP-9 levels have also been reported to play a role in the development of hypertension [26].

We observed an elevation of circulating plasma MMPs as a consequence of the cardiotoxic effects of doxorubicin [27] and after mediastinal irradiation. Moreover, the effects of doxorubicin were in rats associated with up-regulation of 72 kDa MMP-2 activites in left ventricular cardiac tissue. We observed a similar 72 kDa MMP-2 activation in left ventricle in diabetic ZDF rats [28]. Increased activation of this form of MMP-2 can occur in disease conditions associated with redox imbalance. This was supported also by findings that the effects of doxorubicin and diabetes on MMP-2 activation were associated with a reduction of total superoxide dismutase (SOD) activities. Several lines of evidence point to a close association between MMPs and diabetes [28, 29]. Diabetes was found to enhance through increased oxidative stress vascular MMP-9 activity. This gelatinase can hydrolyze the basement membrane and extracellular matrix of the blood vessels and in such way, significantly impact the vascular system [30]. Described was also crucial role of MMP-9 in disruption of blood–brain barrier [31]. Studies have shown MMPs as an essential factor involved in the generation and maintenance of neuroinflammation and pain [32].

Effects of Molecular Hydrogen on Modulation of Matrix Metalloproteinases

Matrix metalloproteinases represent potential targets of H2 action, and regulatory role of H2 in modulation of expression of several MMPs (MMP-2, MMP-3, MMP-9, and MMP-13) was documented [33, 34]. In our studies, we investigated the potential role of several MMPs (MMP-2, MMP-9, and MMP-28) in development of myocardial pathological changes induced by mediastinal irradiation (MI) of rats. The effects of molecular hydrogen on radiation-induced changes in levels/activities of these MMPs were also determined. We found that MI induced significant increase of circulating plasma 72 kDa MMP-2 activities. Effects of irradiation on circulating MMP-2 were transient, with maximum at the day 2, and with prolongation of time, were reduced.

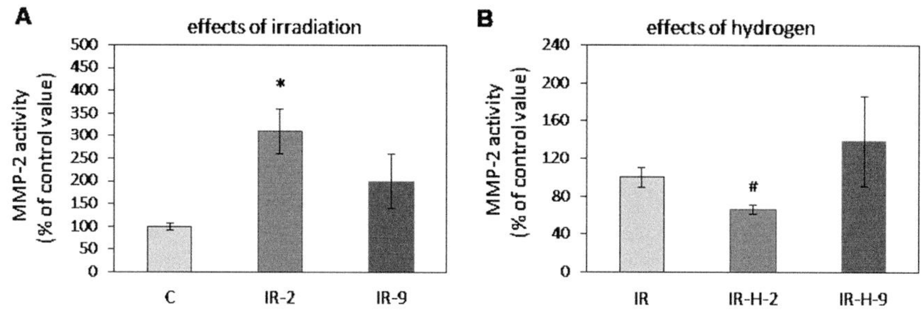

Fig. 11.1 Effects of molecular hydrogen on irradiation and molecular hydrogen on activities of circulating plasma 72 kDa MMP-2. **A)** Graph showing changes in 72 kDa MMP-2 activities 2 and 9 days after mediastinal irradiation, **B)** Graph showing effects of molecular hydrogen on irradiation-induced changes 2 and 9 days after irradiation. MMP-2 activities were determined in plasma samples using gelatine zymography. C- control; IR-2 – two days after irradiation; IR-9 – nine days after irradiation; IR-H-2 – two days after irradiation + application of molecular hydrogen; IR-H-9 – nine days after irradiation + application of molecular hydrogen. $^{*}p < 0.05$ compared to C, $\neq < 0.05$ compared to IR

Inhalation of molecular hydrogen (4% gas, 3 × 30 min.) reduced at the day 2 the adverse effects of irradiation on plasma 72 kDa MMP-2 activities (Fig. 11.1).

Analysis of tissue samples of the left ventricle showed that neither irradiation nor molecular hydrogen significantly influenced the protein levels/activities of cardiac MMP-2 and MMP-9. However, exposure of rats to irradiation of mediastinum was associated with reduction of MMP-28 protein levels at both day 2 and day 9 after exposure to MI (Fig. 11.2). Inhalation of molecular hydrogen reduced the negative effects of irradiation on MMP-28 protein levels in the left ventricle.

MMP-28 is the newest identified member of the MMP family, and decreases in its protein levels in response to pathological conditions have been documented [35]. Our recent results showed significant down-regulation of protein levels of MMP-28 in obese diabetic ZDF rats [28]. In other studies, it was reported that deletion of MMP-28 may accelerate the decline of function of the left ventricle with age and increase dysfunction of the left ventricle post-myocardial infarction [35]. In contrast to MMP-2 and MMP-9, where myocardial infarction increased levels/activities of these MMPs, a decrease of MMP-28 levels was found post-myocardial infarction. The observed data demonstrate negative modulation of MMPs after the mediastinal irradiation. Activation of circulating MMP-2 and down-regulation of cardiac MMP-28 may negatively impact the progression of pathological changes induced by mediastinal irradiation. Molecular hydrogen possessed beneficial effects against MI-induced changes in MMP-2 and MMP-28. From the point of molecular mechanisms involved in the positive effects of molecular hydrogen was important finding that exposure of animals to the effects of molecular hydrogen induced after 2 days activation of cellular antioxidant defense through up-regulation of protein levels of superoxide dismutase-2 (SOD-2) (Fig. 11.3).

The connection between superoxide dismutases and MMPs was demonstrated in a study showing increased superoxide production (decreased SOD activities) and

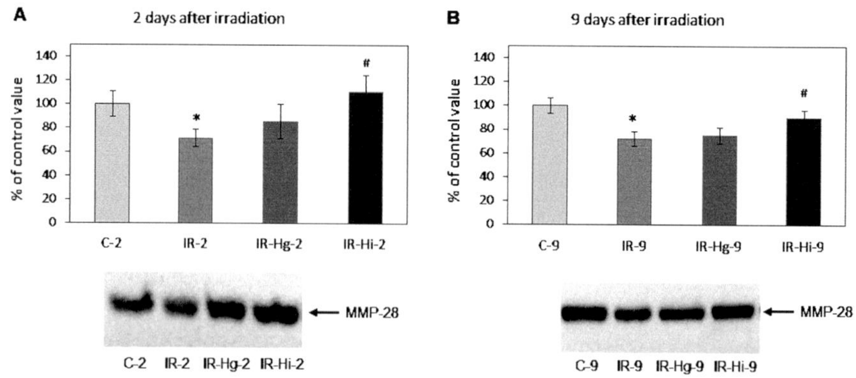

Fig. 11.2 Effects of molecular hydrogen on irradiation-induced changes in MMP-28 protein levels in left ventricle. **A**) Graph and Western blot record showing changes 2 days after irradiation, **B**) Graph and Western blot record showing changes 9 days after irradiation. C- control; IR – rats exposed to irradiation of mediastinum; IR-Hg – irradiation + application of molecular hydrogen in hydrogen-rich water (1 ppm in water, 3 × 3 mL); IR-Hi - irradiation + inhalation of molecular hydrogen (4% gas, 3 × 30 min.). *$p < 0.05$ compared to C, $^{\neq}p < 0.05$ compared to IR

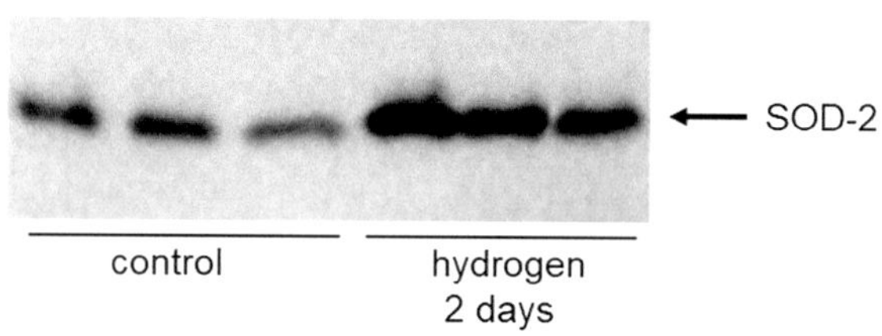

Fig. 11.3 Western blot record showing the effect of molecular hydrogen inhalation on protein levels of superoxide dismutase-2 (SOD-2) in rat tissue of the left ventricle. The SOD-2 protein levels were determined by Western blot analysis using a specific antibody

MMP-2 activation induced by high glucose in bovine retinal endothelial cells [36]. The positive effects of H2 on changes induced by radiation were documented in study showing that H2 can improve the heart and lung fibrosis caused by radiation damage. Another study demonstrated that treatment with hydrogen-rich saline (HRS) reduced the development of myocardial fibrosis in spontaneously hypertensive rats [37]. The positive effects of H2 were associated with down-regulation of the expression of tissue inhibitors of matrix metalloproteinases (TIMPs) and improved antioxidant enzymatic activity of SOD. Other data showed that hydrogen-rich water inhibits the migration of smooth muscle cells into the vein grafts. This effect of molecular hydrogen was partially explained through its effects on lowering of MMP-2 and MMP-9 expression [38]. To the relation between effects of H2 and MMPs point also data showing that HRS treatment attenuates acute obstructive cholangitis (AOC)-induced inflammatory and oxidative damage in rats, and this was associated with a significant reduction of MMP-2 and MMP-9 activities [39]. The authors assumed that molecular hydrogen's antioxidant and anti-inflammatory properties might decrease MMPs activities and reverse the disruption of gap junctions and tight junctions proteins.

In pigs with simulated heart transplantation we observed, after 60 min of spontaneous reperfusion, significantly increased activities of circulating MMP-9. Inhalation of hydrogen-rich air led to a partial reversal of MMP-9 activation after simulated heart transplantation. This indicates the potential role of molecular hydrogen in protection against post-transplant complications. MMP-9 is an enzyme that plays crucial role in extracellular matrix remodeling, and its significant elevation in plasma of patients with pulmonary arterial hypertension has been documented [40]. Moreover, increased MMP-9 was observed in the plasma of patients with coronary artery disease undergoing coronary artery bypass graft surgery with cardiopulmonary bypass [41]. MMP-9 was also found to promote hemorrhagic infarction by disrupting cerebral vessels in a rat model of middle cerebral artery occlusion (MCAO) [34]. Inhalation with H2 significantly reduced the injury though inhibition of MMP-9 expression. Molecular hydrogen has also been reported to inhibit MMP-9 activity in order to reduce brain infarction and improve neurological function [42]. Other data showed the relation between therapeutic effects of H2 and matrix metalloproteinase-3 (MMP-3) [43]. It has been found that a 5-day infusion of hydrogen-enriched saline improved disease activity in patients with rheumatoid arthritis, and the positive effects of hydrogen were confirmed by the reduction of plasma MMP-3 levels.

Conclusion

Matrix metalloproteinases (MMPs) are enzymes with an essential role in processes associated with remodeling extracellular matrix (ECM). Dysregulation of their activities has a crucial role in pathophysiology of several diseases and is associated with disturbed ECM turnover, tissue damage, and negative impact on cell–cell interactions. Several lines of evidence indicate that MMPs could represent potential targets of molecular hydrogen action. In our studies, we demonstrated the potential role of MMP-2 and MMP-28 in development of myocardial pathological changes induced by mediastinal irradiation (MI) of rats and confirmed beneficial effects of molecular hydrogen against MI-induced changes in MMP-2 and MMP-28. The observed data also indicates the potential role of molecular hydrogen in protection against post-transplant complications through inhibition of MMP-9 activities. In summaray, current information indicates that modulation of matrix metalloproteinases using molecular hydrogen could be promising strategy for treatment of several diseases, in which oxidative stress and tissue remodeling is an important partner.

Acknowledgements This work was funded by VEGA SR grant no. 2/0179/21 and grants of Agency for Research and Development APVV-18-0548 and APVV-19-0317, and project of EU Structural Funds ITMS 2001+: 313011AVG3.

References

1. Ohsawa I, Ishikawa M, Takahashi K et al (2007) Hydrogen acts as a therapeutic antioxidant by selectively reducing cytotoxic oxygen radicals. Nat Med 13:688–694. https://doi.org/10.1038/nm1577
2. Fukuda K, Asoh S, Ishikawa M et al (2007) Inhalation of hydrogen gas suppresses hepatic injury caused by ischemia/reperfusion through reducing oxidative stress. Biochem Biophys Res Commun 361:670–674. https://doi.org/10.1016/j.bbrc.2007.07.088
3. Huang C-S, Kawamura T, Toyoda Y, Nakao A (2010) Recent advances in hydrogen research as a therapeutic medical gas. Free Radic Res 44:971–982. https://doi.org/10.3109/10715762.2010.500328
4. Zhang Y, Long Z, Xu J et al (2017) Hydrogen inhibits isoproterenol-induced autophagy in cardiomyocytes in vitro and in vivo. Mol Med Rep 16:8253–8258. https://doi.org/10.3892/mmr.2017.7601
5. Zálešák M, Kura B, Graban J et al (2017) Molecular hydrogen potentiates beneficial anti-infarct effect of hypoxic postconditioning in isolated rat hearts: a novel cardioprotective intervention. Can J Physiol Pharmacol 95:888–893. https://doi.org/10.1139/cjpp-2016-0693
6. Bai X, Liu S, Yuan L et al (2016) Hydrogen-rich saline mediates neuroprotection through the regulation of endoplasmic reticulum stress and autophagy under hypoxia-ischemia neonatal brain injury in mice. Brain Res 1646:410–417. https://doi.org/10.1016/j.brainres.2016.06.020
7. Yang L, Guo Y, Fan X et al (2020) Amelioration of coagulation disorders and inflammation by hydrogen-rich solution reduces intestinal ischemia/reperfusion injury in rats through NF-κ B/NLRP3 pathway. Mediators Inflamm 2020:1–12. https://doi.org/10.1155/2020/4359305
8. LeBaron T, Singh R, Fatima G et al (2020) The effects of 24-week, high-concentration hydrogen-rich water on body composition, blood lipid profiles and inflammation biomarkers in men and women with metabolic syndrome: a randomized controlled trial. Diabetes, Metab Syndr Obes Targets Ther 13:889–896. https://doi.org/10.2147/DMSO.S240122
9. Ming Y, Ma Q, Han X, Li H (2020) Molecular hydrogen improves type 2 diabetes through inhibiting oxidative stress. Exp Ther Med 20:359–366. https://doi.org/10.3892/etm.2020.8708
10. Guan P, Sun Z-M, Luo L-F et al (2019) Hydrogen protects against chronic intermittent hypoxia induced renal dysfunction by promoting autophagy and alleviating apoptosis. Life Sci 225:46–54. https://doi.org/10.1016/j.lfs.2019.04.005
11. Meng J, Liu L, Wang D et al (2020) Hydrogen gas represses the progression of lung cancer via down-regulating CD47. Biosci Rep 40. https://doi.org/10.1042/BSR20192761
12. Kawamura T, Wakabayashi N, Shigemura N et al (2013) Hydrogen gas reduces hyperoxic lung injury via the Nrf2 pathway in vivo. Am J Physiol Cell Mol Physiol 304:L646–L656. https://doi.org/10.1152/ajplung.00164.2012
13. Chen H, Xie K, Han H et al (2015) Molecular hydrogen protects mice against polymicrobial sepsis by ameliorating endothelial dysfunction via an Nrf2/HO-1 signaling pathway. Int Immunopharmacol 28:643–654. https://doi.org/10.1016/j.intimp.2015.07.034
14. Yang C, Wei X-Q, Zheng J et al (2023) A correlative study between IVIM-DWI parameters and VEGF and MMPs expression in hepatocellular carcinoma. Quant Imaging Med Surg 13:1887–1898. https://doi.org/10.21037/qims-22-271
15. Yang D, Su L, Li X et al (2023) Evidence that enolase-phosphatase 1 exacerbates early cerebral ischemia injury and blood–brain barrier breakdown by enhancing extracellular matrix destruction and inhibiting the interaction between ADI1 and MT1-MMP. Exp Neurol 365:114410. https://doi.org/10.1016/j.expneurol.2023.114410
16. Hong O-Y, Jang H-Y, Lee Y-R et al (2022) Inhibition of cell invasion and migration by targeting matrix metalloproteinase-9 expression via sirtuin 6 silencing in human breast cancer cells. Sci Rep 12:12125. https://doi.org/10.1038/s41598-022-16405-x
17. Wu J, Han W, Chen X et al (2017) Matrix metalloproteinase-2 and −9 contribute to functional integrity and noise-induced damage to the blood-labyrinth-barrier. Mol Med Rep 16:1731–1738. https://doi.org/10.3892/mmr.2017.6784

18. Ren C, Li N, Wang B et al (2015) Limb ischemic perconditioning attenuates blood-brain barrier disruption by inhibiting activity of MMP-9 and occludin degradation after focal cerebral ischemia. Aging Dis 6:406. https://doi.org/10.14336/AD.2015.0812
19. Viappiani S, Nicolescu AC, Holt A et al (2009) Activation and modulation of 72kDa matrix metalloproteinase-2 by peroxynitrite and glutathione. Biochem Pharmacol 77:826–834. https://doi.org/10.1016/j.bcp.2008.11.004
20. Banaszkiewicz M, Krzywonos-Zawadzka A, Olejnik A, Bil-Lula I (2020) Tissue expression of atrial and ventricular myosin light chains in the mechanism of adaptation to oxidative stress. Int J Mol Sci 21:8384. https://doi.org/10.3390/ijms21218384
21. Klionsky DJ, Abdelmohsen K, Abe A et al (2016) Guidelines for the use and interpretation of assays for monitoring autophagy, 3rd edn. Autophagy 12. https://doi.org/10.1080/15548627.2015.1100356
22. Spinale FG (2007) Myocardial matrix remodeling and the matrix metalloproteinases: influence on cardiac form and function. Physiol Rev 87:1285–1342. https://doi.org/10.1152/physrev.00012.2007
23. Krebber MM, van Dijk CGM, Vernooij RWM et al (2020) Matrix metalloproteinases and tissue inhibitors of metalloproteinases in extracellular matrix remodeling during left ventricular diastolic dysfunction and heart failure with preserved ejection fraction: a systematic review and meta-analysis. Int J Mol Sci 21:6742. https://doi.org/10.3390/ijms21186742
24. Wang W, Schulze CJ, Suarez-Pinzon WL et al (2002) Intracellular action of matrix metalloproteinase-2 accounts for acute myocardial ischemia and reperfusion injury. Circulation 106:1543–1549. https://doi.org/10.1161/01.CIR.0000028818.33488.7B
25. Altieri P, Brunelli C, Garibaldi S et al (2003) Metalloproteinases 2 and 9 are increased in plasma of patients with heart failure. Eur J Clin Invest 33:648–656. https://doi.org/10.1046/j.1365-2362.2003.01187.x
26. Flamant M, Placier S, Dubroca C et al (2007) Role of matrix metalloproteinases in early hypertensive vascular remodeling. Hypertension 50:212–218. https://doi.org/10.1161/HYPERTENSIONAHA.107.089631
27. Barteková M, Šimončíková P, Fogarassyová M et al (2015) Quercetin improves postischemic recovery of heart function in doxorubicin-treated rats and prevents doxorubicin-induced matrix metalloproteinase-2 activation and apoptosis induction. Int J Mol Sci 16:8168–8185. https://doi.org/10.3390/ijms16048168
28. Boťanská B, Barteková M, Ferenczyová K et al (2021) Matrix metalloproteinases and their role in mechanisms underlying effects of quercetin on heart function in aged Zucker diabetic fatty rats. Int J Mol Sci 22:4457. https://doi.org/10.3390/ijms22094457
29. Camp TM, Tyagi SC, Senior RM et al (2003) Gelatinase B(MMP-9) an apoptotic factor in diabetic transgenic mice. Diabetologia 46:1438–1445. https://doi.org/10.1007/s00125-003-1200-y
30. Trocme C, Deffert C, Cachat J et al (2015) Macrophage-specific <scp>NOX2</scp> contributes to the development of lung emphysema through modulation of <scp>SIRT1</scp> / <scp>MMP</scp> -9 pathways. J Pathol 235:65–78. https://doi.org/10.1002/path.4423
31. Tejima E, Zhao B-Q, Tsuji K et al (2007) Astrocytic Induction of Matrix Metalloproteinase-9 and Edema in Brain Hemorrhage. J Cereb Blood Flow Metab 27:460–468. https://doi.org/10.1038/sj.jcbfm.9600354
32. Kuhad A, Singh P, Chopra K (2015) Matrix metalloproteinases: potential therapeutic target for diabetic neuropathic pain. Expert Opin Ther Targets 19:177–185. https://doi.org/10.1517/14728222.2014.960844
33. Hanaoka T, Kamimura N, Yokota T et al (2011) Molecular hydrogen protects chondrocytes from oxidative stress and indirectly alters gene expressions through reducing peroxynitrite derived from nitric oxide. Med Gas Res 1:18. https://doi.org/10.1186/2045-9912-1-18
34. Chen CH, Manaenko A, Zhan Y et al (2010) Hydrogen gas reduced acute hyperglycemia-enhanced hemorrhagic transformation in a focal ischemia rat model. Neuroscience 169:402–414. https://doi.org/10.1016/j.neuroscience.2010.04.043

35. Ma Y, Chiao YA, Zhang J et al (2012) Matrix Metalloproteinase-28 Deletion Amplifies Inflammatory and Extracellular Matrix Responses to Cardiac Aging. Microsc Microanal 18:81–90. https://doi.org/10.1017/S1431927611012220
36. Kowluru RA, Kanwar M (2009) Oxidative stress and the development of diabetic retinopathy: Contributory role of matrix metalloproteinase-2. Free Radic Biol Med 46:1677–1685. https://doi.org/10.1016/j.freeradbiomed.2009.03.024
37. Liu L, Shi Q, Liu X et al (2022) Attenuation of Myocardial Fibrosis Using Molecular Hydrogen by Inhibiting the TGF-β Signaling Pathway in Spontaneous Hypertensive Rats. Am J Hypertens 35:156–163. https://doi.org/10.1093/ajh/hpab159
38. Sun Q, Kawamura T, Masutani K et al (2012) Oral intake of hydrogen-rich water inhibits intimal hyperplasia in arterialized vein grafts in rats. Cardiovasc Res 94:144–153. https://doi.org/10.1093/cvr/cvs024
39. Zhu Z, Yu J, Lin W et al (2019) Molecular hydrogen accelerates the reversal of acute obstructive cholangitis-induced liver dysfunction by restoring gap and tight junctions. Mol Med Rep. https://doi.org/10.3892/mmr.2019.10179
40. Schäfer M, Ivy DD, Nguyen K et al (2021) Metalloproteinases and their inhibitors are associated with pulmonary arterial stiffness and ventricular function in pediatric pulmonary hypertension. Am J Physiol Circ Physiol 321:H242–H252. https://doi.org/10.1152/ajpheart.00750.2020
41. Sokal A, Zembala M, Radomski A et al (2009) A differential release of matrix metalloproteinases 9 and 2 during coronary artery bypass grafting and off-pump coronary artery bypass surgery. J Thorac Cardiovasc Surg 137:1218–1224. https://doi.org/10.1016/j.jtcvs.2008.11.004
42. Takeuchi S, Nagatani K, Otani N et al (2015) Hydrogen improves neurological function through attenuation of blood–brain barrier disruption in spontaneously hypertensive stroke-prone rats. BMC Neurosci 16:22. https://doi.org/10.1186/s12868-015-0165-3
43. Ishibashi T, Sato B, Shibata S et al (2014) Therapeutic efficacy of infused molecular hydrogen in saline on rheumatoid arthritis: A randomized, double-blind, placebo-controlled pilot study. Int Immunopharmacol 21:468–473. https://doi.org/10.1016/j.intimp.2014.06.001

Chapter 12
Perioperative Mitigation of Oxidative Stress with Molecular Hydrogen During Simulated Heart Transplantation in Pigs

Branislav Kura, Barbara Szeiffova Bacova, Miroslav Barancik, Matus Sykora, Ludmila Okruhlicova, Narcisa Tribulova, Roberto Bolli, Barbora Kalocayova, Tyler W. LeBaron, Katarina Andelova, and Jan Slezak

Abstract Heart transplantation is now a routine method for severe heart failure treatment. It is critical to focus on preventing ischemia–reperfusion damage and mitigating oxidative stress to achieve successful outcomes. However, prolonged anesthesia, hyperoxia, and defibrillations contribute to an increase of ROS/RNS and disrupt the redox homeostasis, which poses a serious risk factor. Numerous

---

B. Kura · B. S. Bacova · M. Barancik · M. Sykora · L. Okruhlicova · N. Tribulova · B. Kalocayova · K. Andelova · J. Slezak (✉)
Centre of Experimental Medicine, Institute for Heart Research, Slovak Academy of Sciences, Dúbravská Cesta 9, Bratislava 84104, Slovak Republic
e-mail: jan.slezak@savba.sk

B. Kura
e-mail: branislav.kura@savba.sk

B. S. Bacova
e-mail: barbara.bacova@savba.sk

M. Barancik
e-mail: miroslav.barancik@savba.sk

M. Sykora
e-mail: matus.sykora@savba.sk

L. Okruhlicova
e-mail: ludmila.okruhlicova@savba.sk

N. Tribulova
e-mail: narcisa.tribulova@savba.sk

B. Kalocayova
e-mail: barbora.kalocayova@savba.sk

K. Andelova
e-mail: katarina.andelova@savba.sk

R. Bolli
Institute of Molecular Cardiology, University of Louisville, Louisville, KY 40202, USA
e-mail: roberto.bolli@louisville.edu

J. Slezak and B. Kura (eds.), *Molecular Hydrogen in Health and Disease*, Advances in Biochemistry in Health and Disease 27, https://doi.org/10.1007/978-3-031-47375-3_12

publications have confirmed the remarkable antioxidant, anti-apoptotic, and anti-inflammatory properties of molecular hydrogen. In our simulated heart transplantation experiment, we demonstrate that administering 2% hydrogen gas during anesthesia and extracorporeal circulation (ECC) significantly alleviates oxidative stress-induced damage. This is evidenced by a significant decrease in markers of ischemia, lipid peroxidation, and inflammation. The restoration of the pumping activity in the implanted pig hearts showed improvement, with a reduced need for repeated defibrillations. The administration of H_2 during graft collection and transplantation significantly enhances the function of the transplanted heart and the overall condition of the recipient. Hydrogen administered by conventional ventilators and ECC oxygenators represents an innovative therapy that can significantly improve current transplantation techniques.

Keywords Heart transplantation · Pigs · Ischemia/reperfusion injury · Oxidative stress mitigation · Inflammation · Molecular hydrogen inhalation · Selective antioxidant

Introduction

Heart transplantation has become a routine treatment for end-stage heart failure. However, achieving full functionality of the implanted graft after reperfusion and warming often requires repeated electrical defibrillation shocks. The critical transient period of ventricular fibrillation is mainly attributed to the increased formation of free oxygen/nitrogen radicals during cumulative action of cardiac ischemia/reperfusion (I/R), anesthesia, and defibrillation after cold asystole storage. I/R injury and oxidative stress have been identified as the primary risk factors for peri- and post-operative adverse events [1]

Ischemia and reperfusion: Mitochondria play an essential physiological role in metabolism by supplying energy for optimal cellular function [2]. Altered cardiac ion homeostasis and structural remodeling are associated with elevated reactive oxygen species (ROS), and subsequent oxidative and metabolic stress [3]. During ischemia, the mitochondrial oxidative chain has been impaired and after reperfusion ROS increases dramatically [4, 5].

The gold standard for graft preservation is cardioplegia with cold storage [6, 7]. The quality of the cardiac graft is directly proportional to the cold ischemic time,

T. W. LeBaron
Molecular Hydrogen Institute, Enoch, UT 84721, USA

Department of Kinesiology and Outdoor Recreation, Southern Utah University, Cedar City, UT 84720, USA

T. W. LeBaron
e-mail: tylerlebaron@suu.edu

which in ideal conditions should not exceed 5–6 h [8]. To get a good quality heart from a donor, fast cardioplegic arrest remains the gold standard of cardioprotection [1]. Hypothermic cardioplegia requires immediate and sustained electromechanical quiescence, rapid myocardial cooling, and the addition of certain therapeutic additives.

The effect of *anesthesia* on oxidative stress is controversial. According to some authors, volatile anesthetics like isoflurane and sevoflurane may reveal promising effects on relieving ischemia via regulating apoptosis-related genes, which in turn reduce apoptosis [5, 9]. However, prolonged exposure to sevoflurane (2%) may lead to increased DNA damage and decreased glutathione contents [10].

During anesthesia, excess O_2 or *hyperoxia*, is known to be harmful. Toxicity arises from the enhanced formation of ROS that, when it exceeds the antioxidant defense, generates oxidative stress. High O_2 administration intraoperatively may increase harm, possibly through increased oxidative damage and inflammation, resulting in more complications and worse outcomes [11].

Repeated *defibrillations* can have adverse effects on the heart and may complicate the success of transplantation. Strong defibrillation shocks can cause temporary or permanent damage to the heart. It is worth noting that weak defibrillation shocks do not cause any harm to the heart but also fail to defibrillate. Electric shocks of sufficient strength to defibrillate immediately cause significant cellular destruction in the ventricular myocardium, major abnormalities in calcium regulation, oxidative stress, and the activation of pro-inflammatory and pro-fibrotic processes, which contribute to contractile dysfunction [12].

Effects of Molecular Hydrogen

Molecular hydrogen (H_2) appears to be an ideal selective scavenger of ROS, especially for the highly reactive hydroxyl and nitrosyl radicals. Thus, H_2 may help prevent long periods of dangerous fibrillation requiring defibrillation. Use of H_2 gas inhalation has been reported to be an effective protector in heart transplantation [13–15] and it has been shown to have various biological benefits, including suppression of I/R injury in animal studies [16–19].

Our hypothesis is that administering H_2 gas can mitigate I/R injury and other sources of oxidative stress by selectively scavenging hydroxyl/nitrosyl radicals, reducing lipid peroxidation and inflammation, and increasing native antioxidant enzymes. This, in turn, can lead to improved outcomes with fewer complications. In this paper, we present an innovative method for mitigating oxidative stress, which is a primary factor influencing transplantation outcomes. We utilized a simulated transplantation model in pigs to eliminate the confounding effects of immune reactions, denervation, and the time and trauma associated with graft implantation, allowing us to focus primarily on the issue of oxidative stress and its suppression.

Material and Methods

Female pigs (Sus scrofa f. domestica) aged 16–17 weeks weighing 55–60 kg and regarded to be sexually mature were used in the experiment. The animals were cared for in accordance with the law pursuant to § 16 of the Act of the Czech National Council No. 246/1992 Sb. on the protection of animals from cruelty, as amended. After pre-medication total anesthesia (propofol i.v. fentanyl i.v., inhaled isoflurane, $\pm H_2$) was used.

Experimental animals were divided into three groups:

1. Control group (C) where markers were measured from blood plasma collected before hydrogen administration.
2. Transplantation group (T) with simulated heart transplantation,
3. Transplantation with hydrogen-rich air ($T + H_2$) (4% of hydrogen in air + > 40–50% of oxygen) in anesthesia administered during the whole experiment.

Simulated transplantation started with a simulated heart extraction, where the heart was completely isolated from the body (except nerves), but it was not severed off and removed. The heart remained in the chest, stored in a crushed ice of saline solution at 4 °C. The venae cavae and pulmonary veins were occluded. The ascending aorta was occluded, and a cardioplegic solution, Custodiol 1000 mL cooled to 4 °C, was administered (cold ischemic arrest storage) and cardiopulmonary bypass (CPB) was connected. For experimental purposes, simulated transplantation (orthotopic autotransplantation) has a number of advantages like minimization of traumatization, elimination of antigenicity, and reduction of sowing time during graft implantation.

The concentration in arterial blood of H_2 gas administered with anesthesia was measured during the whole time of the experiment. All presented results were normalized by subtracting the blank values. The first measurement of the concentration of administered H_2 gas was performed immediately after the pigs' pre-medication (time 0). During anesthesia, the concentration of H_2 in arterial blood was measured and recorded every 5 min. All measurements were performed by Unisense H_2 micro-probe (Unisense, A/S Aarhus, Denmark).

After three hours of cold storage, reperfusion of the heart with warm blood was initiated. Once the temperature reached 30–33 °C, the heart began contracting. If necessary, the heart was defibrillated. When the heart reached a pressure of around 80 mm/Hg and a pulse of around 100/min, and once these parameters were maintained for 60 min, the last blood and tissue samples were taken, and the experiment was terminated.

Samples were analyzed for.

- Markers of inflammation and tissue damage,
- Markers of oxidative stress damage,
- Protein expression levels of total connexin 43,
- Activity of matrix metalloproteinases 2 and 9.

Histochemistry of hypoxic/ischemic injury by sensitive enzymes assessment was performed, and electron microscopy of cardiac tissue was studied.

Statistical Analysis

Experimental data were expressed as means ± SD. Statistical analysis was performed using one-way two-tailed ANOVA test, followed by Bonferroni post hoc analysis (GraphPad Prism 7). A probability value less than 0.05 was considered to reflect statistically significant differences.

Results

Hydrogen Measurement

During anesthesia, the concentration of H_2 in arterial blood was measured and recorded every 5 min. The highest value 0.18 μM was detected during the anesthesia before connection to ECC. In the ECC, molecular hydrogen was administered directly to the ECC machine oxygenator, where oxygen gas (40–50%) was mixed with 4% of H_2 in air during the whole 180 min of the heart arrest reducing the administered H_2 concentration to approximately 2%. At this time, the concentration of H_2 in the arterial blood decreased to 0.169 μM after 30 min and to 0.144 μM after 180 min of the heart arrest (Fig. 12.1). The reperfusion period lasted 60 min after disconnecting the pig from the ECC machine. For this period, the concentration of H_2 in the arterial blood decreased to 0.103 μM (Fig. 12.1).

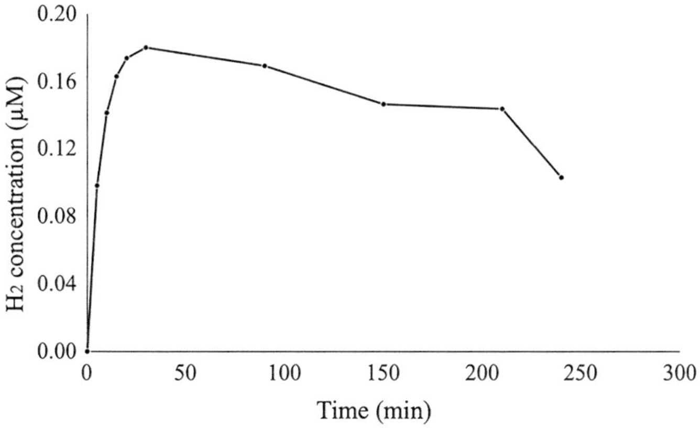

Fig. 12.1 Concentration of arterial blood of pigs throughout the entire experiment. (n = 11)

Table 12.1 Tissue damage and inflammation markers

Parameter	Control	Transplantation	Transplantation with H_2 treatment
Creatine kinase	79.86 ± 88.83	156.33 ± 71.41	108.08 ± 69
Myoglobin	6 ± 0.02	6.17 ± 0.32	5.92 ± 0.07
Hs-Troponin	54.25 ± 5.87	2488 ± 1964.16	579.8 ± 207.46
LDH	7.81 ± 1.85	22.06 ± 4.59****	14.85 ± 4.47****
NF-κB	100 ± 7.67	168.86 ± 30.94*	$103.14 \pm 4.24^{\#}$
TNF-α	100 ± 7.95	110.1 ± 17.43	$87.19 \pm 7.94^{\#\#}$
MPO	100 ± 3.14	104.49 ± 8.67	$86.42 \pm 6.88^{\#}$

$^{*}p \leq 0.05$; $^{****}p \leq 0.0001$; $^{\#}p \leq 0.05$; $^{\#\#}p \leq 0.01$

Tissue Damage and Inflammation Markers

Markers of tissue damage, including creatine kinase, myoglobin, and hs-troponin, showed a non-significant increase in value after the transplant without H_2 treatment. However, with the addition of H_2, there was a non-significant decrease in these markers, and in the case of myoglobin, the values even approached the control levels.

Lactate dehydrogenase (LDH) activity in the blood plasma of experimental pigs significantly increased in the transplantation group, while a significant decrease was observed in the H_2-treated group.

Levels of nuclear factor κ B (NF-κB) and tumor necrosis factor α (TNF-α) in the left ventricle tissue significantly increased in the transplantation group, but a significant decrease was found in the H2-treated groups.

Levels of myeloperoxidase (MPO) from the left ventricle tissue of experimental pigs showed a non-significant increase in the transplantation group compared to the control and H_2-treated groups (Table 12.1).

Oxidative Stress

Levels of SOD activity in blood plasma showed a statistically significant decrease in the H_2-treated group. However, levels of SOD1 protein in the left ventricle tissue demonstrated that transplantation caused a significant upregulation, while H_2 treatment normalized SOD1 protein values.

Levels of 8-hydroxy-2′-deoxyguanosine in the blood plasma showed a significant increase after simulated heart transplantation, but a significant reduction was observed after H_2 treatment.

Glutathione peroxidase (GPx) activity in the blood plasma of experimental pigs significantly decreased in the H_2-treated group. Levels of malondialdehyde (MDA) in the blood plasma of experimental pigs decreased significantly after H_2 treatment.

Table 12.2 Markers of oxidative stress

Parameter	Control	Transplantation	Transplantation with H_2 treatment
SOD activity	78.31 ± 5.7	86.73 ± 2.02*	77.43 ± 1.33#
8-hydroxy-2´-deoxyguanosine	959.75 ± 213.93	1728.28 ± 347.81***	1266.37 ± 170.3#
GPx activity	1384.81 ± 240.14	2643.35 ± 1196.66*	1400.41 ± 275.96#
MDA	7.07 ± 1.99	18.39 ± 5.39****	9.78 ± 0.95##
UA	14.18 ± 3.84	26.6 ± 8.32***	16.67 ± 3.56#
CAT	1.22 ± 0.56	13.97 ± 0.0	12.45 ± 3.0
Keap1	100 ± 0.32	98.73 ± 4.83	98.08 ± 8.59
Nrf2	100 ± 1.98	122.37 ± 24.23	102.54 ± 44.23
SOD1	100.5 ± 3.6	142.18 ± 14.15***	114.34 ± 8.3##

$^{*}p \leq 0.05$; $^{\#}p \leq 0.05$; *** $p \leq 0.0005$; ****$p \leq 0.0001$ $^{\#\#}p \leq 0.01$

Uric acid levels in the blood plasma showed a statistically significant elevation after transplantation, followed by almost normalization of values after H_2 treatment.

Catalase activity levels measured in blood plasma showed a statistically significant increase in activity in the transplantation group and a slight decrease after H_2 treatment.

Levels of Kelch-like ECH-associated protein 1 (Keap1) and nuclear factor-erythroid factor 2-related factor 2 (Nrf2) in the left ventricle tissue of experimental pigs did not change after H_2 treatment (Table 12.2).

Matrix Metalloproteinases 2 and 9

Levels of MMP2 activity in the blood plasma of experimental pigs showed a non-significant increasing tendency in the transplantation group with no H_2 treatment, whereas their normalization effect of H_2 treatment. Levels of MMP9 activity in the blood plasma significantly increased by simulated transplantation of the heart and a less significant increase was found in the activity after H_2 treatment.

Levels of total connexin (Cx)43 in the left ventricle tissue of experimental pigs significantly increased after H_2 treatment. However, levels of phosphorylated Cx43 and PKCε in the left ventricle tissue of experimental pigs practically did not change (Table 12.3).

Histochemistry Results

Histochemistry results of glycogen-dependent phosphorylase, a sensitive marker of ischemic injury, demonstrated normal homogenous staining in the control

Table 12.3 Connexin43, PKCε, and matrix metalloproteinases 2 and 9

Parameter	Control	Transplantation	Transplantation with H_2 treatment
MMP2	69.24 ± 15.83	88.06 ± 5.83	72.01 ± 12.85
MMP9	63.4 ± 15.38	141.75 ± 25.23****	127.97 ± 20.6****
Total Cx43	100 ± 2.46	109.53 ± 3.99**	112.85 ± 5.09***
pCx43	100 ± 3.01	106.29 ± 9.93	105.43 ± 10.93
PKCe	100 ± 8.02	95.75 ± 3.89	93.58 ± 4.2

$p \leq 0.01$; *** $p \leq 0.0005$; * *$p \leq 0.0001$

myocardium (Fig. 12.2A). The transplanted heart showed areas with lower activity and patchy staining (Fig. 12.2B). However, the myocardium of the transplanted heart with H_2 treatment exhibited almost normal and homogenous staining (Fig. 12.2C).

The activity of mitochondrial succinate dehydrogenase was high regardless of H_2 administration. However, compared to the normal control myocardium with homogenous staining (Fig. 12.2D), the myocardium of transplanted and defibrillated heart showed areas of hyper contractions and increased formazan granules (Fig. 12.2E) [20]. Mitochondrial injury was less pronounced in the heart tissue upon treatment with H_2, and the myocardium of the transplanted heart with hydrogen treatment shows almost normal staining (Fig. 12.2F).

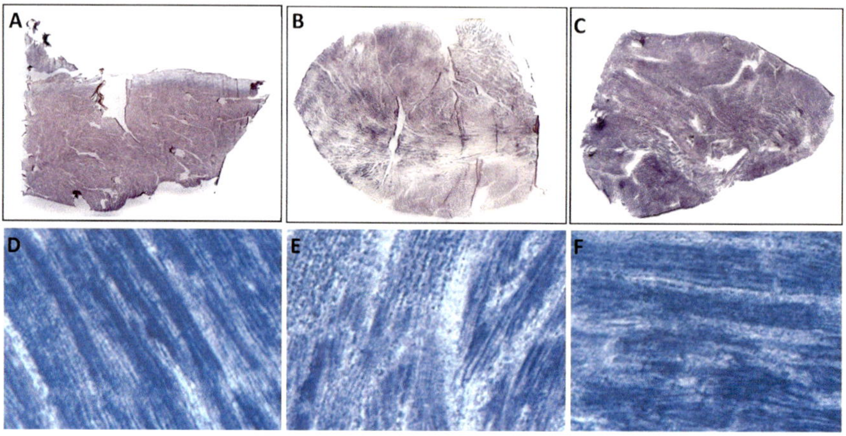

Fig. 12.2 Histochemistry of the heart tissues. Histochemistry of representative glycogen-dependent phosphorylase (sensitive marker of ischemic injury) stained sections. A. normal control myocardium shows homogenous staining indicating normal enzyme activity, B. myocardium of transplanted heart with patchy staining, C. Myocardium of transplanted heart with hydrogen treatment shows almost normal staining

Electron Microscopy

Electron-microscopic examination in our experiments focused on the midmyocardial parts of the left ventricle. Comparison of the individual layers of the left ventricle showed a certain heterogeneity in the degree of damage, where we found the greatest damage in subendocardial layer. We focused on midmyocardial mass because the middle mass of myocardium plays the main role in heart function. The representative results of tissue samples taken after the end of the experiment, (i.e. after 3 h of cardioplegic hypothermic storage, and 1 h of reperfusion ending with normal functioning without fibrillation and reaching a pressure of 80 mm Hg and a pulse of about 100 beats per min.) showed that, after H_2 administration (Fig. 12.3B), there is significantly less manifestations of damage to the mitochondria, endothelium of capillaries and sarcolemma, as well as fewer sarcomere contractures compared to the transplantation group without the use of H_2 (Fig. 12.3A). These findings may be a manifestation, not only of better management of oxidative stress, but also of less frequent use of defibrillation in experiments where H_2 has been used.

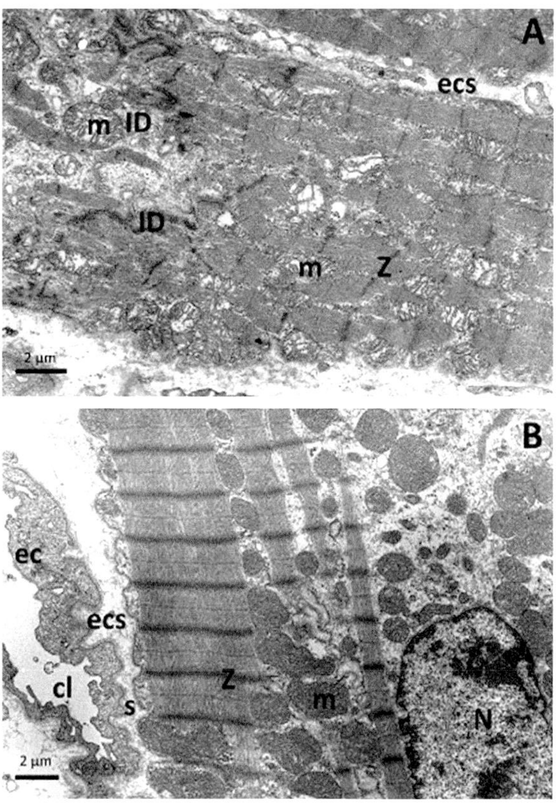

Fig. 12.3 Electron microscopy of the heart tissues. Electron-microscopic examination of experiments focused on the midmyocardial part of the left ventricle myocardium. (A) myocardium of transplanted heart with numerous changes, mainly mitochondrial (m) but also degenerative changes of sarcomeres, intercalated disks (ID), sarcolemma, and endothelial cells of capillaries. (B) Myocardium of transplanted heart with hydrogen treatment shows almost normal ultrastructure

Discussion

Despite the progress in the technique of collection and storage of the heart before its implantation, heart transplantation is always associated with tension and uncertainty when taking over its physiological function (57).

To protect the heart from I/R injury, the multi-target hypothesis for activation of survival pathways (Survivor Activating Factor Enhancement -RISK, Survivor Activating Factor Enhancement –SAFE, and NO/PKG,) and inhibition of deleterious pathways (inflammation, mito/nuclear DNA and DAMPS/PAMPS) was proposed [21, 22].

Since previous methods of cardioprotection focused on improving the potency of the RISK, SAFE and other pathways, our intention is to point out the proven danger of increased ROS formation and particularly of oxidative stress, and to look for its mitigation, which gives the cell its best possible chance of survival [21].

In addition to various factors influencing the critical stage of heart reperfusion after transplantation and the early postoperative course, the excessive production of reactive oxygen and nitrogen species is a key factor that can negatively impact a smooth transplantation process.

Heart transplantation is a complex process that involves reactions at the cellular, organ, and whole-body levels. One common factor in this process, among others, is oxidative stress. It is important to consider not only the condition of the graft but also the overall impact of oxidative stress on the entire organism. Factors such as anesthesia, hyperoxia, and extracorporeal circulation (ECC) contribute to the oxidative stress experienced by the body. However, the administration of free antioxidants has proven to be ineffective (18).

Many reactive oxygen species (ROS) are necessary for normal physiological processes. Therefore, hydrogen gas (H_2), known for its relative inertness, selectively scavenges aggressive hydroxyl and nitrosyl radicals, making it a promising candidate for mitigating functional and structural abnormalities [23].

Currently, there is no consensus on what the optimal concentration of inhaled H_2 should be. For safety considerations, it has been recommended that the concentration of inhaled H_2 ought to be monitored and kept less than 4.6% when mixed with air and less than 4.1% when mixed with oxygen [24]. In our experiments, we used 4% H_2 mixed with air, which was used for inhalation through an anesthesiology apparatus where 40–50% O_2 was added and the same concentration of H_2 was used in the ECC oxygenator mixed with 40%-50% O_2. This would reduce the concentration of H_2 to an effective administered dose closer to 2% A similar concentration of H_2 was used for inhalation by authors for other treatments [25, 26].

During the entire experiment, we managed to maintain the level of H_2 in the arterial blood between 0.1 and 0.18 μM This was achieved by inhalation of H_2 during narcosis and oxygenation of the blood with H_2 and in the ECC oxygenator. However, according to Henry's Law, a 2% H_2 concentration should provide a plasma concentration of 14.2 μM. Thus, there is discrepancy between our measured results and the expected value by nearly a factor of 10x. This may be due to systematic errors in the instrumentation/methodology, or other factors. This will be investigated in future research. The mentioned method of administration of $\approx 2\%$ H_2 with 40–50% O_2 ensures that the arterial blood contains a sufficient amount of H_2 to reduce the oxidative stress of the organism.

In our experiments, we selected markers of graft damage and oxidative stress of the whole organism. After the application of hydrogen, these markers approached normal levels, indicating its positive effect in protecting both the graft and the entire organism. This preventive effect optimized the heart's function and the post-implantation period.

Simulated transplantation without H_2 treatment significantly increased the levels of troponin and creatine kinase, indicating myocardial cell death or damage. Both markers decreased after H_2 treatment. The tissue damage marker lactate dehydrogenase (LDH) in the blood plasma significantly increased after transplantation without H_2 but decreased significantly after H_2 treatment, indicating an improvement in the organism's condition. Inflammatory markers nuclear factor-kappa B (NF-κB) and tumor necrosis factor-alpha (TNF-α) increased after transplantation but significantly decreased in hearts treated with hydrogen, suggesting improved conditions shortly after graft function restoration [27].

Markers of oxidative stress in plasma were also measured. It was observed that levels of malondialdehyde (marker of lipid peroxidation), 8-hydroxy-2'-deoxyguanosine 8-OHdG (marker of DNA damage), glutathione peroxidase, uric acid, and catalase, all significantly decreased in the group with hydrogen, compared to transplantation without hydrogen. This indicates an improvement in the overall redox condition of the entire organism when H_2 is applied [23].

It is reported that uric acid (UA) contributes to $> 50\%$ of the antioxidant capacity of the blood (28). However, there is still no consensus if UA is protective or a risk factor. Nevertheless, it seems that acute elevation is a protective factor, whereas chronic elevation is a risk for disease [28]. Xie et al. [29] showed a protective effect in the prevention of renal injury and can inhibit renal fibrosis after an IR injury.

Superoxide dismutase-1 (SOD-1), (which catalyzes the conversion of superoxide into oxygen and hydrogen peroxide) in the left ventricle tissue in the group with H_2 decreased significantly compared to transplantation without H_2. Although both groups had elevated levels of SOD1 compared to control, the lower level of SOD1 in the H_2 group coupled with the lower levels of oxidative stress markers indicates an improvement in antioxidant status in the transplanted heart with H_2 [30].

Matrix metalloproteinases (MMPs) degrade and deposit both matrix and non-matrix proteins in the blood plasma of experimental pigs and indicate signs of a profibrotic situation [31]. MMP 2 showed a non-significant increase in tendency in the transplantation group with no H_2 treatment and subsequent normalization effect of hydrogen treatment. However, MMP9 activity in the blood plasma of experimental pigs showed a significant increase of the activity of MMP9 caused by simulated transplantation of the heart as a sign of profibrotic situation and a less significant increase of the activity after H_2 treatment [31].

Electrical activation is a key phenomenon in cardiac physiology. Action potential is transferred from one cardiomyocyte to another via the gap junction connexin

(Cx) channels and Cx43 is a prevalent cardiac isoform. The expression of Cx43, as well as its functional phosphorylated state, is highly dynamic and responsive to pathophysiological conditions. While down-regulation of Cx43 is deleterious to the heart function [32], up regulation may be beneficial. It was shown in our experiments that Cx43 is essential for the formation of heart structures and intercellular communication [33]. In present experiments, increased protein expression levels of total connexin 43 (TCx43) indicate increased intercellular communication and decreased tendency to arrhythmias after H_2 was added.

Glycogen phosphorylase, demonstrated on frozen sections of hearts after implantation and resuming normal function, is a very sensitive indicator of hypoxia/ischemia and, in fact, reflects the reduced amount of glycogen in the myocardium [34, 35]. Likewise, succinate dehydrogenase is not as sensitive as the previous reaction, but it reliably demonstrates mitochondrial damage, which is manifested by larger formazan granules. Based on our results, histochemical reactions clearly demonstrate that myocardial tissue is better protected in hydrogen-treated animals. Hypercontraction foci have also been found, but in addition to a disorder of calcium metabolism, they can also be caused by more frequent and repeated shocks during defibrillation [36].

Electron microscopy examination of heart tissue of our experiments focused on the subendocardial, midmyocardial, subepicardial, and apex parts of the left ventricle of the transplanted heart. The most pronounced changes were found in mitochondria, as well as in sarcomeres, intercalated discs (ID), cellular membranes, and capillaries. Samples taken from transplanted hearts with hydrogen treatment showed significantly improved ultrastructure. Our ultrastructural results are consistent with the results of other authors [37, 38].

Based on our presented results and in accordance with the selected results of other authors, it appears that the use of molecular hydrogen throughout the entire experiment has a protective effect and improves the outcome of the transplantation. This can be supported by the following observations: (i) Hydrogen alleviates transplantation-related ischemia–reperfusion injury; (ii) Hydrogen prevents hyperoxia and ROS formation during anesthesia; (iii) Hydrogen protects the heart from damage during defibrillation after transplantation; (iv) Hydrogen administration mitigates oxidative stress in heart transplantation, improving overall results; and (v) Addition of 4% H_2 during anesthesia decreases oxidative stress, ischemia markers, inflammation, and peroxidation. Collectively, H_2 administration led to improved cardiac morphology, reduced ventricular fibrillation severity, and better heart pumping activity after 3 h of cold ischemia and 1 h warm reperfusion in a pig heart.

Taken together, H_2 treatment may be regarded as a novel, efficient and comprehensive therapeutic strategy to successfully abrogate the effects of oxidative stress and can significantly improve the outcome of heart transplantation. Furthermore, these results provide a rationale for designing clinical trials aimed at testing the efficacy of this strategy. Specifically, H_2 administration to patients challenged by ischemia–reperfusion events involving the heart and other organs such as kidney or liver etc. in transplantation surgeries.

Competing Interest We thank the management of IKEM Prague for hiring operating rooms and transplant assistants. We thank the collaborating team of the Slovak Institute of Heart Diseases (Drs. M. Hulman, V. Hudec, J. Luptak, I-Olejarova, M. Ondrusek, I. Gasparovic, R.Sramaty) for performing simulated transplantation.

Funding This research was funded by grants from Slovak Research and Development Agency (APVV-0241–11, APVV-15–0376, APVV-19–0317), grant from the Slovak Academy of Sciences (VEGA 2/0092/22 and 2/0063/18), grant from European Union Structural funds (ITMS 26230120009), grant (2018/7838:1-26C0), and grant from Ministry of Health of The Slovak Republic (2019-CEMSAV-1). Authors declare no conflict of interests.

References

1. Uto K, Sakamoto S, Que W et al (2019) Hydrogen-rich solution attenuates cold ischemia-reperfusion injury in rat liver transplantation. BMC Gastroenterol 19:25. https://doi.org/10.1186/s12876-019-0939-7
2. Valko M, Leibfritz D, Moncol J et al (2007) Free radicals and antioxidants in normal physiological functions and human disease. Int J Biochem Cell Biol 39:44–84. https://doi.org/10.1016/j.biocel.2006.07.001
3. Jeong E-M, Liu M, Sturdy M et al (2012) Metabolic stress, reactive oxygen species, and arrhythmia. J Mol Cell Cardiol 52:454–463. https://doi.org/10.1016/j.yjmcc.2011.09.018
4. Murphy MP (2009) How mitochondria produce reactive oxygen species. Biochem J 417:1–13. https://doi.org/10.1042/BJ20081386
5. Suleiman M-S, Zacharowski K, Angelini GD (2008) Inflammatory response and cardioprotection during open-heart surgery: the importance of anaesthetics. Br J Pharmacol 153:21–33. https://doi.org/10.1038/sj.bjp.0707526
6. Krezdorn N, Tasigiorgos S, Wo L et al (2017) Tissue conservation for transplantation. Innov Surg Sci 2:171–187. https://doi.org/10.1515/iss-2017-0010
7. Shi S, Xue F (2016) Current antioxidant treatments in organ transplantation. Oxid Med Cell Longev 2016:1–9. https://doi.org/10.1155/2016/8678510
8. Hicks M, Hing A, Gao L et al (2006) Organ preservation. Methods Mol Biol 333:331–374. https://doi.org/10.1385/1-59745-049-9:331
9. Lee Y-M, Song BC, Yeum K-J (2015) Impact of volatile anesthetics on oxidative stress and inflammation. Biomed Res Int 2015:1–8. https://doi.org/10.1155/2015/242709
10. Alleva R, Tomasetti M, Solenghi MD et al (2003) Lymphocyte DNA damage precedes DNA repair or cell death after orthopaedic surgery under general anaesthesia. Mutagenesis 18:423–428. https://doi.org/10.1093/mutage/geg013
11. Oldman AH, Martin DS, Feelisch M et al (2021) Effects of perioperative oxygen concentration on oxidative stress in adult surgical patients: a systematic review. Br J Anaesth 126:622–632. https://doi.org/10.1016/j.bja.2020.09.050
12. Rogan F, Funston R, Meenan B, Burke G (2017) 209 Evaluation of acute cardiac damage in a porcine model of defibrillation. Heart 103:A139.1–A139. https://doi.org/10.1136/heartjnl-2017-311726.207
13. Tan M, Sun X, Guo L et al (2013) Hydrogen as additive of HTK solution fortifies myocardial preservation in grafts with prolonged cold ischemia. Int J Cardiol 167:383–390. https://doi.org/10.1016/j.ijcard.2011.12.109
14. Noda K, Shigemura N, Tanaka Y et al (2013) A novel method of preserving cardiac grafts using a hydrogen-rich water bath. J Hear Lung Transplant 32:241–250. https://doi.org/10.1016/j.healun.2012.11.004

15. Tao G, Song G, Qin S (2019) Molecular hydrogen: current knowledge on mechanism in alleviating free radical damage and diseases. Acta Biochim Biophys Sin (Shanghai) 51:1189–1197. https://doi.org/10.1093/abbs/gmz121
16. Ohsawa I, Ishikawa M, Takahashi K et al (2007) Hydrogen acts as a therapeutic antioxidant by selectively reducing cytotoxic oxygen radicals. Nat Med 13:688–694. https://doi.org/10.1038/nm1577
17. Hayashida K, Sano M, Ohsawa I et al (2008) Inhalation of hydrogen gas reduces infarct size in the rat model of myocardial ischemia–reperfusion injury. Biochem Biophys Res Commun 373:30–35. https://doi.org/10.1016/j.bbrc.2008.05.165
18. Hayashida K, Sano M, Kamimura N et al (2014) Hydrogen inhalation during normoxic resuscitation improves neurological outcome in a rat model of cardiac arrest independently of targeted temperature management. Circulation 130:2173–2180. https://doi.org/10.1161/CIRCULATIONAHA.114.011848
19. Matsuoka T, Suzuki M, Sano M et al (2017) Hydrogen gas inhalation inhibits progression to the "irreversible" stage of shock after severe hemorrhage in rats. J Trauma Acute Care Surg 83:469–475. https://doi.org/10.1097/TA.0000000000001620
20. Lojda Z, Gossrau R, Schiebler TH (1976) Enzym-histochemische Methoden, 1st ed. Springer-Verlag Berlin and Heidelberg GmbH & Co. K, Berlin
21. Rossello X, Yellon DM (2018) The RISK pathway and beyond. Basic Res Cardiol 113:2. https://doi.org/10.1007/s00395-017-0662-x
22. Hadebe N, Cour M, Lecour S (2018) The SAFE pathway for cardioprotection: is this a promising target? Basic Res Cardiol 113:9. https://doi.org/10.1007/s00395-018-0670-5
23. Slezak J, Kura B, LeBaron TW et al (2021) Oxidative stress and pathways of molecular hydrogen effects in medicine. Curr Pharm Des 27:610–625. https://doi.org/10.2174/1381612826666200821114016
24. Ordin PM (1997) Safety standard for hydrogen and hydrogen systems guidelines for hydrogen system. Design, Materials Selection, Operations, Storage, and Transportation No Title
25. Buchholz BM, Kaczorowski DJ, Sugimoto R et al (2008) Hydrogen inhalation ameliorates oxidative stress in transplantation induced intestinal graft injury. Am J Transplant 8:2015–2024. https://doi.org/10.1111/j.1600-6143.2008.02359.x
26. Yan M, Yu Y, Mao X et al (2019) Hydrogen gas inhalation attenuates sepsis-induced liver injury in a FUNDC1-dependent manner. Int Immunopharmacol 71:61–67. https://doi.org/10.1016/j.intimp.2019.03.021
27. Liu B, Xie Y, Chen J et al (2021) Protective effect of molecular hydrogen following different routes of administration on D-galactose-induced aging mice. J Inflamm Res 14:5541–5550. https://doi.org/10.2147/JIR.S332286
28. de Oliveira EP, Burini RC (2012) High plasma uric acid concentration: causes and consequences. Diabetol Metab Syndr 4:12. https://doi.org/10.1186/1758-5996-4-12
29. Xie F, Jiang X, Yi Y et al (2022) Different effects of hydrogen-rich water intake and hydrogen gas inhalation on gut microbiome and plasma metabolites of rats in health status. Sci Rep 12:7231. https://doi.org/10.1038/s41598-022-11091-1
30. Fujii J, Homma T, Osaki T (2022) Superoxide radicals in the execution of cell death. Antioxidants 11:501. https://doi.org/10.3390/antiox11030501
31. Barancik M, Kura B, LeBaron TW et al (2020) Molecular and cellular mechanisms associated with effects of molecular hydrogen in cardiovascular and central nervous systems. Antioxidants 9:1281. https://doi.org/10.3390/antiox9121281
32. Dhein S, Salameh A (2021) Remodeling of cardiac gap junctional cell-cell coupling. Cells 10:2422. https://doi.org/10.3390/cells10092422
33. Szeiffova Bacova B, Viczenczova C, Andelova K et al (2020) Antiarrhythmic effects of melatonin and omega-3 are linked with protection of myocardial Cx43 topology and suppression of fibrosis in catecholamine stressed normotensive and hypertensive rats. Antioxidants 9:546. https://doi.org/10.3390/antiox9060546
34. Slezak J, Tribulova N, Ravingerova T, Singal PK (1992) Myocardial heterogeneity and regional variations in response to injury. Lab Invest 67:322–330

35. Tribulova N, Novakova S, Macsaliova A et al (2002) Histochemical and ultrastructural characterisation of an arrhythmogenic substrate in ischemic pig heart. Acta Histochem 104:393–397. https://doi.org/10.1078/0065-1281-00670
36. Slezák J, Klobusická M (1969) To some questions of early changes of ischemic myocardium. Folia Morphol (Warsz) 17:165–170
37. Schaper J (1986) Ultrastructural changes of the myocardium in regional ischaemia and infarction. Eur Heart J 7:3–9. https://doi.org/10.1093/eurheartj/7.suppl_B.3
38. Slezak J, Geller SA, Litwak RS, Smith H (1983) Long-term study of the ultrastructural changes of myocardium in patients undergoing cardiac surgery, with more than 10 years follow-up. Int J Cardiol 4:153–168. https://doi.org/10.1016/0167-5273(83)90129-8

Chapter 13
Application of Hydrogen in Hemodialysis: A Brief Review with Emphasis on the Quantification of Dissolved H_2

Foivos Leonidas Mouzakis, Lal Babu Khadka, Miguel Pereira da Silva, and Khosrow Mottaghy

Abstract Chronic kidney disease patients frequently manifest signs of oxidative and inflammatory stress associated with the lifesaving haemodialysis therapy itself. Lately, hydrogen has gained traction as an antioxidant with a deluge of reports bearing evidence of its potential in treating a plethora of medical conditions. Among the various animal models and clinical studies communicated, a novel hydrogen administration method revolving around H_2-rich dialysate stands out. Over the past decade, hydrogen enriched haemodialysis (E-HD) has steadily been gaining ground thanks to its ameliorating effect on oxidative and inflammatory complications arising during haemodialysis. To complement this modality, a hydrogen water monitoring system (HWMS) has been developed by Pureron Japan Co., Ltd to assist in determining the levels of dissolved hydrogen in liquids. Preliminary investigations have validated the sensitivity and accuracy of the contactless hydrogen sensor and its applicability in various clinical settings. Nevertheless, a few setbacks such as the long response time, and range-specific accuracy prompted further examination and in-depth analysis of its capabilities. To achieve that, three such sensors have been integrated in an in vitro haemodialysis circuit to monitor H_2 concentration at the dialyzer's inlet–outlet. Moreover, experiments have been conducted using a regular dialyzer

F. L. Mouzakis (✉) · L. B. Khadka · M. Pereira da Silva · K. Mottaghy
Institute of Physiology, RWTH Aachen University Hospital, Aachen, Germany
e-mail: foivos.mouzakis@rwth-aachen.de

L. B. Khadka
e-mail: lal.babu.khadka@rwth-aachen.de

M. Pereira da Silva
e-mail: miguel.pereira.da.silva@tecnico.ulisboa.pt

K. Mottaghy
e-mail: kmottaghy@ukaachen.de

M. Pereira da Silva
Laboratory of Membrane Processes CeFEMA, Instituto Superior Técnico, Universidade de Lisboa, Lisbon, Portugal

J. Slezak and B. Kura (eds.), *Molecular Hydrogen in Health and Disease*, Advances in Biochemistry in Health and Disease 27, https://doi.org/10.1007/978-3-031-47375-3_13

as well as a modified one with a purely diffusive coating, in order to discern the nature of hydrogen transfer through the capillary membrane and distinguish between ultrafiltration and diffusion.

Keywords Hemodialysis · Chronic kidney disease · Sensors · Hydrogen · Extracorporeal circulation · Hydrogen water

Hydrogen as Antioxidant and Anti-inflammatory Agent

Chronic Kidney Disease (CKD) is among the ten leading causes of death in the US [1, 2] and already in the early '00s more than 10% of the adult population worldwide was being diagnosed with relevant markers [3]. End stage renal disease (ESRD) patients receive renal replacement therapy in the form of regular hemodialysis (HD) sessions, where blood is continuously purified using a capillary membrane hemodialyzer in an extracorporeal circulation (ECC) system.

Renal disease is associated with the production of reactive oxidative species (ROS), leading to oxidative stress. HD procedures exacerbate this stress due to increased ROS production and systemic inflammation caused by the treatment. This combination of oxidative stress and inflammation poses serious risks for cardiovascular events and death in chronic HD patients [4, 5].

Molecular hydrogen (H_2), known for its therapeutic potential in eliminating cytotoxic radicals and treating various injuries and diseases [6–10], is proposed as a suitable means of mitigating oxidative stress-related and inflammatory diseases. Among the different administration methods (e.g. inhalation, ingestion of hydrogen-rich water) hydrogen delivery during hemodialysis stands out for its originality and non-invasiveness. H_2-enriched dialysate (H_2-HD) allows for optimal hydrogen administration directly into the bloodstream using small dosages over the duration of each HD session.

Several pilot studies have been carried out in Japan with ESRD patients using the H_2-HD system and so far, the outlook is quite auspicious. Initial reports from a short application (12 sessions) revealed a decrease in blood pressure, which is usually elevated in patients undergoing HD regularly [11]. A 6-month implementation corroborated the decline in hypertension and further highlighted the lower levels of oxidative stress markers [12]. Recent studies clearly indicate that H_2-HD successfully combats oxidative stress and prevents further oxidation during therapy, which translates into enhanced hemo- and biocompatibility [13]. Furthermore, it is reported that patients receiving H_2-HD treatment for prolonged periods show reduced signs of fatigue, which depending on the dosage might even warrant complete absence of fatigue in some cases [14].

Indeed little insight has been gained so far regarding the optimum dosage of hydrogen in all these treatments. To address this matter the Hydrogen Water Monitoring System (HWMS) has been developed, offering online contactless measurements of hydrogen concentration in liquids. The system's capabilities and shortcomings have been previously recounted in depth [15], but some updates since are worth shedding light upon.

Calibration of the HWMS Device

One of the most troubling issues with HWMS has been the diversified readings from the different hydrogen sensors used in the aforementioned study. To counteract this phenomenon and guarantee the measurements' accuracy all three HWMS devices have been recalibrated according to the following equation:

$$C = Ae^{bV} \tag{13.1}$$

where *C* is the concentration (in ppb), *V* is the voltage (in V), and *A*, *b* are temperature dependent calibration constants.

Investigations have been carried out in an experimental circuit, where all three sensors (S1, S2, S3) have been appended in series, in order to readily recognize and rectify any measurement discrepancies between them. A membrane gas exchanger has been supplied with a gas admixture of variable H_2: Air ratio. Data pairs of voltage—H_2 concentration data have been obtained from the HWMS devices under room temperature ($\theta_w = 20 \pm 1$ °C) and at constant water flow rate ($Q_w = 500$ ml/min). Data analysis lead to sensor recalibration, and facilitated the assessment of post-calibration estimates, as Table 13.1 indicates.

To minimize the measurement error between the three sensors Eq. 13.2 has been applied in conjunction with Eq. 13.3 to appropriately adjust the calibration constants A, b for each sensor. One significant limitation of the HWMS MkIII device is that parameter A may not receive values higher than 9.99 (A ∈ (0, 9.99)).

$$SSE_i = \sum_j (SV_{j,i} - TV_{j,i})^2, j = 2.7, 3.5, 5.45; i = S1, S2, S3 \tag{13.2}$$

Table 13.1 Induced voltage and expected hydrogen concentration by the HWMS devices at different operating conditions (H_2: Air ratio in the gas mixture)

H_2 (vol%)	Sensor voltage (V)			Estimated hydrogen concentration (ppb)			
	S1	S2	S3	S1	S2	S3	Theoretical
2.7	1.25	1.46	1.33	43.47	41.96	45.14	44
3.5	1.49	1.76	1.51	57.54	56.20	55.63	57
5.3	1.82	2.19	1.89	85.89	87.07	86.28	86

Table 13.2 Optimized calibration constants for the three HWMS devices (S1, S2 and S3)

	A	*b*
S1	9.7567	1.1925
S2	9.7567	0.9971
S3	9.7168	1.1548

where *SV* is the sensor value and *TV* is the theoretical value.

$$\langle \mathrm{A}, \mathrm{b}\rangle_{\mathrm{i}} : \mathrm{SSE_i}(\mathrm{A}, \mathrm{b}) = \min(\mathrm{SSE_i})\text{with } \mathrm{A} \leq 9.98 \tag{13.3}$$

The new, error-adjusted calibration coefficients for each sensor are listed in Table 13.2.

Figure 13.1 illustrates the simultaneous H_2 concentration measurements of the three sensors at gas mixtures with 2.7%, 3.5% and 5.45% H_2:Air ratio. The equilibrium values attained by each sensor are encapsulated in Table 13.3.

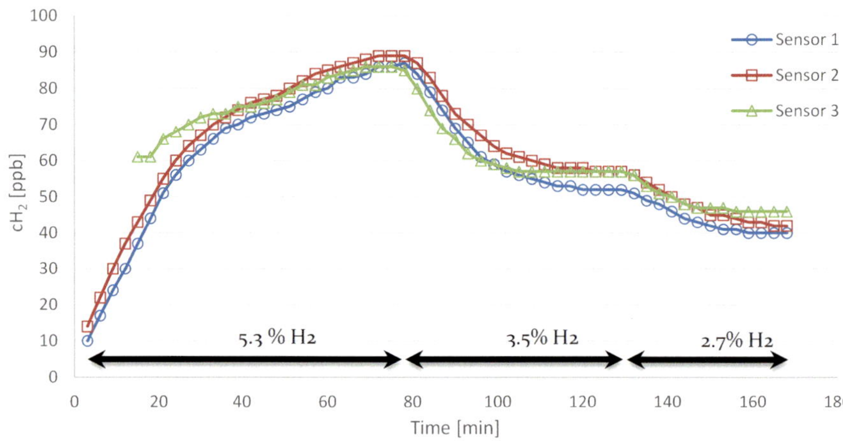

Fig. 13.1 Hydrogen concentration readings by sensors 1, 2 and 3 at different operating conditions

Table 13.3 Hydrogen concentration readings from the three sensors at different operating conditions

H_2 (vol%)	Hydrogen concentration (ppb)			
	S1	S2	S3	Theoretical
0	0	0	0	0
2.7	40	42	46	44
3.5	51	56	56	57
5.4	87	89	85	86

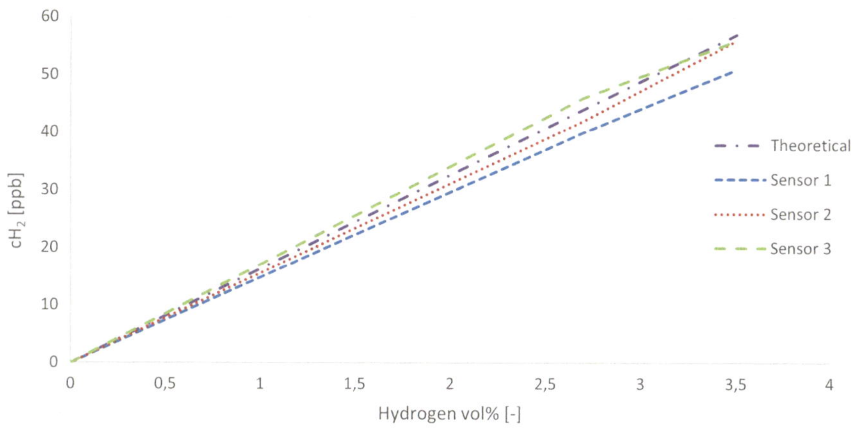

Fig. 13.2 Juxtaposition of experimental and theoretical data for a small range of H_2:Air ratios

Figure 13.2 accentuates the deviations between the sensors' readings as a function of the hydrogen concentration in the gas admixture (manipulated variable).

Hydrogen Mass Transfer During Dialysis

Using the uniformly calibrated sensors, a hemodialysis test setup has been assembled, consisting of two sub-circuits that both converge into a dialyzer module, as depicted in Fig. 13.3. The first one, designated as the blood circuit, includes only a reservoir (*R1*), and a peristaltic pump (*P1*). The reservoir continuously receives fresh water from the tap, which the pump propels through the dialyzer and into the drain at diverse flow rates (Q_B = 150–400 mL/min). A hydrogen sensor (*S3*) at the blood outlet registers hydrogen concentration in the water that may vary depending on the blood flow rate employed.

On the other hand, the dialysate circuit comprises a reservoir (*R2*), a peristaltic pump (P2), and a small gas exchanger (*MGE*) and is primed with water. The gas exchanger unit is constantly supplied with gas to maintain the concentration of dissolved hydrogen, and the dialysate pump runs at a steady flow rate, Q_D = 500 mL/min, to keep the liquid homogeneous. Different ratios of H_2:Air are used to investigate the impact of H_2 concentration on diffusivity. Two hydrogen sensors, *S2* and *S1*, measure the concentration of dissolved hydrogen before and after the dialyzer, respectively. The former tracks the equilibrium state, whereas the latter registers how much hydrogen remains in the liquid after having traversed the dialyzer module.

Pressure sensors at all 4 ports of the dialyzer ($PS_{1\text{-}4}$) enable monitoring of the transmembrane pressure (*TMP*), which can be effortlessly adjusted by placing a constrictor valve (*CV*) downstream of the dialysate's outlet. Maintaining *TMP* at 0 mmHg guarantees zero net ultrafiltration, which reduces the contribution of convection on mass

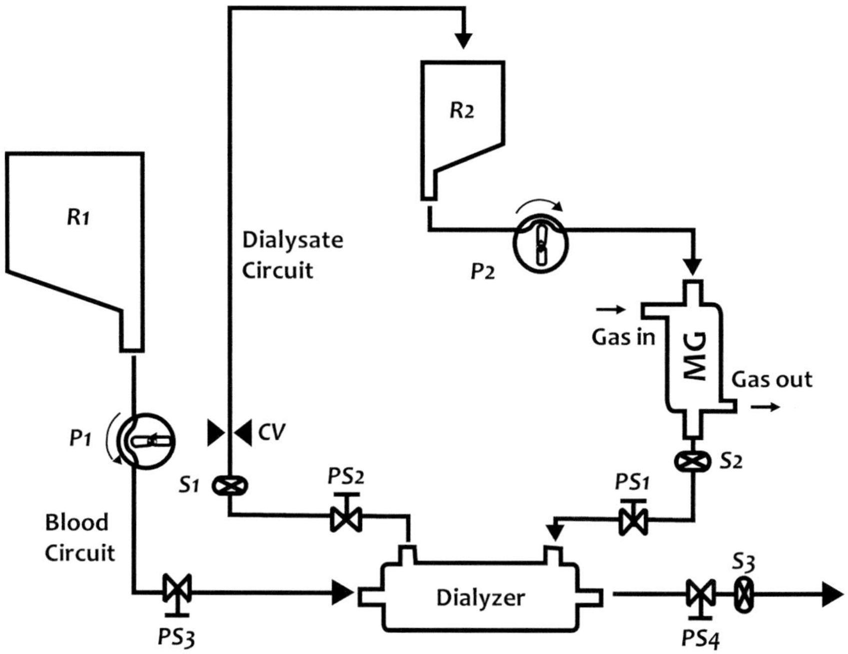

Fig. 13.3 Schematic representation of the dialysis experimental setup

transfer. To further attenuate the effect of convection a low flux dialyzer has been initially selected that ought to keep ultrafiltration to a minimum. Further investigations have been carried out with a modified dialyzer, which through unique treatment has attained a purely diffusive capillary membrane (therefore *TMP* becomes irrelevant in this case). All the experiments have been conducted at room temperature ($\theta_w = 20 \pm 1$ °C).

Low-Flux Dialyzer

The exact characteristics of the low-flux dialyzer employed in this study are listed in Table 13.4.

Table 13.4 Technical specifications of the low-flux dialyzer

Surface area	A	0.8 m^2
Internal diameter of hollow fibers	d_{in}	200 μm
Number of fibers	N	5606
Effective length	L_{eff}	227 mm

The course of the experiment with the low-flux dialyzer is presented in Fig. 13.4. Further details about the operating conditions, and the equilibrium values achieved at each setting can be obtained from Tables 13.5 and 13.6. Based on the above data, an efficiency factor (*EF*), and a hydrogen transfer rate parameter are defined (Eqs. 13.4–13.5).

$$\mathrm{EF} = \frac{H_{Bout} - H_{Bin}}{H_{Din} - H_{Bin}} \tag{13.4}$$

$$\mathrm{HTR} = \dot{Q}_B (H_{Bout} - H_{Bin}) \tag{13.5}$$

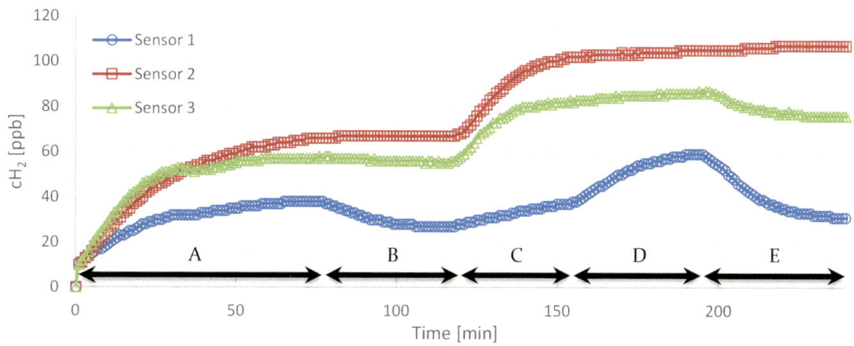

Fig. 13.4 Hydrogen concentration measurements by all three sensors at different operating conditions (S1-D_{out}, S2-D_{in}, S3-B_{out})

Table 13.5 Hydrogen concentration values at the dialyzer's inlet/outlet ports after reaching equilibrium at different operating conditions

	H_2 (vol%)	Q_D (ml/min)	Q_B (ml/min)	Hydrogen concentration (ppb)				EF (%)	HTR (mm^3/min)
				B_{in}	B_{out}	D_{in}	D_{out}		
A	3.5	500	150	0	57	64	38	89	103.39
B	3.5	500	300	0	55	67	27	82.1	199.52
C	5.4	500	300	0	83	102	38	81.4	301.09
D	5.4	500	150	0	86	105	59	81.9	155.99
E	5.4	500	400	0	75	107	31	70.1	362.76

Table 13.6 Pressure at the dialyzer's inlet/outlet ports at different operating conditions

	Q_D (ml/min)	Q_B (ml/min)	Presssure (mmHg)				TMP (mmHg)
			B_{in}	B_{out}	D_{in}	D_{out}	
A	500	150	89	66	78	76	0.5
B	500	300	113	70	103	79	0.5
C	500	300	122	70	106	84	1
D	500	150	99	72	96	74	0.5
E	500	400	138	73	117	93	0.5

Purely Diffusive Dialyzer

Table 13.7 contains the characteristics of the purely diffusive dialyzer tested next.

Figure 13.5 portrays the progress of the experiment with the purely diffusive dialyzer.

Analytical information about the operating conditions, as well as the steady state data are disclosed in Table 13.8.

Table 13.7 Technical specifications of the modified zero-flux dialyzer

Surface area	A	0.7 m^2
Internal diameter of hollow fibers	d_{in}	200 μm
Number of fibers	N	4905
Effective length	L_{eff}	227 mm

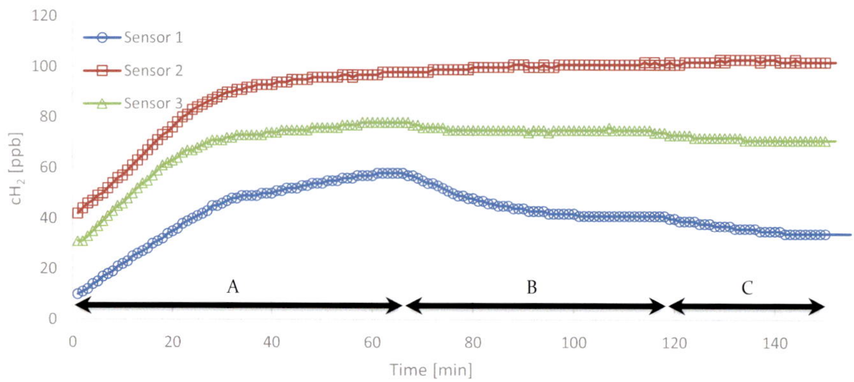

Fig. 13.5 Hydrogen concentration measurements by all three sensors at different operating conditions (S1-D_{out}, S2-D_{in}, S3-B_{out})

Table 13.8 Equilibrium hydrogen concentration values at the dialyzer's inlet/outlet ports at different operating conditions

	H_2 (vol%)	Q_{dia} (ml/min)	Q_{blood} (ml/min)	Hydrogen concentration [ppb]				EF (%)	HTR (mm^3/min)
				B_{in}	B_{out}	D_{in}	D_{out}		
A	5.4	500	150	0	77	100	58	77	139.66
B	5.4	500	300	0	75	100	40	75	272.07
C	5.4	500	400	0	71	100	34	71	343.41

Epilogue

As early as 1975 it was demonstrated that hydrogen is effective against cytotoxic radicals [16]. Ever since, it has been implemented in a plethora of pilot studies and animal models to counteract the oxidative and inflammatory stress induced by various conditions (cerebral/cardiac ischemia, pulmonary oedema, organ transplantation, hepatitis, colitis, pancreatitis etc.). Across these studies, different administration methods (inhalation, ingestion of H_2-rich water, injection with H_2-dissolved saline) have been tested, including the application of H_2-enriched dialysate for dialysis.

Nevertheless, accurate determination of the optimal clinical dosage still eludes scientists. HWMS can assist in this endeavour with its high sensitivity and precise measurements. As demonstrated here, sensor recalibration can be a powerful tool in adjusting the accuracy of HWMS in the desired range and under case-specific operating conditions (Figs. 13.1 and 13.2). Further investigations with an in-vitro HD system confirm the recalibration's success and enable the assessment of mass transfer between the liquid phases (blood-dialysate) in terms of hydrogen diffusion-ultrafiltration. Minimizing the impact of convection by adjusting TMP to 0 permits maintaining a constant liquid volume in the dialysate sub-circuit through reduction of liquid losses-gains in the closed loop. Both the low-flux and the purely diffusive dialyzer permit sufficient amounts of hydrogen to migrate from the dialysate to the blood side (EF > 70%). Whilst on the subject of the efficiency factor, it should be pointed out that only the blood-side equilibrium has been taken into consideration in its definition, as potential hydrogen losses on the dialysate side are not relevant or of note. Moreover, these experiments incontestably show that owing to its minuscule size, hydrogen, migrates to a great extent by diffusion (Figs. 13.4 and 13.5), hinting that membrane porosity, flow rate etc. are not decisive parameters for its application.

The above findings cement the capabilities and applicability of HWMS, which in turn can greatly enhance hydrogen's reputation as a potent antioxidant and anti-inflammatory agent. In view of its already vast implementation (e.g. diabetes melitus, ischaemia–reperfusion injury, rheumatoid arthritis) [17, 18] its application in other medical disciplines (e.g. cardiovascular/pulmonary support) ought to be considered forthwith.

Acknowledgements The authors wish to express their gratitude towards Pureron Japan Co., Ltd, for providing the HWMS devices, and for the successful cooperation on the co-development of the hydrogen sensor.

References

1. Heron M (2021) Deaths: leading causes for 2019. Natl Vital Stat Rep 70(9):1–114
2. Xu J, Murphy SL, Kochanek KD, Arias E (2022) Mortality in the United States, 2021. NCHS Data Brief 456:1–8
3. Levey AS, Atkins R, Coresh J, Cohen EP, Collins AJ, Eckardt KU, Nahas ME, Jaber BL, Jadoul M, Levin A, Powe NR, Rossert J, Wheeler DC, Lameire N, Eknoyan G (2007) Chronic kidney disease as a global public health problem: approaches and initiatives - a position statement from kidney disease improving global outcomes. Kidney Int 72(3):247–259. https://doi.org/10.1038/sj.ki.5002343
4. Zimmermann J, Herrlinger S, Pruy A, Metzger T, Wanner C (1999) Inflammation enhances cardiovascular risk and mortality in hemodialysis patients. Kidney Int 55(2):648–658. https://doi.org/10.1046/j.1523-1755.1999.00273.x
5. Himmelfarb J, Stenvinkel P, Ikizler TA, Hakim RM (2002) The elephant in uremia: oxidant stress as a unifying concept of cardiovascular disease in uremia. Kidney Int 62(5):1524–1538. https://doi.org/10.1046/j.1523-1755.2002.00600.x
6. Hayashida K, Sano M, Ohsawa I, Shinmura K, Tamaki K, Kimura K, Endo J, Katayama T, Kawamura A, Kohsaka S, Makino S, Ohta S, Ogawa S, Fukuda K (2008) Inhalation of hydrogen gas reduces infarct size in the rat model of myocardial ischemia-reperfusion injury. Biochem Biophys Res Commun 373(1):30–35. https://doi.org/10.1016/j.bbrc.2008.05.165
7. Buchholz BM, Kaczorowski DJ, Sugimoto R, Yang R, Wang Y, Billiar TR, McCurry KR, Bauer AJ, Nakao A (2008) Hydrogen inhalation ameliorates oxidative stress in transplantation induced intestinal graft injury. Am J Transplant 8(10):2015–2024. https://doi.org/10.1111/j.1600-6143.2008.02359.x
8. Spulber S, Edoff K, Hong L, Morisawa S, Shirahata S, Ceccatelli S (2012) Molecular hydrogen reduces LPS-induced neuroinflammation and promotes recovery from sickness behaviour in mice. PLoS ONE 7(7):e42078. https://doi.org/10.1371/journal.pone.0042078
9. Kajiya M, Sato K, Silva MJ, Ouhara K, Do PM, Shanmugam KT, Kawai T (2009) Hydrogen from intestinal bacteria is protective for Concanavalin A-induced hepatitis. Biochem Biophys Res Commun 386(2):316–321. https://doi.org/10.1016/j.bbrc.2009.06.024
10. Chen H, Sun YP, Li Y, Liu WW, Xiang HG, Fan LY, Sun Q, Xu XY, Cai JM, Ruan CP, Su N, Yan RL, Sun XJ, Wang Q (2010) Hydrogen-rich saline ameliorates the severity of l-arginine-induced acute pancreatitis in rats. Biochem Biophys Res Commun 393(2):308–313. https://doi.org/10.1016/j.bbrc.2010.02.005
11. Nakayama M, Kabayama S, Nakano H, Zhu WJ, Terawaki H, Nakayama K, Katoh K, Satoh T, Ito S (2009) Biological effects of electrolyzed water in hemodialysis. Nephron Clin Pract 112(1):c9-15. https://doi.org/10.1159/000210569
12. Nakayama M, Nakano H, Hamada H, Itami N, Nakazawa R, Ito S (2010) A novel bioactive haemodialysis system using dissolved dihydrogen (H_2) produced by water electrolysis: a clinical trial. Nephrol Dial Transplant 25(9):3026–3033. https://doi.org/10.1093/ndt/gfq196
13. Satta H, Iwamoto T, Kawai Y, Koguchi N, Shibata K, Kobayashi N, Yoshida M, Nakayama M (2021) Amelioration of hemodialysis-induced oxidative stress and fatigue with a hemodialysis system employing electrolyzed water containing molecular hydrogen. Ren Replace Ther 7:37. https://doi.org/10.1186/s41100-021-00353-9
14. Uemura S, Kegasa Y, Tada K, Tsukahara T, Kabayama S, Yamamoto T, Miyazaki M, Takada J, Nakayama M (2022) Impact of hemodialysis solutions containing different levels of molecular

hydrogen (H_2) on the patient-reported outcome of fatigue. Ren Replace Ther 8:32. https://doi.org/10.1186/s41100-022-00422-7

15. Mouzakis FL, Khadka LB, Pereira da Silva M, Mottaghy K (2022) Quantification of dissolved H_2 and continuous monitoring of hydrogen-rich water for haemodialysis applications: an experimental study. Int J Artif Organs 45(3):254–261. https://doi.org/10.1177/03913988211070588
16. Dole M, Wilson FR, Fife WP (1975) Hyperbaric hydrogen therapy: a possible treatment for cancer. Science 190(4210):152–154. https://doi.org/10.1126/science.1166304
17. Kajiyama S, Hasegawa G, Asano M, Hosoda H, Fukui M, Nakamura N, Kitawaki J, Imai S, Nakano K, Ohta M, Adachi T, Obayashi H, Yoshikawa T (2008) Supplementation of hydrogen-rich water improves lipid and glucose metabolism in patients with type 2 diabetes or impaired glucose tolerance. Nutr Res 28(3):137–143. https://doi.org/10.1016/j.nutres.2008.01.008
18. Ishibashi T, Sato B, Rikitake M, Seo T, Kurokawa R, Hara Y, Naritomi Y, Hara H, Nagao T (2012) Consumption of water containing a high concentration of molecular hydrogen reduces oxidative stress and disease activity in patients with rheumatoid arthritis: an open-label pilot study. Med Gas Res 2(1):27. https://doi.org/10.1186/2045-9912-2-27

Chapter 14
Hydrogen as a Potential Therapeutic Approach in the Treatment of Cancer: From Bench to Bedside

Arian Karimi Rouzbehani, Golnaz Mahmoudvand, Zahra Goudarzi, Arshia Fakouri, Simin Farokhi, Saeideh Khorshid Sokhangouy, Elnaz Ghorbani, Amir Avan, Elham Nazari, and Majid Khazaei

Abstract Cancer is still remained among the leading cause of death worldwide, mainly due to the metastatic spread and chemotherapy resistance. In turn chemotherapy is associated with several side effects. Therefore, there is a need to develop new therapies for the management of this condition, reducing toxicity and drug resistance. Recently, hydrogen molecule has been reported as a novel therapeutic

---

Arian Karimi Rouzbehani, Golnaz Mahmoudvand, Zahra Goudarzi, Arshia Fakouri, Simin Farokhi and Saeideh Khorshid Sokhangouy are equal contribution.

A. K. Rouzbehani · G. Mahmoudvand · A. Fakouri · S. Farokhi
Student Research Committee, Lorestan University of Medical Sciences, Khorramabad, Iran

USERN office, Lorestan University of Medical Sciences, Khorramabad, Iran

A. K. Rouzbehani
e-mail: arian.karimi@sbmu.ac.ir

G. Mahmoudvand
e-mail: GolnazMahmoudvand@sbmu.ac.ir

A. Fakouri
e-mail: ArshiaFakouri@sbmu.ac.ir

S. Farokhi
e-mail: SiminFarokhi@sbmu.ac.ir

Z. Goudarzi
Student Research Center, Tehran University of Medical Sciences, Tehran, Iran
e-mail: GoudarziZ3@mums.ac.ir

S. K. Sokhangouy · E. Ghorbani · A. Avan · M. Khazaei (✉)
Metabolic Syndrome Research Center, Mashhad University of Medical Sciences, Mashhad, Iran
e-mail: khazaeim@mums.ac.ir

A. Avan
e-mail: Avana@mums.ac.ir

A. Avan · M. Khazaei
Basic Sciences Research Institute, Mashhad University of Medical Sciences, Mashhad, Iran

J. Slezak and B. Kura (eds.), *Molecular Hydrogen in Health and Disease*, Advances in Biochemistry in Health and Disease 27, https://doi.org/10.1007/978-3-031-47375-3_14

approach for many diseases including cancer. there is growing body of data showing the anti-tumor, antioxidant, and anti-inflammatory activity of hydrogen therapy in cancer. Also, various studies have reported the effectiveness of hydrogen therapy in cancer in preclinical setting, However, it is important to note that these studies are still preliminary, and more research is needed to determine the safety and efficacy of hydrogen therapy for cancer treatment. Additionally, while some studies suggest that hydrogen therapy may have potential benefits for certain types of cancer, it is not a cure or replacement for conventional cancer treatments such as chemotherapy or radiation therapy. while there is some evidence to suggest that hydrogen molecules may have therapeutic potential for cancer treatment, more research is needed before any definitive conclusions can be drawn. In this chapter, we provide an overview on the antiproliferative activity of hydrogen and its potential molecular mechanisms of actions followed by recent reports on its potential side effects.

Keywords Cancer · Hydrogen therapy · Hydrogen water · Molecular hydrogen

Introduction

In 1766, hydrogen was identified as a chemical element, and its first use in medicine is reported for the treatment of tuberculosis. After several years of its discovery, it was found in the digestive system by natural gut bacteria and then used the hydrogen molecule to prevent decompression syndrome in divers. In 1975, an animal study was conducted on squamous cell carcinoma, which confirmed the therapeutic effects of hydrogen molecules. Another animal study showed that hydrogen inhalation therapy was used to prevent damage after re-ischemia stroke. During the COVID-19 pandemic, there are some studies for treatment of Covid-19 patients, and they found that hydrogen could prevents the development of pulmonary dysfunction in infections and emphysema [1, 2].

Hydrogen is a diatomic, non-polar, and highly flammable molecule that is known as the most abundant chemical element [3]. It is included in the range of medical gases due to its antioxidant and anti-apoptotic characteristics and other therapeutic properties. This gas is highly available and due to its low molecular weight, which can easily cross biological barriers, while other antioxidants usually require special carriers to cross the biological barriers [2]. According to the proposed properties of hydrogen in different studies, this molecule has been shown to be effective in treating

S. K. Sokhangouy · A. Avan
Medical Genetics Research Center, Mashhad University of Medical Sciences, Mashhad, Iran

E. Nazari (✉)
Department of Health Information Technology and Management, School of Allied Medical Sciences, Shahid Beheshti University of Medical Sciences, Tehran, Iran
e-mail: NazariE4001@mums.ac.ir

many diseases, such as diabetes, cancer, viral infectious diseases such as COVID-19, asthma, and other diseases [1]. There are growing body of evidence on the effect of hydrogen. In particular, 82 people with different types of cancer were examined and the results indicate that hydrogen therapy has been effective in 57.5% of people treated with inhaled hydrogen therapy after 55 days of using this treatment, 83% of those with stage 3 cancer and 41% have been in contact with stage 4 cancer patients, which has improved their conditions. The effect of this treatment in stage 3 cancer patients was more than in stage 4 patients. Also hydrogen therapy in cancer patients has improved the physical condition by 41.5%, the highest effect is related to lung cancer patients and the lowest effect is in pancreatic cancer patients. In general, in 36% of patients treated with hydrogen after 23 days, there was a decrease in abnormal markers, the largest decrease was seen in lung cancer patients and the least effect in pancreatic and liver cancer patients. In the patients who were treated with hydrogen, side effects related to hematological toxicity were not seen, and the side effects seen in the patients were very minor and disappeared without the need for treatment [1, 4]. Therefore, considering the importance of hydrogen treatment, the following sections are provided to further explain in this regard.

Hydrogen: As a New Treatment Option for Human Disease

Hydrogen (H_2) molecules have a strong covalent link that makes them non-polar. Hydrogenase is an enzyme found in many bacteria that can reversibly split H_2 into protons and electrons. In addition to iron-sulfur (Fe-S) clusters, the catalytic core of most hydrogenases consists of two metal atoms, often [NiFe] or [FeFe]. Interestingly, H_2 absorption is facilitated by O_2-tolerant membrane-bound [NiFe]-hydrogenases (MBH), which oxidize H_2 in the proximity of electron acceptors, including ubiquinone and menaquinone [5]. These features suggest that MBH-type enzymes are related to the proto-respiratory complex I. Even though human cells do not produce hydrogenases, H_2 is increasingly applied as a medical gas due to its impact on conditions as diverse as neurodegenerative illnesses, metabolic problems, and inflammatory diseases. As a relatively inert gas that does not alter physiological enzymatic processes, therapeutic H_2 is thought to have a positive safety profile and diffuse throughout the human body soon after delivery passively. To this day, it has been theorized that H_2's effects on mammalian cells are primarily attributable to its non-enzymatic interactions with highly reactive oxygen species (ROS) through random collisions [6]. Ohsawa and coworkers attributed H_2's biological impacts and its ability to function as an electron donor reactive oxygen species (ROS) molecule like the highly reactive hydroxyl radical or peroxynitrite [7] in the liquid phase of chemical processes. These scavenging actions may originate from H_2's amphipathic characteristics in the lipid bilayer inner membrane of mitochondria [8]. By reducing oxidative stress and scavenging hydroxyl radicals, they showed that H_2 might protect against ischemia–reperfusion (I/R) damage.

Hydrogen therapy has shown promising results in Parkinson's disease (PD) treatment in both animal models and human studies [9, 10]. In a recent clinical study for PD, Yoritaka et al. [11] illustrated the effectiveness of 1 L/d hydrogen water treatment in Japanese patients using levodopa and they found a significant enhancement in the entire unified PD rating scale, which was in line with several other observations [12–15]. Memory was enhanced when hydrogen-rich saline was injected intraperitoneally [14, 15] via modulation of oxidative stress, cytokines and nuclear factor kappa B. Palladium hydride nanoparticles' persistent release of bioactive hydrogen in AD mice models and reduce cognitive impairment by preventing synaptic and neuronal death, boosting neuronal metabolism [16]. Molecular hydrogen's potential as a treatment for brain damage has been studied in animal models of traumatic brain injury and surgical brain injury [17–19]. It has been shown in a traumatic brain injury (TBI) rat model [18] that inhalation of hydrogen protected the blood–brain barrier, reduced cerebral edema, and prevented the decline in antioxidant enzymes. Neuroprotective mechanisms of hydrogen treatment in TBI [17] was shown by the suppression of hypoxia-inducible factor-1, Matrix metalloproteinase-9, and cyclophilin A, and regulation of cytokines and chemokines. Eckermann et al. [19] found that frontal lobectomy reduced cerebral edema and enhanced neurobehavioral scores in mice models of surgical trauma. A hydrogen-producing dissolving tablet was administered buccally to a concussed athlete immediately after the incident and then every two hours for a whole day. The athlete's concussion score decreased from 16 to 16. More information on the optimal dosage and effectiveness of such medicines in the context of TBI [12] may be forthcoming from future, bigger clinical investigations. In a mouse model of cecal ligation and puncture, hydrogen gas treatment was also shown to alleviate septic encephalopathy symptoms [20]. Liu et al. revealed that administering 2% inhaled hydrogen gas reduced brain edema, blood–brain barrier breakdown, cytokine production, and oxidative stress in the CA1 area of the hippocampus. Neuroprotective effects and enhanced survival rate were seen after low-concentration hydrogen gas inhalation or hydrogen-rich saline intraperitoneal injection in a mouse model of global brain ischemia/reperfusion [17]. Moreover, Huang et al. found that hydrogen produced by electrolyzing water at 67% improved both short- and long-term neurological impairments and reduced neuronal degeneration in the hippocampus [21].

Hydrogen therapy for rheumatoid arthritis (RA) showed positive results in a randomized controlled trial. During the intervention, there was a 21.2 percentage point drop in the disease activity score in 28 joints (DAS28; score ranges from 0 to 10, with higher scores indicating more RA activity), which suggests that H_2 might supplement conventional treatment in RA. Additionally, professional athletes with acute soft-tissue injury had a quicker restoration to a normal joint range of motion of the damaged limb after the H_2 intervention compared to the control intervention [22]. Twenty-two patients with pressure ulcers had better wound healing after receiving 600 mL/day of hydrogen-rich water through tube feeding in a Japanese clinical study. After H_2 intervention, wound size reduced by around 6 cm^2, and patients spent fewer days in the hospital (113.3 vs. 155.4) than those in the control group [23].

Several studies have demonstrated the value of H_2 as an anti-apoptotic agent. H_2 downregulates the expression of pro-apoptotic factors like B-cell lymphoma-2-associated X-protein, caspase-3, caspase-8, and caspase-12, while upregulating the expression of anti-apoptotic factors like B-cell lymphoma-2 and B-cell lymphoma-extra large [24–26]. In 2014, Hong et al. detected that the activation of the anti-apoptotic protein kinase B pathway (also known as the Akt/glycogen synthase kinase 3 (GSK3) pathway) in neurons was responsible for the H_2-triggered neuroprotective effect. H_2 has been found to affect mesenchymal stem cells by increasing their capacity to form colonies in an aplastic anemia model [25]. Hydrogen supplementation also improves mitochondrial health and function in some cancer cell types. The first is activation of genes involved in the production of elements of complex I of the electron transport chain [27]. The second is that H_2 has several positive effects on mitochondria, including increasing their mass, superoxide radical concentration, and membrane potential. This combination of events promotes cell growth, at least in certain cancer cells. As a newly recognized mediator implicated in mitochondrial function and cell survival under harsh settings, the action of H_2 on mtUPR merits special consideration [28]. Even though some research has utilized different cancer cells, the data suggest that H_2 may facilitate the rapid division, differentiation, and expansion of stem cells. Regenerative medicine's primary goal is to provide the safest possible methods of activating the body's natural tissue regeneration processes. Hence H_2's unique properties are particularly relevant to this field. Molecular hydrogen stimulates the proliferation and development of mesenchymal stem cells, allowing them to fill up for missing cells directly. In addition, a group of pro-regenerative cytokines initiates and control this process, and cell migration into a definite niche is aided by the enhanced production of cell adhesion molecules. H_2 has a positive influence on the health of mesenchymal stem cells and the regeneration of tissues [29]. Another study by Akagi et al. investigated its effect on the outcomes of H_2 gas (67% concentration) on immunological function in 55 patients with stage IV colorectal cancer. Results showed that H_2 gas substantially improved PFS and OS by decreasing fatigued terminal PD-1+ CD8+ T cells and increasing activated PD-1$^+$ CD8$^+$ T cells in peripheral blood. They hypothesized that H_2 may be connected with the regeneration of fatigued CD8+ T cells [30]. Chen et al. treated 82 cancer patients with H_2 gas (67%) at least three hours a day for three months. Only one-third of patients (34%) were treated with H_2 gas inhalation alone, while the remaining two-thirds (66%) were also given several low-dose anticancer medications. Quality of life (QOL) indicators such as tiredness, sleeplessness, hunger, and pain were all improved in 41.5% of patients. Disease control was achieved in 57.5% (36) of patients between 21 and 80 days (median 55 days) following H_2 gas intake. The fewest disease control rate was shown in pancreatic cancer patients, whereas 83.0% of stage III patients had their illness under control, compared to 47.7% of stage IV patients [31]. Chen et al. assumed that inhalation of H_2 gas may increase QOL in cancer patients and slow the disease's advancement. Patients with liver-originating metastatic gallbladder cancer were the focus of a case study including H_2 gas (67%) inhalation [32, 33]. The gallbladder and liver malignancies progressed and were worsened by intestinal obstruction. However, the metastases in the abdominal cavity diminished, and symptomatic treatment for the intestinal blockage improved, which is in line with several other studies in other diseases [34–37]. Itoh et al. [37] assessed its activity on allergic response. They showed that ingesting hydrogen-rich water reduced an immediate-type allergic response via suppression of the phosphorylation of the FcRI-associated Lyn and its downstream signal transduction, which in turn suppressed the activity of NADPH oxidase and decreased hydrogen peroxide production. Blocking NADPH oxidase reduced phosphorylation of Lyn and its downstream targets, including Syk, PPAR, PPAR, ERK1/2, Akt, cPLA2, and p38, suggesting that the FcRI-related Lyn was the target of the feed-forward loop [38].

Potential Mechanisms of Hydrogen

Reactive Oxygen Species (ROS)

Oxidative stress, which is caused by an imbalance between reactive oxygen species (ROS) and the antioxidant system is associated with progression and development of several disorders [39, 40]. ROS originate from mitochondrial respiration, NADH/NADPH oxidase, or xanthine oxidoreductase [41] and comprise superoxide anion (O_2^-), hydroxyl (OH) radicals, peroxyl (RO_2) radicals, alkoxyl (RO) radicals, and nitric oxide (NO). Whenever a cell is damaged, electron leakage from mitochondrial oxidative phosphorylation and electron transport results in an abundance of ROS. H_2 is a form of reduction that may cross the cell membrane and neutralize damaging particles (OH and ONOO) without affecting the cell structure or releasing ROS [42]. One possible mechanism is proposed [43] by direct scavenging of the hydroxyl radical through the chemical reaction $H_2 + OH\ H_2O + H$, then $H + O_2\ HO_2$. Superoxide dismutase (SOD) activity and lipid peroxide malondialdehyde (MDA) levels were both reported to be improved by H_2 in a schistosomiasis-associated liver inflammation model [44, 45]. H_2 could protect against I/R damage by lowering oxidative stress and scavenging OH and ONOO. Tissue damage could be attenuated in vasculitis mice when they were treated with 1.3% H_2 gas inhalation [45]. They could be transformed to OH radicals through the Haber–Weiss and Fenton reaction in the proximity of catalytically active metals like Fe^{2+} and Cu^+ [46], and controlling their concentration also prevents the generation of hydroxyl radicals [46–54]. The translocation of nuclear factor erythroid-2 related factor 2 (Nrf2) to the nucleus may regulate genes involved in oxidative stress defense. Furthermore, H_2 lowers intracellular ROS by decreasing the expression of NADPH oxidase and upregulating the transcription of Nrf2, SOD and glutathione (GSH) [55].

Regulation of Mitochondria

The events preceding the malfunction of the electron transport chain, which occurs at the onset of mitochondrial oxidative stress, were also studied. Mitochondria are sometimes called the "powerhouses" of the cell since they produce almost all of the ATP used by the cell. ROS are produced in tandem with oxidative phosphorylation by forward and backward electron transfer [56], respectively. In mitochondria, H_2 reduces dysfunction by blocking electron leakage in the electron transport chain, which has the potential to restore damaged cells. The mitochondrial ATP-sensitive K + channel (mKATP) is a key player in energy homeostasis. H_2 gas has the capacity to activate mKATP and control mitochondrial membrane potential during acute myocardial infarction, resulting in a decrease in myocardial I/R damage by normalizing the level of myocardial NAD^+ (the precursor to ATP synthesis) and the generation of mitochondrial ATP [57]. The mitochondrial electron transport chain relies on coenzyme Q (CoQ). CoQ10 predominates in humans, whereas CoQ9 does so in rats. CoQ delivers electrons from Complex I and Complex II to Complex III, aiding in the formation of NAD^+, a necessary cofactor in ATP synthesis and the source of the proton incentive force required for ATP synthesis [58]. Rat plasma and myocardial CoQ9 levels rose dramatically after H_2 administration. The mitochondrial oxidative phosphorylation route to ATP generation is enhanced by elevated CoQ9 levels. An increase in mitochondrial Coenzyme Q10 has been proposed as a means by which H_2 gas might boost the clinical effectiveness of nivolumab in reviving fatigued $CD8^+$ T lymphocytes [59]. Accordingly, we hypothesize that H_2's ability to enhance mitochondrial activity protects against cellular harm. As mitochondrial dysfunction is corrected, cell death processes like Bax and caspase activity should become less chaotic as a result. Mitophagy is crucial for mitochondrial homeostasis because it removes unhealthy or damaged mitochondria. Administration with 2% H_2 for 3 h enhanced Fundc1-induced mitophagy and guarded mice from sepsis-induced liver injury [60]. Fun 14 domain-containing protein 1 (Fundc1) is one of the mitophagy receptors located on the outer membrane of the mitochondrion and can keep mitochondrial ATP balance by regulating the mitophagy and contracting with LC3 II. Increased expression of mitophagy-related proteins PINK1 and Parkin suggests that H_2 is helpful for ATP production by boosting mitochondrial autophagy [61], and H_2 has a neuroprotective impact on oxygen/glucose deprivation brain injury in rats. Multiple organ failure due to sepsis has been linked to mitochondrial malfunction, according to animal investigations of the disease. To prevent organ damage caused by sepsis, H_2 therapy upregulated HO-1 (also called heat shock protein 32) in cardiac tissues, which scavenged ROS [62]. Many forms of neurodegeneration may be traced back to mitochondrial dysfunction brought on by an abundance of ROS. Intervention with H_2 has been found to have antioxidant benefits in animal models of Parkinson's disease and Alzheimer's disease [10, 63]. Despite the safety of H_2 gas, a double-blind randomized placebo-controlled study of 16 weeks of twice-daily 1-h inhalation did not reveal any positive benefits for individuals with Parkinson's disease inhaling 6.5% H_2 gas at 2 L/min [64]. We suspect a correlation between H_2 concentration and length of therapy. There has to be more research done because of the limited size of this clinical trial. H_2's capacity to scavenge ROS and promote mitochondrial energy metabolism may be explained by its potential to equilibrate mitochondrial electron flow [65].

Anti-inflammatory Effect

By aggregating monocytes, local neutrophils, and other immune cells and releasing inflammatory cytokines, inflammation is thought to be an adaptive response of the body triggered by infection of external pathogens or tissue injury. Because phagocytes are the primary source of growth factors and cytokines, this process is characterized by the migration of mononuclear phagocytes and lymphocytes from the veins to the location of damaged tissue, where they become activated and develop into macrophages. Apoptosis may be triggered when there is an abundance of intracellular ROS [66–68]. This is because ROS can trigger inflammatory transcription factors such as p53, hypoxia-inducible factor-1a (HIF-1a), nuclear factor kB (NF-kB), peroxisome proliferator activated receptor-g, matrix metalloproteinases, and nitrosyl radicals. Therefore, cell damage, inflammation, and apoptosis all accompany and impact one another during the whole pathogenic phase of oxidative stress. Suppressing the expression of intercellular adhesion molecules and chemokines [69], like early pro-inflammatory cytokines IL-1b and TNF-a, H_2 can decrease the infiltration of neutrophils and macrophages, thereby reducing the production of inflammatory cytokines like IL-6 and IFN-g [70]. H_2-rich saline was reported to alleviate the airway inflammatory response generated by a burn in rats by inhibiting the activation of the key inflammatory signaling pathway NF-kB and lowering serum IL-1b, IL-6, and TNF-a levels. Additionally, the H_2 can significantly decrease NF-kB expression in liver injury [71], hematencephalon [72], and skeletal muscle injury caused by acute sports [73, 74], indicating that molecular hydrogen can impact the inflammatory process by regulating nuclear transcription factors and downstream pro-inflammatory cytokines. Furthermore, the balance between anti-inflammation and pro-inflammation should be addressed for the treatment of illnesses of inflammatory dysfunction. By increasing the number of Tregs, which have an immunosuppressive role and reduce the expression of NF-kB, H_2 show an anti-inflammatory impact in animal models of I/R brain damage and allergic rhinitis. The microsomal enzyme heme oxygenase-1 is the rate-limiting enzyme in heme catabolism, and it is a member of the heat-shock protein family. Rapid reduction of biliverdin to bilirubin, a powerful endogenous antioxidant, is the result. It has the potential to inhibit IL-1β and NF-kB expression, hence reducing septic damage [75]. Human umbilical vein endothelial cells treated with LPS and lung tissue from lung-damaged mice showed enhanced HO-1 expression and IL-10 production after H_2 injection [76]. Preventing acute pancreatitis in mice by the early induction of Hsp60 protein expression (a heat stress protein that stimulates production by high temperature to defend itself) was also observed by us [77]. Thus, we hypothesize that H_2 may activate the immune system to significantly contribute to reducing inflammation [54].

Apoptosis

H_2 may control endogenous apoptosis by scavenging ROS or by regulating gene transcription. Intestinal epithelial cells were shown to be resistant to apoptosis, oxidative stress, and caspase-3 and -9 activation in vitro when exposed to H_2-rich media. In addition, H_2 [78] normalized the overabundant expression of Bax and Bcl-2. It has been hypothesized [79] that H_2-rich water achieves this action by blocking the mitochondrial import of the apoptotic markers caspase-3 and Bax. By increasing the expression of Bcl-2, a critical anti-apoptotic factor [80], H_2-rich water may serve an anti-apoptotic effect. In addition, H_2 may protect type II alveolar epithelial cells from hypoxia-induced apoptosis by stimulating the PI3K/Akt signaling pathway [81] and decrease neuronal death by stimulating the MAPK/HO-1 pathway [82]. By increasing the expression of cleaved caspase-3 [83, 84], H_2 showed promise as a potential application in tumor therapy by promoting cell apoptosis and inhibiting the invasion, growth, and migration of lung cancer and esophageal cancer cells in vitro. Thus, we hypothesize that H_2 may have more than one function, including shielding healthy cells from harm and slowing the growth of cancerous ones [54].

Autophagy

By destroying macromolecular compounds, autophagy helps maintain energy homeostasis, but too much of it may exacerbate inflammation and organ damage, as shown in sepsis. Light chain 3 proteins (LC3) and Beclin-1, two autophagy-related proteins, serve critical roles in autophagy detection. Inhibiting autophagy, Zhang et al. showed that H_2 reduced isoproterenol-induced damage to cardiomyocytes [85]. In LPS-induced lung damage, the addition of H_2 to the water dramatically decreased the expression of autophagy-related proteins LC3 and Beclin-1 [86]. By decreasing mTOR expression in glial cells, raising the LC3 II/LC3 I ratio, and boosting autophagy [86], H_2 was shown to be effective in lowering LPS-induced neuroinflammation. Possible explanations include variations in the intensity of LPS-induced inflammation used in the studies. Mitochondrial membrane-localized mitophagy receptor Fundc1 can regulate mitophagy to keep mitochondrial ATP levels stable. Fundc1-induced mitophagy was boosted after 3 h of H_2 administration at 2%, and the mice were protected from sepsis-induced liver damage [63]. Additionally, research has revealed that H_2 is a part of the breakdown process of injured mitochondria intracellular homeostasis [87], as both the LC3 II/LC3 I ratio and Beclin-1 expression of damaged cardiomyocytes increased under the control of H_2-rich water. Stress-related p38 and JNK/MAPK pathways [88], which H_2 may block, are both necessary for autophagy activation. Similarly, A549 and H1975 lung cancer cell lines treated with varying doses of H_2 gas exhibited dramatically increased apoptosis and autophagy [89, 90]. The studies concluded that H_2 might protect cells and tissues from harm and also has a regulating influence on autophagy when autophagy is hyperactivated during inflammation.

Pyrolysis

Protecting macrophages, monocytes, and other invading pathogens, pyrolysis is a programmed death mechanism of pro-inflammatory cells. Although pyrolysis is generally helpful to the host, it may cause sepsis and septic shock if it occurs in excess. Activation of pyrolysis requires caspase-1, and the cytokines IL-1b and IL-18 are the primary downstream inflammatory agents. Several studies have shown that H_2 has a protective effect in septic mice [91, 92]. Subarachnoid hemorrhage brain damage models show that treatment with H_2-rich saline significantly reduces caspase-1 expression and inhibits inflammatory response [93]. In addition, H_2 therapy dramatically decreased the expression of caspase-1 in the injured organ and the levels of IL-1β and IL-18 cytokines in models of organ destruction caused by sepsis [54, 92]. Inflating the lungs with H_2 is a tried and reliable strategy for preventing I/R harm in lung donors. Inhibiting pyroptosis by H_2 decreased lung I/R damage in Wistar rats, as shown by a decrease in the pyroptosis-related proteins NLRP3 caspase-1 and the N-terminal of the gasdermin D (GSDMD-N) after 3% H_2 inflation of the lungs [94]. In cancers, however, H_2 may have a distinct regulatory function. By activating the NLRP3 inflammasome and the caspase-1-mediated classical pyroptosis pathway. Yang et al. demonstrated that H_2-rich water reduced the growth of endometrial cancer cells [95]. Cell pyrolysis [96] may be influenced by the negative regulation of HMGB1 by H_2, an endotoxin transporter protein that is required for activation of the caspase-11 pyroptosis pathway [95]. Through mechanisms including apoptosis, autophagy, and pyrolysis, H_2 may impede the growth of cancer cells. While the precise role of H_2 in cellular pyroptosis remains unknown, the protein's modulation of certain nuclear proteins and inflammatory factors might impede the process. Similar to apoptosis [54], H_2's influence on the pyrolysis route may suppress tumor cells or shield healthy tissues and cells from harm.

Impacts of Hydrogen on the Immune System

The overactivation of immune cells and pro-inflammatory chemicals is a major contributor to inflammation in many inflammatory disorders. EAE has long served as a reliable proxy for human MS. By decreasing $CD4^+$ T cell infiltration and blocking

Th17 cell development in the spinal cord, intervention with H_2-rich water may improve EAE symptoms [97]. H_2 at varying doses may boost antitumor immune activity and ameliorate immunodeficiency by raising the percentage of $CD8^+$ T cells [30]. Many individuals undergoing radiotherapy have negative side effects due to radiation-induced immunological dysfunction. Pretreatment with H_2 protected mice against radiation-induced splenocyte death and enhanced $CD4^+$ and $CD8^+$ T cells, according to a study [98]. In addition, healthy individual peripheral blood cells showed considerable downregulation of inflammation and apoptosis signaling after four weeks of H_2-water intake [99]. Congestion and swelling in the nasal tissues are the symptoms of allergic rhinitis, which results from a type I hypersensitivity response, including the activation of mast cells and eosinophils. Reducing the inflammatory response of Th2 cells [100] allows for a 67% H_2 concentration to provide relief. Furthermore, Xu et al. [59] showed that H_2-rich saline relieves allergic rhinitis by correcting the Th1/Th2 imbalance. Many inflammatory illnesses exhibit pathological characteristics associated with macrophage polarization and an imbalance of M1/M2 cells [101]. Previous research has shown that acute kidney damage [102], rheumatoid arthritis [103], and ischemic stroke [104], all of which are characterized by inflammation, respond favorably to treatments that increase H_2 levels. In a rat model of chronic pancreatitis, the H_2 was the first to be shown to restore Treg loss, suggesting that H_2 also controls inflammation by mediating Treg. Reduced inflammation was achieved by low-dose H_2 intervention's promotion of Treg proliferation and suppression of immunological overactivation [49, 105]. By controlling the growth of immune cells, H_2 may prevent either an overactive or underactive immune system [54]. Potential Mechanisms of hydrogen therapy in disease treatment can be seen in the Fig. 14.1.

Hydrogen Therapy in Cancer

Preclinical Studies on Cancer

The contemporary approach to managing cancer primarily involves surgical intervention, chemotherapy, and targeted therapy. The efficacy of anticancer therapy has led to a marked improvement in the prognosis of individuals with cancer. Nevertheless, some patients exhibit a heightened frequency of unfavorable outcomes resulting from surgical procedures and chemotherapeutic interventions. Hence, it is imperative to investigate novel therapeutic approaches for these individuals [106–108]. Studies have demonstrated that molecular hydrogen possesses antioxidant, anti-inflammatory, and antiapoptotic characteristics [29]. Dole et al. [109] stated the capacity of H_2 for cancer therapy. Ohsawa et al. [7] showed that hydrogen could play an anti-oxidant role, resulting in the removal of ROS (123). The findings indicate that the potential anti-tumor effect of hydrogen may be attributed to its ability to modulate intratumoral ROS levels. Also, the consumption of hydrogen-rich water resulted

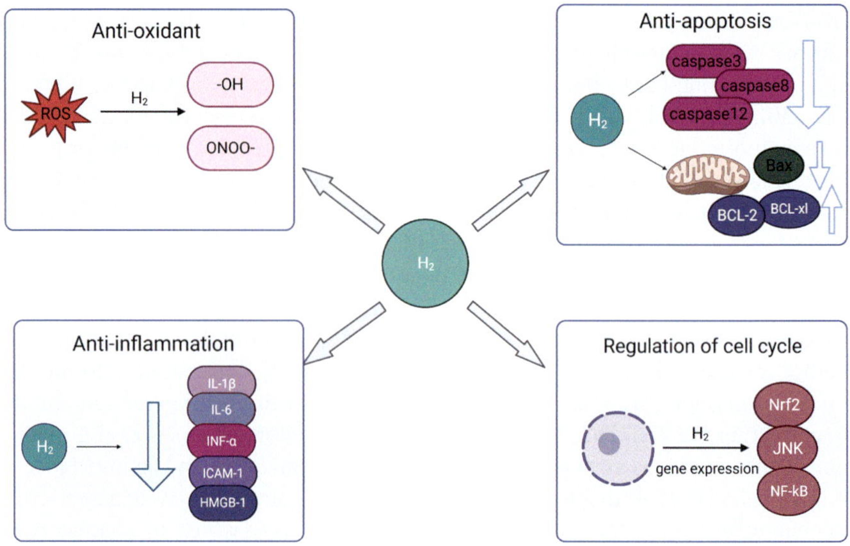

Fig. 14.1 Potential Mechanisms of hydrogen therapy in disease treatment

in a notable increase in the levels of antioxidant enzyme SOD. This suggests that hydrogen-rich water has the potential to stimulate SOD activity [110]. Zhang et al. [113] revealed that the proliferation of colorectal cancer cell lines was suppressed through hydrogen treatment. They indicated that elevated levels of H_2 exhibit a suppressive impact on colorectal cancer by impeding the pAKT/SCD1 signaling pathway. Similarly, Wang et al. [115] investigated the effects of hydrogen gas on cell viability, apoptosis, migration, and invasion in lung cancer cell lines A549 and H1975. The findings demonstrated that H_2 decreased cell viability, migration, and invasion, accelerated cell apoptosis, and caused the G2/M arrest in A549 and H1975 cells. Additionally, H_2 decreased the expression of Cyclin D1, CDK4, and CDK6, as well as NIBPL, SMC3, SMC5, and SMC6. During cell division, H_2 changed SMC3's subcellular localization, lowering its stability and raising its ubiquitination in both A549 and H1975 cells. Overall, they showed that preserving the chromosomal integrity with 60% H_2 might prevent lung cancer migration and invasion. Also, Liu et al. [111] aimed to evaluate the possible impact of molecular hydrogen on glioblastoma multiforme (GBM). Both a mouse subcutaneous xenograft model and a rat orthotopic glioma model were used in the in vivo investigations. hydrogen inhalation inhibitted the development of GBM tumors. Immunohistochemistry and immunofluorescence staining showed that hydrogen treatment elevated GFAP expression, (ki67). Yang et al. showed the upregulation of ROS and pyroptosis-related protein expression, as well as the increase in the number of PI- and TUNEL-positive cells and the release of LDH and IL-1β, following hydrogen pretreatment. However, the release of these factors was reduced upon depletion of GSDMD [112] (Table 14.1).

Table 14.1 Preclinical studies inquiring about the effects of hydrogen on tumor cells

Study	Cancer type	Cell lines	Animals	Hydrogen administration route	Results
Zhang et al. [113]	Colorectal cancer	CRC cell lines RKO, SW480, and HCT116	BALB/c-nude mice	Hydrogen inhalation	High amounts of H_2 showed an inhibitory effect on colorectal cancer via inhibiting the pAKT/SCD1 pathway
Shang et al. [114]	Ovarian cancer	Hs38.T and PA-1	BALB/c nude mice	Hydrogen inhalation	Molecular hydrogen has an anti-tumor role in ovarian cancer via reducing the proliferation of CSCs-like cells and angiogenesis
Wang et al. [115]	Lung cancer	A549 and H1975	BALB/c-nude mice	Hydrogen inhalation	H_2 blocked lung cancer progression via down-regulating SMC3
Yang et al. [116]	Endometrial cancer	Ishikawa, HEC1A, AN3CA	BALB/c-nude mice	Hydrogen inhalation	The results demonstrated the ability of hydrogen to provoke NLRP3 inflammasome/ GSDMD activation in pyroptosis
Asgharzadeh et al. [117]	Colorectal cancer	CT-26	BALB/c mice	Hydrogen-rich water	Administration of hydrogen-rich water, with or without 5-fluorouracil, may serve as a therapeutic for treating colorectal cancer
Liu et al. [118]	Glioblastoma	Rat C6 glioma cells and human U87 cells	Wistar rats	Hydrogen inhalation	Hydrogen reduced the expression of markers involved in stems (CD133, Nestin), proliferation (ki67), and angiogenesis (CD34) and also increased GFAP expression
Zhao et al. [119]	Thymic lymphoma	–	BALB/c mice	H_2-rich saline	H_2 pre-treated mice showed elevated antioxidant levels and a higher survival

(continued)

Table 14.1 (continued)

Study	Cancer type	Cell lines	Animals	Hydrogen administration route	Results
Runtuwene [120]	Colorectal cancer	CT-26	BALB/c mice	Hydrogen-rich water	High-content hydrogen water in combination with 5-fluorouracil suppressed colon cancer
Meng et al. [121]	Lung cancer	A549 and H1975	BALB/c athymic nude mice	Hydrogen inhalation	H_2 inhibits the progression of lung cancer via down-regulating CD47

Clinical Studies

Despite the extensive preclinical evidence supporting the potential of hydrogen therapy in inhibiting tumor progression, there is a dearth of research examining this phenomenon in clinical trial. According to a study conducted in Japan, the inhalation of H_2 for three hours per day has the potential to enhance the prognosis of patients with advanced colorectal cancer [122]. In addition, it was observed that hydrogen-enriched water demonstrated potential in mitigating liver damage associated with mFOLFOX6 chemotherapy in individuals with colorectal cancer undergoing treatment [123]. Kang and colleagues [124] observed that drinking hydrogen-rich water improved the quality-of-life of the patients undergoing radiotherapy and did not need additional hospital visits. In general, the results of a clinical investigation indicate that the inhalation of hydrogen has been shown to enhance the overall well-being of individuals. Following two weeks of hydrogen inhalation, notable amelioration of patients' symptoms has been observed. Furthermore, the inhalation of hydrogen has been shown to enhance physical fitness. Thirdly, the inhalation of hydrogen has been shown to potentially decrease the levels of tumor markers. Hydrogen inhalation therapy has been found to have the potential to regulate the progression of cancer. Following three months, patients diagnosed with stage III cancer exhibited a significantly higher rate of tumor control in comparison to those diagnosed with stage IV cancer. Notably, patients with lung cancer demonstrated the most favorable response, while those with pancreatic cancer exhibited the least favorable response [30].

Despite the lack of clear elucidation regarding the mechanisms underlying the advantageous impacts of hydrogen-rich water in cancer, the consumption of water supplemented with hydrogen has been observed to decrease levels of reactive oxidative metabolites derivatives and sustain levels of biological antioxidant power in the serum. This suggests that hydrogen-rich water possesses potent systemic antioxidant activity [29]. A recent randomized controlled trial revealed that the group receiving hydrogen-water exhibited a significant decrease in apoptosis of peripheral blood mononuclear cells, a significant reduction in the frequency of $CD14^+$

cells, and a significant down-regulation of transcriptional networks associated with inflammatory responses and NF-κB signaling [125]. A recent study demonstrated that the inhalation of hydrogen gas may have a positive impact on the prognosis of patients with advanced colorectal cancer. This effect is attributed to an increase in the number of PD-L1/$CD8^+$ cells [29]. The findings of Lian et al. [122] also illustrate that hydrogen-enriched water has the potential to mitigate hyperalgesia induced by oxaliplatin, modify the composition of gut microbiota, reduce microbial diversity, restore the balance of inflammatory cytokines and oxidative stress, and lower the expression of LPS and TLR4.

Side Effects of Hydrogen Therapy

The side effects of hydrogen therapy are still not fully understood. In March 2019, 79 human publications with a total of 1676 participants used hydrogen therapy. several other clinical trial studies have been published and reported minor side effects [126].

Gastrointestinal Side Effects

Diarrhea and increased stool frequency and bloating are complications that have occurred in more than one person in some studies, and this bloating is due to the accumulation of hydrogen gas in the intestines. Also, increased exhaled hydrogen and irritable bowel syndrome are related because lactose intolerance causes increased hydrogen in the gut, which can be associated with bloating and diarrhea, but based on some evidence, hydrogen therapy can also have a positive effect on the gut. Improvement of some intestinal inflammatory diseases, colitis, and ulcerative [126, 127].

Hypoglycemia

A report of a case of hypoglycemia of diabetes following the consumption of hydrogenated water has been recorded. Studies show that taking hydrogen tablets increases insulin sensitivity in people with nonalcoholic fatty liver disease, decreases insulin in overweight people, and improves metabolic syndrome [126, 128].

Heartburn

In a clinical trial, heartburn was mentioned as a common problem among the participants of hydrogen therapy. Another study showed that hydrogen therapy was effective in relieving gastric reflux and heartburn, which are the main symptoms of gastric reflux [126].

Heart Failure

In a study, one person suffered heart failure after treatment with hydrogen, but intravenous hydrogen salt on patients recovering from acute ischemic stroke, who are at risk of cardiac arrest was at a rate of 3.9% [126–128].

Headache

A few reports have shown that headache is the most prominent complication of hydrogen therapy. The participants stated that they get headaches after the first hydrogen therapy, and after receiving several more high-dose hydrogens, their headaches disappear [129].

Conclusion

Hydrogen molecules have been studied for their potential therapeutic effects in various diseases, including cancer. Hydrogen gas has been shown to have anti-inflammatory and antioxidant properties that may help reduce oxidative stress and inflammation, which are known to contribute to the development and progression of cancer. The role of hydrogen molecules in cancer has been studied extensively over the past decade. Several studies have shown that hydrogen molecules can inhibit tumor growth and metastasis by regulating various cellular processes. Inhalation of hydrogen gas reduced oxidative stress and inflammation in mice with lung cancer. The researchers concluded that hydrogen gas may be a promising therapeutic agent for lung cancer treatment. Another study showed that drinking water enriched with hydrogen molecules inhibited tumor growth and metastasis in mice with colon cancer. The researchers suggested that hydrogen water may be a safe and effective adjuvant therapy for colon cancer patients. In addition to its anti-tumor effects, hydrogen molecules have also been shown to enhance the efficacy of chemotherapy and radiation therapy. Another study published in 2014 showed that drinking water enriched with hydrogen molecules enhanced the anti-tumor effects of radiation therapy in mice with liver cancer. The mechanisms by which hydrogen molecules exert their anti-cancer effects are not fully understood. However, several studies have suggested that hydrogen molecules may regulate various cellular processes, including oxidative stress, inflammation, and apoptosis. Despite promising preclinical studies, there are currently limited clinical trials investigating the use of hydrogen molecules as a therapeutic agent for cancer treatment. However, several clinical trials are ongoing or planned. In conclusion, while there is some evidence to suggest that hydrogen molecules may have therapeutic potential for cancer treatment, more research is needed before any definitive conclusions can be drawn. It is important for individuals with cancer to consult with their healthcare providers before considering any alternative or complementary therapies.

References

1. Schut GJ, Zadvornyy O, Wu C-H, Peters JW, Boyd ES, Adams MWW (2016) The role of geochemistry and energetics in the evolution of modern respiratory complexes from a proton-reducing ancestor. Biochim Biophys Acta Bioenerg 1857(7):958–970. Available from: https://doi.org/10.1016/j.bbabio.2016.01.010
2. Ishibashi T (2019) Therapeutic efficacy of molecular hydrogen: a new mechanistic insight. Curr Pharm Des 25(9):946–955
3. Ohsawa I, Ishikawa M, Takahashi K, Watanabe M, Nishimaki K, Yamagata K, et al (2007) Hydrogen acts as a therapeutic antioxidant by selectively reducing cytotoxic oxygen radicals. Nat Med 13(6):688–694. Available from: https://doi.org/10.1038/nm1577
4. Ohta S (2014) Molecular hydrogen as a preventive and therapeutic medical gas: initiation, development and potential of hydrogen medicine. Pharmacol Ther 144(1):1–11. Available from: https://doi.org/10.1016/j.pharmthera.2014.04.006
5. Ito M, Hirayama M, Yamai K, Goto S, Ito M, Ichihara M et al (2012) Drinking hydrogen water and intermittent hydrogen gas exposure, but not lactulose or continuous hydrogen gas exposure, prevent 6-hydorxydopamine-induced Parkinson's disease in rats. Med Gas Res 2(1):15. Available from: https://pubmed.ncbi.nlm.nih.gov/22608009
6. Fu Y, Ito M, Fujita Y, Ito M, Ichihara M, Masuda A et al (2009) Molecular hydrogen is protective against 6-hydroxydopamine-induced nigrostriatal degeneration in a rat model of Parkinson's disease. Neurosci Lett 453(2):81–85. Available from: https://doi.org/10.1016/j.neulet.2009.02.016
7. Yoritaka A, Takanashi M, Hirayama M, Nakahara T, Ohta S, Hattori N (2013) Pilot study of H_2therapy in Parkinson's disease: a randomized double-blind placebo-controlled trial. Mov Disord 28(6):836–839. Available from: https://doi.org/10.1002/mds.25375
8. Ramanathan D, Huang L, Wilson T, Boling W (2023) Molecular hydrogen therapy for neurological diseases: a review of current evidence. Med Gas Res 13(3):94–98
9. Gu Y, Huang C-S, Inoue T, Yamashita T, Ishida T, Kang K-M et al (2010) Drinking hydrogen water ameliorated cognitive impairment in senescence-accelerated mice. J Clin Biochem Nutr 46(3):269–276. Available from: https://pubmed.ncbi.nlm.nih.gov/20490324
10. Wang C, Li J, Liu Q, Yang R, Zhang JH, Cao Y-P et al (2011) Hydrogen-rich saline reduces oxidative stress and inflammation by inhibit of JNK and NF-κB activation in a rat model of amyloid-beta-induced Alzheimer's disease. Neurosci Lett 491(2):127–132. Available from: https://doi.org/10.1016/j.neulet.2011.01.022
11. Li J, Wang C, Zhang JH, Cai J-M, Cao Y-P, Sun X-J (2010) Hydrogen-rich saline improves memory function in a rat model of amyloid-beta-induced Alzheimer's disease by reduction of

oxidative stress. Brain Res 1328:152–61. Available from: https://doi.org/10.1016/j.brainres.2010.02.046
12. Zhang L, Zhao P, Yue C, Jin Z, Liu Q, Du X et al (2019) Sustained release of bioactive hydrogen by Pd hydride nanoparticles overcomes Alzheimer's disease. Biomaterials 197:393–404. Available from: https://doi.org/10.1016/j.biomaterials.2019.01.037
13. Dohi K, Kraemer BC, Erickson MA, McMillan PJ, Kovac A, Flachbartova Z et al (2014) Molecular hydrogen in drinking water protects against neurodegenerative changes induced by traumatic brain injury. PLoS One 9(9):e108034–e108034. Available from: https://pubmed.ncbi.nlm.nih.gov/25251220
14. Ji X, Liu W, Xie K, Liu W, Qu Y, Chao X et al (2010) Beneficial effects of hydrogen gas in a rat model of traumatic brain injury via reducing oxidative stress. Brain Res 1354:196–205. Available from: https://doi.org/10.1016/j.brainres.2010.07.038
15. Eckermann JM, Chen W, Jadhav V, Hsu FP, Colohan AR, Tang J et al (2011) Hydrogen is neuroprotective against surgically induced brain injury. Med Gas Res 1(1):7. Available from: https://pubmed.ncbi.nlm.nih.gov/22146427
16. Javorac D, Stajer V, Ostojic SM (2019) Case report: buccal administration of hydrogen-producing blend after a mild traumatic brain injury in a professional athlete. F1000Research 8:1024. Available from: https://pubmed.ncbi.nlm.nih.gov/32595937
17. Liu L, Xie K, Chen H, Dong X, Li Y, Yu Y et al (2014) Inhalation of hydrogen gas attenuates brain injury in mice with cecal ligation and puncture via inhibiting neuroinflammation, oxidative stress and neuronal apoptosis. Brain Res 1589:78–92. Available from: https://doi.org/10.1016/j.brainres.2014.09.030
18. Zhang J, Liu C, Tai M, Qu K (2012) Effect of hydrogen gas on the survival rate of mice following global cerebral ischemia (Shock 37(6), 645–652, 2012). Shock 38(4):444. Available from: https://doi.org/10.1097/shk.0b013e318268deb2
19. Huang L, Applegate 2nd RL, Applegate PM, Boling W, Zhang JH (2018) Inhalation of high concentration hydrogen gas improves short-term outcomes in a rat model of asphyxia induced-cardiac arrest. Med Gas Res 8(3):73–78. Available from: https://pubmed.ncbi.nlm.nih.gov/30319760
20. Yu Y, Feng J, Lian N, Yang M, Xie K, Wang G, Wang C, Yu Y (2020) Hydrogen gas alleviates blood-brain barrier impairment and cognitive dysfunction of septic mice in an Nrf2-dependent pathway. Int Immunopharmacol 85:106585 1567–5769. https://doi.org/10.1016/j.intimp.2020.106585
21. Huang L, Applegate Ii RL, Applegate PM, Gong L, Ocak U, Boling W, et al (2019) Inhalation of high-concentration hydrogen gas attenuates cognitive deficits in a rat model of asphyxia induced-cardiac arrest. Med Gas Res 9(3):122–6. Available from: https://pubmed.ncbi.nlm.nih.gov/31552874
22. Ishibashi T (2013) Molecular hydrogen: new antioxidant and anti-inflammatory therapy for rheumatoid arthritis and related diseases. Curr Pharm Des 19(35):6375–81. https://doi.org/10.2174/13816128113199990507. PMID: 23859555; PMCID: PMC3788323
23. Li Q, Kato S, Matsuoka D, Tanaka H, Miwa N (2013) Hydrogen water intake via tube-feeding for patients with pressure ulcer and its reconstructive effects on normal human skin cells in vitro. Med Gas Res Sep 10:3(1):20. https://doi.org/10.1186/2045-9912-3-20. PMID: 24020833; PMCID: PMC3843550
24. Cai J, Kang Z, Liu WW, Luo X, Qiang S, Zhang JH et al (2008) Hydrogen therapy reduces apoptosis in neonatal hypoxia–ischemia rat model. Neurosci Lett 441(2):167–172. Available from: https://doi.org/10.1016/j.neulet.2008.05.077
25. Hong Y, Shao A, Wang J, Chen S, Wu H, McBride DW et al (2014) Neuroprotective effect of hydrogen-rich saline against neurologic damage and apoptosis in early brain injury following subarachnoid hemorrhage: possible role of the Akt/GSK3β signaling pathway. PLoS One 9(4):e96212–e96212. Available from: https://pubmed.ncbi.nlm.nih.gov/24763696
26. Adzavon YM, Xie F, Yi Y, Jiang X, Zhang X, He J et al (2022) Long-term and daily use of molecular hydrogen induces reprogramming of liver metabolism in rats by modulating NADP/NADPH redox pathways. Sci Rep 12(1):3904. Available from: https://pubmed.ncbi.nlm.nih.gov/35273249

27. Hasegawa T, Ito M, Hasegawa S, Teranishi M, Takeda K, Negishi S et al (2022) Molecular hydrogen enhances proliferation of cancer cells that exhibit potent mitochondrial unfolded protein response. Int J Mol Sci 23(5):2888. Available from: https://pubmed.ncbi.nlm.nih.gov/35270030
28. Fiorese CJ, Schulz AM, Lin Y-F, Rosin N, Pellegrino MW, Haynes CM (2016) The transcription factor ATF5 mediates a mammalian mitochondrial UPR. Curr Biol 26(15):2037–2043. Available from: https://pubmed.ncbi.nlm.nih.gov/27426517
29. Artamonov MY, Martusevich AK, Pyatakovich FA, Minenko IA, Dlin SV, LeBaron TW (2023) Molecular hydrogen: from molecular effects to stem cells management and tissue regeneration. Antioxid (Basel) 12(3):636. https://doi.org/10.3390/antiox12030636. PMID: 36978884; PMCID: PMC10045005
30. Akagi J, Baba H (2018) Hydrogen gas restores exhausted CD8+ T-cells in patients with advanced colorectal cancer to improve prognosis. Oncol Rep. Available from: https://doi.org/10.3892/or.2018.6841
31. Chen J-B, Kong X-F, Lv Y-Y, Qin S-C, Sun X-J, Mu F et al (2019) "Real world survey" of hydrogen-controlled cancer: a follow-up report of 82 advanced cancer patients. Med Gas Res 9(3):115–121. Available from: https://pubmed.ncbi.nlm.nih.gov/31552873
32. Chen J-B, Pan Z-B, Du D-M, Qian W, Ma Y-Y, Mu F et al (2019) Hydrogen gas therapy induced shrinkage of metastatic gallbladder cancer: a case report. World J Clin cases 7(15):2065–2074. Available from: https://pubmed.ncbi.nlm.nih.gov/31423439
33. Chen J, Mu F, Lu T, Ma Y, Du D, Xu K (2019) A gallbladder carcinoma patient with pseudo-progressive remission after hydrogen inhalation. Onco Targets Ther 12:8645–8651. Available from: https://pubmed.ncbi.nlm.nih.gov/31695424
34. Chen J, Mu F, Lu T, Du D, Xu K (2019) Brain metastases completely disappear in non-small cell lung cancer using hydrogen gas inhalation: a case report. Onco Targets Ther 12:11145–11151. Available from: https://pubmed.ncbi.nlm.nih.gov/31908482
35. Gloire G, Legrand-Poels S, Piette J (2006) NF-κB activation by reactive oxygen species: fifteen years later. Biochem Pharmacol 72(11):1493–505. Available from: https://doi.org/10.1016/j.bcp.2006.04.011
36. Moe KT, Aulia S, Jiang F, Chua YL, Koh TH, Wong MC et al (2006) Differential upregulation of Nox homologues of NADPH oxidase by tumor necrosis factor-alpha in human aortic smooth muscle and embryonic kidney cells. J Cell Mol Med 10(1):231–239. Available from: https://pubmed.ncbi.nlm.nih.gov/16563235
37. Ryter SW, Alam J, Choi AMK (2006) Heme oxygenase-1/carbon monoxide: from basic science to therapeutic applications. Physiol Rev 86(2):583–650. Available from: https://doi.org/10.1152/physrev.00011.2005
38. Itoh T, Fujita Y, Ito M, Masuda A, Ohno K, Ichihara M, Kojima T, Nozawa Y, Ito M (2009) Molecular hydrogen suppresses FcepsilonRI-mediated signal transduction and prevents degranulation of mast cells. Biochem Biophys Res Commun 389(4):651–656. https://doi.org/10.1016/j.bbrc.2009.09.047
39. Zhang J-Y, Liu C, Zhou L, Qu K, Wang R, Tai M et al (2012) A review of hydrogen as a new medical therapy. Hepatogastroenterology 59(116):1026–1032
40. Burton GJ, Jauniaux E (2010) Oxidative stress. Best Pract Res Clin Obstet Gynaecol 25(3):287–299. Available from: https://pubmed.ncbi.nlm.nih.gov/21130690
41. Dan Dunn J, Alvarez LA, Zhang X, Soldati T (2015) Reactive oxygen species and mitochondria: a nexus of cellular homeostasis. Redox Biol 6:472–485. Available from: https://pubmed.ncbi.nlm.nih.gov/26432659
42. Wang J, Hossain M, Thanabalasuriar A, Gunzer M, Meininger C, Kubes P (2017) Visualizing the function and fate of neutrophils in sterile injury and repair. Science (80-) 358(6359):111–116. Available from: https://doi.org/10.1126/science.aam9690
43. Iuchi K, Nishimaki K, Kamimura N, Ohta S (2019) Molecular hydrogen suppresses free-radical-induced cell death by mitigating fatty acid peroxidation and mitochondrial dysfunction. Can J Physiol Pharmacol 97(10):999–1005. Available from: https://doi.org/10.1139/cjpp-2018-0741

44. LeBaron TW, Kura B, Kalocayova B, Tribulova N, Slezak J (2019) A new approach for the prevention and treatment of cardiovascular disorders. Molecular hydrogen significantly reduces the effects of oxidative stress. Molecules 24(11):2076. Available from: https://pubmed.ncbi.nlm.nih.gov/31159153
45. Gharib B, Hanna S, Abdallahi OMS, Lepidi H, Gardette B, De Reggi M (2001) Anti-inflammatory properties of molecular hydrogen: investigation on parasite-induced liver inflammation. Comptes Rendus l'Académie des Sci - Ser III - Sci la Vie 324(8):719–724. Available from: https://doi.org/10.1016/s0764-4469(01)01350-6
46. Kiyoi T, Liu S, Takemasa E, Nakaoka H, Hato N, Mogi M (2020) Constitutive hydrogen inhalation prevents vascular remodeling via reduction of oxidative stress. PLoS One 15(4):e0227582–e0227582. Available from: https://pubmed.ncbi.nlm.nih.gov/32302306
47. Huang L (2016) Molecular hydrogen: a therapeutic antioxidant and beyond. Med Gas Res 6(4):219–222. Available from: https://pubmed.ncbi.nlm.nih.gov/28217294
48. Dixon BJ, Tang J, Zhang JH (2016) The evolution of molecular hydrogen: a noteworthy potential therapy with clinical significance. Med Gas Res 3(1):10. Available from: https://pubmed.ncbi.nlm.nih.gov/23680032
49. Tonelli C, Chio IIC, Tuveson DA (2018) Transcriptional regulation by Nrf2. Antioxid Redox Signal 29(17):1727–1745. Available from: https://pubmed.ncbi.nlm.nih.gov/28899199
50. Xu F, Yu S, Qin M, Mao Y, Jin L, Che N et al (2017) Hydrogen-Rich saline ameliorates allergic rhinitis by reversing the imbalance of Th1/Th2 and Up-Regulation of CD4+CD25+Foxp3+Regulatory T Cells, Interleukin-10, and Membrane-Bound transforming growth factor-β in Guinea Pigs. Inflammation 41(1):81–92. Available from: https://doi.org/10.1007/s10753-017-0666-6
51. Su N, Wu Q, Chen H, Huang Y, Zhu Z, Chen Y et al (2019) Hydrogen gas alleviates toxic effects of cadmium in Brassica campestris seedlings through up-regulation of the antioxidant capacities: possible involvement of nitric oxide. Environ Pollut 251:45–55. Available from: https://doi.org/10.1016/j.envpol.2019.03.094
52. Zhao Y-S, An J-R, Yang S, Guan P, Yu F-Y, Li W et al (2019) Hydrogen and oxygen mixture to improve cardiac dysfunction and myocardial pathological changes induced by intermittent hypoxia in rats. Oxid Med Cell Longev 7415212. Available from: https://pubmed.ncbi.nlm.nih.gov/30984338
53. Poljsak B, Šuput D, Milisav I (2013) Achieving the balance between ROS and antioxidants: when to use the synthetic antioxidants. Oxid Med Cell Longev 2013:956792. Available from: https://pubmed.ncbi.nlm.nih.gov/23738047
54. Singh K, Bhori M, Kasu YA, Bhat G, Marar T (2018) Antioxidants as precision weapons in war against cancer chemotherapy induced toxicity—exploring the armoury of obscurity. Saudi Pharm J SPJ Off Publ Saudi Pharm Soc 26(2):177–190. Available from: https://pubmed.ncbi.nlm.nih.gov/30166914
55. Ma Q (2013) Role of nrf2 in oxidative stress and toxicity. Annu Rev Pharmacol Toxicol 53:401–26. https://doi.org/10.1146/annurev-pharmtox-011112-140320. PMID: 23294312; PMCID: PMC4680839
56. Tian Y, Zhang Y, Wang Y, Chen Y, Fan W, Zhou J et al (2021) Hydrogen, a novel therapeutic molecule, regulates oxidative stress, inflammation, and apoptosis. Front Physiol 12:789507
57. Annesley SJ, Fisher PR (2019) Mitochondria in health and disease. Cells 8(7):680. Available from: https://pubmed.ncbi.nlm.nih.gov/31284394
58. Yoshida A, Asanuma H, Sasaki H, Sanada S, Yamazaki S, Asano Y et al () H2 mediates cardioprotection via involvements of KATP channels and permeability transition pores of mitochondria in dogs. Cardiovasc Drugs Ther 26(3):217–226. Available from: https://doi.org/10.1007/s10557-012-6381-5
59. Gvozdjáková A, Kucharská J, Kura B, Vančová O, Rausová Z, Sumbalová Z et al (2020) A new insight into the molecular hydrogen effect on coenzyme Q and mitochondrial function of rats. Can J Physiol Pharmacol 98(1):29–34. Available from: https://doi.org/10.1139/cjpp-2019-0281

60. Liu QU, Li B-S, Song Y-J, Hu M-G, Lu J-Y, Gao ANG et al (2016) Hydrogen-rich saline protects against mitochondrial dysfunction and apoptosis in mice with obstructive jaundice. Mol Med Rep 13(4):3588–3596. Available from: https://doi.org/10.3892/mmr.2016.4954
61. Yan M, Yu Y, Mao X, Feng J, Wang Y, Chen H et al (2019) Hydrogen gas inhalation attenuates sepsis-induced liver injury in a FUNDC1-dependent manner. Int Immunopharmacol 71:61–67. Available from: https://doi.org/10.1016/j.intimp.2019.03.021
62. Wu X, Li X, Liu Y, Yuan N, Li C, Kang Z et al (2018) Hydrogen exerts neuroprotective effects on OGD/R damaged neurons in rat hippocampal by protecting mitochondrial function via regulating mitophagy mediated by PINK1/Parkin signaling pathway. Brain Res 1698:89–98. Available from: https://doi.org/10.1016/j.brainres.2018.06.028
63. Zhang Y, Dong A, Xie K, Yu Y (2020) Protective effects of hydrogen on myocardial mitochondrial functions in septic mice. Biomed Res Int 2020:1568209. Available from: https://pubmed.ncbi.nlm.nih.gov/32083123
64. Hirayama M, Ito M, Minato T, Yoritaka A, LeBaron TW, Ohno K (2019) Inhalation of hydrogen gas elevates urinary 8-hydroxy-2'-deoxyguanine in Parkinson's disease. Med Gas Res 8(4):144–149. Available from: https://pubmed.ncbi.nlm.nih.gov/30713666
65. Yoritaka A, Kobayashi Y, Hayashi T, Saiki S, Hattori N (2021) Randomized double-blind placebo-controlled trial of hydrogen inhalation for Parkinson's disease: a pilot study. Neurol Sci 42(11):4767–4770. Available from: https://pubmed.ncbi.nlm.nih.gov/34319514
66. Eming SA, Wynn TA, Martin P (2017) Inflammation and metabolism in tissue repair and regeneration. Science (80-) 356(6342):1026–1030. Available from: https://doi.org/10.1126/science.aam7928
67. Mittal M, Siddiqui MR, Tran K, Reddy SP, Malik AB (2014) Reactive oxygen species in inflammation and tissue injury. Antioxid Redox Signal 20(7):1126–67. Available from: https://pubmed.ncbi.nlm.nih.gov/23991888
68. Rimessi A, Previati M, Nigro F, Wieckowski MR, Pinton P (2016) Mitochondrial reactive oxygen species and inflammation: molecular mechanisms, diseases and promising therapies. Int J Biochem Cell Biol 81:281–93. Available from: https://doi.org/10.1016/j.biocel.2016.06.015
69. Forrester SJ, Kikuchi DS, Hernandes MS, Xu Q, Griendling KK (2018) Reactive oxygen species in metabolic and inflammatory signaling. Circ Res 122(6):877–902. Available from: https://pubmed.ncbi.nlm.nih.gov/29700084
70. Chen M, Zhang J, Chen Y, Qiu Y, Luo Z, Zhao S et al (2018) Hydrogen protects lung from hypoxia/re-oxygenation injury by reducing hydroxyl radical production and inhibiting inflammatory responses. Sci Rep 8(1):8004. Available from: https://pubmed.ncbi.nlm.nih.gov/29789753
71. Wang X, Yu P, YongYang, Liu X, Jiang J, Liu D et al (2015) Hydrogen-rich saline resuscitation alleviates inflammation induced by severe burn with delayed resuscitation. Burns 41(2):379–385. Available from: https://doi.org/10.1016/j.burns.2014.07.012
72. Tan YC, Xie F, Zhang HL, Zhu YL, Chen K, Tan HM et al (2014) Hydrogen-rich saline attenuates postoperative liver failure after major hepatectomy in rats. Clin Res Hepatol Gastroenterol 38(3):337–345. Available from: https://doi.org/10.1016/j.clinre.2013.11.007
73. Zheng Z-G, Sun W-Z, Hu J-Y, Jie Z-J, Xu J-F, Cao J et al (2021) Hydrogen/oxygen therapy for the treatment of an acute exacerbation of chronic obstructive pulmonary disease: results of a multicenter, randomized, double-blind, parallel-group controlled trial. Respir Res 22(1):149. Available from: https://pubmed.ncbi.nlm.nih.gov/33985501
74. Nogueira JE, Amorim MR, Pinto AP, da Rocha AL, da Silva ASR, Branco LGS (2021) Molecular hydrogen downregulates acute exhaustive exercise-induced skeletal muscle damage. Can J Physiol Pharmacol 99(8):812–20. Available from: https://doi.org/10.1139/cjpp-2020-0297
75. Li Q, Yu P, Zeng Q, Luo B, Cai S, Hui K et al (2016) Neuroprotective effect of hydrogen-rich saline in global cerebral ischemia/reperfusion rats: up-regulated tregs and down-regulated miR-21, miR-210 and NF-κB expression. Neurochem Res 41(10):2655–65. Available from: https://pubmed.ncbi.nlm.nih.gov/27386874

76. Fujioka K, Kalish F, Zhao H, Lu S, Wong S, Wong RJ et al (2017) Induction of Heme Oxygenase-1 attenuates the severity of sepsis in a non-surgical preterm mouse model. Shock 47(2):242–50. Available from: https://doi.org/10.1097/shk.0000000000000689
77. Chen H, Xie K, Han H, Li Y, Liu L, Yang T et al (2015) Molecular hydrogen protects mice against polymicrobial sepsis by ameliorating endothelial dysfunction via an Nrf2/HO-1 signaling pathway. Int Immunopharmacol 28(1):643–654. Available from: https://doi.org/10.1016/j.intimp.2015.07.034
78. Obeng E (2021) Apoptosis (programmed cell death) and its signals—a review. Brazilian J Biol 81(4):1133–1143. Available from: https://doi.org/10.1590/1519-6984.228437
79. Qiu X, Dong K, Guan J, He J (2020) Hydrogen attenuates radiation-induced intestinal damage by reducing oxidative stress and inflammatory response. Int Immunopharmacol 84:106517. Available from: https://doi.org/10.1016/j.intimp.2020.106517
80. Zhang Q, Ge Y, Li H, Bai G, Jiao Z, Kong X, et al (2018) Effect of hydrogen-rich saline on apoptosis induced by hepatic ischemia reperfusion upon laparoscopic hepatectomy in miniature pigs. Res Vet Sci 119:285–291. Available from: https://doi.org/10.1016/j.rvsc.2018.07.005
81. Wang S-T, Bao C, He Y, Tian X, Yang Y, Zhang T et al (2020) Hydrogen gas (XEN) inhalation ameliorates airway inflammation in asthma and COPD patients. QJM 113(12):870–875. Available from: https://pubmed.ncbi.nlm.nih.gov/32407476
82. Mo X-Y, Li X-M, She C-S, Lu X-Q, Xiao C-G, Wang S-H et al (2019) Hydrogen-rich saline protects rat from oxygen glucose deprivation and reperusion-induced apoptosis through VDAC1 via Bcl-2. Brain Res 1706:110–115. Available from: https://doi.org/10.1016/j.brainres.2018.09.037
83. Wu D, Liang M, Dang H, Fang F, Xu F, Liu C (2018) Hydrogen protects against hyperoxia-induced apoptosis in type II alveolar epithelial cells via activation of PI3K/Akt/Foxo3a signaling pathway. Biochem Biophys Res Commun 495(2):1620–1627. Available from: https://doi.org/10.1016/j.bbrc.2017.11.193
84. Li Q, Tanaka Y, Miwa N (2017) Influence of hydrogen-occluding-silica on migration and apoptosis in human esophageal cells in vitro. Med Gas Res 7(2):76–85. Available from: https://pubmed.ncbi.nlm.nih.gov/28744359
85. Zhang Y, Long Z, Xu J, Tan S, Zhang N, Li A et al (2017) Hydrogen inhibits isoproterenol-induced autophagy in cardiomyocytes in vitro and in vivo. Mol Med Rep 16(6):8253–8258. Available from: https://doi.org/10.3892/mmr.2017.7601
86. Zhang Y, Liu Y, Zhang J (2015) Saturated hydrogen saline attenuates endotoxin-induced lung dysfunction. J Surg Res 198(1):41–49. Available from: https://doi.org/10.1016/j.jss.2015.04.055
87. Zhuang X, Yu Y, Jiang Y, Zhao S, Wang Y, Su L et al (2020) Molecular hydrogen attenuates sepsis-induced neuroinflammation through regulation of microglia polarization through an mTOR-autophagy-dependent pathway. Int Immunopharmacol 81:106287. Available from: https://doi.org/10.1016/j.intimp.2020.106287
88. Yao L, Chen H, Wu Q, Xie K (2019) Hydrogen-rich saline alleviates inflammation and apoptosis in myocardial I/R injury via PINK-mediated autophagy. Int J Mol Med 44(3):1048–62. Available from: https://pubmed.ncbi.nlm.nih.gov/31524220
89. Guan P, Sun Z-M, Luo L-F, Zhou J, Yang S, Zhao Y-S et al (2019) Hydrogen protects against chronic intermittent hypoxia induced renal dysfunction by promoting autophagy and alleviating apoptosis. Life Sci 225:46–54. Available from: https://doi.org/10.1016/j.lfs.2019.04.005
90. Liu L, Yan Z, Wang Y, Meng J, Chen G (2020) Suppression of autophagy facilitates hydrogen gas-mediated lung cancer cell apoptosis. Oncol Lett 20(4):112. Available from: https://pubmed.ncbi.nlm.nih.gov/32863925
91. Man SM, Karki R, Kanneganti T-D (2017) Molecular mechanisms and functions of pyroptosis, inflammatory caspases and inflammasomes in infectious diseases. Immunol Rev 277(1):61–75. Available from: https://pubmed.ncbi.nlm.nih.gov/28462526

92. Zhai Y, Zhou X, Dai Q, Fan Y, Huang X (2015) Hydrogen-rich saline ameliorates lung injury associated with cecal ligation and puncture-induced sepsis in rats. Exp Mol Pathol 98(2):268–276. Available from: https://doi.org/10.1016/j.yexmp.2015.03.005
93. Xie K, Wang Y, Yin L, Wang Y, Chen H, Mao X et al (2020) Hydrogen gas alleviates sepsis-induced brain injury by improving mitochondrial biogenesis through the activation of PGC-α in mice. Shock 55(1):100–109. Available from: https://doi.org/10.1097/shk.0000000000001594
94. Zheng P, Kang J, Xing E, Zheng B, Wang X, Zhou H (2021) Lung inflation with hydrogen during the cold ischemia phase alleviates lung ischemia-reperfusion injury by inhibiting pyroptosis in rats. Front Physiol 12:699344. Available from: https://pubmed.ncbi.nlm.nih.gov/34408660
95. Yang Y, Liu PY, Bao W, Chen SJ, Wu FS, Zhu PY (2020) Hydrogen inhibits endometrial cancer growth via a ROS/NLRP3/caspase-1/GSDMD-mediated pyroptotic pathway. BMC Cancer 20(1):28. https://doi.org/10.1186/s12885-019-6491-6. https://pubmed.ncbi.nlm.nih.gov/31924176. PMID: 31924176; PMCID: PMC6954594
96. Deng M, Tang Y, Li W, Wang X, Zhang R, Zhang X et al (2018) The endotoxin delivery protein HMGB1 mediates caspase-11-dependent lethality in sepsis. Immunity 49(4):740–753.e7. Available from: https://pubmed.ncbi.nlm.nih.gov/30314759
97. Yu Y, Yang Y, Yang M, Wang C, Xie K, Yu Y (2019) Hydrogen gas reduces HMGB1 release in lung tissues of septic mice in an Nrf2/HO-1-dependent pathway. Int Immunopharmacol 69:11–8. Available from: https://doi.org/10.1016/j.intimp.2019.01.022
98. Zhao M, Liu M-D, Pu Y-Y, Wang D, Xie Y, Xue G-C et al (2016) Hydrogen-rich water improves neurological functional recovery in experimental autoimmune encephalomyelitis mice. J Neuroimmunol 294:6–13. Available from: https://doi.org/10.1016/j.jneuroim.2016.03.006
99. Zhao S, Yang Y, Liu W, Xuan Z, Wu S, Yu S et al (2014) Protective effect of hydrogen-rich saline against radiation-induced immune dysfunction. J Cell Mol Med 18(5):938–946. Available from: https://pubmed.ncbi.nlm.nih.gov/24618260
100. Sim M, Kim C-S, Shon W-J, Lee Y-K, Choi EY, Shin D-M (2020) Hydrogen-rich water reduces inflammatory responses and prevents apoptosis of peripheral blood cells in healthy adults: a randomized, double-blind, controlled trial. Sci Rep 10(1):12130. Available from: https://pubmed.ncbi.nlm.nih.gov/32699287
101. Huang P, Wei S, Huang W, Wu P, Chen S, Tao A et al (2019) Hydrogen gas inhalation enhances alveolar macrophage phagocytosis in an ovalbumin-induced asthma model. Int Immunopharmacol 74:105646. Available from: https://doi.org/10.1016/j.intimp.2019.05.031
102. Shapouri-Moghaddam A, Mohammadian S, Vazini H, Taghadosi M, Esmaeili S, Mardani F et al (2018) Macrophage plasticity, polarization, and function in health and disease. J Cell Physiol 233(9):6425–6440. Available from: https://doi.org/10.1002/jcp.26429
103. Yao W, Guo A, Han X, Wu S, Chen C, Luo C et al (2019) Aerosol inhalation of a hydrogen-rich solution restored septic renal function. Aging (Albany NY) 11(24):12097–113. Available from: https://pubmed.ncbi.nlm.nih.gov/31841441
104. Meng J, Yu P, Jiang H, Yuan T, Liu N, Tong J et al (2016) Molecular hydrogen decelerates rheumatoid arthritis progression through inhibition of oxidative stress. Am J Transl Res. 8(10):4472–4477
105. Kheir JN, DiNardo JA (2022) Commentary: hydrogen: lightweight molecule takes on a heavyweight problem. J Thorac Cardiovasc Surg 164(6):e286–e287
106. Mattiuzzi C, Lippi G (2019) Current cancer epidemiology. J Epidemiol Glob Health. 9(4):217–222
107. Seebacher N, Stacy A, Porter G, Merlot A (2019) Clinical development of targeted and immune based anti-cancer therapies. J Exp Clin Cancer Res 38(1):1–39
108. Bahadur S, Sahu AK, Baghel P, Saha S (2019) Current promising treatment strategy for glioblastoma multiform: a review. Oncol Rev 13(2)
109. Dole M, Wilson FR, Fife WP (1975) Hyperbaric hydrogen therapy: a possible treatment for cancer. Science 190(4210):152–154

110. Nakao A, Toyoda Y, Sharma P, Evans M, Guthrie N (2010) Effectiveness of hydrogen rich water on antioxidant status of subjects with potential metabolic syndrome-an open label pilot study. J Clin Biochem Nutr. 46(2):140–149
111. Liu M-Y, Xie F, Zhang Y, Wang T-T, Ma S-N, Zhao P-X et al (2019) Molecular hydrogen suppresses glioblastoma growth via inducing the glioma stem-like cell differentiation. Stem Cell Res Ther 10:1–10
112. Shang L, Xie F, Li J, Zhang Y, Liu M, Zhao P et al (2018) Therapeutic potential of molecular hydrogen in ovarian cancer. Transl Cancer Res 7(4):988–995
113. Zhang X, Tao G, Zhao Y, Xing S, Jiang J, Liu B, Qin S (2022) Molecular hydrogen inhibits colorectal cancer growth via the AKT/SCD1 signaling pathway. Biomed Res Int 26:8024452. https://doi.org/10.1155/2022/8024452. PMID: 35528164; PMCID: PMC9071919
114. Shang L, Xie F, Li J, Zhang Y, Liu M, Zhao P, et al (2018) Therapeutic potential of molecular hydrogen in ovarian cancer. Transl Cancer Res 7(4):988–95
115. Wang D, Wang L, Zhang Y, Zhao Y, Chen G (2018) Hydrogen gas inhibits lung cancer progression through targeting SMC3. Biomed Pharmacother 104:788–797. https://doi.org/10.1016/j.biopha.2018.05.055
116. Yang Y, Liu PY, Bao W, Chen SJ, Wu FS, Zhu PY (2020) Hydrogen inhibits endometrial cancer growth via a ROS/NLRP3/caspase-1/GSDMD-mediated pyroptotic pathway. BMC Cancer 10:20(1):28. https://doi.org/10.1186/s12885-019-6491-6. PMID: 31924176; PMCID: PMC6954594
117. Asgharzadeh F, Tarnava A, Mostafapour A, Khazaei M, LeBaron TW (2022) Hydrogen-rich water exerts anti-tumor effects comparable to 5-fluorouracil in a colorectal cancer xenograft model. World J Gastroint Oncol 14(1):242
118. Liu MY, Xie F, Zhang Y, Wang TT, Ma SN, Zhao PX, Zhang X, Lebaron TW, Yan XL, Ma XM (2019) Molecular hydrogen suppresses glioblastoma growth via inducing the glioma stem-like cell differentiation. Stem Cell Res Ther May 21:10(1):145. https://doi.org/10.1186/s13287-019-1241-x. PMID: 31113492; PMCID: PMC6528353
119. Zhao L, Zhou C, Zhang J, Gao F, Li B, Chuai Y et al (2011) Hydrogen protects mice from radiation induced thymic lymphoma in BALB/c mice. Int J Biol Sci 7(3):297–300
120. Runtuwene J, Amitani H, Amitani M, Asakawa A, Cheng KC, Inui A (2015) Hydrogen-water enhances 5-fluorouracil-induced inhibition of colon cancer. PeerJ Apr 7:3:e859. https://doi.org/10.7717/peerj.859. PMID: 25870767; PMCID: PMC4393812
121. Meng J, Liu L, Wang D, Yan Z, Chen G (2020) Hydrogen gas represses the progression of lung cancer via down-regulating CD47. Biosci Rep 40(4)
122. Lian N, Shen M, Zhang K, Pan J, Jiang Y, Yu Y et al (2021) Drinking hydrogen-rich water alleviates chemotherapy-induced neuropathic pain through the regulation of gut microbiota. J Pain Res 681–691
123. https://hho-bulgaria.com/en/hydrogen-therapy-in-cancer-treatment/
124. Liu C, Cui J, Sun Q, Cai J (2010) Hydrogen therapy may be an effective and specific novel treatment for acute radiation syndrome. Med Hypotheses 74(1):145–146
125. Chen JB, Kong XF, Lv YY, Qin SC, Sun XJ, Mu F, Lu TY, Xu KC (2019) "Real world survey" of hydrogen-controlled cancer: a follow-up report of 82 advanced cancer patients. Med Gas Res 9(3):115
126. Beer AM, Uebelhack R, Pohl U (2016) Efficacy and tolerability of hydrogen carbonate-rich water for heartburn. World J Gastrointest Pathophysiol 15:7(1):171–80. https://doi.org/10.4291/wjgp.v7.i1.171. PMID: 26909240; PMCID: PMC4753184
127. Wu Y, Yuan M, Song J, Chen X, Yang H (2019) Hydrogen gas from inflammation treatment to cancer therapy. ACS Nano 13(8):8505–8511
128. Ji H, Zhao Z, Liu Z, Sun R, Li Y, Ding X, Ni T (2023) Real-World effectiveness and safety of hydrogen inhalation in chinese patients with type 2 diabetes: a single-arm, retrospective study. Diabetes Metab Syndr Obes Jul 5:16:2039–2050. https://doi.org/10.2147/DMSO.S412898. PMID: 37431394; PMCID: PMC10329830
129. Ge L, Yang M, Yang NN, Yin XX, Song WG (2017) Molecular hydrogen: a preventive and therapeutic medical gas for various diseases. Oncotarget Sep 21:8(60):102653–102673. https://doi.org/10.18632/oncotarget.21130. PMID: 29254278; PMCID: PMC5731988

Chapter 15
The Role of the Smallest Molecule Hydrogen Overcoming Ageing-Related Disease

Wenjing He, Md. Habibur Rahman, Chaodeng Mo, Arounnapha Vongdouangchanh, Cheol-Su Kim, and Kyu-Jae Lee

Abstract Ageing is an inevitable process that increases the probability of chronic disease that involves an evolving decline in organism function over time. The length of human life is closely related to ageing, which can cause a deterioration of physiological functions and lead to the development of various chronic diseases. Many theories have been put forward to illustrate the ageing mechanisms, with the free oxidative stress theory being the most well-known. Aging is caused by the excessive accumulation of cells or tissues related to oxidative damage caused by ROS. As everyone knows that oxidative stress is closely related to aging, antioxidants may be potentially valuable in the treatment of senescence disease. Hydrogen (H_2) is a colorless, tasteless small molecule that plays a major role in eliminating harmful free radicals and reducing oxidative damage in vivo, such as anti-oxidation, anti-inflammation, and anti-apoptosis. Additionally, H_2 can be utilized to prevent and cure several aging-related diseases, including cancer, Alzheimer's disease, and gastrointestinal diseases. Understanding the aging process is crucial for comprehending how to slow the aging process and the development of aging-related diseases. This chapter summarized the preventive and treatment applications of molecular H_2 in anti-ageing and underlying mechanisms in aging-related diseases.

Keywords Aging-related diseases · Molecular hydrogen · Reactive oxygen species · Oxidative stress · Redox mechanism · Antioxidant · Longevity

---

W. He · Md. H. Rahman · C. Mo · A. Vongdouangchanh · C.-S. Kim · K.-J. Lee (✉)
Department of Convergence Medicine, Yonsei University Wonju College of Medicine, Gangwon-Do, Wonju 26426, Republic of Korea
e-mail: medbio@yonsei.ac.kr

W. He
e-mail: hwjing20@gmail.com

C.-S. Kim
e-mail: cs-kim@yonsei.ac.kr

J. Slezak and B. Kura (eds.), *Molecular Hydrogen in Health and Disease*, Advances in Biochemistry in Health and Disease 27, https://doi.org/10.1007/978-3-031-47375-3_15

Introduction

The global population's age structure is continuing to shift in an unprecedented and continuous manner due to rising life expectancy and falling fertility rates. The share of the world population of older persons is predicted to increase from 9.3 to 16.0% from 2020 to 2050 [1]. As lifespans lengthen, ageing diseases are thriving [2]. Aging could cause physiological integrity loss and damage function and increased death ratio which will enhance the risk of human disease, including cardiovascular disorders, diabetes, cancer, and neurodegenerative diseases [3].

Aging is a complex with many factors process that has experienced progress over recent years. Although many theories put forward to explain this process, the mechanism is still unclear. Denham Harman's free oxidative stress theory has been getting more attention [4, 5]. This theory proved that reactive oxygen species (ROS) is the main factor causing aging, when ROS cannot be controlled by tissue re-oxygenation, oxidative stress occurs which lead to the damage to lipid, DNA damage, protein oxidation, and mitochondrial dysfunction [6]. In addition, due to metabolic and pathophysiological changes, exposure to external stressors, and an imbalance in the production and removal of ROS, also lead to oxidative stress [7]. Particularly, many countries have seen an increase in the total number of elderly individuals [8]. It is crucial to comprehend how the aging process works in order to postpone its progression and the onset of age-related illnesses. According to this theory, anti-oxidant might have potential therapeutical value in aging-related diseases. To prevent aging-related diseases, it is important to find a valid antioxygen for the remission of oxidative stress.

Hydrogen has biological effects such as anti-inflammation, anti-oxidation, and anti-apoptosis [9–11]. H_2 treatment may provide an anti-aging intervention [12–14]. H_2 is not only safe and effectivity as a disease treatment mechanism but also as a therapeutic method for aging disease. It is safer and more convenient that has a broad application prospect in anti-aging. The objective of this present chapter is to highlight the multiple functions of the smallest H_2 molecule in overcoming aging-related diseases.

The Source of ROS

When free radicals and reactive metabolites are out of equilibrium, ROS are produced, which could be harmful to the organism [15]. Although organisms have anti-oxidant defenses, overproduction or failure will lead to the stage of "oxidative stress" in the anti-oxidant system of ROS which is a key hallmark of aging [3, 16–18]. ROS including the hydroxyl radicals (OH·), superoxide anion (O_2^-), and hydrogen peroxide (H_2O_2) that are normally related to oxidative stress theory, all have inherent properties of reactivity to different biological targets [19, 20]. Harman posits that ROS promotes aging via its responsiveness towards macromolecules in cellular, especially in mitochondria [4]. In addition, damaged mitochondria produced increasing ROS which inevitably impair cellular function [21].

ROS could be divided into endogenous and exogenous sources, NADPH oxidases (NOX) and mitochondria act as intracellular sources, and several factors such as chemical poisons, ultraviolet radiation ionization, pollution, alcohol, drugs, etc. that act as exogenous sources [22] (Fig. 15.1). Mitochondria is the primary source of ROS [23]. The electron transport chain (ETC) produced $O_2^{\bullet-}$ massively and is responsible for adenosine triphosphate (ATP) generation [24, 25]. The electrons in ETC may interact with oxygen to produce hydrogen peroxide or superoxide directly [26, 27]. NOX is a great source of ROS that overexpressed or has high activity will be implicated in aging diseases [28–30]. Many factors could stimulate ROS production as well as mitochondria in stressful conditions that will increase inflammatory cytokines [31, 32].

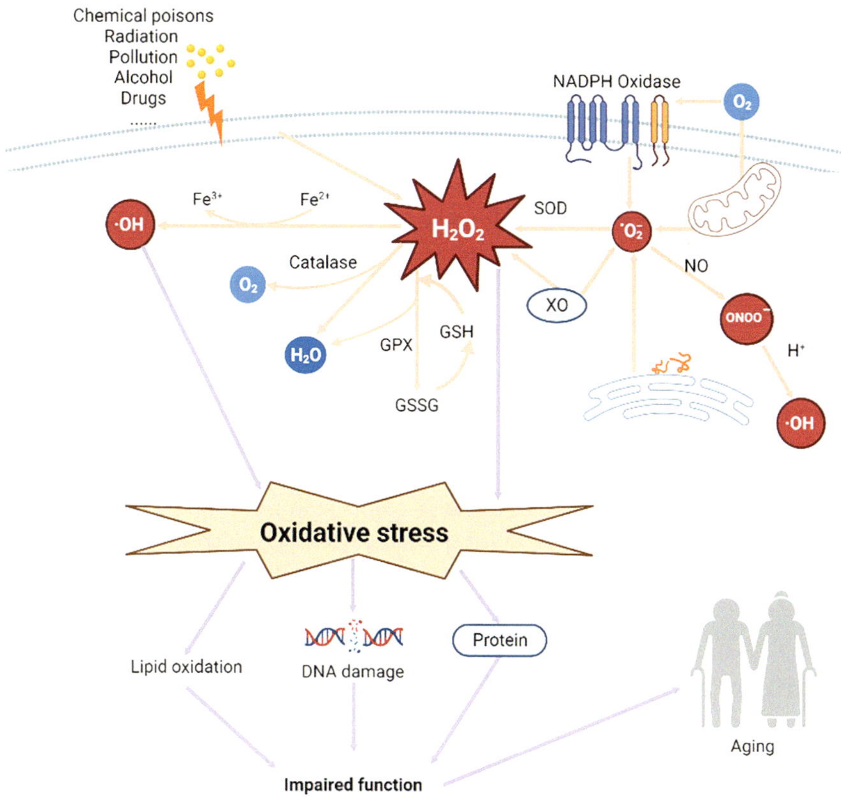

Fig. 15.1 Schematic of ROS (Drawn by using Biorender software)

ROS as an inescapable byproduct of metabolic activity has several intracellular and exogenous sources. NAPH oxidase enzymes (NOX) produce intracellular superoxide (O_2^-) mainly by transferring an electron leakage in mitochondria aerobic respiration. O_2^- is rapidly transferred into hydrogen peroxide (H_2O_2) by SOD. Hydroxyl radicals (OH·) which are formed by metal cations (Fe^{2+}) will cause irreversible damage when H_2O_2 levels uncontrollably. If H_2O_2 could not completely controlled intracellular and extracellular, oxidative damage to proteins, DNA damage, and lipids oxidation may occur. Oxidative stress themselves potent ROS, autocatalyzing this process will cause typical aging disease.

ROS and Regulation of Aging

The aging process is very complicate and no theories will explain these features to be entirely satisfactory. ROS-mediated redox signaling is associated with aging in many factors. Besides, one of the features of aging is inflammation [33]. The loss of Nrf2 produced a high level of ROS leading to the increase in proinflammatory cytokine levels that aggravate pneumonia and sepsis in mice [34, 35]. Telomerase activity and telomere length are also important factors in aging-related diseases. Promoting telomere shortening is a key hallmark of various diseases associated with ageing [36–39]. One source of endogenous telomeric damage is ROS. Oxidative stress can damage in DNA and induce telomere loss directly [40]. ROS can injure cells via DNA damage, proteins, and lipids oxidation to the short the telomeres which could damage to genomic DNA through p53 activation [41]. Additionally, oxidative stress expedites telomere shortening [42].

Over 50 years ago the free radical theory of senescence put forward that ROS promotes senescence [4]. Damaged mitochondria produced ROS by low-efficiency oxidative phosphorylation that damaged cellular function [21] (Fig. 1). It has been demonstrated that activation of mitogen-activated protein kinase (MAPK), possibly lead to ROS overproduction will cause the damage to heart failure [43]. ROS will interfere with the protein synthesis in skeletal muscle cells increases the expression of disease regulator myomiRs and promotes apoptosis of skeletal muscle cells [44].

Hydrogen and Aging

Aging is a typical programmed process that we want to slow down aging in organisms. In the field of anti-aging, the previous research strategy is to use drug molecules to intervene in the whole body. For example, metformin extends lifespan in mice [45, 46]. Calorie restriction (CR) prolongs life-span in yeast, worms, mice, and perhaps human beings [47, 48]. Resveratrol could activate the silent information regulator 1 (Sirt1) that extends life span in diverse species [49, 50]. Rapamycin and other famous anti-aging tools have been discovered [51]. Currently, H_2 is the smallest molecule we have found which is colorless, odorless, and highly flammable. Although Dole et al. in 1975 study raised the possibility of H_2 having an antioxidative effect, its biological effects were long disregarded [52]. In 2007, Ohsawa et al. found that hydrogen is a great source of antioxidant in biological systems and that inhaling H_2 gas reduced oxidative stress [53]. Various research has reported that H_2 could prevent oxidative stress damage through its antioxidant functions. Zhang et al. reported that H_2 can effectively reduce the body's ROS levels and prolonged Caenorhabditis elegans model lifespan which is associated with oxidative stress and reducing oxidative stress damage [54]. Moreover, the drosophila model confirms the effect of H_2 on anti-aging at the organismal level. Klichko reported that the ingestion of H_2 could increase the survival of aging or short-lived drosophila, delay the intestinal barrier dysfunction development, and improved physical activity significantly [55]. Zhang et al. found using hydrogen water could reduce the cognitive disorder in aging mice via the H_2 anti-oxidant properties that proved the protection effects of neurons [56]. Thus, H_2 could be a new therapeutic for slower aging and chronic disease.

Signal Pathway on Hydrogen and Aging

At present, there are not too many studies directly related to aging and anti-aging research on H_2. But most of them are related to the signaling pathways. Hydrogen decreases cellular aging in vivo by regulating the ROS/p53/p21 pathway [57]. It has been proved the effect that hydrogen-rich saline (HRS) plays an important role in the anti-apoptotic mechanism protecting against ischemia/reperfusion (I/R) injury by the way to increase insulin-like growth factor 1 (IGF-1,) phosphoinositide 3-kinase (PI3K) and mTOR mRNA expression and decreased miR-199a-3p level [58]. Molecular H_2 through mTOR-autophagy pathway improve and ameliorates sepsis-induced neuroinflammation in mice and BV-2 cells and protects against sepsis-induced cognitive impairment in mice [59]. Fu et al. confirmed HRS could protect against lung injury in lipopolysaccharide-induced by regulating apoptosis through control of the mTOR/TFEB signaling pathway [60]. Sirt1 is a histone deacetylase that reportedly participates in stress response pathways [61, 62], apoptosis [63], inflammation [64], and metabolism [65]. Sirt1 expression may be connected with human health and longevity [66–69]. Resveratrol, the hypothesized small molecule activator of Sirt1, prolongs the life in yeast, Caenorhabditis elegans, and fruit fly [65, 70, 71]. Du et al. report that 66.7% H_2 protected against lipopolysaccharide-induced lung injury by nuclear factor kappa B (NF-κB) and catalase mediated in a Sirt1-dependent manner inhibiting inflammatory response and oxidative stress [72].

Molecule Hydrogen and Aging-Related Disease

Cancer

Oxidative stress interacts with all three stages of the multi-stage cancer process, which is made up of initiation, promotion, and progression. By activating cell signal transduction pathways in both animal models and people, ROS has been shown to promote tumor cell proliferation, survival, and migration [73]. On the one hand, because chronic inflammation is a key mediator of cancer and ROS controls its incidence, low levels of ROS increase the survival of cancer cells. In addition, excessive amounts of ROS can stop the growth of tumors by permanently activating cell cycle inhibitors, causing cell senescence, and killing cells by degrading macromolecules [74]. Most cancers affect cells through the production of ROS, a risk factor [75]. In addition, ROS also regulates the activity of many tumor suppressor genes, and genomic instability is considered to be a major driver of tumorigenesis, which may explain the genetic diversity in many cancers [76].

So far, many studies have shown that H_2's role in cancer does not target specific proteins, but rather modulates several key roles. [NiFe]-hydrogenase and mitochondrial complex I share a common evolutionary history and a high degree of homology [77]. Because H_2 changes ubiquitin ketones into ubiquitin alcohols, it may correct mitochondrial electron flow, preventing excessive electron leakage and the formation of ROS. It was also noted that consumption of HRW raised the concentration of oxidized coenzyme Q (ubiquitin) in rat cardiac tissue and enhanced the level of mitochondrial ATP via raising the concentration of complex I and II substrates [78]. Recent research has demonstrated that complex I is directly related to the evolutionary activity of H_2 in eukaryotic mitochondria, and that this activity takes place close to the completely oxidized ubiquitin ketone binding site. These findings support the idea that H_2 could control the formation of ROS from the source and successfully lowering oxidative stress damage to control chronic inflammation and the incidence and growth of malignancies.

Alzheimer's Disease

Alzheimer's disease (AD) is the most common disease in elderly persons, and is also the most prevalent neurodegenerative illness worldwide. The pathogenesis and injury of AD are heavily influenced by oxidative stress which plays a crucial role in AD [79]. As a result, using antioxidants to lower ROS levels offers a viable plan for finding anti-AD medications. Recently, the use of antioxidant regulation of ROS in animal models for the treatment of neurological diseases has made significant progress.

H_2 can protect to excessive oxidative stress in an organism that was completed by the raising level of anti-oxidant enzymes and redox-related gene expression.

H_2 therapy boosted the activity of natural anti-oxidant enzymes such SOD, CAT, and GPx in the brain and spinal cord [80]. The expression of HO-1 and nNOS might be greatly increased in the hypoxia/reoxygenation damage model when H_2 is present. The expression of p-p38 MAPK, HO-1, and Nrf2 was found to be increased by H_2 treatment in numerous in vitro and in vivo studies [81]. Nrf2 is a crucial neuroprotective effect in a variety of neurological illnesses. By stimulating the Nrf2 antioxidant defense mechanism, H_2 therapy may cause the nervous system to adapt to oxidative stress. HRW could drastically lower the level of MDA and raise T-SOD and GSH activity in APP/PS1 mice. By promoting AMP-activated protein kinase (AMPK) and the downstream Sirt1-FoxO3a axis in vitro, H_2 administration improved the antioxidant system of A-stimulated human neuroblastoma SK-N-MC cells, avoiding mitochondrial malfunction and ROS generation and eventually preserving cell viability [82].

Gastrointestinal Diseases

Our gut microbiota has only recently become the subject of intensive research, having a considerable impact on human health and the onset or progression of diseases. Therefore, its crucial functions in warding off infections, controlling metabolic, endocrine, and immunological systems, and affecting drug metabolism and absorption have begun to be clarified [83, 84]. In the human colon, H_2 gas is naturally created during microbial fermentation, Since H_2 has been found to have antioxidant characteristics. Indeed, the balance between H_2-producing (hydrogen genic) and H_2-consuming organisms is reflected in H_2 metabolism [85]. Despite ample evidence of the significance of microbial hydrogen metabolism, their little attempts to modify H_2 associate with microbe in the colon. The effectiveness of gut microbial fermentation depends on H_2 cycling. H_2 is assumed to be reoxidized mostly via anaerobic respiratory microbes and is primarily produced by fermentative bacteria [86]. The firmicutes members Bacteroidetes and Clostridia mediate fermentative activities, which are the main mode of H_2 evolution in this ecosystem [87].

H_2 has the effect to improve abnormal fermentation in the colon inflammatory bowel disease (IBD), which includes ulcerative colitis (UC) and Crohn's disease (CD). H_2 saline/water has been displayed in studies to reduce colitis in mouse models, although the underlying process is still not fully understood [88]. Surprisingly, oral lactulose, digested by intestinal flora, boosted endogenous H_2 synthesis, which also significantly reduced inflammation in DSS-induced mouse UC. Although lactulose's anti-inflammatory effects can be reversed by antibiotics, the molecular mechanism by which H_2 controls gut flora is not well understood [89]. This research suggests a new function for H_2 in the control of the gut flora [90]. Curcumin as an anti-inflammatory in ancient times, and it has also been used to treat gastrointestinal diseases. Curcumin also suppresses oxidative stress via regulation of ROS/NF-κB signaling pathway [91].

Conclusion

We review the functions and efficacy of ROS and aging and propose a new idea for anti-aging using hydrogen in this chapter. Meanwhile, we also explore the signal pathway of hydrogen and aging. H_2, a new therapeutic antioxidant that is known to be non-toxic, has the potential to effectively alleviate oxidative stress. The selective antioxidant effect of hydrogen may be attributed to its ability to reduce oxidative stress, decrease cell apoptosis, and regulate cell signaling pathways. Increasing reports proved that H_2 could be a therapeutic method for aging-related diseases. Therefore, as a kind of antioxidant compound, H_2 has many advantages that are strong clinical applications in many aging diseases. Its potent effect and various new concepts make it a potential application for a wide range of aging-related diseases. Therefore, additional long-term researches are necessary to explain how H_2 affects the process of physiological aging. However, we believe that H_2 is a crucial factor in aging and aging-related disorders, providing hopeful prospects for treatment in this field.

References

1. Affairs DoEaS, Division P (2020) World population ageing 2020 highlights. https://www.un.org/development/desa/pd/sites/www.un.org.development.desa.pd/files/undesa_pd-2020_world_population_ageing_highlights.pdf. Accessed
2. Carmona JJ, Michan S (2016) Biology of healthy aging and longevity. Rev Invest Clin 68(1):7–16
3. López-Otín C, Blasco MA, Partridge L, Serrano M, Kroemer G (2013) The hallmarks of aging. Cell 153(6):1194–1217
4. Rahman MH, Bajgai J, Fadriquela A, Sharma S, Trinh Thi T, Akter R, Goh SH, Kim CS, Lee KJ (2021) Redox effects of molecular hydrogen and its therapeutic efficacy in the treatment of neurodegenerative diseases. Processes 9(2):308
5. Harman D (1972) The biologic clock: the mitochondria? J Am Geriatr Soc 20(4):145–147
6. Polidori MC, Mecocci P (2022) Modeling the dynamics of energy imbalance: the free radical theory of aging and frailty revisited. Free Radical Biol Med
7. Gupta A, Jagadis K (2015) Reactive oxygen and nitrogen species signaling and communication in plants. Signal Commun Plants
8. de Haan G, Lazare SS (2018) Aging of hematopoietic stem cells Blood. J Am Soc Hematol 131(5):479–487
9. Huang C-S, Kawamura T, Toyoda Y, Nakao A (2010) Recent advances in hydrogen research as a therapeutic medical gas. Free Radical Res 44(9):971–982
10. Sakai T, Kurokawa R, Hirano S-I, Imai J (2019) Hydrogen indirectly suppresses increases in hydrogen peroxide in cytoplasmic hydroxyl radical-induced cells and suppresses cellular senescence. Int J Mol Sci 20(2):456
11. Wang C, Li J, Liu Q, Yang R, Zhang JH, Cao Y-P et al (2011) Hydrogen-rich saline reduces oxidative stress and inflammation by inhibit of JNK and NF-κB activation in a rat model of amyloid-beta-induced Alzheimer's disease. Neurosci Lett 491(2):127–132
12. Liu B, Xie Y, Chen J, Xue J, Zhang X, Zhao M et al (2021) Protective effect of molecular hydrogen following different routes of administration on D-galactose-induced aging mice. J Inflamm Res 14:5541

13. Rahman MH, Jeong E-S, You HS, Kim C-S, Lee K-J (2023) Redox-Mechanisms of molecular hydrogen promote healthful longevity. Antioxidants 12(5):988
14. Rahman MH, Bajgai J, Fadriquela A, Sharma S, Trinh Thi T, Akter R et al (2021) Redox effects of molecular hydrogen and its therapeutic efficacy in the treatment of neurodegenerative diseases. Processes 9(2):308
15. Lushchak VI (2014) Free radicals, reactive oxygen species, oxidative stress and its classification. Chem Biol Interact 224:164–175
16. Almeida AJPOD, Ribeiro TP, Medeiros IAD (2017) Aging: molecular pathways and implications on the cardiovascular system. Oxidative Med Cell Longevity 2017
17. Rahman MH, Jeong ES, You HS, Kim CS, Lee KJ (2023) Redox-Mechanisms of molecular hydrogen promote healthful longevity. Antioxidants 12(5):988
18. Riera CE, Merkwirth C, De Magalhaes Filho CD, Dillin A (2016) Signaling networks determining life span. Ann Rev Biochem 85
19. Rahman MH, Bajgai J, Sharma S, Jeong ES, Goh SH, Jang YG, Kim CS, Lee KJ (2023) Effects of hydrogen gas inhalation on community-dwelling adults of various ages: a single-arm, open-label, prospective clinical trial. Antioxidants 12(6):1241
20. Li R, Jia Z, Trush MA (2016) Defining ROS in biology and medicine. React Oxygen Species (Apex, NC) 1(1):9
21. Balaban RS, Nemoto S, Finkel T (2005) Mitochondria, oxidants, and aging. Cell 120(4):483–495
22. Holmström KM, Finkel T (2014) Cellular mechanisms and physiological consequences of redox-dependent signalling. Nat Rev Mol Cell Biol 15(6):411–421
23. Dai D-F, Chiao YA, Marcinek DJ, Szeto HH, Rabinovitch PS (2014) Mitochondrial oxidative stress in aging and healthspan. Longevity Healthspan 3(1):1–22
24. Zhao RZ, Jiang S, Zhang L, Yu ZB (2019) Mitochondrial electron transport chain, ROS generation and uncoupling. Int J Mol Med 44(1):3–15
25. Letts JA, Sazanov LA (2017) Clarifying the supercomplex: the higher-order organization of the mitochondrial electron transport chain. Nat Struct Mol Biol 24(10):800–808
26. Turrens JF (2003) Mitochondrial formation of reactive oxygen species. J Physiol 552(2):335–344
27. Cadenas E, Davies KJ (2000) Mitochondrial free radical generation, oxidative stress, and aging. Free Radical Biol Med 29(3–4):222–230
28. Touyz RM, Anagnostopoulou A, Rios F, Montezano AC, Camargo LL (2019) NOX5: molecular biology and pathophysiology. Exp Physiol 104(5):605–616
29. Harvey A, Montezano AC, Touyz RM (2015) Vascular biology of ageing—implications in hypertension. J Mol Cell Cardiol 83:112–121
30. Montezano AC, Nguyen Dinh Cat A, Rios FJ, Touyz RM (2014) Angiotensin II and vascular injury. Current Hypertens Rep 16:1–11
31. Missiroli S, Genovese I, Perrone M, Vezzani B, Vitto VA, Giorgi C (2020) The role of mitochondria in inflammation: from cancer to neurodegenerative disorders. J Clin Med 9(3):740
32. Rimessi A, Previati M, Nigro F, Wieckowski MR, Pinton P (2016) Mitochondrial reactive oxygen species and inflammation: Molecular mechanisms, diseases and promising therapies. Int J Biochem Cell Biol 81:281–293
33. Franceschi C, Campisi J (2014) Chronic inflammation (inflammaging) and its potential contribution to age-associated diseases. J Gerontol Series A: Biomed Sci 69(Suppl_1):S4–S9
34. Thimmulappa RK, Lee H, Rangasamy T, Reddy SP, Yamamoto M, Kensler TW et al (2016) Nrf2 is a critical regulator of the innate immune response and survival during experimental sepsis. J Clin Investig 116(4):984–995
35. Kong X, Thimmulappa R, Craciun F, Harvey C, Singh A, Kombairaju P et al (2011) Enhancing Nrf2 pathway by disruption of Keap1 in myeloid leukocytes protects against sepsis. Am J Respir Crit Care Med 184(8):928–938
36. Blasco MA (2005) Telomeres and human disease: ageing, cancer and beyond. Nat Rev Genet 6(8):611–622

37. Shawi M, Autexier C (2008) Telomerase, senescence and ageing. Mech Ageing Dev 129(1–2):3–10
38. Collado M, Gil J, Efeyan A, Guerra C, Schuhmacher AJ, Barradas M et al (2005) Senescence in premalignant tumours. Nature 436(7051):642
39. Moon IK, Jarstfer MB (2007) The human telomere and its relationship to human disease, therapy, and tissue engineering. Front Biosci-Landmark 12(7):4595–4620
40. Von Zglinicki T (2002) Oxidative stress shortens telomeres. Trends Biochem Sci 27(7):339–344
41. Cataldi A, Zara S, Rapino M, Zingariello M, Di Giacomo V, Antonucci A (2009) p53 and telomerase control rat myocardial tissue response to hypoxia and ageing. Eur J Histochem 53(4):e25-e
42. Uziel O, Reshef H, Ravid A, Fabian I, Halperin D, Ram R et al (2008) Oxidative stress causes telomere damage in Fanconi anaemia cells–a possible predisposition for malignant transformation. Br J Haematol 142(1):82–93
43. Heusch P, Canton M, Aker S, Van De Sand A, Konietzka I, Rassaf T et al (2010) The contribution of reactive oxygen species and p38 mitogen-activated protein kinase to myofilament oxidation and progression of heart failure in rabbits. Br J Pharmacol 160(6):1408–1416
44. Di Filippo ES, Mancinelli R, Pietrangelo T, La Rovere RML, Quattrocelli M, Sampaolesi M et al (2016) Myomir dysregulation and reactive oxygen species in aged human satellite cells. Biochem Biophys Res Commun 473(2):462–470
45. Dilman VM, Anisimov VN (1980) Effect of treatment with phenformin, diphenylhydantoin or L-dopa on life span and tumour incidence in C3H/Sn mice. Gerontology 26(5):241–246
46. Anisimov VN, Berstein LM, Egormin PA, Piskunova TS, Popovich IG, Zabezhinski MA et al (2008) Metformin slows down aging and extends life span of female SHR mice. Cell Cycle 7(17):2769–2773
47. Heilbronn LK, Ravussin E (2003) Calorie restriction and aging: review of the literature and implications for studies in humans. Am J Clin Nutr 78(3):361–369
48. Kennedy B, Steffen K, Kaeberlein M (2007) Ruminations on dietary restriction and aging. Cell Mol Life Sci 64:1323–1328
49. Howitz KT, Bitterman KJ, Cohen HY, Lamming DW, Lavu S, Wood JG et al (2003) Small molecule activators of sirtuins extend Saccharomyces cerevisiae lifespan. Nature 425(6954):191–196
50. Bauer JH, Goupil S, Garber GB, Helfand SL (2004) An accelerated assay for the identification of lifespan-extending interventions in Drosophila melanogaster. Proc Natl Acad Sci 101(35):12980–12985
51. Wilkinson JE, Burmeister L, Brooks SV, Chan CC, Friedline S, Harrison DE et al (2012) Rapamycin slows aging in mice. Aging Cell 11(4):675–682
52. Dole M, Wilson FR, Fife WP (1975) Hyperbaric hydrogen therapy: a possible treatment for cancer. Science 190(4210):152–154
53. Ohsawa I, Ishikawa M, Takahashi K, Watanabe M, Nishimaki K, Yamagata K et al (2007) Hydrogen acts as a therapeutic antioxidant by selectively reducing cytotoxic oxygen radicals. Nat Med 13(6):688–694
54. Zhang M, Li Z, Gao D, Gong W, Gao Y, Zhang C (2020) Hydrogen extends Caenorhabditis elegans longevity by reducing reactive oxygen species. PLoS ONE 15(4):e0231972
55. Klichko VI, Safonov VL, Safonov MY, Radyuk SN (2019) Supplementation with hydrogen-producing composition confers beneficial effects on physiology and life span in Drosophila. Heliyon. 5(5):e01679
56. Gu Y, Huang C-S, Inoue T, Yamashita T, Ishida T, Kang K-M et al (2010) Drinking hydrogen water ameliorated cognitive impairment in senescence-accelerated mice. J Clin Biochem Nutr 46(3):269–276
57. Zhang W, Huang C, Sun A, Qiao L, Zhang X, Huang J et al (2018) Hydrogen alleviates cellular senescence via regulation of ROS/p53/p21 pathway in bone marrow-derived mesenchymal stem cells in vivo. Biomed Pharmacother 106:1126–1134
58. Yao W, Lin X, Han X, Zeng L, Guo A, Guan Y et al (2020) MicroRNA files in the prevention of intestinal ischemia/reperfusion injury by hydrogen rich saline. Biosci Rep 40(1)

59. Zhuang X, Yu Y, Jiang Y, Zhao S, Wang Y, Su L et al (2020) Molecular hydrogen attenuates sepsis-induced neuroinflammation through regulation of microglia polarization through an mTOR-autophagy-dependent pathway. Int Immunopharmacol 81:106287
60. Fu Z, Zhang Z, Wu X, Zhang J (2020) Hydrogen-rich saline inhibits lipopolysaccharide-induced acute lung injury and endothelial dysfunction by regulating autophagy through mTOR/TFEB signaling pathway. BioMed Res Int 2020
61. Zhu W, Chen S, Li Z, Zhao X, Li W, Sun Y et al (2014) Effects and mechanisms of resveratrol on the amelioration of oxidative stress and hepatic steatosis in KKAy mice. Nutr Metab 11(1):1–11
62. Brunet A, Sweeney LB, Sturgill JF, Chua KF, Greer PL, Lin Y et al (2004) Stress-dependent regulation of FOXO transcription factors by the SIRT1 deacetylase. Science. 303(5666):2011–2015
63. Cohen HY, Miller C, Bitterman KJ, Wall NR, Hekking B, Kessler B et al (2004) Calorie restriction promotes mammalian cell survival by inducing the SIRT1 deacetylase. Science 305(5682):390–392
64. Tamaki N, Orihuela-Campos RC, Inagaki Y, Fukui M, Nagata T, Ito H-O (2014) Resveratrol improves oxidative stress and prevents the progression of periodontitis via the activation of the Sirt1/AMPK and the Nrf2/antioxidant defense pathways in a rat periodontitis model. Free Radical Biol Med 75:222–229
65. Bordone L, Cohen D, Robinson A, Motta MC, Van Veen E, Czopik A et al (2007) SIRT1 transgenic mice show phenotypes resembling calorie restriction. Aging Cell 6(6):759–767
66. Li S, Fujino M, Ichimaru N, Kurokawa R, Hirano S, Mou L et al (2018) Molecular hydrogen protects against ischemia-reperfusion injury in a mouse fatty liver model via regulating HO-1 and Sirt1 expression. Sci Rep 8(1):14019
67. Sun Q, Han W, Hu H, Fan D, Li Y, Zhang Y et al (2017) Hydrogen alleviates hyperoxic acute lung injury related endoplasmic reticulum stress in rats through upregulation of SIRT1. Free Radical Res 51(6):622–632
68. Liu Y, Li R, Xie J, Hu J, Huang X, Ren F et al (2018) Protective effect of hydrogen on sodium iodate-induced age-related macular degeneration in mice. Front Aging Neurosci 10:389
69. Yan W-M, Chen T, Wang X-C, Qi L-S, Zhao G-H, Yang G-Q et al (2017) The reason for the amelioration of N-methyl-N-nitrosourea-induced retinitis pigmentosa in rats by hydrogen-rich saline. Int J Ophthalmol 10(10):1495
70. Lamming DW, Wood JG, Sinclair DA (2004) MicroReview: small molecules that regulate lifespan: evidence for xenohormesis. Mol Microbiol 53(4):1003–1009
71. Wood JG, Rogina B, Lavu S, Howitz K, Helfand SL, Tatar M et al (2004) Sirtuin activators mimic caloric restriction and delay ageing in metazoans. Nature 430(7000):686–689
72. Du J, Li J, Li R, Yan X (2022) High concentration of hydrogen ameliorates lipopolysaccharide-induced acute lung injury in a sirt1-dependent manner. Respir Physiol Neurobiol 296:103808
73. Aggarwal V, Tuli HS, Varol A, Thakral F, Yerer MB, Sak K et al (2019) Role of reactive oxygen species in cancer progression: molecular mechanisms and recent advancements. Biomolecules 9(11):735
74. Ramsey MR, Sharpless NE (2006) ROS as a tumour suppressor? Nat Cell Biol 8:1213–1215
75. Rahman MH, Bajgai J, Cho Y, Fadriquela A, Sharma S, Thuy TT, Cho SH, Jeong YJ, Goh SH, Kim Y, Kim CS (2023) Immune redox modulation effects of non-electrolyzed hypochlorous acid water on helicobacter pylori-Infected C57BL/6 mouse model. Processes 11(5):1474
76. Loeb LA, Loeb KR, Anderson JP (2003) Multiple mutations and cancer. Proc Natl Acad Sci 100(3):776–781
77. Marreiros BC, Batista AP, Duarte AM, Pereira MM (2013) A missing link between complex I and group 4 membrane-bound [NiFe] hydrogenases. Biochimica et Biophysica Acta (BBA)-Bioenergetics 1827(2):198–209
78. Gvozdjáková A, Kucharská J, Kura B, Vančová Og, Rausová Z, Sumbalová Z et al (2020) A new insight into the molecular hydrogen effect on coenzyme Q and mitochondrial function of rats. Can J Physiol Pharmacol 98(1):29–34
79. Butterfield DA, Lange MLB, Sultana R (2010) Involvements of the lipid peroxidation product, HNE, in the pathogenesis and progression of Alzheimer's disease. Biochimica et Biophysica Acta (BBA)-Mol Cell Biol Lipids 1801(8):924–929

80. Xu L, Liang J, Jin T, Zhou F (2016) Neuroprotective effects of mangiferin on acute spinal cord injury in rats and its mechanism. Zhongguo xiu fu Chong Jian wai ke za zhi= Zhongguo Xiufu Chongjian Waike Zazhi= Chin J Repar Reconstr Surg 30(8):1019–1025
81. Shi H, Jing X, Wei X, Perez RG, Ren M, Zhang X et al (2015) S-allyl cysteine activates the Nrf2-dependent antioxidant response and protects neurons against ischemic injury in vitro and in vivo. J Neurochem 133(2):298–308
82. Lin C-L, Huang W-N, Li H-H, Huang C-N, Hsieh S, Lai C et al (2015) Hydrogen-rich water attenuates amyloid β-induced cytotoxicity through upregulation of Sirt1-FoxO3a by stimulation of AMP-activated protein kinase in SK-N-MC cells. Chem Biol Interact 240:12–21
83. Cho I, Blaser MJ (2012) The human microbiome: at the interface of health and disease. Nat Rev Genet 13(4):260–270
84. Palmer C, Bik EM, DiGiulio DB, Relman DA, Brown PO (2007) Development of the human infant intestinal microbiota. PLoS Biol 5(7):e177
85. Carbonero F, Benefiel AC, Gaskins HR (2012) Contributions of the microbial hydrogen economy to colonic homeostasis. Nat Rev Gastroenterol Hepatol 9(9):504–518
86. Nakamura N, Lin HC, McSweeney CS, Mackie RI, Gaskins HR (2010) Mechanisms of microbial hydrogen disposal in the human colon and implications for health and disease. Annu Rev Food Sci Technol 1:363–395
87. Wolf PG, Biswas A, Morales SE, Greening C, Gaskins HR (2016) H_2 metabolism is widespread and diverse among human colonic microbes. Gut microbes 7(3):235–245
88. Ge L, Qi J, Shao B, Ruan Z, Ren Y, Sui S et al (2022) Microbial hydrogen economy alleviates colitis by reprogramming colonocyte metabolism and reinforcing intestinal barrier. Gut Microbes 14(1):2013764
89. Cheng C-Y, Kung C-T, Wu K-H, Chen F-C, Cheng H-H, Cheng F-J et al (2021) Liver cirrhosis affects serum lactate level measurement while assessing disease severity in patients with sepsis. Eur J Gastroenterol Hepatol 33(9):1201
90. Hasegawa S, Goto S, Tsuji H, Okuno T, Asahara T, Nomoto K et al (2015) Intestinal dysbiosis and lowered serum lipopolysaccharide-binding protein in Parkinson's disease. PLoS ONE 10(11):e0142164
91. Huang J, Yi Q, You Y, Chen Y, Niu T, Li Y, Zhang J, Ji X, Xu G, Zou W, Ji F (2021) Curcumin suppresses oxidative stress via regulation of ROS/NF-κB signaling pathway to protect retinal vascular endothelial cell in diabetic retinopathy. Mol Cell Toxicol 17(3):367–376

Chapter 16
Dihydrogen and Hepatic Function: Systematic Review and Meta-analysis

Nikola Todorovic and Sergej M. Ostojic

Abstract Molecular hydrogen (H_2, dihydrogen) is an innovative experimental agent that could foster beneficial effects in a plethora of human conditions, including liver disorders. Recent trials demonstrated the positive effects of dihydrogen in several acute and chronic liver diseases, including chronic hepatitis B, non-alcoholic fatty liver disease, and liver cancer. Still, no systematic review or meta-analysis investigated the effects of dihydrogen intake on the hepatic function panel. For this report, we searched three relevant databases (PubMed, Web of Science, Scopus) from inception until December 24, 2022, using PRISMA guidelines. A literature search yielded 365 publications in total. After removing duplicates and studies that did not meet the inclusion criteria, 12 studies (published from 2008 to 2022) were included in this analysis. We found that serum malondialdehyde levels were significantly decreased following dihydrogen intake (pooled standardized mean difference = − 0.97 [95% confidence interval [CI], from − 1.65 to − 0.19; $P = 0.01$), with results indicating a large effect of dihydrogen intervention. A strong trend for a reduction in the liver function tests (including aspartate aminotransferase and alanine transaminase) has also been observed after dihydrogen administration ($P < 0.20$). Our findings suggested that dihydrogen can favorably affect the hepatic function panel; additional well-sampled interventional trials are highly warranted to corroborate our results.

Keywords Dihydrogen · Hepatic panel · Malondialdehyde · Aspartate aminotransferase · Hydrogen-rich water

Introduction

Chronic liver diseases are a significant public health concern and one of the leading causes of mortality and morbidity [1]. Worldwide, it is estimated that ~ 1.5 billion people suffer from the chronic liver disease [2]. In terms of prevalence, non-alcoholic

N. Todorovic · S. M. Ostojic (✉)
Applied Bioenergetics Lab, Faculty of Sport and Physical Education, University of Novi Sad, 21000 Novi Sad, Serbia
e-mail: sergej.ostojic@chess.edu.rs

J. Slezak and B. Kura (eds.), *Molecular Hydrogen in Health and Disease*, Advances in Biochemistry in Health and Disease 27, https://doi.org/10.1007/978-3-031-47375-3_16

fatty liver disease (NAFLD) accounts for 59% of illnesses, followed by hepatitis B (29%), hepatitis C (9%), and alcoholic liver disease (2%) [3, 4]. The hepatic function panel is the most common method to assess liver health, which measures blood levels of different *enzymes*, proteins, and other substances made by the liver. Aspartate aminotransferase (AST) and alanine transaminase (ALT) are the key liver enzymes and critical elements of the hepatic function panel. AST catalyzes a reaction between the amino acids aspartate and glutamate, while ALT facilitates the conversion of proteins into energy for the liver cells [5]. Blood levels of these two enzymes rise when liver cells are damaged or stressed. Besides these enzymes, the serum levels of gamma-glutamyl transferase (GGT) are commonly measured as a marker of hepatic health. GGT is a non-specific marker of hepatobiliary disease found in the liver and biliary epithelial cells [6], with high GGT levels may indicate liver disease or bile duct damage. Recently, the levels of malondialdehyde (MDA), a final product of polyunsaturated fatty acids peroxidation, were used as a novel biomarker of hepatic health [7]. Oxidative stress plays a critical role in the pathogenesis of many acute and chronic liver diseases [8], with plasma MDA providing a reliable biomarker of free radical damage. In addition, increased ALT and AST levels are tightly linked to plasma MDA [7], advancing the above compounds as potential biomarkers in liver diseases.

The liver is linked to various metabolic pathways responsible for producing and using nutrients and other substances [9]. It is one of the primary organs that incorporate nutrients into the body's cell mass and plays a vital role in lipid metabolism. Many nutritional strategies are used as supplementary therapy in managing liver disease [10]. It appears that various medical gases could modulate hepatic function. Nitrite oxide (NO), carbon monoxide (CO_2), and hydrogen sulfide (H_2S), also known as gasotransmitters, were considered endogenous mediators in many metabolic pathways [11]. For example, it was shown that NO plays an integral role in metabolic health and lipid metabolism that might involve hepatocytes [12]. A fourth member of the medical gases family, molecular hydrogen (dihydrogen, H_2), has shown promising potential as a novel medical gas, with the beneficial effects of supplemental dihydrogen confirmed in a plethora of disease models and human studies over the last two decades [13]. The rationale for medical applications and H_2 intake is primarily based on its fine-tuning properties in energy-demanding tissues rich in mitochondria, including the liver [14]. The effects of dihydrogen on clinical end-points and surrogate markers have been demonstrated in several clinical trials, ranging from metabolic disorders to chronic inflammatory bowel disease [for a detailed review, see Ref. [15].

Dihydrogen has recently been identified as a potential protective molecule that may improve liver health [16]. For instance, dihydrogen improved the liver function of colorectal cancer patients treated with chemotherapy [17], chronic hepatitis B [18], and non-alcoholic fatty liver disease [19]. In addition, it was demonstrated that dihydrogen was effectively accumulated in the liver after supplementation, with continuous inhalation of 3% dihydrogen exhibited the greatest mean maximum concentration of the gas in the liver compared to other organs [20]. Furthermore, exogenous hydrogen might reduce blood lipid profiles, hepatic oxidative stress, and insulin resistance in type 1 and type 2 diabetes and prospective metabolic syndrome patients [21–24]. The most recent meta-analysis by Todorovic and co-workers [25] found that hydrogen-rich water (HRW) could impact serum triglycerides, low-density lipoproteins, and high-density lipoproteins, indicating that H_2 could be a possible therapy in regulating metabolic health. However, despite evidence of its beneficial effects on liver health and metabolism, to the best of the authors' knowledge, there is no systematic review and meta-analysis (SRMA) considering the effects of dihydrogen on the hepatic function panel. Therefore, this SRMA aims to evaluate the effects of dihydrogen intake on liver enzymes and related biomarkers in both healthy and clinical populations.

Methods

The PRISMA® (Preferred Reporting Items for Systematic Reviews and Meta-Analyses) statement standards [26] were followed to investigate dihydrogen intake's impact on the hepatic function panel. Our study was reported in the prospective register of systematic reviews (PROSPERO) with study ID: CRD42023386791.

Eligibility Criteria

Only trials involving human subjects were considered for the inclusion in our study. In addition, the following criteria were utilized: (1) dihydrogen intake; (2) clinical population (metabolic diseases and/or obesity and/or NAFLD); (3) healthy population; (4) dihydrogen effects concerning the following hepatic function tests: AST, ALT, GGT, MDA; (5) human experimental trials, (6) controlled with a placebo group/or other methods of medical care, and (7) original and peer-reviewed English-language studies. In contrast, the exclusion criteria included: (1) animal studies, (2) in vitro studies, (3) co-administration of dihydrogen with other supplements, (4) absence of a placebo group for comparison of findings, and (5) studies without pre- and post-supplementation results.

Information Sources and Search Strategy

The following databases were systematically searched for pertinent articles published prior to December 24, 2022: PubMed, Web of Science, and Scopus. The search was conducted independently by the two authors (NT and SMO), and disagreements were clarified throughout the discussion. The combinations of the following Boolean search phrases were utilized to locate records: ("molecular hydrogen" OR "dihydrogen") AND ("hepatic function" OR "liver diseases ") AND ("H_2 supplementation")

AND ("AST" OR "ALT" OR "GGT" OR "MDA"). In addition, the references of the related articles were manually examined using the snowball approach to identify publications that were not recognized by the original search.

Selection Process

In the initial round of the screening, the lead author evaluated the eligibility of all titles and abstracts (NT). Afterward, the second author (SMO) independently reviewed the existing literature. The same authors then screened the full texts to determine which experimental trials should be included in the SRMA. The authors discussed uncertainty regarding eligibility to limit the number of studies that did not meet the inclusion and exclusion criteria.

Data Extraction and Study Coding

The following data were extracted from each study that satisfied the inclusion criteria: essential study and participants information (authors, year of publication), study design, supplementation procedure, duration of the supplementation treatment, and hepatic function test results (e.g., pre- and post-data). The authors were approached personally to get the raw data when values were plotted as figures. If there was no response, the means and standard deviation values were calculated using the Image J program ® (National Institutes of Health, Bethesda, MD, USA) by measuring the pixel length of each magnitude.

Study Risk of Bias Assessment

The process was conducted separately by two researchers (NT and SMO), and differences between authors were resolved through discussion. The Physiotherapy Evidence Database (PEDro) scale was used to assess the methodological quality of the included research [27]. This scale consists of eleven distinct components. However, only items 2 through 11 can be rated. Depending on fulfilling specific criteria, studies are rated with 0 or 1; therefore, the maximal possible score on the PEDro scale is 10 (representing the low risk of bias), and the lowest is 0 (representing the high risk of bias). Studies with scores of 0–3 points were considered poor quality, with scores of 4–5 points of acceptable quality, with scores of 6–8 points of good quality, and with excellent quality when the score was 9 to 10 points.

Statistical Analysis

Using a random effects modeling approach, a meta-analysis was carried out using the R computer language. The weighted estimation of standardized mean differences (SMD) across studies was pooled using the inverse variance method. The SMD was used to determine the effect's magnitude, with a value of 0.2 indicating a trivial effect, a value of 0.2–0.3 small effect, a value of 0.4–0.8 moderate effect, and a value of > 0.8 representing a large effect [28]. In the meta-analysis, statistical heterogeneity between various trials was evaluated using the I^2 statistic, where 25% denotes a low risk of heterogeneity, 25–75% denotes a moderate risk of heterogeneity, and > 75% denotes a high risk of heterogeneity [29]. The I^2 statistic was calculated using the two constrained maximum-likelihood estimators. Any potential funnel plot asymmetry could be visually assessed for the included studies by plotting standard errors against Hedges' G values [30]. Additionally, the Trim and Fill method and Egger's regression test were employed to evaluate the asymmetry of funnel plots [31, 32].

Results

Literature Search

A literature search yielded 365 publications from relevant databases. After removing duplicates and studies that did not meet the inclusion criteria, a total of 12 studies were included in the SRMA. The study search was conducted by December 24, 2022. The PRISMA flow diagram is displayed in Fig. 16.1.

Study Characteristics

Characteristics of all included studies are presented in Table 16.1. All studies were published from 2008 to 2022. The duration of the interventions ranged from 5 days to 24 weeks. Two studies investigated the effect of dihydrogen on subjects with potential metabolic syndrome, two on subjects with NAFLD, two studies on healthy subjects, one in subjects with gastroesophageal reflux disease, rheumatoid arthritis, malignant liver tumors, chronic hepatitis B, and elderly population, respectively. In terms of hydrogen intake, in nine out of twelve studies, participants consumed hydrogen-rich water (HRW), while in two studies, the patients received hydrogen-rich saline or inhaled dihydrogen gas. In terms of the effect of dihydrogen on the liver panel, three studies reported a significant reduction in AST levels after the intervention. Three studies reported a significant decrease in ALT test outcomes, while one reported a decrease in GGT levels. Finally, the MDA levels significantly decreased in five studies, while only one showed increases in MDA levels.

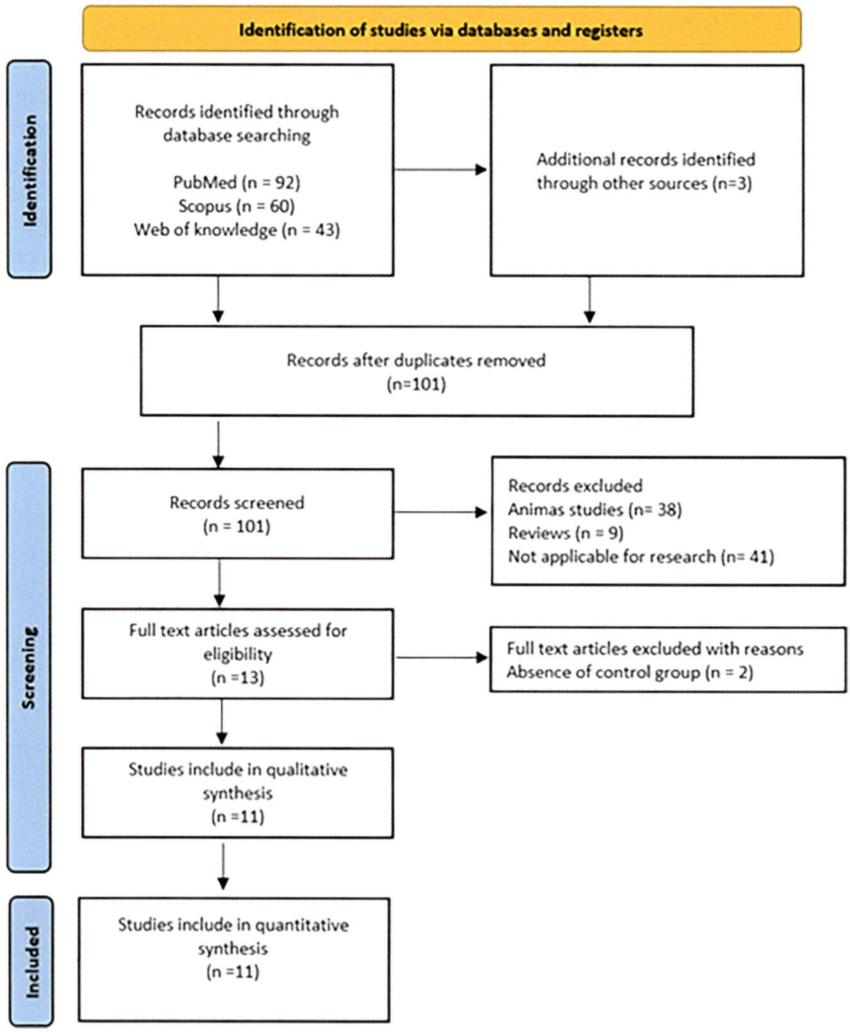

Fig. 16.1 PRISMA flow diagram

Level of Quality of the Studies

The mean PEDro scale score for the included studies was 8.27, which is considered an excellent average score. Six studies were rated excellent quality, while five were rated good quality. The results of the study quality analysis are depicted in Table 16.2.

Table 16.1 Effects of molecular hydrogen on hepatic function panel

Study	Population	Intervention	Variables	Outcomes
[33]	49 patients with malignant liver tumors	*Study design:* Randomized controlled trial *Intervention:* 500 mL of HRW per day *Duration:* 6 weeks	ALT AST GGT	↔ ↔ ↔
[18]	60 patients with chronic hepatitis B	*Study design:* Randomized controlled trial *Intervention:* 1.2–1.8 L of HRW per day *Duration:* 6 weeks	ALT MDA GGT	↓ ↓ ↔
[34]	26 patients with rheumatoid arthritis	*Study design:* Randomized controlled trial *Intervention:* 500 ml of hydrogen-rich saline *Duration:* 5 days	AST ALT GGT	↓ ↓ ↓
[35]	84 gastroesophageal reflux disease patients	*Study design:* Clinical trial *Intervention:* 1.5 L of HRW per day *Duration:* 12 weeks	MDA	↓
[36]	16 healthy participants	*Study design:* Randomized controlled trial *Intervention:* 300 mL of HRW per day *Duration:* 4 weeks	MDA	↓
[37]	38 juvenile female football players	*Study type:* Randomized controlled trial *Intervention:* 1.5–2 L of HRW per day *Duration:* 12 weeks	MDA	↓
[19]	10 overweight middle-aged women	*Study design*: Randomized controlled trial *Intervention*: 1 L of HRW per day *Duration*: 4 weeks	ALT AST GGT	↔ ↓ ↔
[38]	60 metabolic syndrome patients	*Study design:* Randomized controlled trial *Intervention:* 750 ml HRW *Duration:* 24 weeks	MDA	↓
[39]	40 elderly individuals	*Study design:* Randomized controlled trial *Intervention:* 500 ml of HRW per day *Duration: 24 weeks*	MDA	↔

(continued)

Table 16.1 (continued)

Study	Population	Intervention	Variables	Outcomes
[40]	62 patients diagnosed with NAFLD	*Study design:* Randomized controlled trial *Intervention:* Inhalation of 66% H_2 and 33% O_2 for 1 h/d *Duration:* 13 weeks	ALT AST GGT MDA	↓ ↓ ↔ ↓
[41]	30 subjects with NAFLD	*Study design:* Randomized controlled trial *Intervention:* 1L of HRW or placebo water *Duration*: 8 weeks	AST ALT MDA	↔ ↔ ↑

Abbreviations: ↔ no change; ↑ increase; ↓ decrease; *ALT* alanine aminotransferase; *AST* aspartate aminotransferase; *GGT* gamma-glutamyltransefase; *MDA* malondialdehyde; *NAFLD* non-alcoholic fatty liver disease; *HRW* hydrogen-rich water

Table 16.2 Table PEDro ratings of the included studies

Study	1	2	3	4	5	6	7	8	9	10	11	Total
[33]	Yes	1	0	1	0	0	1	1	1	1	1	7
[18]	Yes	1	0	1	0	0	1	1	1	1	1	7
[34]	Yes	1	1	1	1	1	1	1	1	1	1	10
[35]	Yes	1	0	1	0	0	0	1	1	1	1	6
[36]	No	1	0	1	0	0	0	1	1	1	1	6
[37]	Yes	1	0	1	1	1	1	1	1	1	1	9
[19]	Yes	1	0	1	1	1	1	1	1	1	1	9
[38]	Yes	1	1	1	1	1	1	1	1	1	1	10
[39]	Yes	1	0	1	1	0	1	1	1	1	1	8
[40]	Yes	1	1	1	1	0	1	1	1	1	1	9
[41]	Yes	1	1	1	1	1	1	1	1	1	1	10

Legend: 1, eligibility criteria were specified; 2, volunteers were randomly allocated to groups; 3, the allocation was concealed; 4, no baseline differences between the groups; 5, blinding of all participants; 6, the clinicians involved in providing the therapy were blinded; 7, blinding of all assessors who measured at least one key outcome; 8, measures of primary outcomes were acquired from 85% of participants; 9, all participants with accessible outcome measures received the treatment or control condition as assigned (if this was not possible, data for at least one primary outcome was analyzed using "intention to treat."); 10, at least one key outcome has between-group statistical comparisons; 11, the study provides both point measures and measures of variability for at least one key outcome

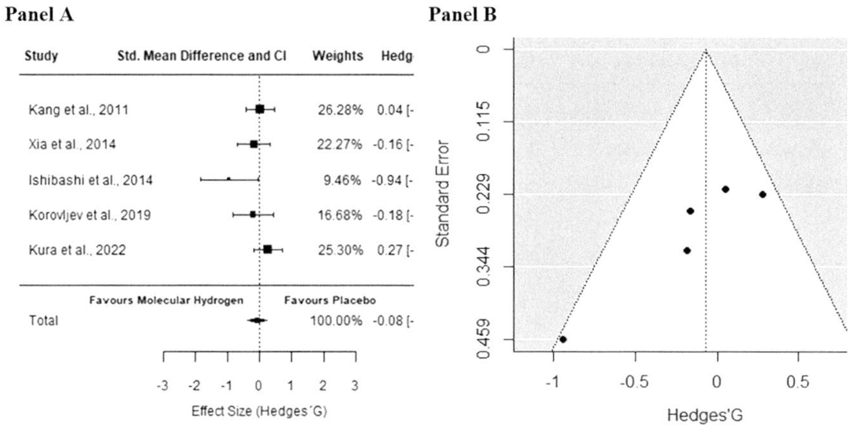

Fig. 16.2 Forest plot and funnel plot for meta-analysis of changes in serum alanine aminotransferase (ALT) levels after dihydrogen intervention

Meta-analysis

Effects of Dihydrogen on ALT Levels

The results of the pooled meta-analysis indicated a non-significant change (pooled SMD = – 0.08 [95% CI, from – 0.38 to 0.23]) in ALT levels following dihydrogen intake ($P = 0.42$) (Fig. 16.2, Panel A). Concerning the risk of bias, ALT levels of the I^2 square test demonstrated a significant heterogeneity between studies ($P = 0.04$). In addition, the I^2 statistics demonstrated a moderate risk of heterogeneity ($I^2 = 36.28\%$). Visual inspection of the funnel plot demonstrated no publication bias asymmetry (Fig. 16.2, Panel B). In addition, Egger's regression test yielded no significant results (df = 4; $P = 0.18$), while Duval and Tweedie Trim and Fill methods failed to locate missing studies.

Effect of Dihydrogen on AST Levels

AST levels did not change significantly following dihydrogen intake (pooled SMD = – 0.15 [95% CI; from -0.48 to 0.19]; $P = 0.38$) (Fig. 16.3, Panel A). Based on the I^2 square, there was no heterogeneity between studies ($P = 0.11$). The I^2 statistic indicated a moderate likelihood of heterogeneity ($I^2 = 41.33\%$). However, the funnel plot showed no publication bias asymmetry (Fig. 16.3, Panel B). Additionally, Egger's regression test did not encounter significant results (df = 4; $P = 0.15$), while Duval and Tweedie Trim and Fill methods failed to locate missing studies.

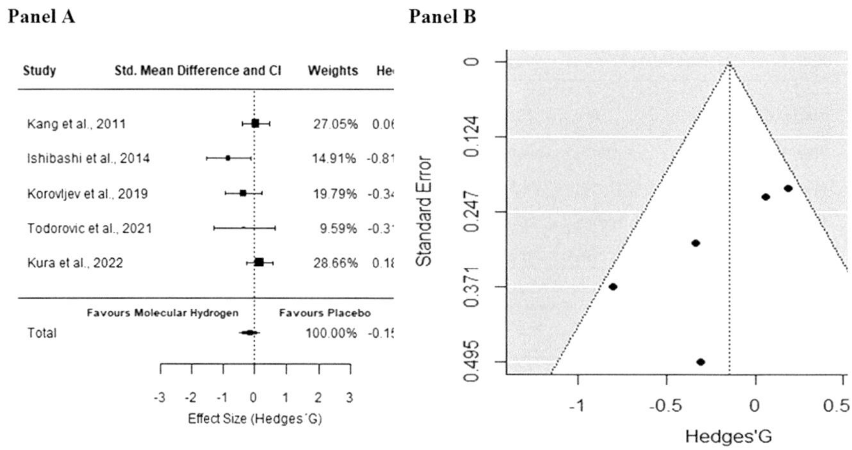

Fig. 16.3 Forest plot and funnel plot for meta-analysis of changes in serum aspartate aminotransferase (AST) level after dihydrogen intervention

Effect of Dihydrogen on GGT Levels

Results of the meta-analysis showed no significant change for GGT after dihydrogen intake (pooled SMD = 0.17 [95% CI, from − 0.65 to 0.99]; $P = 0.39$) (Fig. 16.4, Panel A), with values slightly evaluated at follow-up. However, the I^2 square test demonstrated significant heterogeneity between studies ($P < 0.01$) and a high risk of heterogeneity ($I^2 = 89.6\%$). The funnel plot's visual examination revealed asymmetry, which implied publishing bias (Fig. 16.4, Panel B); this was confirmed by a significant asymmetry found in Egger's regression test (df = 3; $P < 0.01$). Moreover, Duval and Tweedie Trim and Fill's method identified one missing study on the plot's right side. Heterogeneity was diminished after excluding the not evenly distributed study, signifying a low risk of heterogeneity ($I^2 = 0\%$; $P = 0.94$). After bias correction, Egger's regression test revealed no funnel plot asymmetry (df = 5; $P = 0.65$), whereas Duval and Tweedie Trim and Fill's approach did not detect missing studies on either side of the plot (Fig. 16.4, Panel D). This had influence also on the statistical part of the analysis. Although still not significant, the results indicate a slight decrease of GGT after reducing the possible risk of bias (pooled SMD = -0.19 [95% CI, from -0.52 to 0.15]; $P = 0.39$) (Fig. 16.4, Panel C).

Effect of Dihydrogen on MDA Levels

MDA levels were significantly decreased following dihydrogen intake (pooled SMD = − 0.97 [95% CI, from − 1.65 to − 0.19]; $P = 0.01$), with results indicating a large effect (Fig. 16.5, Panel A). However, the I^2 square test of heterogeneity points to significant heterogeneity between studies ($P = 0.01$), with an extremely high risk of heterogeneity ($I^2 = 90.6\%$). The visual analysis of the funnel plot showed significant

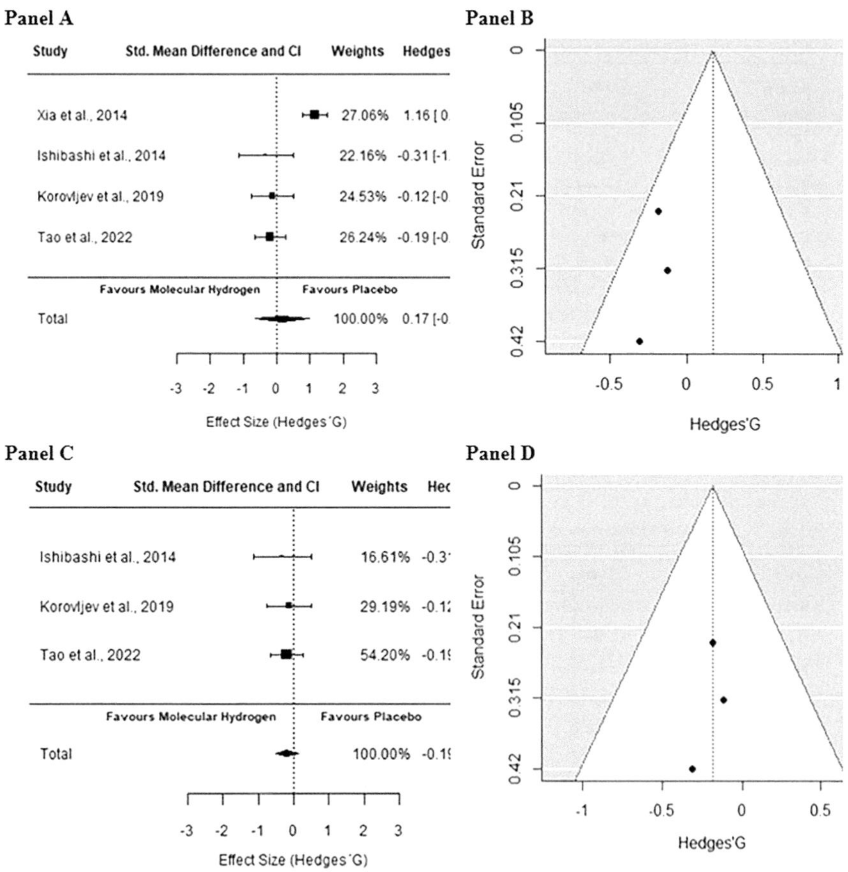

Fig. 16.4 Forest plot and funnel plot for meta-analysis of changes in serum gamma-glutamyltransefase (GGT) level after dihydrogen intervention

asymmetry and publication bias (Fig. 16.5, Panel B). There were no significant results from Egger's regression test (df = 8; $P = 0.33$); however, Duval and Tweedie Trim and Fill found that one study was missing on the left side of the funnel plot and two on the right. By excluding studies distributed unevenly around the funnel plot base, the heterogeneity between studies was significantly decreased, leading to a minimal heterogeneity risk ($I^2 = 0\%$; $P = 0.83$). After bias adjustment, Egger's regression test revealed no funnel plot asymmetry (df = 5; $P = 0.88$), whereas the Duval and Tweedie Trim and Fill methods did not reveal any missing studies (Fig. 16.5, Panel D). However, this did not influence the statistical part of the analysis, where the results showed a large effect of dihydrogen intake on MDA levels after an adjustment (pooled SMD = − 0.65 [95% CI, from − 0.86 to − 0.15]; $P = 0.01$) (Fig. 16.5, Panel C).

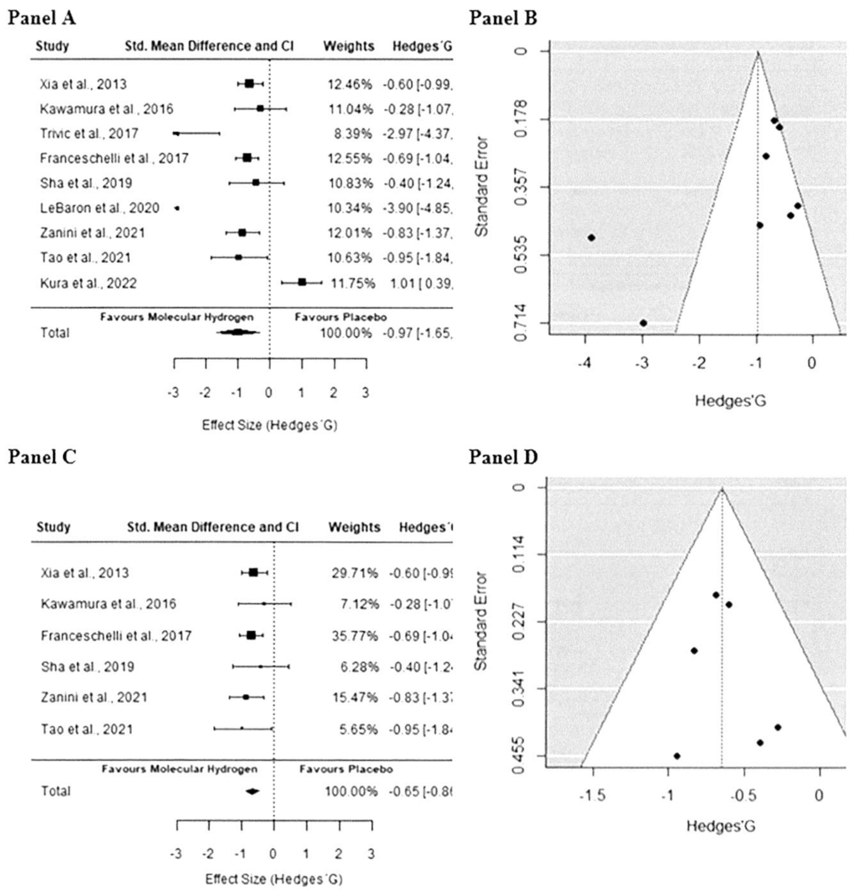

Fig. 16.5 Forest plot and funnel plot for meta-analysis of changes in serum malondialdehyde (MDA) level after dihydrogen intervention

Discussion

The primary objective of this SRMA was to synthesize and evaluate the existing scientific literature on whether dihydrogen supplementation affects the hepatic function panel. Our results suggest that consuming dihydrogen may improve liver health by affecting several indicators in the hepatic panel, including lowering MDA levels and tending to reduce ALT, AST, and GGT levels.

Although MDA is considered a marker of oxidative stress, it can also be associated with NAFLD and hepatic viability [7, 42]. The pathophysiology of both acute and chronic liver diseases is influenced by oxidative stress [8, 43]. The oxidation of various cellular components such as DNA, proteins, and lipids/fatty acids could cause damage and potential cell death [44]. In particular, lipid peroxidation results in the generation of many different pro-inflammatory products, all of which contribute to the development of the disease. In addition, the liver stellate cells can be prompted to produce collagen by MDA, leading to fibrosis [45]. Prior research examined MDA levels in peripheral blood, hepatic tissue, hepatic vein, peripheral vein, and urine samples as a marker for lipid peroxidation in the liver [46]. In patients with NAFLD, an increase in MDA is accompanied by a decrease in antioxidant activity [47]. Moreover, high levels of plasma MDA have been linked to diabetes and obesity/NAFLD [48]. In the present analysis, seven out of eight studies showed a significant decrease in MDA levels after dihydrogen intake compared to placebo. The only study that showed the adverse effect was a trial conducted by Kura et al. [41] on patients with NAFLD, where a mild non-significant increase in MDA (~ 17.2%) occurred. Despite the capability of dihydrogen to be primarily responsible for MDA reduction, its therapeutic effects are occasionally associated with temporarily increased levels of MDA. This is one of the possible explanations for the observed increase. In addition, there is a question of dosage, low dihydrogen concentration (~ 4 ppm), the long-term duration of the trial (6 months), and compliance with dihydrogen intake among participants throughout the study. Therefore, these results should not be interpreted as an adverse effect of dihydrogen intake. The reduction in MDA levels reported in our SRMA was also seen in animal studies. For example, the effects of dihydrogen on preventing early steatohepatitis in mice were analyzed by Lin and colleagues [49]. Elevations in hepatic lipid accumulation and inflammatory cytokines, including tumor necrosis factor-alpha and interleukin-6, were caused by ethanol consumption; however, HRW therapy significantly reduced these effects. Furthermore, the liver's superoxide dismutase (SOD), glutathione peroxidase (GSH-Px), and catalase activities were all enhanced by dihydrogen, while MDA levels were reduced. In addition, ALT and AST levels were also significantly reduced.

Small but insignificant effects of dihydrogen were seen on ALT, AST, and GGT levels across all trials we included in the analysis. Although multiple animal studies showed the possible effect of dihydrogen in reducing hepatic function panel biomarkers [50, 51], our SRMA did not find any significance in terms of these specific liver enzymes. The possible explanation for such results might be an insufficient number of studies, diversity among the population evaluated in studies included, a dosage of dihydrogen administration, varying duration of the treatment, and influence of other accompanying diseases. However, despite this, some positive trends are observed, and future studies should evaluate the effects of dihydrogen on hepatic enzymes and the possible beneficial effect of these therapies on liver health.

Several mechanisms could explain the beneficial effects of dihydrogen on liver health and hepatic function panel. Dihydrogen is often considered a biological modulator of target cells, similar to other medical gases. It appears that several antioxidant enzymes, including heme oxygenase [52], SOD [53], catalase [54], and myeloperoxidase [53], could play a role in dihydrogen indirect reduction of oxidative stress [55]. Guan and co-workers found that dihydrogen could protect against renal dysfunction induced by chronic intermittent hypoxia [56]. In this study, dihydrogen has been shown to reduce oxidative stress by elevating the reduced glutathione/oxidized glutathione ratio and regulating SOD and GSH-Px activity. MDA, another end product of oxidative stress, is also reduced by dihydrogen. Possible antioxidative effects of dihydrogen are likely mediated via the nuclear factor erythroid 2–related factor 2/antioxidant response element pathway [57, 58]. This pathway could be crucial for preventing damage caused by free radicals and controlling the production of other antioxidant and cytoprotective proteins. An essential role for the transcription factor Nrf2 can be seen in the redox-sensitive regulation of the expression of a number of endogenous antioxidants and detoxification enzymes [59]. In addition, dihydrogen may influence peroxisome proliferator-activated receptor gamma coactivator 1-alpha (PGC-1) activation [60]. PGC-1 plays a critical role in controlling energy metabolism in cells by stimulating mitochondrial biogenesis [61]. In addition, PGC-1 raises the liver-derived and systemic fibroblast growth factor 21 (FGF-21). One possible mechanism by which dihydrogen regulates lipid metabolism, leading to better fuel utilization and less body fat accumulation, is via induction of hepatic FGF-21 [60]. Furthermore, FGF-21 can also effectively regulate metabolic homeostasis and cellular senescence. Moreover, FGF-21 analogs and FGF-21 receptor agonists have therapeutic potential for treating age-related metabolic diseases [62].

This SRMA has several limitations. The publication bias in our study was mitigated by using stringent inclusion criteria and blinded RCTs. However, the study still has inherent limitations shared by all meta-analyses [63]. Moreover, as a result, only a handful of studies were included in this SRMA, which may weaken the reliability of the findings. Second, there was a possible risk of bias due to differences in the following categories: baseline participant characteristics; supplement intake protocols; the form of supplemental intake; the study duration; accompanying diseases; and potential interference with participants' regular therapies. Finally, there is a large gap between evidence obtained from external outcome measures, such as reductions in hepatic function panel, and fundamental basic research that would confirm the exact and direct mechanisms of dihydrogen action. The effects on gene expression and cellular/mitochondrial bioenergetics are commonly attributed to dihydrogen and explained by its involvement in a cascade of cellular signaling pathways. Specifically, there is a lack of potential mechanisms, enzymes, and genes that transport dihydrogen directly throughout cells. Due to the absence of the hydrogen-synthesizing enzyme hydrogenase in humans, the precise effect of dihydrogen remains unknown.

Conclusion

The present systematic review and meta-analysis demonstrated that dihydrogen supplementation improved individual elements of the hepatic function panel regardless of supplementation protocol. Strong evidence suggests that dihydrogen reduces the oxidative stress biomarker MDA in the serum, thereby enhancing hepatic health. A strong trend for a reduction in liver function tests (including aspartate aminotransferase and alanine transaminase) has also been observed after dihydrogen administration. To confirm and expand the current knowledge of the potential effects of dihydrogen on hepatic health, well-structured and organized randomized controlled trials are highly warranted.

References

1. Cheemerla S, Balakrishnan M (2021) Global epidemiology of chronic liver disease. Clin Liver Dis 17:365–370
2. Moon AM, Singal AG, Tapper EB (2020) Contemporary epidemiology of chronic liver disease and cirrhosis. Clin Gastroenterol Hepatol 18:2650–2666
3. Sepanlou SG, Safiri S, Bisignano C et al (2020) The global, regional, and national burden of cirrhosis by cause in 195 countries and territories, 1990–2017: a systematic analysis for the global burden of disease study 2017. Lancet Gastroenterol Hepatol 5:245–266
4. Asrani SK, Devarbhavi H, Eaton J, Kamath PS (2019) Burden of liver diseases in the world. J Hepatol 70:151–171
5. Thapa BR, Walia A (2007) Liver function tests and their interpretation. Indian J Pediatr 74:663–671
6. van Beek JH, de Moor MH, de Geus EJ et al (2013) The genetic architecture of liver enzyme levels: GGT, ALT and AST. Behav Genet 43:329–339
7. Atiba AS, Abbiyesuku FM, Oparinde DP, Temitope A, Akindele RA (2016) Plasma malondialdehyde (MDA): an indication of liver damage in women with pre-eclamsia. Ethiop J Health Sci 26:479–486
8. Comporti M, Arezzini B, Signorini C, Vecchio D, Gardi C (2009) Oxidative stress, isoprostanes and hepatic fibrosis. Histol Histopathol 24:893–900
9. Pimpin L, Cortez-Pinto H, Negro F et al (2018) Burden of liver disease in Europe: epidemiology and analysis of risk factors to identify prevention policies. J Hepatol 69:718–735
10. Nompleggi DJ, Bonkovsky HL (1994) Nutritional supplementation in chronic liver disease: an analytical review. Hepatol 19:518–533
11. Cirino G, Vellecco V, Bucci M (2017) Nitric oxide and hydrogen sulfide: the gasotransmitter paradigm of the vascular system. Br J Pharmacol 174:4021–4031
12. Sansbury BE, Hill BG (2014) Regulation of obesity and insulin resistance by nitric oxide. Free Radic Biol Med 73:383–399
13. Ohta S (2011) Recent progress toward hydrogen medicine: potential of molecular hydrogen for preventive and therapeutic applications. Curr Pharm Des 17:2241–2252
14. Ostojic SM (2015) Targeting molecular hydrogen to mitochondria: barriers and gateways. Pharmacol Res 94:51–53
15. Ostojic SM (2015) Molecular hydrogen: an inert gas turns clinically effective. Ann med 47:301–304
16. Zhang Y, Xu J, Yang H (2020) Hydrogen: An endogenous regulator of liver homeostasis. Front Pharmacol 11:877
17. Yang Q, Ji G, Pan R, Zhao Y, Yan P (2017) Protective effect of hydrogen-rich water on liver function of colorectal cancer patients treated with mFOLFOX6 chemotherapy. Mol Clin Oncol 7:891–896
18. Xia C, Liu W, Zeng D, Zhu L, Sun X, Sun X (2013) Effect of hydrogen-rich water on oxidative stress, liver function, and viral load in patients with chronic hepatitis B. Clin Transl Sci 6:372–375
19. Korovljev D, Stajer V, Ostojic J, LeBaron TW, Ostojic SM (2019) Hydrogen-rich water reduces liver fat accumulation and improves liver enzyme profiles in patients with non-alcoholic fatty liver disease: a randomized controlled pilot trial. Clin Res Hepatol Gastroenterol 43:688–693
20. Yamamoto R, Homma K, Suzuki S, Sano M, Sasaki J (2019) Hydrogen gas distribution in organs after inhalation: real-time monitoring of tissue hydrogen concentration in rat. Sci Rep 9:1–7

21. Kajiyama S, Hasegawa G, Asano M et al (2008) Supplementation of hydrogen-rich water improves lipid and glucose metabolism in patients with type 2 diabetes or impaired glucose tolerance. Nut Res 28:137–143
22. Nakao A, Toyoda Y, Sharma P, Evans M, Guthrie N (2010) Effectiveness of hydrogen rich water on antioxidant status of subjects with potential metabolic syndrome—an open label pilot study. J Clin Biochem Nutr 46:140–149
23. Koyama Y, Taura K, Hatano E et al (2014) Effects of oral intake of hydrogen water on liver fibrogenesis in mice. Hepatol Res 44:663–677
24. Reeder SB, Cruite I, Hamilton G, Sirlin CB (2011) Quantitative assessment of liver fat with magnetic resonance imaging and spectroscopy. J Magn Reson Imaging 34:729–749
25. Todorovic N, Fernández-Landa J, Santibañez A et al (2023) The effects of hydrogen-rich water on blood lipid profiles in clinical populations: a systematic review and meta-analysis. Pharmaceuticals 16:142
26. Liberati A, Altman DG, Tetzlaff J et al (2009) The PRISMA statement for reporting systematic reviews and meta-analyses of studies that evaluate health care interventions: explanation and elaboration. J Clin Epidemiol 62:e1–e34
27. De Morton NA (2009) The PEDro scale is a valid measure of the methodological quality of clinical trials: a demographic study. Aust J Physiother 55:29–133
28. Primer AP (1992) Quantitative methods in psychology. Psychol Bull 112:155–159
29. Higgins JP, Thompson SG, Deeks JJ, Altman DG (2003) Measuring inconsistency in meta-analyses. BMJ 327:557–560
30. Hedges LV (1981) Distribution theory for Glass's estimator of effect size and related estimators. J Educ Stat 6:107–128
31. Egger M, Smith GD, Schneider M, Minder C (1997) Bias in meta-analysis detected by a simple, graphical test. BMJ 315:629–634
32. Duval S, Tweedie R (2000) Trim and fill: a simple funnel-plot–based method of testing and adjusting for publication bias in meta-analysis. Biometrics 56:455–463
33. Kang KM, Kang YN, Choi IB et al (2011) Effects of drinking hydrogen-rich water on the quality of life of patients treated with radiotherapy for liver tumors. Med Gas Res 1:1–8
34. Ishibashi T, Sato B, Shibata S et al (2014) Therapeutic efficacy of infused molecular hydrogen in saline on rheumatoid arthritis: a randomized, double-blind, placebo-controlled pilot study. Int Immunopharmacol 21:468–473
35. Franceschelli S, Gatta DMP, Pesce M et al (2018) Modulation of the oxidative plasmatic state in gastroesophageal reflux disease with the addition of rich water molecular hydrogen: a new biological vision. J Cell Mol Med 22:2750–2759
36. Trivic T, Vojnovic M, Drid P, Ostojic SM (2017) Drinking hydrogen-rich water for 4 weeks positively affects serum antioxidant enzymes in healthy men: a pilot study. Curr Top Nutraceutical Res 15:45–48
37. Sha JB, Zhang SS, Lu YM et al (2018) Effects of the long-term consumption of hydrogen-rich water on the antioxidant activity and the gut flora in female juvenile soccer players from Suzhou. China. Med Gas Res 8:135
38. LeBaron TW, Singh RB, Fatima G et al (2020) The effects of 24-week, high-concentration hydrogen-rich water on body composition, blood lipid profiles and inflammation biomarkers in men and women with metabolic syndrome: a randomized controlled trial. Diabetes Metab Syndr Obes 13:889–896
39. Zanini D, Todorovic N, Korovljev D et al (2021) The effects of 6-month hydrogen-rich water intake on molecular and phenotypic biomarkers of aging in older adults aged 70 years and over: a randomized controlled pilot trial. Exp Gerontol 155:111574
40. Tao G, Zhang G, Chen W et al (2022) A randomized, placebo-controlled clinical trial of hydrogen/oxygen inhalation for non-alcoholic fatty liver disease. J Cell Mol Med 26:4113–4123
41. Kura B, Szantova M, LeBaron TW et al (2022) Biological effects of hydrogen water on subjects with NAFLD: a randomized placebo-controlled trial. Antioxidants 11:1935
42. Zelber-Sagi S, Ivancovsky-Wajcman D, Fliss-Isakov N et al (2020) Serum malondialdehyde is associated with non-alcoholic fatty liver and related liver damage differentially in men and women. Antioxidants 9:578

43. Lee KC, Yang YY, Wang YW et al (2010) Increased plasma malondialdehyde in patients with viral cirrhosis and its relationships to plasma nitric oxide, endotoxin, and portal pressure. Dig Dis Sci 55:2077–2085
44. Negre-Salvayre A, Coatrieux C, Ingueneau C, Salvayre R (2008) Advanced lipid peroxidation end products in oxidative damage to proteins. Potential role in diseases and therapeutic prospects for the inhibitors. Br J Pharmacol 153:6–20
45. Hadizadeh F, Faghihimani E, Adibi P (2017) Non-alcoholic fatty liver disease: diagnostic biomarkers. World J Gastrointest Pathophysiol 8:11–26
46. Vuppalanchi R, Ghabril M, Chalasani N et al (2011) Oxidative stress in chronic liver disease: relationship between peripheral and hepatic measurements. Am J Med Sci 342:314–317
47. Yesilova Z, Yaman H, Oktenli C et al (2005) Systemic markers of lipid peroxidation and antioxidants in patients with non-alcoholic fatty liver disease. Am J Gastroenterol 100:850–855
48. Varma M, Makwane H, Kare P, Jha R, Parmar A (2016) Study of serum ferritin, serum uric acid and plasma malondialdehyde (MDA) levels in non-alcoholic fatty liver disease. Int J Biomed Adv Res 7:4
49. Lin CP, Chuang WC, Lu FJ, Chen CY (2017) Antioxidant and anti-inflammatory effects of hydrogen-rich water alleviate ethanol-induced fatty liver in mice. World J Gastroenterol 23:4920
50. Kawai D, Takaki A, Nakatsuka A et al (2012) Hydrogen-rich water prevents progression of non-alcoholic steatohepatitis and accompanying hepatocarcinogenesis in mice. Hepatology 56:912–921
51. Kamimura N, Nishimaki K, Ohsawa I, Ohta S (2011) Molecular hydrogen improves obesity and diabetes by inducing hepatic FGF21 and stimulating energy metabolism in db/db mice. Obesity 19:1396–1403
52. Huang CS, Kawamura T, Toyoda Y, Nakao A (2010) Recent advances in hydrogen research as a therapeutic medical gas. Free Radic Res 44:971–982
53. Ming Y, Ma QH, Han XL, Li HY (2020) Molecular hydrogen improves type 2 diabetes through inhibiting oxidative stress. Exp Ther Med 20:359–366
54. Ge YS, Zhang QZ, Li H, Bai G, Jiao ZH, Wang HB (2019) Hydrogen-rich saline protects against hepatic injury induced by ischemia-reperfusion and laparoscopic hepatectomy in swine. Hepatobiliary Pancreat Dis Int 18:48–61
55. Zhao Z, Pu Y (2019) Lixisenatide enhances mitochondrial biogenesis and function through regulating the CREB/PGC-1α pathway. Biochem Biophys Res Commun 508:1120–1125
56. Guan P, Sun ZM, Luo LF et al (2019) Hydrogen protects against chronic intermittent hypoxia induced renal dysfunction by promoting autophagy and alleviating apoptosis. Life Sci 225:46–54
57. Kura B, Bagchi AK, Singal PK et al (2019) Molecular hydrogen: Potential in mitigating oxidative-stress-induced radiation injury. Can J Physiol Pharmacol 97:287–292
58. Kawamura T, Wakabayashi N, Shigemura N et al (2013) Hydrogen gas reduces hyperoxic lung injury via the Nrf2 pathway in vivo. Am J Physiol Lung Cell Mol 304:646–656
59. Huang Y, Li W, Su ZY, Kong ANT (2015) The complexity of the Nrf2 pathway: beyond the antioxidant response. J Nutr Biochem 26:1401–1413
60. Kamimura N, Ichimiya H, Iuchi K, Ohta S (2016) Molecular hydrogen stimulates the gene expression of transcriptional coactivator PGC-1α to enhance fatty acid metabolism. NPJ Aging Mech Dis 2:16008
61. Miller KN, Clark JP, Anderson RM (2019) Mitochondrial regulator PGC-1a—modulating the modulator. Curr Opin Endocr Metab Res 5:37–44
62. Yan J, Nie Y, Cao J et al (2021) The roles and pharmacological effects of FGF21 in preventing aging-associated metabolic diseases. Front Cardiovasc Med 8:655575
63. LeLorier J, Gregoire G, Benhaddad A, Lapierre J, Derderian F (1997) Discrepancies between meta-analyses and subsequent large randomized, controlled trials. N Engl J Med 337:536–542

Chapter 17
Hydrogen-Rich Water Using as a Modulator of Gut Microbiota and Managing the Inflammatory Bowel Disease

Atieh Yaghoubi, Saman Soleimanpour, and Majid Khazaei

Abstract Molecular hydrogen (H_2) is well known as a colorless gas, and water enriched with H_2 (HRW) is an innovative, beneficial beverage for human health that improves gut microbiota management and the viability of the intestinal. Drinking water enriched with hydrogen has been found to have therapeutic effects on inflammatory bowel diseases (IBDs). The low molecular weight of H_2 enables it to easily diffuse and permeate cell membranes to exert various biological effects. In addition, H_2 may control the immune system, antioxidant and anti-inflammatory activities (metabolism of mitochondrial energy), and cell death processes (apoptosis, autophagy, and pyroptosis) by reducing excessive reactive oxygen species generation and altering nuclear transcription factors. The fundamental mechanism of H_2 is still not fully understood. Given its safety and possible usefulness, H_2 has a promising future as a treatment for a variety of illnesses, including IBD. This review aimed to comprehensively highlight the current knowledge in the fields of H_2 function in antioxidative, anti-inflammatory, and anti-apoptotic effects as well as its underlying mechanism, with a focus on IBD, and also offer recommendations for using H_2 medically to treat IBD.

Keywords Molecular hydrogen · H_2 · Hydrogen-rich water · HRW · Inflammatory bowel diseases · IBD

A. Yaghoubi · S. Soleimanpour
Department of Microbiology and Virology, Faculty of Medicine, Mashhad University of Medical Sciences, Mashhad, Iran
e-mail: YaghoubiA2@mums.ac.ir

S. Soleimanpour
e-mail: SoleimanpourS@mums.ac.ir

M. Khazaei (✉)
Department of Physiology, Faculty of Medicine, Mashhad University of Medical Sciences, Mashhad, Iran
e-mail: Khazaeim@mums.ac.ir

J. Slezak and B. Kura (eds.), *Molecular Hydrogen in Health and Disease*, Advances in Biochemistry in Health and Disease 27, https://doi.org/10.1007/978-3-031-47375-3_17

Introduction

Hydrogen (H_2) is well-known as a colorless and abundant gas in the atmosphere of the earth. During the last decade, this gas attracts much attention as a natural antioxidant with potential benefits for medical [1, 2]. Molecular hydrogen with at least capacity for interaction with most of the biomolecules, for the first time used in combination with H_2, helium, and oxygen (O_2) called hydreliox for the prevention of decompression illness (DCI) and depth intoxication (narks), occurs due to the very deep technical diving [3].

In 1975, molecular hydrogen was for the first time used in therapy that exhibits great ability in tumor regression in the animal model of skin squamous carcinoma [4]. After that, in 2007, H_2 was used as a breathing gas in rats with cerebral ischemia–reperfusion (I/R) injury and stroke. The result of this investigation shows that H_2 can significantly inhibit this injury through affecting the oxidative stress imbalance [5]. Since then, H_2 explored and now it is well understood that this gas can be considered an ideal therapy for different diseases due to its anti-apoptotic, anti-oxidative, and anti-inflammatory activity [6].

H_2 is able to penetrate into the cellular membrane and easily spread into the cytosol of the target cell then specifically decrease the hydroxide radicals and peroxynitrite [7]. Clinical trials have demonstrated that H_2 treatment was safe and effective in patients with asthma and chronic obstructive pulmonary disease (COPD) [8, 9]. Different clinical trials had launched to discover the safety and efficacy of H_2 and it is well demonstrated this gas is ideal therapy in patients who had COPD, also can decrease intestinal ischemia-reperfusion (I/R) injury, and inflammatory bowel diseases (IBD) [8–12]. Here we aimed to comprehensively review the current knowledge in the field of therapeutic application of molecular hydrogen specifically in inflammatory bowel diseases.

Molecular Hydrogen

Physiology

As we know H_2 is an odorless and colorless gas with a low molecular weight (2.016 g/mol) that is flammable and with volumes of 4 and 75% it can be burned in air. Within the body, this gas can rapidly diffuse from alveoli into the blood and then circulate all over the body. H_2 due to its low molecular weight and lack of polarity can rapidly penetrate into the membrane of the cells and reach the cytoplasm then separate to the nucleus and other organelles of the target cells to induce its biological effects. In addition, molecular hydrogen can be across the blood-brain barrier, while usually most of the antioxidant agents cannot. In addition, there is no report on the toxicity of this gas. In addition, it is exhibited that H_2 has no adverse effect on body physiology like blood pressure, the temperature of the body, pH, and partial pressure of oxygen (PO_2) [13].

But it is well known that H_2 induces an anti-inflammatory and anti-oxidative effect by manipulating intracellular homeostasis, neutralizing the biomarker of oxidative stress, affecting transporting the of mitochondrial electrons, and effect on the transcription of key inflammatory regulatory proteins [7]. Mammalian cells are not able to produce H_2 due to the lack of hydrogenases. However, some of the anaerobic metabolism microorganism that belong to the gut microbial community are able to produces H_2 following the breakdown of the lactulose which would be about 50–1000 mg daily [14].

Biological Activities

Anti-oxidative Activities

Molecular hydrogen induces its antioxidant effects through different mechanisms like the neutralization of reactive oxygen species (ROS) that are produced due to mitochondrial respiration, xanthine oxidoreductase, or even oxidase of NADH/NADPH. ROS is a general term used for describing the wide range of oxidant molecules like nitric oxide (NO•), hydroxyl (•OH), peroxyl (RO2•), superoxide anion (O2•−), and alkoxyl (RO•) radicals [15]. Cell damage is well known as an initiator trigger for blocking the transport of electron to the mitochondria and oxidative phosphorylation that result in using the electrons for a generation of more ROS. This imbalance between ROS and antioxidant biomarkers leads to oxidative stress that is allowed to initiate different pathological processes following the damaging of the cell membrane or even the membrane of the organelle due to the overgeneration of ROS [16]. These membrane damages lead to detached of the lipids form membrane that cused excess peroxide then prodcution of further leukotrienes and arachidonic acid which all involved in the inflammatory pain. In addition, still there is no known enzyme which exclusively responsible for removing and dealing with hydroxyl (•OH) because this molecule nonspecifically interacts with the neighbor nucleophilic biomolecules, while molecular hydrogen is well-known as a molecule with the ability for penetrating to the damaged membrane of cells then reductant [17, 18]. H_2 is able to directly scavenge the hydroxyl radical via reaction with •OH that results in the production of H_2O and H• then the resulting H• reacting with O_2^- that results in the production of HO_2^- [18, 19]. The previous investigations demonstrated that H_2 is able to radically raise the intracellular activity of superoxide dismutase (SOD), while it decreased the level of the malondialdehyde (MDA) in the inflammatory model that exhibits its protective benefit toward reperfusion oxidative injury [5, 20]. In addition, the protective benefit of H_2 was also observed in I/R injury via decreasing the oxidative stress and scavenging the •OH and $ONOO^-$ which involve in the over generation of ROS via donating the extra electrons. Furthermore, shifting the nuclear factor erythroid-2 related factor 2 (Nrf2) into the nucleus can regulate the expression of the genes that have a role in defense toward oxidative stress [21]. Evidence suggests that gavage of the H_2-rich saline can induce the activation of the Nrf2-ARE signaling pathway in mice models of autoimmune encephalomyelitis (EAE) [22]. Therefore, H_2 upregulate intracellular transcription of Nrf2 that lead to overexpression of SOD and glutathione (GSH) and decrease the expression of NADPH oxidase which all result in decreasing the intracellular ROS [23, 24].

Mitochondria regulation is another antioxidant activity of H_2 through affecting disorders on the electron transport chain that would be the initial step of mitochondrial oxidative stress and excessive ROS generation. The term cell powerhouse is well defined by the mitochondria's responsibility that generates 90% of the cell's energy in the form of ATP [25]. In this process H_2 manage the dysfunction of the electron transport chain of mitochondrial through inhibiting the uncontrolled leakage of electrons from the chain, also it's able to regenerate this disorder in the cells and then prevent excessive ROS generation. Mitochondrial ATP-sensitive K^+-channel (mKATP-channel) that involve in the regulation of energy produced via mitochondria. Evidence suggests that therapy with H_2 in patients with acute myocardial infarction can activate mKATP and through that regulate the membrane potential of mitochondrial to equilibrate the myocardial NAD^+ level as a precursor for the synthesis of ATP then lead to ATP production from mitochondria which all results improving the myocardial I/R injury [26]. All this evidence suggests the protective ability of H_2 against dysfunction of mitochondrial that also affects signal transduction of crucial processes involved in cell death like Bax and caspase pathway [27]. Mitophagy has a key role in homeostasis by removing the mitochondria that are dysfunctional or damaged. One of the receptors of mitophagy is called Fun 14 domain-containing protein 1 (Fundc1) which is present on the mitochondrion outer membrane and regulates the balance of mitochondrial ATP via interacting with LC3 II. An investigation suggests that therapy with H_2 (2%) for 3 h can enhance the mitophagy induced by Fundc1 and then protect recipient mice from liver injury-mediated sepsis [28]. Previous reports demonstrate the antioxidant effects of H_2 on the treatment of animal models with Parkinson's disease as well as Alzheimer's disease [29, 30]. After that, a randomized double-blind placebo-controlled trial on Parkinson's disease patients used the 6.5% H_2 gas as inhaled with a dose of 2 L/min twice a day for 16 weeks, but this trial didn't find any beneficial effects [31]. However, the beneficial effects of H_2 are may due to the relation between its concentration and duration time of treatment, also the low number of participants in that trial probably affects the results.

Anti-inflammatory Activities

Inflammation is well-known as a component of adaptive immune responses elicited against pathogens or even tissue damage that serves as a warning signal for the aggregation or migration of lymphocytes, neutrophils, and mononuclear phagocytes (MNPs) to the site of infection and then differentiation into macrophages as a major source of releasing pro-inflammatory cytokines [32]. As previously stated, excessive ROS production causes upregulation of transcription from inflammatory factors such as p53, hypoxia-inducible factor-1α (HIF-1α), nuclear factor-kB (NF-kB), matrix metalloproteinases (MMPs), and peroxisome proliferator-activated receptor gamma (PPAR- or PPARG), which are all considered as triggers for the initiation of apoptosis processes [33–35]. This evidence makes it clear that all of the pathological processes like inflammation, oxidative stress, apoptosis, and damage to cells act together and mutually influence each other. According to reports, H_2 blocks neutrophil and macrophage infiltration by reducing the intercellular expression of adhesion molecules and chemokines like IL-1β and TNF-α. This, combined with a decrease in inflammatory cytokines like IL-6 and IFN-γ, is thought to be how H_2 inhibits inflammation at the primary stage [36, 37]. According to Wang et al. (2015), H_2-rich saline lowered serum IL-1β, IL-6, and TNF-α levels and blocked the activation of the critical inflammatory signaling pathway NF- kB, easing the airway inflammatory response brought on by a burn in rats [38]. Moreover, it is exhibited that molecular hydrogen can improve the hematencephalon, injury of the liver, and even injury of skeletal muscle following acute sports by downregulating the expression of NF-kB which suggests that H_2 is involved in healing through regulating the inflammatory process and affecting the transcription and cytokines production [39–41]. Upregulating regulatory T cells (Tregs) is also another anti-inflammatory effect of H_2 that lead to downregulating the expression of NF-kB [42, 43].

Anti-apoptotic Activities

One of the ways that molecular hydrogen induces its antiapoptotic effect is through gene transcription regulation or scavenging ROS, and via them, it affects endogenous apoptosis. In addition to preventing ROS generation, H_2 can preserve cell viability by blocking the caspase-3 and -9 pathways in intestine epithelial cells [44]. It is well understood that there are two types of signals that can cause apoptosis: intrinsic and extrinsic. The receptors of the extrinsic signal that medicated the apoptosis are present on the cell surface that interact with the Fas and TNF then leading to activating downstream caspase-8 that mediates the apoptosis. The second signal, or intrinsic signal, that mediates apoptosis is due to the expression level of B-cell lymphoma 2 (Bcl-2) as an antiapoptotic protein and Bax as a proapoptotic protein. However, the common endpoint for both pathways is the activation of caspase-3 and the fragmentation of DNA [45]. Evidence shows that H_2 is also able to affect the overexpression of Bax and Bcl-2 [46]. It also was suggested, H_2-rich water probebly achieved its medication effects by preventing the translocation of the apoptotic biomarkers, caspase-3, and Bax to the mitochondria [44]. In addition to these, H_2-rich water also induces its antiapoptotic effects via increasing the main antiapoptotic factor called Bcl-2 expression [47]. In addition to the apoptosis pathways, H_2 regulates cell death by affecting the autophagy pathways. The term "autophagy" is used to define the set of actions that balance the cell's energy via the degradation of macromolecular substances, while the same as all processes, excessive activity in this case also leads to tissue damage and acute inflammation that results in sepsis. Light chain 3 protein (LC3) and Beclin-1 are autophagy-related proteins that are widely used in its detection. Previous reports suggest that H_2-rich water can protect tissues from damage caused by LPS and excessive autophagy by decreasing the expression levels of LC3 and Beclin-1 [48]. H_2 is also able to decrease the mTOR expression in glial cells that mediate the raising of the LC3 II/LC3 I ratio to promote autophagy, thus molecular hydrogen improves the neuro-inflammation induced by LPS through blocking autophagy [49]. Moreover, 2% H_2 in a mouse model can prevent sepsis mediated by liver injury by increasing Fundc1, which is a mitophagy receptor present at the membrane of the mitochondrion and responsible for maintaining ATP balance [50]. Pyrolysis is another pathway of programmed cell death that protects monocytes, macrophages, and other invading pathogens; thus, this pathway is beneficial for us, although, like the other pathways, the excessive activity of this pathway also causes sepsis and septic shock [51]. The key marker of pyrolysis activation is caspase-1, and IL-1β and IL-18 are considered critical downstream inflammatory factors for this pathway. Evidence suggests that therapy with H_2-rich saline in mouse models of early subarachnoid hemorrhage brain injury can significantly decrease caspase-1 expression and block the inflammatory response [52]. Moreover, H_2 therapy markedly decreases the caspase-1 expression as well as the intracellular level of IL-1β and IL-18 in the sepsis models mediated by organ injury [28, 53]. In addition, it is established that therapy with 3% H_2 can decrease the NLRP3, which is a protein related to pyroptosis, as well as caspase-1 and the N-terminal of gasdermin D (GSDMD-N), in a lung I/R injury [54].

Hydrogen-Rich Water (HRW) and Gut Microbiota

There are numerous microorganisms in the human body, and since the gastrointestinal (GI) tract represents as one of the largest (250–400 m^2) areas of the body, it is expected to colonising with the numerous bacteria, archaea, and eukarya that all define the term "gut microbiota" [55]. It seems that this microbial community has, over thousands of years, co-evolved with the host to create a complex and advantageous partnership [56, 57]. Gut microbiota act as a huge network with more than 1014 microorganisms, which seems to be 10 times more than the number of human cells, and microbiome genomic content exceeds the human genome by over 100 times [58]. As mentioned, the microbiota act as a huge network to benefit the host and are involved in a wide range of physiological activities like enhancing the integrity of the gut, structuring the epithelium of the intestinal tract, gaining energy, acting as a part of the host's defense, and even regulating the immune system [59–62]. Therefore, as expected, any disturbance in this system leads to host intestinal and extraintestinal diseases. Dysbiosis is the term used to describe a disruption in the microbiota community [63, 64]. There are a few reports that focus on the effect of drinking HRW on the microbiota of the gut, which is the first report related to a study in 2018. This study investigated the effect of drinking HRW in an animal model of small intestine toxicity induced by radiation and demonstrated that using the water enriched with 0.80 mM of H_2 for 5 days could improve the gastrointestinal toxicity mediated by radiation by stabilising the gastrointestinal level of MyD88. Myeloid differentiation primary response gene 88 (MyD88) is a modulator that is necessary for responses of the innate immune system toward gut pathogens. In addition, this study found that therapy with HRW can ameliorate the functions of the intestinal tract through counterbalanced gut microbiota. The result of this study demonstrated that radiation affects the microbiota community in a way that decreases the abundance of *Bacteroidia*, *Betaproteobacteria*, and *Coriobacteria*, while increasing the abundance of *Phycisphaerae*, *Planctomycetia*, and *Sphingobacteria* spp [65]. After that, another study investigates the effect of drinking HRW on the microbiota community of intestinal and short-chain fatty acid (SCFA) contents after receiving the 0.32 mM H_2 for 6 weeks. They found that this therapy significantly raised the weight of cecal contents as well as the production of SCFAs like isobutyric acid, propionic acid, and isovaleric acid. In addition, this study exhibit that drinking HRW leads to altering the microbiota community with a lower abundance of *Bifidobacterium*, *Clostridiaceae*, *Coprococcus*, *Ruminococcus*, and *Sutterella*, while it leads to increasing the abundance of *Parabacteroides*, *Rikenellaceae*, *Butyricimonas*, *Prevotella*, and *Candidatus arthromitus* [66]. Another study exhibited therapy with drinking water rich in 0.6 mM H_2 for 25 days at 10 mL/kg in female piglets that were fed maize contaminated with *Fusarium mycotoxin*. The result demonstrated that therapy with HRW reduces the rate of diarrhea, and leads to increasing the butyrate level of the colon, acetate, and SCFAs in animal models. Moreover, this investigation demonstrated that microbiota populations in different segments of the intestine are also affected via H_2 therapy. They found that microbiota populations of ileum were altering with a lower abundance of *E. coli* and a higher abundance of *Bifidobacterium*. However, microbiota populations of the colon were altering with a higher abundance of methanogenic Archaea and sulfate-reducing bacteria after therapy with HWR [67]. Another investigation in 2019 demonstrated that drinking water rich in H_2 0.4–0.9 mM for 15 days can ameliorate the integrity of the intestinal barrier that induced by permethrin pesticide in a rat model of Parkinson's disease. In addition, they found that this therapy enhances the levels of occludin in the ileum which is a necessary biomarker for the integrity of tight junctions. Moreover, this study demonstrated that drinking the HRW for 15 days raised the butyric acid level in feces, preserved the population of *Lachnospira* and *Defluviitaleaceae*, and increase the abundance of *Blautia*, *Lachnospiraceae*, *Ruminococcaceae*, *Papillibacter* that are well known as a butyrate-producing bacteria [68]. The first clinical trial in 2019 had been launched on investigating the beneficial effects of therapy with drinking water rich in H_2. This trial used the HRW (1.5–2.0 L/day) for 2 months in 38 adolescent female football players. They suggest drinking the HRW for 2 months altering the diversity of gut microbiota in favor of establishing a balance of microflora [69, 70]. However, most of the available studies are focused on the protective effect of HRW on the integrity of the gut barrier or increasing the abundance of bacteria that produce butyrate, while still there are no studies for assessment the HRW effectiveness as therapy in modulating the gut microbiota in different diseases (Table 17.1; Fig. 17.1).

Table 17.1 Review of studies examining the relationship between gut microbiota and hydrogen-rich water (HRW)

Author	Year	Sample size (n)	Study model	Dosage of HRW	Control group	Experiment length (days)	Key finding (s)	Refs.
Hui-wen Xiao	2018	12	Male C57BL/6 mice model of gastrointestinal toxicity induced by radiation	0.80 mM	Normal water (0.2 ml)	5	Upregulated the level of miR-1968-5p; Downregulation expression of MyD88; Alleviate enteric damages induced via radiotherapy	[65]
Yasuki Higashimura	2018	16	Male C57BL/6N mice model of the intestinal environment	0.32 mM	Normal water	30	Reducing the intraserum level of LDL-C and ALT; Higher intraserum level of propionic, isobutyric, and isovaleric acids; Higher abundance of *Parabacteroides*, *Rikenellaceae*, *Butyricimonas*, *Prevotella*, *Mucispirillum*, *Candidatus arthromitus*, *Erysipelotrichaceae*, *Allobaculum*, *Desulfovibrionaceae*; Lower abundance of *Bifidobacterium*, *Adlercreutzia*, *Clostridiaceae*, *Lachnospiraceae*, *Coprococcus*, *Ruminococcus*, *Sutterella*	[66]
Mitsunori Ikeda	2018	36	Male C57/BL6 mice model of sepsis	3.5 mM	Normal saline	7	Improved the survival rates; Inhibit the translocation of *Enterobacteriaceae*; Alleviate intestinal injury; Decreasing the pro-inflammatory cytokines (MDA, TNF-α, IL-1β, IL-6)	[119]
Weijiang Zheng	2018	24	Female piglets model of Fusarium toxin-contaminated diet	0.6 mM	Lactulose	25	Alleviate rate of diarrhea; increasing the level of acetate, butyrate, SCFAs; Higher abundance of *E. coli*; Lower abundance of *Bifidobacterium* and *Archaea*	[67]
Xu Ji	2019	24	Female piglets model of Fusarium toxin-contaminated diet	0.6–0.8 mM	Hydrogen-free water	25	Higher intraserum levels of DAO and D-lactic acid; Deacresing the apoptosis index and small intestine leak induced by mycotoxin; Alleviate the intestinal damages; increasing the expression and distribution of CLDN3	[120]
Laura Bordoni	2019	58	Male and female Wistar rats model of Parkinson's disease	0.4–0.9 mM	Permethrin Vehicle	15	Improve the integrity of the intestinal barrier; Higher level of butyric acid; Increase the abundance of bacteria that produce butyrate like *Blautia*, *Lachnospiraceae* family, *Ruminococcaceae* family, *Papillibacter*, *Roseburia*, *Intestinimonas*, *Shuttleworthia*	[68]
Ji-Bin Sha	2019	38	female football players	NF	Normal water	60	Higher intrasrum level of MDA, SOD, TAC, and hemoglobin; Lower abundance of *Bifidobacterium*, *Oscillibacter*, *lostridia*, and *Coriobacteria*; Higher abundance *Prevotella* and *Bacteroides*	[69]
Zitao Guo	2020	30	Male BALB/c mice model of Intestinal environment	NF	Deionized water N2 nanobubble water	150	Higher abundance of *Clostridium* and *Coprococcus*; Lower abundance of *Mucispirillum* and *Helicobacter*	[121]

Abbreviations *LDL-C* Low-density lipoprotein cholesterol; *ALT* Alanine aminotransferase; *SCFAs* Short-chain fatty acids; *DAO* Diamine oxidase; *CLDN3* Claudin-3; *NF* Not defining

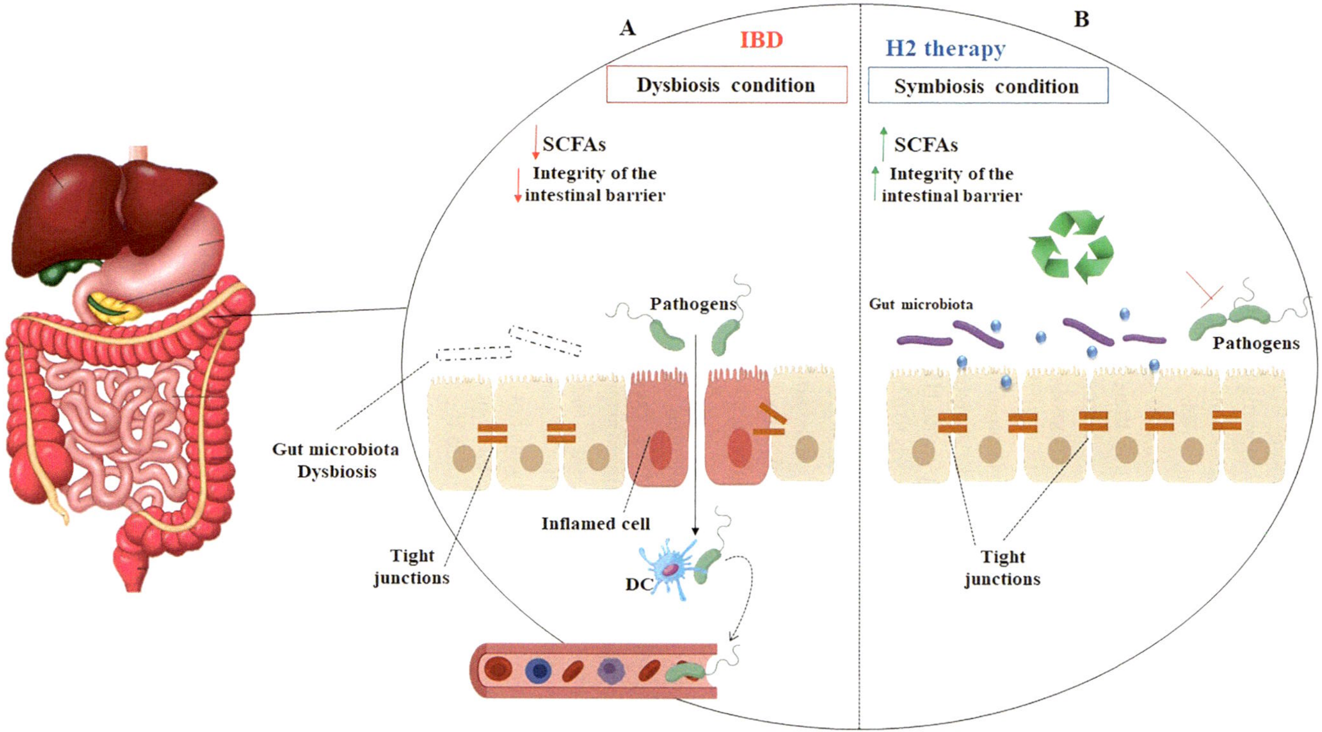

Fig. 17.1 Effect of the microenvironment on the intestinal and microbiota communities (**a**) Inflammatory diseases, including inflammatory bowel disease (IBD), cause three major pathophysiological manifestations: dysbiosis (decreased gut-microbial diversity, which then deactivates SCFAs), leaky gut (loosening of tight-junction proteins), inflammation, and translocation of bacterial endotoxins from the gut lumen to the bloodstream (LPS, ASCA, etc.). (**b**) Therapy with H_2 ameliorates the integrity of the intestinal barrier (inhibiting the bacterial leakage), leading to a symbiosis condition (restoring the gut-microbial diversity that raises the level of SCFAs production). [DC = dendritic cells, SCFAs = short-chain fatty acids]

HRW Mechanisms of Action on Gut Microbiota

There is evidence that suggested drinking water enriched with H_2 could modulate the microbial community of the gut. Following the HRW drinking, extra H_2 is delivered to the intestinal metabolite environment. The milieu of the intestinal is well known as a dynamic environment that is rich in H_2 due to the hydrogen-producing microflora like *Bacteroidetes* and *Firmicutes* with the output of about 13 L of this gas per day [14]. Meanwhile, in the gut, there are bacteria like hydrogenotrophic, methanogens, acetogens, and sulfate-reducing that used this released H_2 in their growth or metabolism process which are mostly located at the distal part of the intestine [71, 72]. Here drinking water enriched with H_2 prepares the substrate which leads to increasing the abundance of hydrogenotrophic bacteria among the community of microbiota to raise their metabolites like acetate, H_2S, and CH_4 in the intestinal. The result of the study confirms that therapy with HRW for 25 days' results in increasing the levels of acetate which means H_2 leads to a higher abundance of methanogens and sulfate-reducing bacteria like *Methanobrevibacter smithii*, *Desulfovibrio spp.* [67]. However, another investigation demonstrated that gavage the water enriched with H_2 for 5 days had no considerable effects on the abundance of intestinal microbiota thus it means that the HRW effect is probably dependent on the duration of therapy since its short-term usage had no significant effect [65]. Using the HRW in addition to its effect on the abundance of hydrogenotrophic bacteria, increased the partial pressure hydrogen that decreases the redox potential in the intestinal lumen to favor anaerobes and butyrate-releasing bacteria [73]. Previous reports confirm that HRW also affects the fermentation in the intestinal through raising the abundance of bacteria like *Deferribacteres*, *Bacteroidetes*, and *Firmicutes* which are anaerobic phyla and facilitate the fermentation [66, 68]. This all can be considered as a primary effect of H_2 that happen at once but a secondary upshot can be the initial trigger for producing numerous compounds that are active biologically like propionic acid, butyric acid, acetate, and H_2S, then via them regulate the metabolism of gut microflora. These biologically active compounds like H_2S and propionic acid are well known for their physiological activity that is probably relevant to the immunomodulation, expression of genes, and signaling pathways within cells [74, 75]. These effects can be both intestinal and systemic that probably occurs following the passing through the intestine and entering the circulation then transported to the other organs [76]. Furthermore, H_2 drive from drinking the HRW also can act as a signaling agent to affect the expression of intestinal-specific metabolic genes like fibroblast growth factor 21 (FGF21) and proliferator-activated receptor gamma coactivator 1-alpha (PGC-1α) that regulates crucial metabolic pathways [77, 78]. However, this is a justly simplified overview of the way that HRW modulates the gut microbiota and still, there remained many unanswered questions.

HRW Be Used as a Therapy (Hydrobiotic)

There is an unanswered question about the HRW and that is whether water enriched with hydrogen can be considered a prebiotic or not. Before answering this question, we should mention that prebiotics is food sours and promote the activity and growth of microflora in the gut. As an example plant fibers are common prebiotics that is non-digestible and improves fermentation in the colon they are also able to be a substrate for *Bifidobacteria* and *Lactobacillus* which are protective endogenous in the intestinal with the ability to effecting the host immune responses and digestion [79]. Although evidence suggests that drinking water enriched with H_2 can modulate the community of gut microbiota, it should not be considered a prebiotic or probiotic because it has complex acts in the intestine. May not be able to categorize HRW as a prebiotic or probiotic based on the evidence, but we can give it a special name "hydrobiotic" which is a unique term referring to the compound that aimed to stabilize or even compensate the levels of H_2 in the intestinal. Now it is well known that an imbalance in the H_2 concentration of gut cycling is a risk factor for the onset of different diseases like inflammatory bowel disease (IBD), syndrome of irritable bowel, obesity, and Parkinson's disease [71, 80]. Reports suggest that patients with irritable bowel syndrome have a lower abundance of H_2-producing bacteria, while they have a higher abundance of *Bacteroides*, *Ruminoccocus*, and *Prevotella* [81]. According to Shen finding therapy with water-enriched H_2 can protect against IBD in experimental models. Although profiles of the gut microflora were not evaluated for confirmation of any modulation in gut microbiota, it was observed that drinking HRW for 7 days can improve the symptoms induced by IBD including changing the body weight, decreasing the blood in the stool, consistency of stool, alleviated diarrhea, ameliorated both macroscopic and microscopic damages of the colon, and rebalance the oxidative stress and inflammation [82]. In addition, the results of the other study also demonstrated that therapy with HRW for 8 weeks can enhance the life quality of irritable bowel syndrome patients, probably by increasing the generation of anaerobic bacteria like *Lactobacilli* and *Bifidobacteria* [83].

Administration Routes for H_2

H_2 gas inhalation, drinking water enriched with H_2, and administration of saline enriched with H_2 are the three main administration routes. In addition, currently using the nanoparticles also attract attention as systems for the delivery of molecular hydrogen. However, the delivery abilities of all routes depended on H_2 solubility in blood, saline, and water. H_2 gas inhalation is the simplest therapeutic route and has been utilized extensively from the first report. H_2 inhalation determined the dose and retention time in the body. H_2 that has been inhaled diffuses via the alveoli into the plasma and travels to the body through circulation. Results of the trial exhibit that 72 h of exposure to 2.4% H_2 gas had no negative impacts on any physiological markers, indicating that H_2 may not have any side effects [84]. Additionally, the research suggests that the concentration of H_2 in blood and tissues depended on the time and intensity of inhalation thus H_2 had dose-dependent antioxidant activity [85–87]. High H_2 gas concentrations may be used to provide more beneficial effects. H_2 can be consumed through drinking water, which is a portable, more secure, and simple method of administration [88]. In addition, evidence suggests a low H_2 concentration in the rats' brains after therapy with H_2 water [89]. Therefore, to increase the effectiveness of H_2 intervention, it is crucial to administer H_2 at a high payload to targeted sites. Microbubbles (MBs) are the new delivery system that is used for H_2 transfer and monitoring. In this manner, H_2 gas is loaded on the MB shell then it would be able to travel to the body by blood circulation. H_2-MBs may be a more effective way to prevent myocardial I/R damage in a rat model because they have a higher H_2 content per unit volume of solution than saline enriched with H_2 [90]. The ability of H_2 to diffuse from the cell membrane into the cytoplasm caused its advanced therapeutic usage. Bathing with H_2 water was effective in treating skin conditions [91, 92]. In addition, H_2 usage has been expanded in the preservation of organ transplants [93]. Administration of saline enriched with H_2 is a technique that allows for the direct use of H_2 in the afflicted area and can quickly deliver a significant volume of H_2 into the body. But this approach is intrusive, challenging for patients to accept, and might result in cross-infection. Direct admission of H_2 into the skin or a vein could also be extremely harmful. In a study utilising a rat model, H_2 was given orally, intravenously, intraperitoneally, as water or saline enriched with H_2, or as H_2 gas by inhalation. They found that H_2 was in various tissues with different concentrations after being measured using high-quality sensor gas chromatography [89]. As a result, H_2 can independently enter most organs or blood through the three mechanisms. However, the outcomes of various administration techniques could vary. Although hydrogen peaks via oral delivery and inhalation at the same time, drinking H_2 water has a longer sustained period of time. Compared to H_2 inhalation, drinking H_2 water greatly reduced the expression of NF-kB in rat liver tissue, and the combined impact was much more effective [94]. H_2 concentrations peaked 5 minutes after oral or intraperitoneal delivery, whereas intravenous therapy only required one minute. After peaking at 5 minutes, the H_2 concentrations in the blood and tissues began to fall, but they did so more quickly after intraperitoneal treatment than after oral administration. H_2 gas inhalation caused a slower increase in H_2 concentration than intraperitoneal, intravenous, or oral treatment did. However, the increased H_2 from inhalation persisted for at least 60 min. Because high-quality sensor gas chromatography was used in this study along with special hermetically sealed tubes filled with pure air to prevent H_2 from leaking from the sampling tissue during processing, it is counterintuitive that inhalation takes longer than ingestion of H_2 water for tissue H_2 concentrations to saturate [89]. Additionally, different H_2 inhalation techniques, such as masks or nasal tubes, provide various outcomes. To ensure the best benefits that are acceptable to the user, several H_2 administration routes need to be taken into account.

HRW as IBD Therapy

The term inflammatory bowel disease (IBD) is used for defining the disorders that are contributing to chronic or long-standing inflammation in gastrointestinal tract tissues. According to the European Federation of Crohn's and Ulcerative Colitis (EFCCA) reports, there are about 10 million people around the world who suffer from IBD [95, 96]. The most prevalent types of IBD are Crohn's disease (CD) and ulcerative colitis (UC) which both lead to digestive tract inflammation but also have some differences. One of the key difference is related to their effect site, UC commonly affected the colon, while CD affect from terminal of the ileum to the colon. Another difference is related to the ability of CD in the formation of non-caseating granulomas that has a role in inducing the transmural and long-standing inflammation in the mucosa. However, the primary target of inflammation caused by UC is the rectum then can extend all over the colon and involve mucosal and submucosal areas [97, 98]. Evidence suggests that the accumulation of oxide in the colon mucosa due to the strong oxidizing effect of reactive oxygen species (ROS) is one of the main UC pathogenesis [99]. Oxide accumulation through disrupting the transcriptional and phosphorylation signaling pathways leads to the persistence of inflammation. In addition, excess levels of lipoperoxide (LPO) also can induce UC symptoms through damaging the endothelial cells in the colon [100]. It means that endogenous antioxidant enzymes determine the severity of UC symptoms. Since these enzymes are involved in the redox status of muscle tissue, they can neutralize the oxidants activity of ROS. These enzymes with the ability in balancing between oxidants and antioxidants markers are superoxide dismutase (SOD), catalase (CAT), and glutathione peroxidase (GPx). According to the reports, UC spontaneously develops in mice models with GPx deficiency [101]. In addition, it is observed that overexpressing the SOD and CAT can alleviate the UC symptoms in mice models [102–104]. Colon inflammation, changing visceral sensation, abdominal pain, bleeding from the rectal, and losing weight are the main symptoms of IBDs [105]. IBDs are well-known as relapsing and chronic illnesses caused due to the extended production of inflammatory cytokines and infiltration of the leukocyte that all together lead to damage in the structure or activity of the intestinal. However, it should mention that there are some factors associated with increasing the incidence risk like genetics, microbiota community of the digestive tract, and environmental conditions [106–108]. The main goal in the treatment of patients who live with IBDs is the alleviation of the disease's symptoms through the attenuation of inflammation. Immunosuppressive therapies, antibiotics, anti-inflammatory agents as well as probiotics are the conventional therapy for IBD that are currently used along with biological therapy like anti-TNF-α [109]. As we mentioned above, oxidative stress can be considered one of the primary triggers as well as exacerbation markers for UC, thus finding the appropriate agents with antioxidant effects can likely serve as a beneficial novel therapy for patients suffering from IBD.

Over the past ten years, molecular hydrogen as a potential anti-oxidative with anti-apoptotic and anti-inflammatory abilities attract much attention in medicine and as an ideal treatment for different diseases [6]. The hydrogen molecule is of interest due to its benefit like penetrating ability into the membrane of cells, easily diffusing from the cytosol into the target organelles, and potential in specifically targeting and decreasing the hydroxide radicals and peroxynitrite without affecting the ROS that contribute in the signaling of normal cells [7]. In addition to these beneficial effects, different clinical trials on therapy with H_2 confirm the safety and effectiveness of this molecule [76, 110]. Reports suggest that hydrogen-rich water (HRW) as an affective and suitable delivery way for H_2 with the same potency as inhaled route and is convenient in medical applications. In 2009, Kajiya and colleagues, with the aim of investigating the anti-inflammatory activity of H_2 as a novel antioxidant medication, used dextran sulfate sodium (DSS) to induce the mouse model of human IBD. After modelling the IBD, mice received (i) DSS at 5%; (ii) DSS at 5% with H_2; (iii) H_2 at 0.78 mM in their drinking water for 7 days. They found that the addition of H_2 greatly reduced the DSS-induced symptoms of UC, improved weight loss, and reduced colon shortening induced by DSS. In addition, therapy with H_2 markedly reduces the levels of IL-12, TNF-α, and IL-1β in the colon tissues. Additionally, histological outcomes also demonstrated that H_2 significantly reduced the intestinal damage induced by DSS that was accompanied by macrophage infiltration [111]. Shen et al. in 2017, investigate the effect of water enriched with H_2 as therapy for IBD. To induce the IBD model they used DSS and after modeling the IBD, mice received the DSS along with HRW. In addition, one group of mice along with DSS and HRW also received ZnPP as an inhibitor of hemeoxygenase-1 (HO-1) at 25 mg/kg/day for 7 days. The result of this study exhibit that therapy with HRW significantly ($P < 0.05$) improves the losing weight induced by DSS. In addition, H_2 ameliorates the disease activity index (DAI) of colitis, and also protects from the shortening of the colon induced by DSS in comparison to the group that only received DSS ($P < 0.05$). The histological investigation of this study also confirms the effect of HRW on improving the damage induced by inflammation in comparison to the colitis group. They also assessed the biomarker of oxidative stress and found that therapy with HRW significantly reduce the malonaldehyde (MDA) and myeloperoxidase (MPO) levels in colon tissue ($P < 0.05$). In addition, the results of this study suggest that HRW probably attenuates inflammation via decreasing the TNF-α, IL-6, and IL-1β. Furthermore, HRW therapy also affects the endoplasmic reticulum (ER) stress by dramatically decreasing the p-eIF2α, ATF4, XBP1s, and CHOP ($P < 0.05$). Therapy with H_2 also up-regulated the expression level of HO-1, while its protective effect was reversed in the group that received ZnPP. This investigation same as the previous knowledge confirms the anti-inflammatory effect of HRW therapy, and also added new results like its effect on ER stress and regulated the expression level of HO-1 [82]. After that, in 2021, another study with the aim of investigating the therapeutic effect of water enriched with H_2 also used DSS to induce the UC model. In this study, after modeling the IBD, mice in addition to the H_2 therapy also received sulfasalazine as a standard therapy of UC alone or in combination with H_2. Water is enriched with H_2 through dissolving tablet-generated hydrogen, and mice receive 200 μL of HRW through oral gavage as well as at will. To investigate the antioxidant and anti-inflammatory activity of H_2, they evaluated the DAI, biomarkers of oxidative stress, and any changes in histological, and inflammation markers. They found that the co-therapy of H_2 with sulfasalazine markedly ameliorated the colitis symptoms and losing weight induced by DSS. Moreover, this combination therapy ameliorated damage in the mucosal layer and also improved crypt loss in comparison to the colitis group. In addition, the result demonstrated that HRW therapy alone or in combination with sulfasalazine significantly reduces inflammation by decreasing the DSS-mediated rise in high-sensitivity C-reactive protein (hsCRP). The result of this study also demonstrated that HRW therapy ameliorated the colitis symptom by restoring the balance of oxidant and antioxidant markers via improving the antioxidant status. This study suggests that H_2 therapy is as effective as therapy with a standard drug like sulfasalazine, which makes it a potential candidate in the case of IBD therapy [112]. Song et al. (2022) also investigate the therapeutic effect of molecular hydrogen on dysbiosis of the intestinal microflora and oxidative injury, which are significant factors in the development of chronic ulcerative colitis (UC). This investigation, also induced acute colitis in C57BL/6 mice by using DSS. In the current study, mice received DSS in 3 cycles to establish a chronic colitis model, including the 2 primary cycles in which they received 2.5% DSS for 5 days, then 16 days after the second cycle, and the third cycle in which they received 2% DSS for 4 days. As in the previous studies, in the current investigation, after modeling the UC, mice received the DSS along with water-enriched H_2 (0.8 ppm). The results of this study demonstrated that H_2 therapy improved the symptoms of acute colitis induced by DSS. In addition, treatment with HRW ameliorated the histopathological damages mediated via acute colitis [113]. Moreover, they found that this therapy significantly raised the concentration of tripeptide glutathione (GSH), which is a crucial intracellular antioxidant, to a high level that enhanced the mucosal activity in the acute colitis models [114]. They discovered that HRW exerts anti-inflammatory effects by lowering the intracellular levels of pro-inflammatory mediators such as TNF-α. In addition, this study demonstrated that H_2 therapy significantly prevents the proliferation of *Enterococcus faecalis*, *Clostridium perfringens*, and *Bacteroides fragilis* in comparison to the DSS group ($P < 0.05$). Aside from the previously discovered anti-inflammatory and anti-oxidative effects of H_2 therapy, this study discovered the effect of HRW on dysbiosis intestinal microflora caused by acute colitis [113]. Another study with the aim of investigating the therapeutic effect of water enriched with H_2 on chronic UC used a generator to enrich the water with H_2, which is called electrolyzed hydrogen water (EHW), which electrochemically decreases water that was enriched with molecular hydrogen [115–118]. They induce chronic UC models in rats via colorectal injection of 2,4,6-trinitrobenzene sulfonic acid (TNBS). After UC modeling, all rates received EHW with a concentration of about 600–700 ppb of dissolved hydrogen that refreshed every 30 min. Results of this study indicated that EHW was also markedly able to improve the inflammation and symptoms induced by TNBS. Furthermore, EHW therapy inhibits the release of pro-inflammatory cytokines such as IL-1β, TNF-α, IL-6, and the chemokine MCP-1 in colonic tissue inflamed by TNBS. They also observed that EHW induces its anti-inflammatory effect by inhibiting the overexpression of ROS, which is responsible for inflammation. Furthermore, this finding indicates that EHW can inhibit S100A9 expression, which is well known as an S100 calcium-binding protein A9 and a biomarker of inflammation in IBD. They found that therapy with EHW suppresses inflammation of the colon by blocking the overexpression of ROS due to its ability to scavenge free radicals and rebalance the oxidant and antioxidant markers [115] (Table 17.2; Fig. 17.2).

Conclusion Remarks

It seems that molecular hydrogen can display different phenotypes that improve several pathological conditions by controlling the expression of different genes. Numerous studies have demonstrated that H_2 has a therapeutic effect on acute or chronic inflammatory bowel desires by regulating pro-inflammatory cytokines, free radicals scavenging and reducing cell apoptotic damages. In conclusion, we take into account that H_2 diffuses into cells and controls any possible imbalance caused by mitochondrial membrane damage through electron transfer to decrease the production of free radicals in the mitochondria. Additionally, it affects the occurrence of oxidative stress and the transcriptional control of Nrf2. Another possibility is that H_2 may suppress NF-kB and Foxp3 transcription in the nucleus, which would then impact how apoptotic proteins like Bax and caspase-3 are put together and ultimately prevent the start of apoptosis. Furthermore, H_2 is a strong antioxidant, and, we think that drinking water enriched with H_2 could be a useful strategy for alleviating IBD. Studies on how H_2 modulates the gut microbiota and its therapeutic effects are still in the early stages. It is suggested that knowledge about the impacts of excess accumulation, reduction potential, dosage, dose duration, and the safety of the antioxidant should all be determined to produce better results in clinical trials. However, despite its potency, H_2 has to be further verified and described in clinical and animal experiments. All the findings point to it being a novel and potential antioxidant agent.

Table 17.2 The summary of studies evaluating the therapeutic effect of hydrogen-rich water (HRW) on IBDs

Author	Year	Sample size (n)	Species	Study model	Dosage of HRW	Control group	Experiment length (days)	Key finding(s)	Refs.
Kajiya	2009	18	Male BALB/c mice	DSS-induced IBD	0.78 mM	Normal water	7	Greatly reduced DSS-induced symptoms of UC; Alleviate histological damages; Decrease the generation of IL-1β, TNF-α, IL-12	[111]
Shen	2017	24	Male C57BL/6 J mice	DSS-induced IBD	NF	Normal saline ip (5 mL/kg)	7	Alleviate DSS -induce symptoms of colitis; Reduce the level of MDA and MPO; Decrease marker of ER stress (p-eIF2α, ATF4, XBP1s, and CHOP)	[82]

(continued)

Table 17.2 (continued)

Author	Year	Sample size (n)	Species	Study model	Dosage of HRW	Control group	Experiment length (days)	Key finding(s)	Refs.
LeBaron	2021	30	Male C57BL/6 mice	DSS-induced IBD	1.5 mM	Normal water	10	Alleviate the DAI induced by DSS; Improve the intestinal damage; Reduce the level of hs-CRP; Rebalance the stress oxidative markers	[112]

(continued)

Table 17.2 (continued)

Author	Year	Sample size (n)	Species	Study model	Dosage of HRW	Control group	Experiment length (days)	Key finding(s)	Refs.
Song	2022	18	Male C57BL/6 mice	DSS-induced UC	0.8 mM	Normal water	16	Ameliorated DSS -induce symptoms of colitis; Alleviate histopathological damages of intestinal; Increased GSH; Decrease TNF-α; Blocking the proliferation of *Enterococcus faecalis, Clostridium perfringens* and *Bacteroides fragilis*	[113]
Hu	2022	18	Male Wistar rats	Colorectal TNBS administration induce IBD	NF	Normal water	14	Decrease the generation of IL-1β, TNF-α, IL-6, MCP-1; Reduce the overexpression of ROS; inhibit expression of S100A9	[115]

Abbreviations *Ip* Intraperitoneal; *ER* Endoplasmic reticulum; *DAI* Disease Activity Index; *hs-CRP* high-sensitivity C-reactive protein; *IBD* Inflammatory bowel disease; *UC* ulcerative colitis; *GSH* glutathione; *NF* Not defining

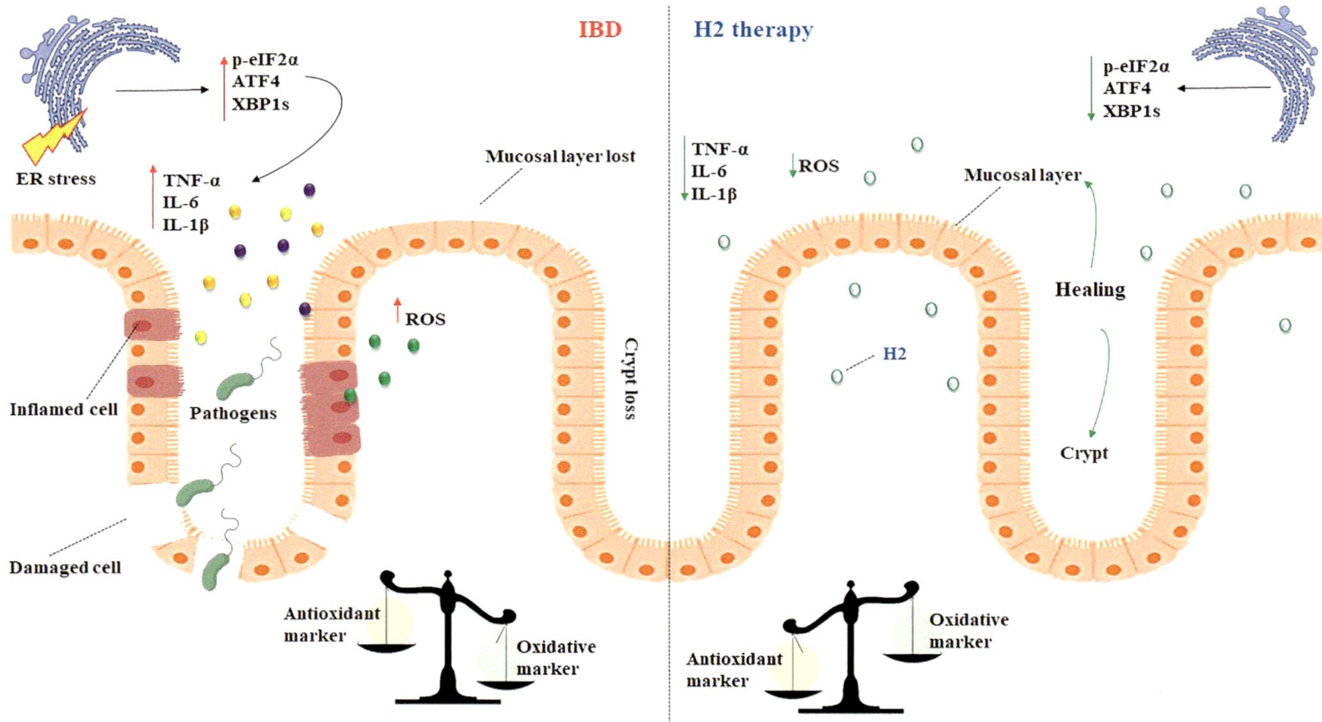

Fig. 17.2 The pathophysiological activity of H_2 therapy for inflammatory bowel disease (IBD). Therapy with H_2-riched water (HRW) significantly ameliorated mucosal damage and crypt loss induced by IBD. H_2 reduces the levels of pro-inflammatory mediators (TNF-α, IL-6, and IL-1β). Moreover, this therapy dramatically reduces the endoplasmic reticulum (ER) stress markers (p-eIF2α, ATF4, XBP1s). H_2 therapy also restored redox balance by reducing the antioxidant marker and ROS

References

1. Kamal Hossain M, Ahmed DQ (2022) Noble metal-decorated nanostructured zinc oxide: strategies to advance chemiresistive hydrogen gas sensing. Chem Rec 22(7):e202200090
2. LeBaron TW, Laher I, Kura B, Slezak J (2019) Hydrogen gas: from clinical medicine to an emerging ergogenic molecule for sports athletes. Can J Physiol Pharmacol 97(9):797–807
3. Lanphier E (1972) Human respiration under increased pressures
4. Dole M, Wilson FR, Fife WP (1975) Hyperbaric hydrogen therapy: a possible treatment for cancer. Science 190(4210):152–154
5. Ohsawa I, Ishikawa M, Takahashi K, Watanabe M, Nishimaki K, Yamagata K et al (2007) L'idrogeno agisce come antiossidante terapeutico riducendo selettivamente i radicali di ossigeno citotossici. Nat Med 13:688–694
6. Xia C, Liu W, Zeng D, Zhu L, Sun X, Sun X (2013) Effect of hydrogen-rich water on oxidative stress, liver function, and viral load in patients with chronic hepatitis B. Clin Transl Sci 6(5):372–375
7. Ostojic SM (2015) Targeting molecular hydrogen to mitochondria: barriers and gateways. Pharmacol Res 94:51–53
8. Wang S-T, Bao C, He Y, Tian X, Yang Y, Zhang T et al (2020) Hydrogen gas (XEN) inhalation ameliorates airway inflammation in asthma and COPD patients. QJM: Int J Med 113(12):870–875
9. Zheng Z-G, Sun W-Z, Hu J-Y, Jie Z-J, Xu J-F, Cao J et al (2021) Hydrogen/oxygen therapy for the treatment of an acute exacerbation of chronic obstructive pulmonary disease: results of a multicenter, randomized, double-blind, parallel-group controlled trial. Respir Res 22(1):1–12
10. Abe T, Li X-K, Yazawa K, Hatayama N, Xie L, Sato B et al (2012) Hydrogen-rich University of Wisconsin solution attenuates renal cold ischemia–reperfusion injury. Transplantation 94(1):14–21
11. He J, Xiong S, Zhang J, Wang J, Sun A, Mei X et al (2013) Protective effects of hydrogen-rich saline on ulcerative colitis rat model. J Surg Res 185(1):174–181
12. Nguyen GC, Tuskey A, Dassopoulos T, Harris ML, Brant SR (2007) Rising hospitalization rates for inflammatory bowel disease in the United States between 1998 and 2004#. Inflamm Bowel Dis 13(12):1529–1535
13. Zhou Z-Q, Zhong C-H, Su Z-Q, Li X-Y, Chen Y, Chen X-B et al (2019) Breathing hydrogen-oxygen mixture decreases inspiratory effort in patients with tracheal stenosis. Respiration 97(1):42–51
14. Hylemon PB, Harris SC, Ridlon JM (2018) Metabolism of hydrogen gases and bile acids in the gut microbiome. FEBS Lett 592(12):2070–2082
15. Dan Dunn J, Alvarez L, Zhang X, Soldati T (2015) Reactive oxygen species and mitochondria: A nexus of cellular homeostasis. Redox Biol. 6:472–485. Epub 2015/10/04. https://doi.org/10.1016/j.redox.2015.09.005. PMID: 26432659
16. Jauniaux GB (2011) Oxidative stress. Best Pract Res Clin Obstet Gynaecol 25:28
17. Wang J, Hossain M, Thanabalasuriar A, Gunzer M, Meininger C, Kubes P (2017) Visualizing the function and fate of neutrophils in sterile injury and repair. Science 358(6359):111–116
18. Ohta S (2014) Molecular hydrogen as a preventive and therapeutic medical gas: initiation, development and potential of hydrogen medicine. Pharmacol Ther 144(1):1–11
19. LeBaron TW, Kura B, Kalocayova B, Tribulova N, Slezak J (2019) A new approach for the prevention and treatment of cardiovascular disorders. Molecular hydrogen significantly reduces the effects of oxidative stress. Molecules. 24(11):2076
20. Gharib B, Hanna S, Abdallahi OM, Lepidi H, Gardette B, De Reggi M (2001) Anti-inflammatory properties of molecular hydrogen: investigation on parasite-induced liver inflammation. Comptes Rendus de l'Académie des Sciences-Series III-Sciences de la Vie 324(8):719–724
21. Kang Q, Yang C (2020) Oxidative stress and diabetic retinopathy: Molecular mechanisms, pathogenetic role and therapeutic implications. Redox Biol 37:101799. Hass DT, Barnstable CJ (2020) Uncoupling proteins in the mitochondrial defense against oxidative stress

22. Liu Y, Dong F, Guo R, Zhang Y, Qu X, Wu X et al (2019) Hydrogen-rich saline ameliorates experimental autoimmune encephalomyelitis in C57BL/6 mice via the Nrf2-ARE signaling pathway. Inflammation 42(2):586–597
23. Su N, Wu Q, Chen H, Huang Y, Zhu Z, Chen Y et al (2019) Hydrogen gas alleviates toxic effects of cadmium in Brassica campestris seedlings through up-regulation of the antioxidant capacities: Possible involvement of nitric oxide. Environ Pollut 251:45–55
24. Zhao Y-S, An J-R, Yang S, Guan P, Yu F-Y, Li W et al (2019) Hydrogen and oxygen mixture to improve cardiac dysfunction and myocardial pathological changes induced by intermittent hypoxia in rats. Oxidative Med Cell Longevity 2019
25. Annesley S, Fisher P (2019) Mitochondria in health and disease. Cells 8(7):680
26. Yoshida A, Asanuma H, Sasaki H, Sanada S, Yamazaki S, Asano Y et al (2012) H_2 mediates cardioprotection via involvements of KATP channels and permeability transition pores of mitochondria in dogs. Cardiovasc Drugs Ther 26(3):217–226
27. Liu Q, Li BS, Song YJ, Hu MG, Lu JY, Gao A et al (2016) Hydrogen-rich saline protects against mitochondrial dysfunction and apoptosis in mice with obstructive jaundice. Mol Med Rep 13(4):3588–3596
28. Yan M, Yu Y, Mao X, Feng J, Wang Y, Chen H et al (2019) Hydrogen gas inhalation attenuates sepsis-induced liver injury in a FUNDC1-dependent manner. Int Immunopharmacol 71:61–67
29. Wang C, Li J, Liu Q, Yang R, Zhang JH, Cao Y-P et al (2011) Hydrogen-rich saline reduces oxidative stress and inflammation by inhibit of JNK and NF-κB activation in a rat model of amyloid-beta-induced Alzheimer's disease. Neurosci Lett 491(2):127–132
30. Hirayama M, Ito M, Minato T, Yoritaka A, LeBaron TW, Ohno K, Hydrogen inhalation, patients with Parkinson's disease. Original Res
31. Yoritaka A, Kobayashi Y, Hayashi T, Saiki S, Hattori N (2021) Randomized double-blind placebo-controlled trial of hydrogen inhalation for Parkinson's disease: a pilot study. Neurol Sci 42(11):4767–4770
32. Eming I. Metabolism in Tissue Repair and Regeneration. Sci (NY NY). (356).
33. Shankar S, Mahadevan A, Satishchandra P, Uday Kumar R, Yasha T, Santosh V et al (2005) Neuropathology of HIV/AIDS with an overview of the Indian scene. Indian J Med Res 121:468–488
34. Vickers NJ (2017) Animal communication: when i'm calling you, will you answer too? Curr Biol 27(14):R713–R715
35. Lerner TR, Borel S, Greenwood DJ, Repnik U, Russell MR, Herbst S et al (2017) Mycobacterium tuberculosis replicates within necrotic human macrophages. J Cell Biol 216(3):583–594
36. Chen M, Zhang J, Chen Y, Qiu Y, Luo Z, Zhao S et al (2018) Hydrogen protects lung from hypoxia/re-oxygenation injury by reducing hydroxyl radical production and inhibiting inflammatory responses. Sci Rep 8(1):1–12
37. Zhao S, Mei K, Qian L, Yang Y, Liu W, Huang Y et al (2013) Therapeutic effects of hydrogen-rich solution on aplastic anemia in vivo. Cell Physiol Biochem 32(3):549–560
38. Wang X, Yu P, Liu X, Jiang J, Liu D, Xue G (2015) Hydrogen-rich saline resuscitation alleviates inflammation induced by severe burn with delayed resuscitation. Burns 41(2):379–385
39. Liu Y, Wu W, Hong W, Sun X, Wu J, Huang Q (2014) Raltitrexed-based chemotherapy for advanced colorectal cancer. Clin Res Hepatol Gastroenterol 38(2):219–225
40. Zhuang K, Zuo Y-C, Sherchan P, Wang J-K, Yan X-X, Liu F (2020) Hydrogen inhalation attenuates oxidative stress related endothelial cells injury after subarachnoid hemorrhage in rats. Front Neurosci 13:1441
41. Nogueira JE, Amorim MR, Pinto AP, da Rocha AL, da Silva AS, Branco LG (2021) Molecular hydrogen downregulates acute exhaustive exercise-induced skeletal muscle damage. Can J Physiol Pharmacol 99(8):812–820
42. Li Q, Yu P, Zeng Q, Luo B, Cai S, Hui K et al (2016) Neuroprotective effect of hydrogen-rich saline in global cerebral ischemia/reperfusion rats: up-regulated tregs and down-regulated miR-21, miR-210 and NF-κB expression. Neurochem Res 41(10):2655–2665

43. Xu F, Yu S, Qin M, Mao Y, Jin L, Che N et al (2018) Hydrogen-rich saline ameliorates allergic rhinitis by reversing the imbalance of Th1/Th2 and up-regulation of CD4+ CD25+ Foxp3+ regulatory T cells, interleukin-10, and membrane-bound transforming growth factor-β in guinea pigs. Inflammation 41(1):81–92
44. Zhang N, Deng C, Zhang X, Zhang J, Bai C (2018) Inhalation of hydrogen gas attenuates airway inflammation and oxidative stress in allergic asthmatic mice. Asthma Res Pract 4(1):1–9
45. Obeng E (2020) Apoptosis (programmed cell death) and its signals—a review. Braz J Biol 81:1133–1143
46. Qiu X, Dong K, Guan J, He J (2020) Hydrogen attenuates radiation-induced intestinal damage by reducing oxidative stress and inflammatory response. Int Immunopharmacol 84:106517
47. Hayes SC, Luoma JB, Bond FW, Masuda A, Lillis J (2006) Acceptance and commitment therapy: Model, processes and outcomes. Behav Res Ther 44(1):1–25
48. Zhang Y, Liu Y, Zhang J. Saturated hydrogen saline attenuates endotoxin-induced lung dysfunction. J Surg Res 198(1):41–49
49. Zhang Y, Dong A, Xie K, Yu Y (2020) Protective effects of hydrogen on myocardial mitochondrial functions in septic mice. BioMed Res Int 2020
50. Guan J, Wei X, Qin S, Liu X, Jiang Y, Chen Y et al (2020) Continuous tracking of COVID-19 patients' immune status. Int Immunopharmacol 89:107034
51. Man SM, Karki R, Kanneganti TD (2017) Molecular mechanisms and functions of pyroptosis, inflammatory caspases and inflammasomes in infectious diseases. Immunol Rev 277(1):61–75
52. Shao A, Wu H, Hong Y, Tu S, Sun X, Wu Q et al (2016) Hydrogen-rich saline attenuated subarachnoid hemorrhage-induced early brain injury in rats by suppressing inflammatory response: possible involvement of NF-κB pathway and NLRP3 inflammasome. Mol Neurobiol 53(5):3462–3476
53. Xie K, Zhang Y, Wang Y, Meng X, Wang Y, Yu Y et al (2020) Hydrogen attenuates sepsis-associated encephalopathy by NRF2 mediated NLRP3 pathway inactivation. Inflamm Res 69(7):697–710
54. Zheng P, Kang J, Xing E, Zheng B, Wang X, Zhou H (2021) Lung inflation with hydrogen during the cold ischemia phase alleviates lung ischemia-reperfusion injury by inhibiting pyroptosis in rats. Front Physiol 1202
55. Bengmark S (1998) Ecological control of the gastrointestinal tract The role of probiotic flora. Gut 42(1):2–7
56. Backhed F, Ley RE, Sonnenburg JL, Peterson DA, Gordon JI (2005) Host-bacterial mutualism in the human intestine. Science 307(5717):1915–1920
57. Neish AS (2009) Microbes in gastrointestinal health and disease. Gastroenterology 136(1):65–80
58. Gill SR, Pop M, DeBoy RT, Eckburg PB, Turnbaugh PJ, Samuel BS et al (2006) Metagenomic analysis of the human distal gut microbiome. Science 312(5778):1355–1359
59. Natividad JM, Verdu EF (2013) Modulation of intestinal barrier by intestinal microbiota: pathological and therapeutic implications. Pharmacol Res 69(1):42–51
60. Den Besten G, Van Eunen K, Groen AK, Venema K, Reijngoud D-J, Bakker BM (2013) The role of short-chain fatty acids in the interplay between diet, gut microbiota, and host energy metabolism. J Lipid Res 54(9):2325–2340
61. Bäumler AJ, Sperandio V (2016) Interactions between the microbiota and pathogenic bacteria in the gut. Nature 535(7610):85–93
62. Gensollen T, Iyer SS, Kasper DL, Blumberg RS (2016) How colonization by microbiota in early life shapes the immune system. Science 352(6285):539–544
63. Chang C, Lin H (2016) Dysbiosis in gastrointestinal disorders. Best Pract Res Clin Gastroenterol 30(1):3–15
64. Schroeder BO, Bäckhed F (2016) Signals from the gut microbiota to distant organs in physiology and disease. Nat Med 22(10):1079–1089
65. Xiao H-W, Li Y, Luo D, Dong J-l, Zhou L-X, Zhao S-Y et al (2018) Hydrogen-water ameliorates radiation-induced gastrointestinal toxicity via MyD88's effects on the gut microbiota. Exp Mol Med 50(1):e433-e

66. Higashimura Y, Baba Y, Inoue R, Takagi T, Uchiyama K, Mizushima K et al (2018) Effects of molecular hydrogen-dissolved alkaline electrolyzed water on intestinal environment in mice. Med Gas Res 8(1):6
67. Zheng W, Ji X, Zhang Q, Yao W (2018) Intestinal microbiota ecological response to oral administrations of hydrogen-rich water and lactulose in female piglets fed a Fusarium toxin-contaminated diet. Toxins 10(6):246
68. Bordoni L, Gabbianelli R, Fedeli D, Fiorini D, Bergheim I, Jin CJ et al (2019) Positive effect of an electrolyzed reduced water on gut permeability, fecal microbiota and liver in an animal model of Parkinson's disease. PLoS ONE 14(10):e0223238
69. Sha J-B, Zhang S-S, Lu Y-M, Gong W-J, Jiang X-P, Wang J-J et al (2018) Effects of the long-term consumption of hydrogen-rich water on the antioxidant activity and the gut flora in female juvenile soccer players from Suzhou, China. Med Gas Res 8(4):135
70. Valdes AM, Walter J, Segal E, Spector TD (2018) Role of the gut microbiota in nutrition and health. BMJ 361
71. Smith NW, Shorten PR, Altermann EH, Roy NC, McNabb WC (2019) Hydrogen cross-feeders of the human gastrointestinal tract. Gut Microbes 10(3):270–288
72. Rey FE, Gonzalez MD, Cheng J, Wu M, Ahern PP, Gordon JI (2013) Metabolic niche of a prominent sulfate-reducing human gut bacterium. Proc Natl Acad Sci 110(33):13582–13587
73. Million M, Raoult D (2018) Linking gut redox to human microbiome. Human Microbiome Journal. 10:27–32
74. Sa'ad H, Peppelenbosch MP, Roelofsen H, Vonk RJ, Venema K (2010) Biological effects of propionic acid in humans; metabolism, potential applications and underlying mechanisms. Biochim Biophysica Acta (BBA)-Mol Cell Biol Lipids 1801(11):1175–1183
75. Blachier F, Davila A-M, Mimoun S, Benetti P-H, Atanasiu C, Andriamihaja M et al (2010) Luminal sulfide and large intestine mucosa: friend or foe? Amino Acids 39(2):335–347
76. McCarty MF (2015) Potential ghrelin-mediated benefits and risks of hydrogen water. Med Hypotheses 84(4):350–355
77. Kamimura N, Ichimiya H, Iuchi K, Ohta S (2016) Molecular hydrogen stimulates the gene expression of transcriptional coactivator PGC-1α to enhance fatty acid metabolism. npj Aging Mech Dis 2(1):1–8
78. Nishiwaki H, Ito M, Negishi S, Sobue S, Ichihara M, Ohno K (2018) Molecular hydrogen upregulates heat shock response and collagen biosynthesis, and downregulates cell cycles: meta-analyses of gene expression profiles. Free Radical Res 52(4):434–445
79. Holscher HD (2017) Dietary fiber and prebiotics and the gastrointestinal microbiota. Gut microbes. 8(2):172–184
80. Ostojic SM (2018) Inadequate production of H_2 by gut microbiota and Parkinson disease. Trends Endocrinol Metab 29(5):286–288
81. Pozuelo M, Panda S, Santiago A, Mendez S, Accarino A, Santos J et al (2015) Reduction of butyrate-and methane-producing microorganisms in patients with Irritable Bowel Syndrome. Sci Rep 5(1):1–12
82. Shen N-Y, Bi J-B, Zhang J-Y, Zhang S-M, Gu J-X, Qu K et al (2017) Hydrogen-rich water protects against inflammatory bowel disease in mice by inhibiting endoplasmic reticulum stress and promoting heme oxygenase-1 expression. World J Gastroenterol 23(8):1375
83. Shin DW, Yoon H, Kim HS, Choi YJ, Shin CM, Park YS, et al. Effects of alkaline-reduced drinking water on irritable bowel syndrome with diarrhea: a randomized double-blind, placebo-controlled pilot study. Evidence-Based Complementary and Alternative Medicine. 2018;2018.
84. Cole AR, Sperotto F, DiNardo JA, Carlisle S, Rivkin MJ, Sleeper LA et al (2021) Safety of prolonged inhalation of hydrogen gas in air in healthy adults. Critical Care Explor 3(10)
85. Kurokawa R, Hirano S-i, Ichikawa Y, Matsuo G, Takefuji Y (2019) Preventing explosions of hydrogen gas inhalers. Med Gas Res 9(3):160
86. Li H, Chen O, Ye Z, Zhang R, Hu H, Zhang N et al (2017) Inhalation of high concentrations of hydrogen ameliorates liver ischemia/reperfusion injury through A2A receptor mediated PI3K-Akt pathway. Biochem Pharmacol 130:83–92

87. Chen J-B, Kong X-F, Qian W, Mu F, Lu T-Y, Lu Y-Y et al (2020) Two weeks of hydrogen inhalation can significantly reverse adaptive and innate immune system senescence patients with advanced non-small cell lung cancer: a self-controlled study. Med Gas Res 10(4):149
88. Shimouchi A, Nose K, Shirai M, Kondo T (2012) Estimation of molecular hydrogen consumption in the human whole body after the ingestion of hydrogen-rich water. Oxygen Transport to Tissue XXXIII: Springer, pp 245–250
89. Liu C, Kurokawa R, Fujino M, Hirano S, Sato B, Li X-K (2014) Estimation of the hydrogen concentration in rat tissue using an airtight tube following the administration of hydrogen via various routes. Sci Rep 4(1):1–11
90. He Y, Zhang B, Chen Y, Jin Q, Wu J, Yan F et al (2017) Image-guided hydrogen gas delivery for protection from myocardial ischemia–reperfusion injury via microbubbles. ACS Appl Mater Interfaces 9(25):21190–21199
91. Zhu Q, Wu Y, Li Y, Chen Z, Wang L, Xiong H et al (2018) Positive effects of hydrogen-water bathing in patients of psoriasis and parapsoriasis en plaques. Sci Rep 8(1):1–8
92. Asada R, Saitoh Y, Miwa N (2019) Effects of hydrogen-rich water bath on visceral fat and skin blotch, with boiling-resistant hydrogen bubbles. Med Gas Res 9(2):68
93. Oharazawa H, Igarashi T, Yokota T, Fujii H, Suzuki H, Machide M et al (2010) Protection of the retina by rapid diffusion of hydrogen: administration of hydrogen-loaded eye drops in retinal ischemia–reperfusion injury. Invest Ophthalmol Vis Sci 51(1):487–492
94. Sobue S, Yamai K, Ito M, Ohno K, Ito M, Iwamoto T et al (2015) Simultaneous oral and inhalational intake of molecular hydrogen additively suppresses signaling pathways in rodents. Mol Cell Biochem 403(1):231–241
95. Holko P, Kawalec P, Sajak-Szczerba M, Avedano L, Mossakowska M (2022) Out-of-pocket expenses of patients with inflammatory bowel disease: a comparison of patient-reported outcomes across 12 European countries. Eur J Health Econ 1–11
96. Holko P, Kawalec P, Sajak-Szczerba M, Avedano L, Mossakowska M (2022) Indirect costs of inflammatory bowel diseases: a comparison of patient-reported outcomes across 12 European countries. Inflamm Bowel Dis
97. Wallace KL, Zheng L-B, Kanazawa Y, Shih DQ (2014) Immunopathology of inflammatory bowel disease. World J Gastroenterol WJG 20(1):6
98. Kornbluth A, Sachar DB, Gastroenterology PPCotACo (2010) Ulcerative colitis practice guidelines in adults: American college of gastroenterology, practice parameters committee. Official J Am Coll Gastroenteroll ACG 105(3):501–523
99. Liu T-C, Stappenbeck TS (2016) Genetics and pathogenesis of inflammatory bowel disease. Annu Rev Pathol 11:127
100. Korenaga D, Takesue F, Kido K, Yasuda M, Inutsuka S, Honda M et al (2002) Impaired antioxidant defense system of colonic tissue and cancer development in dextran sulfate sodium-induced colitis in mice. J Surg Res 102(2):144–149
101. Esworthy RS, Aranda R, Martín MG, Doroshow JH, Binder SW, Chu F-F (2001) Mice with combined disruption of Gpx1 and Gpx2 genes have colitis. Am J Physiol-Gastrointest Liver Physiol 281(3):G848–G55
102. Watterlot L, Rochat T, Sokol H, Cherbuy C, Bouloufa I, Lefèvre F et al (2010) Intragastric administration of a superoxide dismutase-producing recombinant Lactobacillus casei BL23 strain attenuates DSS colitis in mice. Int J Food Microbiol 144(1):35–41
103. Barrett KE, McCole DF (2016) Hydrogen peroxide scavenger, catalase, alleviates ion transport dysfunction in murine colitis. Clin Exp Pharmacol Physiol 43(11):1097–1106
104. Kruidenier L, van Meeteren ME, Kuiper I, Jaarsma D, Lamers CB, Zijlstra FJ et al (2003) Attenuated mild colonic inflammation and improved survival from severe DSS-colitis of transgenic Cu/Zn-SOD mice. Free Radical Biol Med 34(6):753–765
105. Tsapralis D, Panayiotides I, Peros G, Liakakos T, Karamitopoulou E (2012) Human epidermal growth factor receptor-2 gene amplification in gastric cancer using tissue microarray technology. World J Gastroenterol: WJG 18(2):150
106. Levine JS, Burakoff R (2011) Extraintestinal manifestations of inflammatory bowel disease. Gastroenterol Hepatol 7(4):235

107. de Lemos JA, McGuire DK, Drazner MH (2003) B-type natriuretic peptide in cardiovascular disease. The Lancet 362(9380):316–322
108. Strober W, Fuss I, Mannon P (2007) The fundamental basis of inflammatory bowel disease. J Clin Investig 117(3):514–521
109. Triantafillidis JK, Merikas E, Georgopoulos F (2011) Current and emerging drugs for the treatment of inflammatory bowel disease. Drug Des Dev Therapy 185–210
110. Ohta S (2015) Initiation, development and potential of hydrogen medicine: toward therapeutic and preventive applications of molecular hydrogen against a variety of diseases. Seikagaku J Jpn Biochem Soc 87(1):82–90
111. Kajiya M, Silva MJ, Sato K, Ouhara K, Kawai T (2009) Hydrogen mediates suppression of colon inflammation induced by dextran sodium sulfate. Biochem Biophys Res Commun 386(1):11–15
112. LeBaron TW, Asgharzadeh F, Khazaei M, Kura B, Tarnava A, Slezak J (2021) Molecular hydrogen is comparable to sulfasalazine as a treatment for DSS-induced colitis in mice. EXCLI J 20:1106
113. Song L, Zhang Y, Zhu C, Ding X, Yang L, Yan H (2022) Hydrogen-rich water partially alleviate inflammation, oxidative stress and intestinal flora dysbiosis in DSS-induced chronic ulcerative colitis mice. Adv Med Sci 67(1):29–38
114. Narayanankutty A, Job JT, Narayanankutty V (2019) Glutathione, an antioxidant tripeptide: dual roles in carcinogenesis and chemoprevention. Curr Protein Pept Sci 20(9):907–917
115. Hu D, Huang T, Shigeta M, Ochi Y, Kabayama S, Watanabe Y et al (2022) Electrolyzed hydrogen water alleviates abdominal pain through suppression of colonic tissue inflammation in a rat model of inflammatory bowel disease. Nutrients 14(21):4451
116. Zhu W-J, Nakayama M, Mori T, Hao K, Terawaki H, Katoh J et al (2013) Amelioration of cardio-renal injury with aging in dahl salt-sensitive rats by H_2-enriched electrolyzed water. Med Gas Res 3(1):1–8
117. Xue J, Shang G, Tanaka Y, Saihara Y, Hou L, Velasquez N et al (2014) Dose-dependent inhibition of gastric injury by hydrogen in alkaline electrolyzed drinking water. BMC Complement Altern Med 14(1):1–9
118. Yoon Y-S, Sajo MEJV, Ignacio RMC, Kim S-K, Kim C-S, Lee K-J (2014) Positive effects of hydrogen water on 2, 4-dinitrochlorobenzene-induced atopic dermatitis in NC/Nga mice. Biol Pharm Bull 37(9):1480–1485
119. Ikeda M, Shimizu K, Ogura H, Kurakawa T, Umemoto E, Motooka D et al (2018) Hydrogen-rich saline regulates intestinal barrier dysfunction, dysbiosis, and bacterial translocation in a murine model of sepsis. Shock 50(6):640–647
120. Ji X, Zhang Q, Zheng W, Yao W (2019) Morphological and molecular response of small intestine to lactulose and hydrogen-rich water in female piglets fed Fusarium mycotoxins contaminated diet. J Anim Sci Biotechnol 10(1):1–14
121. Guo Z, Hu B, Han H, Lei Z, Shimizu K, Zhang L et al (2020) Metagenomic insights into the effects of nanobubble water on the composition of gut microbiota in mice. Food Funct 11(8):7175–7182

Chapter 18
Effects of Molecular Hydrogen in the Pathophysiology and Management of Metabolic and Non-communicable Diseases

Ram B. Singh, Alex Tarnava, Jan Fedacko, Gizal Fatima, Sunil Rupee, and Zuzana Sumbalova

Abstract The sustainable development goals (SDG) of the UNO would be difficult to achieve without prevention of non-communicable diseases (NCDs). Western diet and lifestyle, which are major risk factors of NCDs, are known to cause oxidative stress and decrease in production of molecular hydrogen in the intestines which leads to a decline in endogenous antioxidant status in the body, leading to increase in systemic inflammation. There are gaps in the knowledge about the role molecular hydrogen plays, in the treatment of these diseases This review aims to discuss the role of hydrogen in the pathogenesis and prevention of NCDs. Molecular hydrogen (H_2) has been studied extensively as a therapeutic gas, with an estimated 2000 publications to date, exploring its potential therapeutic use in 170 disease models across every organ in the mammalian body. Hydrogen therapy can be administered through several methods, such as H_2 inhalation, dissolving H_2 gas in water to make hydrogen-rich water (HRW) for oral consumption or topical application, or hydrogen-rich saline. The exact mechanism of action of molecular hydrogen is not known but it is in itself a potential antioxidant that can also inhibit hydroxyl and nitrosyl radicals in the cells

R. B. Singh (✉)
Halberg Hospital and Research Institute, Moradabad, India
e-mail: rbs@tsimtsoum.net

A. Tarnava
Drink HRW, Vancouver, Canada

J. Fedacko
University Research Park, PJ Safarik University, Kosice, Slovakia

G. Fatima
Era Medical College, Lucknow, India

S. Rupee
School of Natural Science, University of Central Lancashire, Preston, UK

Z. Sumbalova
Pharmacobiochemical Laboratory, 3rd Department of Internal Medicine, Faculty of Medicine, Comenius University in Bratislava, Bratislava, Slovakia
e-mail: zuzana.sumbalova@fmed.uniba.sk

J. Slezak and B. Kura (eds.), *Molecular Hydrogen in Health and Disease*, Advances in Biochemistry in Health and Disease 27, https://doi.org/10.1007/978-3-031-47375-3_18

and tissues. Hydrogen is known to cause a marked decline in oxidative stress, and inflammation that are crucial in the pathogenesis of NCDs. Hydrogen therapy has been found protective against NCDs, including, metabolic diseases, cardiovascular diseases (CVDs), neurodegenerative diseases, chronic kidney disease, cancer, and chronic lung diseases.

Keywords Hydrogen · Cardiovascular · Metabolic diseases · Diet · Oxidative stress · Inflammation

Introduction

Cohort studies have demonstrated that metabolic diseases; obesity, type 2 diabetes mellitus (T2DM) and other chronic diseases; cardiovascular diseases (CVDs), cancer, osteoporosis and neurodegenerative diseases have become a public health problem in both developed and developing countries of the world [1–3]. The increase in risk of these non-communicable diseases (NCDs), may be due to western type diet, sedentary behavior, increase intake of tobacco and alcoholism [1]. Nutrition in transition from poverty to affluence may be associated with increased intake of wester foods and lower intake traditional foods with increased use of automobiles resulting into physical inactivity leading to obesity and metabolic syndrome [1–3].

Western diet and lifestyle are known to cause obesity with an increase in the reactive oxygen species (ROS), along with possible decrease in the production of molecular hydrogen in the intestines with an accompanying decline in endogenous antioxidant status in the body, leading to an increase in systemic inflammation [4, 5]. The increase in inflammation in the adipocytes predisposes obesity and central obesity, and increased inflammation in the beta cells of the pancreas, predisposes T2DM. Inflammation of the LDL receptors in the hepatocytes, endothelium, neurons, osteocytes and gut may predispose related NCDs; CVDs, neuro-degenerative diseases, bone and joint diseases and cancer that are major causes of mortality [6, 7] (Fig. 18.1).

It seems that the existing gaps in knowledge do not allow us to understand how free radical-induced damage in the tissues and concerned organs predispose metabolic diseases and other NCDs. It is possible that increased intake of antioxidants can cause decline in free radical generation and reactive oxygen species (ROS) leading to decline in oxidative stress and inflammation [4–7]. The proposed hypothesis is that the Mediterranean type of diet which is rich in antioxidants can act as potential antioxidant, due to its protective effects on gut microbiota and increased production of molecular hydrogen, due to degradation of dietary fiber in the gut. Molecular hydrogen is known to regulate oxidative stress and inflammation and may maintain cell and tissue homeostasis, with decline in inflammation and risk of NCDs, including metabolic diseases [8–12]. The objective of this article is to highlight the role of molecular hydrogen, in the management of NCDs.

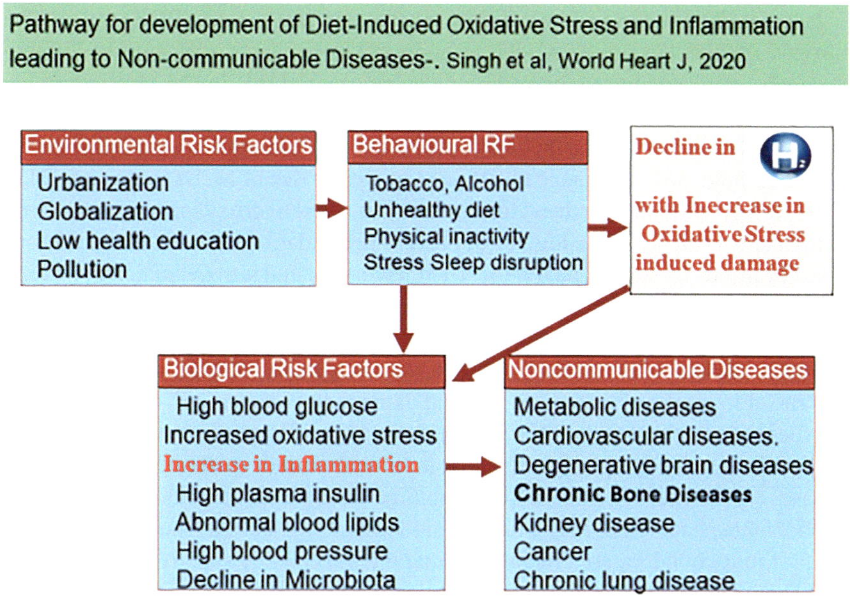

Fig. 18.1 Pathways for development of non-communicable diseases

Oxidative Stress and Antioxidants in the Pathogenesis of Chronic Diseases

Several unhealthy substances and health behaviors such as western diet, fried and fast foods, alcoholism, tobacco smoke, pesticides, radiation, and pollutants can all generate free radicals [4, 5, 12–14]. The body tissues are under constant attack from these free radicals, because inhaled oxygen undergoes single electron reduction to form superoxide radicals (O_2^-). This radical can initiate radical propagation, and also be converted to hydrogen peroxide (H_2O_2) and hydroxyl radicals ($\cdot OH$) [15]. Free oxygen radicals are characterized as having an unpaired electron, which makes them very reactive as they seek another electron to have a stable pair. These free oxygen radicals scavenge the body tissue to seek out other electrons, so they can become a pair, which causes damage to cells, proteins, lipoproteins and DNA resulting in diseases [15]. There is evidence that unhealthy diet and lifestyle may increase the production of free radicals and ROS in the tissues due to high levels of protein carbonylation and lipid peroxidation products [4, 5]. There is decline in the antioxidant defense status, such as decreased production of molecular hydrogen in the gut, as well as antioxidant enzymes; in the tissues. However, certain foods as component of the Mediterranean type of diets, and nutrients, such as dietary fiber, flavonoids and omega-3 fatty acids [10], may produce protective molecules in the gut [11], such as molecular hydrogen, which is a potential free radical scavenger [12, 14].

It seems that numerous free radicals, reactive oxygen species (ROS) and reactive nitrogen species are produced due to endogenous oxidants, exposure to different physiochemical conditions or pathological processes [15]. In the physiology and metabolism, there is a need to have balance between free radical generation and antioxidant status for proper metabolic function, so that there is no increase in oxidative stress-induced cell damage resulting in to greater risk of NCDs including T2DM [8, 9, 12–14]. Butylated hydroxytoluene and butylated hydroxyanisole are synthetic antioxidants used as food additives to prevent peroxidation of foods that are known to be hazardous for human health [4, 5, 15]. Hydroxyl and nitrosyl radicals represent the major cause of the destruction of body tissues, either by a direct reaction, or by triggering a chain reaction of free radicals [4, 5, 15]. It seems that physiological levels of free radicals are protective for cells, hence endogenous antioxidants are crucial to prevent free radical-induced damage to tissues [4, 5]. Therefore, there is a focused attention toward the role of free radical biochemistry and free radical biology. in the physiology and metabolism as well as in the pathogenesis of chronic diseases [4, 5]. There is evidence that xanthine oxidase, nicotinamide adenine dinucleotide phosphate (NADPH) oxidase, and leakage of electron from the respiratory chain of mitochondria, are crucial in damaging the cells by superoxide radical [16, 17]. These free radicals attenuate the biological presence of NO by neutralizing it via conversion to a more detrimental peroxy-nitrite radical. The oxidases that enzymatically generate H2O2 and superoxide, are accumulated in the proteins related to Nox family, the main source of vascular free radicals [16–19]. The shear stress in the vessels activates the Nox proteins; Nox 1, Nox2 and Nox3 that are considered to have crucial role in vascular function [18, 19]. The superoxide radicals are particularly produced via Nox1 and Nox2 via a single transfer of electron to hydrogen molecule. The superoxide radical reacts rapidly in the cells, to inactivate excess of NO, that produces perozxynitrite, known to have adverse effects on the vasodilation induced by nitric oxide [19]. In such situations of oxidative dysfunction in the presence of peroxynitrite, it may inhibit the release of endothelial nitric oxide synthase (eNOS) enzymes causing decline in the production of NO. The cofactor of eNOS, tetrahydrobiopterin (BH4), responsible for oxidation may inactive it to, 7,8-dihydropterin (BH2), causing uncoupling of eNOS. and this mechanism, generates the superoxide. Molecular hydrogen has been found to be useful in protecting against free radical induced damage due to its potential antioxidant effects in various conditions [20–29]. However, the exact mechanisms, how molecular hydrogen protects the body tissues is not clear. Research has found that increased supplementation of molecular hydrogen can inhibit free radical generation with decline in inflammation, which may be protective against NCDs including metabolic diseases [10–15].

Physiology of Molecular Hydrogen and the Gut Microbiota

The microbiota in the gastrointestinal tract, is crucial in the prevention of NCDs including metabolic diseases. Indo-Mediterranean-style foods such as butter milk, curd and yoghurt, are used to alter the composition and improve function of the communities of the microbes that develop colonies in the gut for improvement of health, and protect from NCDs [8, 9]. Increased production of molecular hydrogen is one of the important mechanisms by which healthy foods provide the beneficial effects in health and diseases [8–11]. The mechanisms by which molecular hydrogen provides the benefits are not well understood. Increased production of intestinal hydrogen along with recovery in muscle function has been reported following intensive exercises [30]. A previous study reported increased production of acetate which mediates a microbiome–brain–β- cell axis to have influence on metabolic syndrome [31]. However, increased generation of metabolites such as short chain fatty acids, BDNF and hydrogen from gut microbiota may enhance benefits in the metabolism via gut-brain neural circuits [32]. The microbes present in the gut can metabolize several of these polysaccharides into more useful fatty acids. These short-chain fatty acids (SCFAs); propionate, butyrate, acetate, and gases; methane and hydrogen have potential anti-inflammatory effects [8–11, 31, 32].

There are several biochemical pathways by which microbes convert the complex polysaccharides into monosaccharides, that are mediated by the enzymatic actions [32, 33]. Hydrogen gas in the body is also produced through fermentation of carbohydrates; lactose, lactulose, and fructose by intestinal bacteria. The bacteria normally present in the large intestine are mainly Bacteroides fragilis group, Clostridium perfringens and Pseudomonas that possess hydrogenases to produce hydrogen. The presence of SCFAs in the gut indicates that fermentation due to microbes, occurs in the colon; with a greater level in the proximal colon but lower in the distal colon. The region of distal colon may have the highest number of microbes and with highest level of gases. Since all the Mediterranean type of diets promote the growth of gut microbiota, hydrogen is produced in liter quantities by the intestinal bacteria, in particular, if the individual has a healthy bacterial population in the gut and high fibers and probiotics in the diets [33–35].

The hydrogen production in the gastrointestinal tract in man is primarily dependent upon the delivery of ingested, fermentable fibrous substrates to an abundant intestinal flora [33]. This is normally present mainly in the colon and produce relatively large amounts of H_2. The removal of excess hydrogen through methanogenesis from the gut is not the only microbial mechanism to remove excess hydrogen from the gut; this can also be mediated through the reduction of sulphate to sulphide by sulphate-reducing bacteria. At least one explanation is possible applying the pathogens that can provoke inflammation by remodeling the microbiota [34]. Certain low-abundance microbial pathogens may orchestrate inflammatory disease by remodeling a normally benign microbiota into a dysbiotic one [35]. There is evidence that, Citrobacter rodentium causes in mice global changes in microbial community structure, apparently dependent upon the ability of this pathogen to cause inflammation [35]. Induction of gut

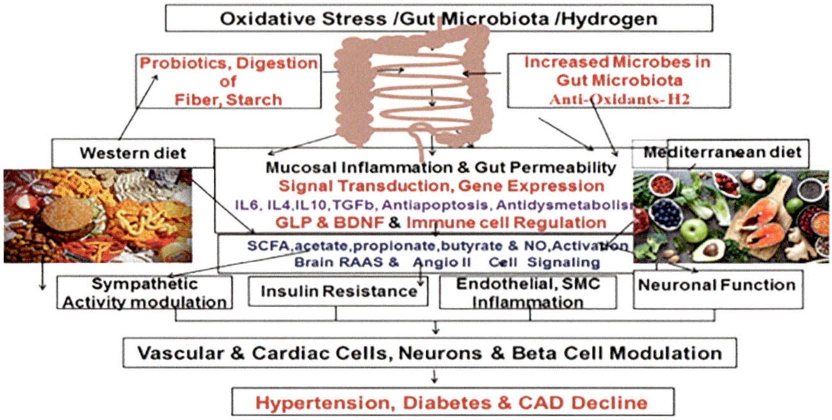

Fig. 18.2 Mechanism of production and inhibition of molecular hydrogen due to diets via microbiota in the gut, and its effects on anti-inflammatory molecules and risk of chronic diseases

inflammation by administration of dextran sodium sulfate leading to a dysbiotic microbiota, and improvement in gut microbiom by healthy diets suggest, an intimate relationship between the inflammatory status of the intestine and gut microbiota [33–36] (Fig. 18.2).

The research on molecular hydrogen is intensified in search of an effective, nontoxic natural compound, with potential antioxidant activity for prevention of chronic NCDs [12–15]. It seems that after the discovery of the biological significance of the molecular hydrogen produced in the intestines, there has been a dramatic shift to the belief that hydrogen may have a critical protective role in health and disease [14–17]. In the last 20 years, large number of studies have been published indicating that molecular hydrogen, either produced endogenously via consuming healthy foods or via any releasing agents, administered exogenously via inhalation or hydrogen rich water (HRW), acts as potential antioxidant in a wide range of physiological and pathophysiological processes [16, 17]. Molecular hydrogen in itself appears to be a potential antioxidant that can also inhibit hydroxyl and nitrosyl radicals in the cells and tissues, causing a marked decline in oxidative stress, leading to a decline in the inflammation that is marker in the pathogenesis of NCDs. Slezak et al. and other experts [12–15] have demonstrated that hydrogen can also rapidly diffuse into tissues and cells without affecting metabolic redox reactions and signaling reactive species [12–15] Apart from its direct neutralizing effects on reactive oxidants, hydrogen, directly decreases free radical stress by regulation of expression of genes [16, 17]. It seems that molecular hydrogen not only regulate gene expression but also does epigenetic modulation, which could be alternative mechanisms for decline in oxidative stress-induced damage to genes, resulting in to increase in its anti-inflammatory and anti-apoptotic potential [16, 17]. It is proposed that apart from improvement in gut microbiota, molecular hydrogen may also activate the production of bioactive

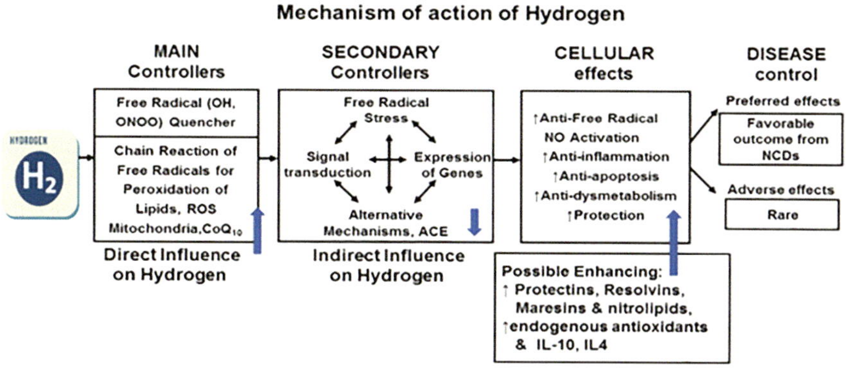

Fig. 18.3 Molecular mechanisms of hydrogen action resulting in to beneficial effects (Modified from Ichihara et al. [14])

lipids that are potential anti-inflammatory metabolites. This potential of molecular hydrogen may be utilized for preventive and therapeutic applications (Fig. 18.3).

The emergence of gut microbiome in the signaling of neuronal cells and neurogenesis poses the possibility that targeting the gut–brain axis could be a new strategy for treatment of neurodegenerative diseases and other NCDs [37]. Apart from SCFA, there may be increased production of bioactive lipids; protectins, maresins, resolvins, and nitrolipids that are also anti-inflammatory molecules. Several studies concerned with diseases of gastroentestinal, metabolic, cardiovascular and neurodegeneration support the existence of a bidirectional communication network of signaling mediators, such as bioactive lipids, that can interact with molecular hydrogen in modulation of inflammation, gut permeability, microbiota composition, and the gut–brain axis[37]. The crosstalk between the gut–brain axis, microbiome, molecular hydrogen and bioactive lipids may emerge as the basis of a promising therapeutic strategy to counteract the pathogenesis of many of the NCDs. The results from recent studies, can contribute to the explanation of a new beneficial mechanism of hydrogen on a part of antioxidant protection in organism [38–40]. The hydrogen produced in the large intestine is metabolized, and excreted in the large intestine [38], by the microbes [39]. The largest amount of hydrogen is produced by *Blautia coccoides* and *Clostridium leptum.* [39]. However, depending on microbiota composition, substantial amounts of hydrogen molecules are consumed by methanogenic and sulfate-reducing bacteria [39, 40].

Molecular Hydrogen; A New Therapeutic Agent

Hydrogen is the most abundant element in the universe, which typically exists as molecular hydrogen in its diatomic form (H_2). It has been considered as a biologically inert gas initially, although later on, in 1975, bio-therapeutic effects of hydrogen were observed, by Dole and co-workers, for the first time [41]. It is only recently, with the discovery of the antioxidant properties of hydrogen, its beneficial effects have been reported in the peer-reviewed journals [42]. The new discoveries drew increasing attention on new therapeutic strategies such as drinking HRW, inhalation of hydrogen gas, injecting the saline rich in hydrogen, along with topical administration via hydrogen baths, and its buccal consumption. It has been reviewed that in all the NCDs, in which oxidative stress-related damage is the major problem, molecular hydrogen may be administered as an effective antioxidant for the prevention of these chronic diseases [37, 43–46]. Hydrogen also alleviates blood-brain barrier impairment and improves cognitive dysfunction [23]. Hydrogen therapy has been found to ameliorate cardiac remodeling [24], dyslipidemia and metabolic syndrome [25] oxygen saturation in chronic lung disease [26] and in NSAID-induced enteropathy [27]. Hydrogen therapy as combine treatment showed beneficial effects in facial paralysis [28], on physical fatigue via prefrontal cortex activation during and after high intensity exercise [29]. The oxidation of LDL by myeloperoxidase produces ROS along with free radical generation which may be protected by antioxidants such as hydrogen [45–48].

Molecular hydrogen therapy has been used among patients with metabolic diseases by the several experts [49–56]. Obesity is the first metabolic disease which begins with hypertrophy of adipocyte causing overweight and central obesity leading to metabolic syndrome and T2DM [51–53]. Metabolic syndrome may be defined as an accumulation of at least three risk factors out of obesity, diabetes, hypertension, hyperlipidemia and low HDL. It is possible that free radical stress with or without inflammation, with decrease in high density lipoprotein (HDL) cholesterol are crucial in the development of metabolic syndrome which is a risk factor of T2DM and CVDs [49, 51–55]. Treatment with HRW may improve metabolism of glucose and lipids in T2DM patients or those with glucose intolerance because both conditions are associated with oxidative stress [49, 50]. The effectiveness of HRW (1.5–2 L/day) was examined in an open label, 8-week study in 20 subjects with potential metabolic syndrome [51]. HRW was produced by placing a metallic magnesium stick into drinking water (hydrogen concentration; 0.55–0.65 mM) by the following chemical reaction: $Mg + 2H_2O \rightarrow Mg(OH)_2 + H_2$. The consumption of hydrogen-rich water for 8 weeks resulted in a 39% increase ($p < 0.05$) in antioxidant enzyme superoxide dismutase (SOD) and a 43% decrease ($p < 0.05$) in thiobarbituric acid reactive substances (TBARS) in urine [49]. Further, subjects showed an 8% increase in high-density lipoprotein (HDL)-cholesterol and a 13% decrease in total cholesterol/HDL-cholesterol from reference to week 4. There was no change in fasting glucose concentrations during the 8-week study. Drinking hydrogen-rich water may represent a potentially novel therapeutic and preventive strategy for metabolic syndrome.

In a randomized, controlled trial, Singh group administered HRW showing favorable effects on multiple parameters of metabolic syndrome after treatment for 24-week [25]. Compared to control group ($P < 0.05$), ($P = 0.309$), HRW group showed significant decline in the parameters of central obesity BMI and WHR after treatment with HRW ($P < 0.05$) [25]. In addition, treatment with HRW caused a significant decline in blood lipids as given in the Table 18.1.

Table 18.1 Effects of hydrogen rich water on blood lipoproteins in patients with metabolic syndrome [25]

Data	Hydrogen rich water (n-30)		Placebo (n = 30)	
Data, mg/dl	Baseline	After 24 weeks	Baseline	After 24 weeks
Cholesterol	187.7 ± 32.4	169.2 ± 26.1***	184.3 ± 37.4	184.4 ± 38.6
LDL-Cholesterol	109.0 ± 34.4	102.5 ± 28.0	105.5 ± 42.0	106.0 ± 43.3
HDL cholesterol	41.7 ± 4.2	40.4 ± 1.8	41.8 ± 2.3	42.3 ± 2.4
VLDL cholesterol	37.3 ± 17.9	28.0 ± 11.3**	36.8 ± 20.6	37.3 ± 20.5
Triglycerides	189.8 ± 93.3	142.4 ± 65.0**	184.4 ± 102.8	185.6 ± 101.3
C-reactive proteins	0.5 ± 0.2	0.5 ± 0.1*	0.6 ± 0.5	0.6 ± 0.5

*** p value < 0.0001, ** p value < 0.01, * p value < 0.05, by comparison of baseline and after follow up using analysis of variance (Modified from Ref. [25])

Treatment with HRW also showed decline in fasting blood glucose after 24-week, in the active treatment group, along with a significant decline in HbA1C (12%, $P <$ 0.05) compared to baseline levels and placebo group [25]. Treatment with HRW also reduced the markers of inflammation; TNF-α, and IL-6 ($P < 0.05$). Interestingly, markers of oxidation showed significant decline, while vitamins C and E showed rise in the hydrogen group. Serum levels of angiotensin converting enzyme showed significant decline whereas serum nitrite level showed significant increase, which may cause decline in blood pressures (Table 18.2).

Table 18.2 Effect of hydrogen rich water on glycaemia, oxidative stress and cytokines in patients with metabolic syndrome [25]

	Hydrogen rich water (n = 30)		Placebo (n = 30)	
Data, mg/dl	Baseline	After 24 weeks	Baseline	After 24 weeks
Fasting blood glucose	121.5 ± 61.0	103.1 ± 33.0*	123.9 ± 43.4	126.4 ± 42.3
HbA1c, %	5.8 ± 0.9	5.1 ± 0.2***	6.2 ± 1.2	6.1 ± 1.2
TNF-α	4.8 ± 1.2	3.9 ± 0.6***	4.8 ± 1.3	4.8 ± 1.3
IL-6	1.9 ± 0.7	1.6 ± 0.2**	1.6 ± 0.6	1.7 ± 0.6
TBARS	2.5 ± 0.3	1.6 ± 0.3*	2.5 ± 0.3	2.5 ± 0.3
Melondialdehyde	3.4 ± 0.2	2.7 ± 0.2***	3.4 ± 0.2	3.5 ± 0.2
Diene conjugates	27.8 ± 1.0	26.7 ± 0.5***	28.3 ± 0.8	28.3 ± 0.8
Vitamin E	23.0 ± 2.3	26.8 ± 1.9***	23.0 ± 1.5	23.1 ± 1.1
Vitamin C	20.7 ± 2.5	24.2 ± 1.8***	20.7 ± 2.5	20.8 ± 2.4
Nitrite	0.63 ± 0.06	0.68 ± 0.06***	0.66 ± 0.04	0.65 ± 0.03
Angiotensin converting enzyme	85.2 ± 7.8	80.7 ± 5.8***	84.5 ± 8.8	83.8 ± 8.7

*** p value < 0.0001, ** p value < 0.01, * p value < 0.05, by comparison of baseline and after follow up using analysis of variance (Modified from Ref. [25])

In a cross-over, randomized, controlled, trial in 30 patients with T2DM and 6 patients with impaired glucose tolerance, patients took either 900 mL/d of HRW or 900 mL of placebo water for 8 weeks, with a 12-week period of washout [50]. Intake of HRW was associated with a significant decline in the concentrations of modified LDL cholesterol, small dense LDL, and urinary 8-isoprostanes by 15.5% ($P < 0.01$), 5.7% ($P < 0.05$), and 6.6% ($P < 0.05$), respectively. Intake of HRW was also associated with a trend of decreased serum concentrations of oxidized LDL and free fatty acids, and increased plasma concentrations of adiponectin and extracellular-superoxide dismutase [50]. These results suggest that supplementation with HRW may have a beneficial role in the prevention of T2DM and insulin resistance, without any adverse effects [51]. This observational, non-interventional, retrospective, double-arm, 6-month clinical study included T2DM patients (n = 1088), receiving anti-diabetes drugs with or without hydrogen initiation from 2018 to 2021 [52]. Interestingly, subjects in hydrogen group maintained greater improvement in the level of HbA1c (− 0.94% vs. − 0.46%), fasting blood glucose (− 22.7 mg/dL vs. − 11.7 mg/dL), total cholesterol (− 12.9 mg/dL vs. − 4.4 mg/dL), HOMA-IR (− 0.76 vs. − 0.17) and HOMA-β (8.2% vs. 1.98%) with all $p < 0.001$ post the treatment, compared to control group. Logistics regression analysis revealed that the likelihood of reaching HbA1c < 7%, ≥ 7% to < 8% and > 1% reduction at the follow-up period was higher in the hydrogen group, while patients in the control group were more likely to attain HbA1c ≥ 9%. Patients in hydrogen group had a lower incidence of several adverse effects including hypoglycemia (2.0% vs. 6.8%), vomiting (2.6% vs. 7.4%), constipation (1.7% vs. 4.4%) and giddiness (3.3% vs. 6.3%) with significance in comparison to the control group [52]. It is clear that hydrogen therapy as an adjunct, may benefit glycemic control, lipid metabolism, insulin resistance and incidence of adverse events in T2DM patients after 6-month treatment.

Hydrogen may activate ATP-binding cassette transporter A1-dependent efflux, enhance HDL anti-atherosclerotic functions, and have beneficial lipid-lowering effects. In a more recent randomized trial among 30 subjects with nonalcoholic fatty liver disease (NAFLD), treatment with HRW was administered for eight weeks [57]. Despite short period of therapy, beneficial trends ($p > 0.05$) were observed in decreased weight (≈1 kg) and body mass index as well as a trend of improved lipid profile and reduced lactate dehydrogenase levels, in the HRW group. HRW tended to non-significantly decrease levels of nuclear factor kappa B, heat shock protein 70 and matrix metalloproteinase-9. In a recent controlled trial, including 43 subjects, inhalation of hydrogen/oxygen improved serum lipid and liver enzymes [56]. There was a significantly improvement in liver fat content detected by ultrasound and CT scans after hydrogen/oxygen inhalation depending upon severity. In the same study, including diet-induced mice model, the effect of hydrogen on mouse NASH showed that hydrogen/oxygen inhalation improved systemic inflammation and liver histology. Molecular hydrogen also inhibited lipid accumulation in AML-12 cells indicating that hydrogen/oxygen inhalation can inhibit NAFLD which may be due to hepatic autophagy [56]. Since metabolic syndrome has become a worldwide problem, hydrogen therapy may be a new approach for the prevention of metabolic disease [51–57].

Hydrogen therapy may positively affect mitochondrial bioenergetics. In an experimental study in rats, Gvozdjakova et al. reported, increased state 3 respiration with both CI and CII-linked substrates in cardiac mitochondria after H2 administration [58]. It is proposed that H2 may convert the quinone intermediates in the Q-cycle to the fully reduced ubiquinol, increasing this way antioxidant capacity of the quinone pool and preventing mitochondrial ROS generation [59].

The increase in the rate of ATP production was accompanied with increased concentration of CoQ9 in heart tissue and mitochondria and decreased plasma levels of malondialdehyde [59]. Beneficial effects of molecular hydrogen in various experimental models of human diseases and in many clinical studies was documented. H2 can be administered by various ways, as a gas inhalation, drinking of hydrogen-enriched water, or taking a hydrogen-dissolved bath as well as in saline infusions [58–60]. As antioxidant, hydrogen selectively scavenges hydroxyl and peroxynitrite radicals, and decreases oxidative stress, however, the hydrogen effect on antioxidant–coenzyme Q information is lacking. A recent review has reemphasized, that Indo-Mediterranean type of foods or DASH diet that are known to decrease blood pressure, may also produce greater amount of hydrogen in the gut, which may be responsible for prevention of hypertension [61].

Effects of Hydrogen Therapy on Blood Lipids

High LDL cholesterol is known to predispose atherosclerosis which is recognized as a chronic inflammatory condition that begins with the dysfunction or activation of arterial endothelium. Low-density lipoprotein (LDL) and especially its oxidized form play a key role in endothelial dysfunction and atherogenesis. Several small-scale studies evaluating the effects of HRW, indicated an advantageous effects in regulating blood lipid profiles. In a recent meta-analysis, including seven studies, the finding reported a significant decline in total cholesterol, LDL, and triglycerides after HRW intake ($p = 0.01$), with small to moderate effects [47]. The results indicate that drinking HRW can cause significant benefit in the lipid profile of the patients. The Forest Plot of studies shows total cholesterol with a small negative effect [pooled SMD = − 0.23 (from − 0.40 to 0.05); $p \leq 0.01$], (b) LDL with small effect [pooled SMD = − 0.22 (from − 0.39 to 0.04); $p \leq 0.01$], and (c) triglycerides, with a moderate beneficial effect [pooled SMD = − 0.38 (from − 0.59 to 0.18), $p \leq 0.01$] (Fig. 18.4).

Increase in triglycerides with decline in HDL cholesterol are important characteristics of metabolic syndrome, which are common in most of the clinical trials, using HRW among patients with metabolic syndrome [25, 52–57]. Treatment with HRW was associated with beneficial effects on blood glucose and blood lipids with significant decline in blood lipoproteins [25, 52–56] and a beneficial trend in some of the trials [57].

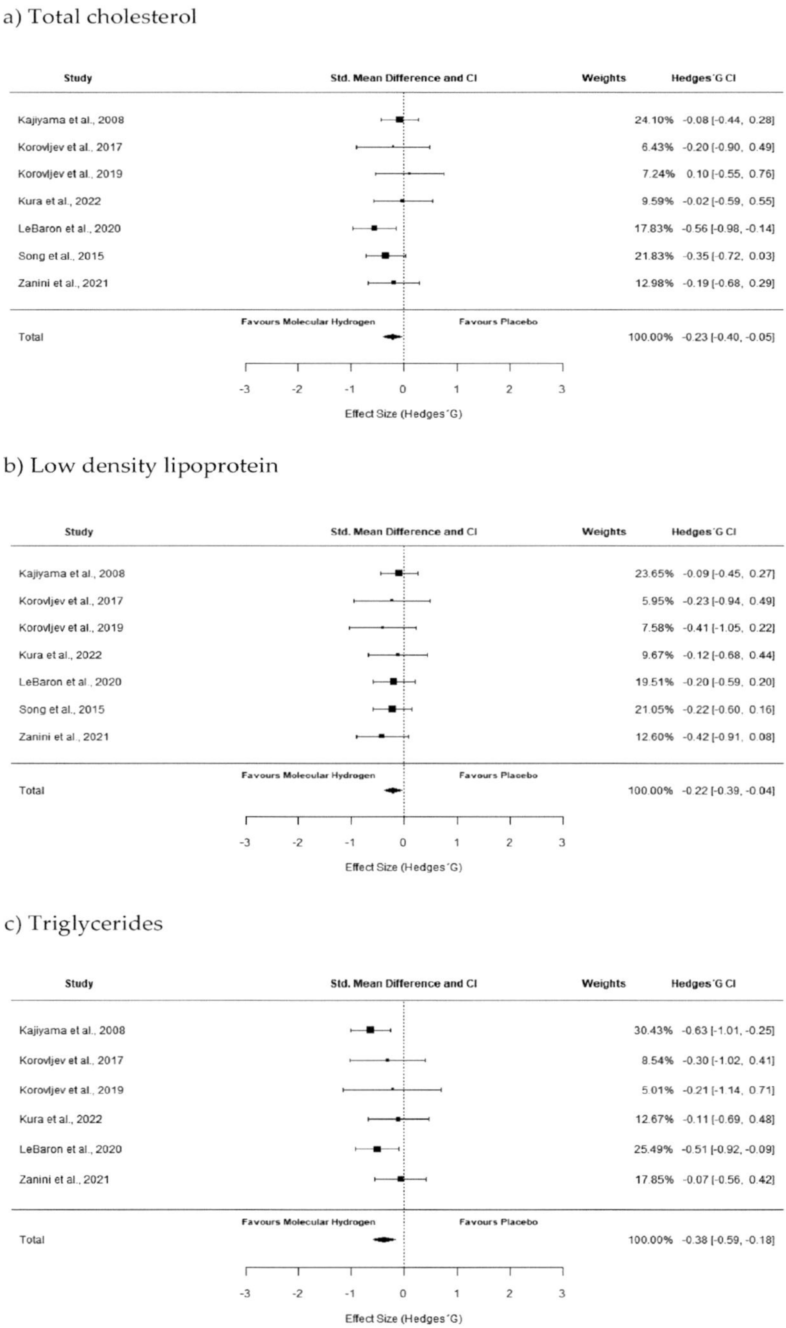

Fig. 18.4 Forest plots of included studies comparing the effects of HRW supplementation on blood lipids (Adapted from Ref. [47])

Effects of Hydrogen Therapy on Blood Pressures

Previous studies showed that hydrogen supplementation m a decrease blood pressures [62, 63]. Nakayama et al. developed a hydrogen enriched dialysate by reverse osmosis via the electrolysis of purified tap water. He and found that compared to hemodialysis with standard dialysate, dialysis with hydrogen-enriched dialysate reduces blood pressure, in particular systolic blood pressure, in patients on chronic maintenance hemodialysis [62]. In an experimental study, 5/6 nephrectomized rats given inhalation of either hydrogen (1.3% H_2 + 21% O_2 + 77.7% N_2) or control (21% O_2 + 79% N_2) gas mixture for 1 h per day [63]. Addition of hydrogen significantly suppressed the rise in blood pressure after 5/6 nephrectomy. Interestingly, blood pressure lowering effect of hydrogen was also confirmed in rats in a stable hypertensive state 3 weeks after nephrectomy. Implantation of telemetry system for continuous blood pressure monitoring revealed, that an anti-hypertensive effect was observed during daytime rest, as well as during night-time activities. It seems that hydrogen therapy caused improvement in blood pressure variability and autonomic imbalance, by suppressing the overly active sympathetic nervous system and augmenting parasympathetic nervous system activity [63]. It is possible that 1-h daily exposure to hydrogen supplementation exerts an anti-hypertensive effect in an animal model of hypertension. This experiment provide further support that decline in blood pressure variability via DASH diet, may be because increased production of hydrogen in the gut, due to high content of fiber, flavonoids and omega-3 fatty acids in the diet [61]. In a randomized, controlled trial [25], supplementation with HRW was associated with significant decline in systolic and diastolic blood pressures, as well as heart rate (Personal communication, by Singh et al.). This study also found that hydrogen therapy was associated with significant decline in angiotensin converting enzyme (ACE) and nitrite, a precursor of nitric oxide (NO) [25]. Decline in ACE and increase in NO can cause decrease in blood pressures. Larger randomized, controlled trials would be necessary to confirm the blood pressure controlling role of hydrogen therapy in hypertension.

In a controlled trial, of 60 patients randomized to treatment, 56 patients completed the study (27 in the Air group and 29 in the H_2-O_2 group). The right and left arm systolic blood pressure (SBP) were significantly declined in H_2-O_2 group compared with the baseline levels (151.9 ± 12.7 mmHg to 147.1 ± 12.0 mmHg, and 150.7 ± 13.3 mmHg to 145.7 ± 13.0 mmHg, respectively; all $p < 0.05$). Interestingly, treatment with hydrogen, significantly reduced diastolic nighttime ambulatory blood pressure by 2.7 ± 6.5 mmHg ($p < 0.05$). In the placebo group, all blood pressures showed no affect (all $p > 0.05$). In addition, the angiotensin II, aldosterone, and cortisol levels as well as the aldosterone-to-renin ratio in plasma were significantly lower in H_2-O_2 group compared with baseline ($p < 0.05$). No significant differences were observed in the Air group before and after the intervention. It is clear, that inhalation of a low-dose H_2-O_2 mixture exerts a favorable effect on blood pressure and reduces the plasma levels of hormones associated with hypertension on renin–angiotensin–aldosterone system and stress in midlife/older adults with hypertension.

Effects of Hydrogen Therapy on Endothelial Function

NO is a signaling molecule involved in many physiological and pathological processes and a vasodilator in the vascular system. The endothelium is the major source of NO, under healthy conditions, Release of NO from the endothelium may be examined by finding out the flow-mediated dilation (FMD) that occur due to hyperemia caused by atherosclerotic occlusion of brachial artery [13]. It seems that the effects of cardiovascular risk factors; tobacco, aging, obesity, hyperlipidemia, diabetes and hypertension, as well as hyperglycemia after meals on endothelial dysfunction may be examined by this method. These risk factors may increase the oxidative stress resulting into endothelial dysfunction due to redox imbalance. The role of ROS is crucial in the development of vascular disease, but physiological generation of ROS is necessary for vasomotor function in most of the vessels [13]. Thus the imbalance in the redox state between NO and superoxide radical produced in the endothelial cells may be crucial in causing endothelial dysfunction. For example, another ROS; H_2O_2 may cause either detrimental or beneficial influence on vascular function, depending on presence endogenous antioxidants in the arterial cells. The role of hydroxyl radical, produced as byproduct of decay of hydrogen peroxide is unclear, although it is known that it causes impairment in the endothelium, which can be neutralized by molecular hydrogen [13]. In a clinical study, administration of hydrogen was associated with significant rise in FMD, from (mean $\pm$ SD) 6.80% $\pm$ 1.96% to 7.64% $\pm$ 1.68%, whereas in the control group, there was a decline from 8.07% $\pm$ 2.41% to 6.87% $\pm$ 2.94% [13]. There was a significant beneficial effect in the ratio of FMD in the high-H_2 group than to the control group [13]. It is possible that treatment with hydrogen may protect the vasculature from shear stress-derived detrimental hydroxyl radical by maintaining the nitric oxide-mediated vasomotor response.

Effects of Molecular Hydrogen in Stroke

In experimental studies, molecular hydrogen therapy was found to act as an antioxidant, and inhalation of hydrogen gas (1–4%) was markedly effective for the improvement of cerebral infarction [64–67]. Inhalation hydrogen during normoxic resuscitation was also found to improve neurological outcomes in a cardiac arrest. In a controlled trial, 50 patients were randomized to (25 each in the hydrogen group and the control group) a therapeutic time window of 6–24 h [67]. The hydrogen group inhaled 3% hydrogen gas (1 h twice a day), and the control group received conventional intravenous medications for the initial 7 days. Treatment with hydrogen was associated with no significant adverse effects but with improvements in oxygen saturation. The relative signal intensity of MRI of brain was significant, indicating the severity of the infarction site, National Institute of Health (NIH) Stroke Scale scores for clinically quantifying stroke severity, and physical therapy evaluation, as judged by the Barthel Index were also significant in the hydrogen group [67]. It is clear that hydrogen treatment was safe and beneficial in patients with cerebral infarction. In another controlled trial, the safety and effectiveness of H2 hydrogen therapy was examined in patients with cerebral infarction in an acute stage with mild-to moderate-severity (NIH) stroke scale scores (NIHSS = 2–6) [68]. No significant adverse effects were noted in the hydrogen group, with improvements in oxygen saturation. The following significant effects were found: the relative signal intensity of MRI, which indicated the severity of the infarction site, NIHSS scores for clinically quantifying stroke severity, and physical therapy evaluation, as judged by the Barthel Index were significant changes in the hydrogen group compared to control group. It is possible that hydrogen acts by reducing cerebral artery vasospasm and direct beneficial effects on neurons, by improving mitochondrial function and over all vascular function, due to its anti-inflammatory, and improved signal transduction and gene function effects [58, 69–71].

In a case study among 200 patients, those treated with H_2 gas had significantly greater improvement in NIH stroke scale score than patients treated with standard therapy [69]. The neurological improvement was significantly greater in patients treated with hydrogen gas than those that received standard therapy for all the days tested ($p < 0.05$). The onset of NIH stroke scale score reduction was faster in patients treated with hydrogen gas when compared to standard therapy. Hydrogen therapy also showed significant improvement in MRI signal Intensity score, which indicates that hydrogen was effective in recovery of infraction site [69]. The findings of this study show that hydrogen gas administered via the inhalation route is effective and safe in Chinese patients with acute cerebral infarction, and is therefore, superior to standard therapy.

Effects of Molecular Hydrogen on Ischemia and Reperfusion Injury

There is evidence that oxidative and pro-inflammatory stress are underlying causative factors in myocardial ischemia/reperfusion (I/R) injury, in which hydrogen therapy may be useful [43, 44, 59]. It seems that increased amounts of ATP are required by the cardiac myocytes, for their physiological function, which needs high density of mitochondria are to allow their high energy needs. These mitochondria, filled with reactive intermediates and pro-apoptotic signals, are intimately involved in I/R injury [65]. The inner mitochondrial membrane is responsible for maintaining mitochondrial transmembrane potential, and is usually impermeable to ions and proteins [71]. However, under stress, the opening of the mitochondrial permeability transition pore (mPTP) creates a non-selective channel between the inner membrane of the mitochondrion and the sarcoplasm. This results in to a loss of the electrochemical gradient, production of ROS, Ca^{2+} overload, and formation of apoptosomes responsible for apoptosis [60]. It is well known that that there is free radical generation through partial reduction of oxygen during I/R injury, that can rapidly overwhelm the cell's endogenous antioxidant self-defense system [59, 60, 72, 73]. These reactions may cause cellular injury by damaging lipids, proteins, DNA, and RNA. The xanthine oxidase substrates, xanthine and hypoxanthine, accumulate during [72]. Ischemia, which triggers xanthine oxidase activation and consequently increased production of ROS. These ROS can also elicit the opening of the mPTP resulting in a positive feedback loop of increased ROS production from the mitochondria [60, 73].

It seems that restoration of blood flow in an acutely obstructed coronary artery, represents the most effective, long-term clinical therapy for AMI, which is also responsible for I/R injury in the myocardial cells. The myocardial remodeling and fibrosis, as component of healing are the most important targets of hydrogen therapy that can influence the outcomes in patients with myocardial infarction [24]. In presence of metabolic diseases and CVDs, the amount of fibrosis and apoptosis are crucial for the development of heart failure. The novel therapeutic approaches for the management of cardiac remodeling and fibrosis of myocardium are important for increased survival of patients with CVDs. In a rat model of myocardial infarction, the treatment included inhalation of 2% hydrogen daily for 3 h, for a duration of 28 days [24]. The results revealed that treatment with hydrogen may be associated with improved function of the heart with decrease in the area of fibrosis. It seems that treatment with hydrogen provides special beneficial effect, in the I/R injury of the cardiomyocyte, via reducing pyroptosis mediated by NLRP3, that occur after restoration of blood flow in the coronary artery (Table 18.3) [24].

Many cellular and molecular processes contribute to ventricular remodeling in response to myocardial irradiation, myocardial I/R injury, myocardial infarction, hypertension, neuro-humoral activation or other pathophysiological stimuli [74]. Increased production of endothelin-1 (ET-1), angiotensin II, catecholamines and pro-inflammatory cytokines, bind to their cognate receptors and activate the downstream signaling events, results in the remodeling. These pathophysiological alterations may cause either necrosis, apoptosis, autophagy, or hypertrophy of the cardiomyocytes. There may be activation of fibroblast to produce collagen and other proteins that cause fibrosis [74, 75]. Although restoration of blood flow in the coronary artery, is critical, the perfusion of oxygen-rich blood induces cytotoxic ROS production. This eruption of ROS leads to cellular necrosis and apoptosis, a process may be called lethal reperfusion injury [76]. Such injury accounts for up to 50% of the final size of the infarct [77]. In addition, the cascade of events occurring within cardiomyocytes, the endothelium is also actively involved in the I/R injury.

NO elicits vasodilation, which provides protective effects during I/R, in part by influencing oxygen consumption, platelet aggregation, leukocyte adhesion, and free radical metabolism [78]. High concentrations of NO exacerbate I/R injury largely by producing highly reactive peroxynitrite. In addition to NO, the coronary endothelium has several other pathophysiological roles in I/R, such as serving as a source of non-NO vasoactive substances, activating the immune system, and an increased expression of cytokines, chemokines, and various adhesion molecules [60]. It is possible that increased availability of hydrogen due to inhalation of HRW, may blunt

Table 18.3 Mechanisms underlying pathophysiology of myocardial ischemia/reperfusion injury (Modified from Ref. [60])

Alterations due I/R injury	Mechanisms
Ion flux	Accumulation of intracellular calcium, a2 + -induced "stone-heart", increased Na influx, Abnormal K flux. Drop in pH followed by normalization upon reperfusion
Loss of myocardial membrane potential	Opening of myocardial permeability transition pore (mPTP)
Reactive Oxygen Species (ROS)	Substrate-level induction of xanthine oxidase resulting in more ROS. Impaired mitochondrial function, worsen with Q10 deficiency, Neutrophil infiltration, ROS induced ROS chain
Dysregulation of Nitric Oxide (NO) metaboilism	Loss of NO vasodilation, Production of Peroxynitrite, Abnormal S-nitrosation,
Apoptosis	JNK Pathway, ceramide generation, cytoplasm acidification, caspase activation
Autophagic cell death	Excessive AMPK activation, Excess of induction of HIF-1α
Endothelial dysfunction	Cytokine, myokine, chemokine signaling. Expression of cellular adhesion markers, Impaired vasodilation
Platelet aggregation Auto-immune activation	Innate immunity; complement activation, induction of TLR Neutrophil accumulation

JNK c-Jun N-terminal kinase, *AMPK* AMP-activated protein kinase, *HIF-1α* Hypoxia-inducible factor 1-alpha

the adverse effects of NO, resulting in to overall beneficial effects. In an experimental study, treatment with hydrogen improved the size of myocardial infarction, cardiac function, apoptosis and cytokine release following myocardial ischemia (MI)/ Reperfusion® in rats [79]. Hydrogen also improved cell viability and LDH release following hypoxia/reoxygenation in myocardial cells in- vitro. Apart from these effects, hydrogen exerted an anti-inflammatory and anti-apoptotic effect in myocardial cells induced by H/R via PINK1/Parkin mediated autophagy. It is clear that hydrogen-rich saline alleviated the inflammation response and apoptosis induced by MI/R or H/R in vivo or in vitro, and that hydrogen-rich saline contributed to the increased expression of proteins associated with autophagy. The present study indicated that treatment with hydrogen-rich saline improved the inflammatory response and apoptosis in MI/R. In a clinical study among 20 patients with ST segment elevation myocardial infarction, hydrogen inhalation during angioplasty was associated with good feasibility and safety, which may have also promoted left ventricular reverse remodeling at 6 months after infarction [80].

Effects of Hydrogen Therapy in Neurodegenerative Diseases

As a recognized reducing gas, hydrogen has shown great anti-oxidative stress and anti-inflammatory effect in many cerebral disease models [81–85]. It can ameliorate neuronal damage, maintain the number of neurons, prolong the lifespan of neurons, and ultimately inhibit disease progression. Hydrogen exhibits the potential to treat Alzheimer's disease [81]. Stereotactic injection has been previously used as an invasive method of administering active hydrogen, but this method has limitations in clinical practice. In this study, triple transgenic (3 × Tg) Alzheimer's disease mice were treated with HRW for 7 months. The results showed that treatment with HRW prevented synaptic loss and neuronal death, inhibited senile plaques, and reduced hyperphosphorylated tau and neurofibrillary tangles in 3 × Tg Alzheimer's disease mice. In addition, hydrogen-rich water improved Interestingly, disorders of brain energy metabolism disorders, intestinal flora imbalances and reduction in inflammatory reactions were also observed on treatment with HRW. The results indicated that HRW therapy is an effective hydrogen donor that can treat Alzheimer's disease [81, 82], anxiety like behavior [83], Parkinson's disease [84] and ageing [85].

Effects of Molecular Hydrogen on Bone and Joint Diseases

There is evidence that compared to common antioxidants vitamin C and vitamin E, molecular hydrogen has two unique characteristics; hydrogen selectively reduces cytotoxic ROS such as the hydroxyl radical and does not react with other ROS that possesses physiological roles. Secondly, since molecular hydrogen is small and electrically neutral, it easily penetrates membranes and enters cells and organelles including nucleus and mitochondria, where normal antioxidant cannot reach [22]. In an experimental study in ovariectomized rats, treatment with HRW was associated with reduction of bone mass including bone mineral content and bone mineral density in femur and vertebrae, and preserved mechanical strength including ultimate load, stiffness, and energy, and bone structure. Interestingly, trabecular bone volume fraction, trabecular number, and trabecular thickness in femur as well as mechanical strength including ultimate load and stiffness, and bone structure including trabecular bone volume fraction and trabecular number in vertebrae also showed benefit on hydrogen administration. In addition, treatment with HW abated oxidative stress and suppressed IL-6 and TNF-α mRNA expressions in femur of ovariectomized rats; treatment with HW increased femur endothelial NOS activity and enhanced circulating NO level in ovariectomized rats, indicating that intake of hydrogen can prevent osteopenia by decreasing oxidative stress and inflammation. The results of this study were also confirmed in another study among rats having bone loss induced by microgravity in which, treatment with hydrogen molecule abated oxidative stress and alleviated bone loss [86].

Effects of Hydrogen on Cancer

Hydrogen therapy has been found to have positive effects in terms of quality of life in patients with cancer [87–93]. It was examined and suggested to use hydrogen therapy in cancer in 1975 [90]. Treatment with hydrogen may improve liver function in patients who were administered chemotherapy, as well as reduce side effects for those receiving radiation therapy, and has protective effects against radiation-induced bone marrow damage in cancer patients [87–89].

Hydrogen has an anti-proliferative, anti-oxidative, pro-apoptotic and anti-tumor effects in carcinogenesis [88]. In a meta-analysis, out of the 677 studies, 27 fulfilled the eligibility criteria, where data was compiled into a table, outlining the general characteristics and findings. Analysis of data found that hydrogen plays a promising therapeutic role as an independent therapy as well as an adjuvant in combination therapy, resulting in an overall improvement in survival, quality of life, blood parameters, and tumor reduction [88].

Previous studies showed that hydrogen display anti-cancer properties when administered on its own. In mice with squamous cell carcinoma, hyperbaric hydrogen therapy has been examined as a potential cancer therapy, revealing potent anti-tumor effects [90]. In another experimental study, study in mice with colon cancer, treatment with HRW was found to cause benefit dose-dependently potentiating the tumor-inhibitory activity of 5-furourasil (5-FU) by enhancing cellular apoptosis of the cancer cells [91]. In a recent experimental study including 24 mice bearing tumors were randomly divided into four groups ($n = 6$ per group) [92]. Combined treatment with HRW and 5-FU or HRW alone, significantly improved tumor size and weight, collagen content and fibrosis as compared to the colorectal cancer control group. Interestingly, HRW alone also attenuated oxidative stress and potentiated antioxidant activity, whereas treatment with 5-FU exacerbated oxidative stress and blunted antioxidant effects due to tissue damage. These findings from experimental study indicate that treatment with HRW, with or without 5-FU, may serve as a therapeutic agent for treating colorectal cancer. Hydrogen enhances proliferation of four out of seven human cancer cell lines indicating response to therapy [93]. The effects showing proliferation-promotion may not correlate with basal levels of cellular ROS. The expression profile of the seven cells showed that the responders have higher gene expression of mitochondrial electron transport chain (ETC) molecules than the non-responders.

Effects of Hydrogen Therapy in Kidney Diseases

Common types of kidney disease include acute kidney injury (AKI), drug toxicity, renal fibrosis, polycystic kidney disease, ischemia reperfusion injury and renal cell carcinoma [51, 94–100]. In most of the kidney diseases also, oxidative stress, inflammatory mediators and inflammatory effector cells and the imbalance between pro-inflammatory cytokines and anti-inflammatory cytokines play an important role in the occurrence and progression of kidney diseases [94]. It seems that the role of hydrogen in regulating and maintaining homeostasis in inflammatory kidney disease and the treatment of inflammatory kidney diseases from the perspective of pro-inflammatory cytokines and anti-inflammatory cytokines may be an interesting route for further research [51, 94–101]. Nephropathy is a serious complication of obesity and diabetes mellitus. Metabolic disorders can occur and manifest as local inflammation of the kidney that can lead to fibrosis and structural remodeling of the organ [101]. Therefore, tackling the immune-mediated inflammation is very significant for the treatment of metabolic nephropathy [102]. There is an activation of immune cells, in AKIs, such as renal artery infarction or toxin-mediated kidney injury [96–98].

It seems that renal epithelial cells damage, activate the stress response pathways, leading to the secretion of cytokines and vasoactive factors, resulting in immune-pathological damage [102]. Hydrogen, on the other hand, can suppress the production of immune-reactive substances [94]. Treatment with HRW has been found to inhibit cyclosporine A-induced nephrotoxicity via the Keap1/Nrf2 signaling pathway [51], kidney fibrosis due to AKI by retaining Klotho expression [96], after ischemia/reperfusion injury in rats via anti-oxidative stress, anti-apoptosis and anti-inflammation [97]. Hydrogen also acts by increased expression of heme-oxygenase-1 in aged rats [98, 99], in mice with calcium oxalate-induced renal injury [100] and immune cell damage in diabetic nephropathy [101] as well as in patients on peritoneal dialysis with peritoneal damage in chronic kidney disease [102]. It seems that the endoplasmic reticulum (ER) stress is characterized with pathological stress inducing an accumulation of unfolded proteins in the ER [94]. It has been observed that the inhalation of hydrogen significantly decreased the ER stress-related protein levels with decline in the tissue damage in myocardial IRI [102]. Similar mechanisms may be involved showing that hydrogen is able to ameliorate chronic intermittent hypoxia (CIH)-induced kidney injury by decreasing ER stress and activating autophagy.

Effects of Hydrogen in Chronic Lung Diseases

Chronic lung diseases affect the airways; alveoli, bronchioles and bronchi as well as other structures of the lungs. Some of the most common lung diseases are; chronic obstructive pulmonary diseases (COPD), bronchial asthma, occupational lung diseases, tuberculosis, cancer, and pulmonary hypertension [1]. All the chronic lung diseases such as bronchial asthma and COPD, as well as post COVID-19 pulmonary fibrosis, may be associated with lung damage along with fibrosis and emphysema [103]. Major pathogenic factors that contribute to developing lung disease include infection and inflammation, protease and antiprotease imbalance, and oxidative stress, which overwhelms the antioxidant defenses. Moreover, ROS play a pivotal role in the incidence of acute exacerbations of diseases [103]. The major feature of COPD and other united airway diseases is characterized with an abnormal response to infection or allergen–induced injury, causing oxidative stress and chronic inflammation, and subsequent activation of macrophages, eosinophils, neutrophils, T lymphocytes, and fibroblasts in the lung [103].

Recently, molecular hydrogen has been suggested for the treatment of COVID-19 [104, 105], as well as COPD [106, 107] due to its potential antioxidant and anti-inflammatory effects. In a randomized, double-blinded, placebo-controlled trial involving 10 centers, patients (n = 54 in each group) with acute exacerbation of COPD (AECOPD) and a Breathlessness, Cough, and Sputum Scale (BCSS) score of at least 6 points were randomly assigned to receive a hydrogen/oxygen therapy or oxygen only [107]. Improvements in the BCSS score in the hydrogen/oxygen group was significantly greater compared to the oxygen only group (− 5.3 vs. − 2.4 point; difference: − 2.75 [95% CI − 3.27 to − 2.22], indicating superiority. Other time points from day 2 through day 6, had similar findings. Cough Assessment Test score also showed significant decline in the hydrogen/oxygen group compared to control (− 11.00 vs. − 6.00, $p < 0.001$). Changes in pulmonary function, arterial blood gas and noninvasive oxygen saturation did not differ significantly between groups as well as other endpoints. However, similar to our findings, analysis of the per-protocol set did find significant group interactions with respect to the changes from baseline in SpO_2 ($P < 0.0001$). Acute exacerbations were reported in 34 (63.0%) patients in hydrogen/oxygen group and 42 (77.8%) in oxygen group. This trial demonstrated that hydrogen/oxygen therapy is superior to oxygen therapy in patients with acute exacerbation of COPD with acceptable safety and tolerability profile [107]. Inhalation of hydrogen has been found to protect against acute lung injury induced by hemorrhagic shock and resuscitation [108], as well as in adjuvant therapy in COVID-19 [109, 110] and chronic heart failure [27]. In a clinical observation among 10 patients with chronic lung diseases, HRW administration was associated with a significant increase in oxygen saturation (SpO_2) and decrease in TBARS, MDA, and diene conjugates, with an increase in vitamin E and nitrite levels, compared to baseline levels [111]. Physical training done after HRW therapy appeared to increase exercise tolerance and decrease hypoxia, as well as delay the need for oxygen therapy. Treatment with HRW in patients with hypoxia from chronic lung diseases may decrease oxidative stress and improve oxygen saturation in some patients. Mechanisms of benefit via hydrogen therapy are given in Fig. 18.5.

Apart from above diseases, molecular hydrogen may be protective against peripheral arterial disease [112, 113], gene expression [112, 113], acute cerebral ischemia [114], sleep deprivation [115], traumatic brain injury [116], aneurismal sub-arachnoid hemorrhage [117], radiation induced bone-marrow damage [118], non-small-cell lung cancer [119], restore exhausted $CD8^+$ T cells in lung cancer

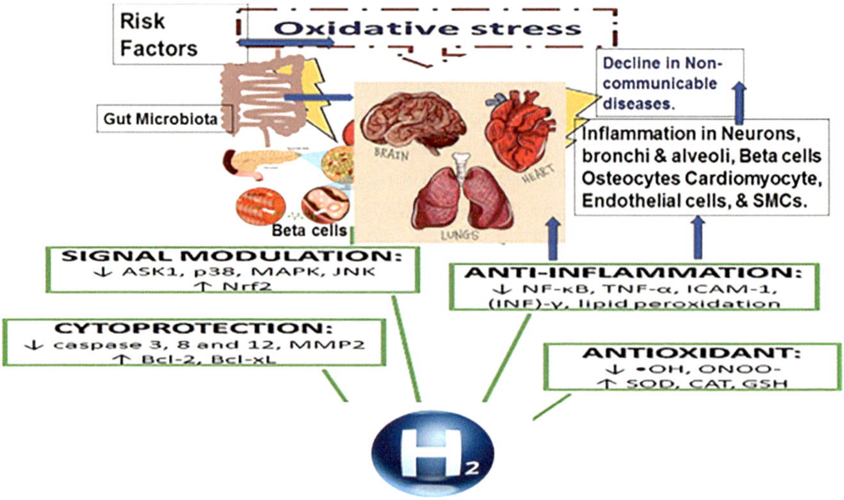

Fig. 18.5 Mechanisms of the effects of hydrogen therapy on pro-inflammatory biomarkers and non-communicable diseases (Modified from Ref. [17])

[120], cognitive function in women [121], alcohol induced hangover [122], idiopathic sudden sensorineural hearing loss[123], post-cardiac arrest syndrome [124], gut microbiota [125], cardiopulmonary bypass surgery [126], stimulation of cardiac autonomic activity [127] and dry eye syndrome [128]. Vascular aging of the aorta [129], radiation therapy for cancer [130].

Conclusions

All the risk factors of NCDs, in particular diet, act by damaging the concerned organ which is mainly involved in producing a disease by causing oxidative stress and inflammations in the cells and tissues. In obesity, adipocytes are the main target but systemic inflammation also occur in the beta cells of pancreas, endothelial cells, smooth muscle cells and hepatocytes resulting in to metabolic syndrome and T2DM. Inflammation in the cardiomyocytes, may predispose cardiac hypertrophy resulting in to heart failure, whereas inflammation in the vasculature, including endothelium and smooth muscle cells, may predispose atherosclerosis and hypertension respectively. Similarly, cellular damage in neurons, osteocytes and osteoblast, bronchial tree, glomerular cells in kidney, may be responsible for development of related diseases. Since oxidative stress enhances systemic inflammation, it may damage the function of cardiomyocyte, beta cells, kidney cells and bone marrow, as well as neurons, apart from endothelium, predisposing development of concerned diseases. Treatment with hydrogen may inhibit the free radical generation with decline in inflammation, resulting in to restoration of cell function, causing improvement in the related organ and concerned NCDs.

Competing Interests

We thank International College of Nutrition for providing logistic support to write this article. AT is involved in the commercial industry pertaining to molecular hydrogen for health, RBS has accepted travel grant to present this work in a conference. All other authors declare no conflict of interest.

References

1. Roth GA, Mensah GA, Johnson CO, Addolorato G, Ammirati E, Baddour LM et al (2020) Global burden of cardiovascular diseases and risk factors, 1990–2019: update from the GBD 2019 study. J Am Coll Cardiol 76:2982–3021. https://doi.org/10.1016/j.jacc.2020.11.010
2. Roth GA, Abate D, Abate KH, Abay SM, Abbafati C, Abbasi N et al (2018) Global, regional, and national age-sex-specific mortality for 282 causes of death in 195 countries and territories, 1980–2017: a systematic analysis for the Global Burden of Disease Study 2017. Lancet 392:1736–1788. https://doi.org/10.1016/S0140-6736(18)32203-7
3. Vaduganathan M, Mensah GA, Varieur Turco J et al (2022) The global burden of cardiovascular diseases and risk: a compass for future health. J Am College Cardiol. 12 Dec 2022. https://doi.org/10.1016/j.jacc.2022.11.005
4. Jiang S, Liu H, Li C (2021) Dietary regulation of oxidative stress in chronic metabolic diseases. Foods 10(8):1854. https://doi.org/10.3390/foods10081854
5. Aleksandrova K, Koelman L, Rodrigues CE (2021) Dietary patterns and biomarkers of oxidative stress and inflammation: a systematic review of observational and intervention studies. Redox Biol 42:101869
6. Micha R, Penalyo JL, Cidhea F, Imamura F, Rehm CD, Mozaffarian D (2017) Association between dietary factors and mortality from heart disease, stroke, and type 2 diabetes in the United States. JAMA 317(9):912–924. https://doi.org/10.1001/jama.2017.0947
7. Sacco RL, Roth GA, Reddy KS, Arnett DK, Bonita R, Gaziano TA, Heidenreich PA, Huffman MD, Mayosi BM, Mendis S, Murray CJL, Perel P, Piñeiro DJ, Smith SC Jr, Taubert KA, Wood DA, Zhao D, Zoghbi WA (2016) The heart of 25 by 25: achieving the goal of reducing global and regional premature deaths from cardiovascular diseases and stroke: a modeling study from the American Heart Association and World Heart Federation. Circulation 133:e674–e690. https://doi.org/10.1161/CIR.0000000000000395
8. Singh RB, Fedacko J, Fatima G, Magomedova A, Watanabe S, Elkilany G (2022) Why and how the indo-mediterranean diet may be superior to other diets: the role of antioxidants in the diet. Nutrients 14:898. https://doi.org/10.3390/nu14040898
9. Singh RB, Fedacko J, Pella D, Fatima G, Elkilany G, Moshiri M, Hristova K, Jakabcin P, Vanova N (2022) High exogenous antioxidant, restorative treatment (heart) for prevention of the six stages of heart failure: the heart diet. Antioxidants 11:1464. https://doi.org/10.3390/antiox11081464
10. Mojto V, Singh RB, Gvozdjakova A, Pella D, Fedacko J, Pella D (2018) Molecular hydrogen: a new approach for the management of cardiovascular diseases. World Heart J 10:83–93

11. Ichikawa Y, Yamamoto H, Hirano Si, Sato B, Takefuji Y, Satoh F (2022) The overlooked benefits of hydrogen-producing bacteria. Med Gas Res 12 [cited 2022 Dec 20]. Available from: https://www.medgasres.com/preprintarticle.asp?id=344977
12. Slezák J, Kura B, Frimmel K, Zálešák M, Ravingerová T, Viczenczová C, Okruhlicová Ľ, Tribulová N (2016) Preventive and therapeutic application of molecular hydrogen in situations with excessive production of free radicals. Physiol Res 65(Suppl. 1):S11–S28
13. Sakai T, Sato B, Hara K, Hara Y, Naritomi Y, Koyanagi S, Hara H, Nagao T, Ishibashi T (2014) Consumption of water containing over 3.5 mg of dissolved hydrogen could improve vascular endothelial function. Vasc Health Risk Manage 10:591–597
14. Ichihara M, Sobue S, Ito M, Ito M, Hirayama M, Ohno K (2015) Beneficial biological effects and the underlying mechanisms of molecular hydrogen- Comprehensive review of 321 original articles. Med Gas Res 5:1–21
15. Lobo V, Patil A, Phatak A, Chandra N (2010) Free radicals, antioxidants and functional foods: Impact on human health. Pharmacogn Rev 4(8):118–126. https://doi.org/10.4103/0973-7847.70902
16. Takac I, Schroder K, Brandes RP (2012) The Nox family of NADPH oxidases: friend or foe of the vascular system. Curr Hypertens Rep 14:70–78
17. Montezano AC, Touyz RM (2012) Reactive oxygen species and endothelial function role of nitric oxide synthase uncoupling and Nox family nicotinamide adenine dinucleotide phosphate oxidases. Basic Clin Pharmacol Toxicol 110:87–94
18. Al Ghouleh I, Khoo NK, Knaus UG (2011) Oxidases and peroxidases in cardiovascular and lung disease: new concepts in reactive oxygen species signaling. Free Radic Biol Med 51:1271–1288
19. Bedard K, Krause KH (2007) The NOX family of ROS-generating NADPH oxidases: physiology and pathophysiology. Physiol Rev 87:245–313
20. Deryugina AV, Danilova DA, Brichkin YD, Taranov EV, Nazarov EI, Pichugin VV, Medvedev AP, Riazanov MV, Fedorov SA, Andrej YS, Makarov EV (2023) Molecular hydrogen exposure improves functional state of red blood cells in the early postoperative period: a randomized clinical study. Med Gas Res 13(2):59–66. https://doi.org/10.4103/2045-9912.356473
21. Yang M, Dong Y, He Q et al (2020) Hydrogen: a novel option in human disease treatment. Oxid Med Cell Longev 2020:8384742. https://doi.org/10.1155/2020/8384742
22. Ohsawa I, Ishikawa M, Takahashi K et al (2007) Hydrogen acts as a therapeutic antioxidant by selectively reducing cytotoxic oxygen radicals. Nat Med 13(6):688–694. https://doi.org/10.1038/nm1577
23. Yu Y, Feng J, Lian N et al (2020) Hydrogen gas alleviates blood-brain barrier impairment and cognitive dysfunction of septic mice in an Nrf2-dependent pathway. Int Immunopharmacol 85:106585
24. Nie C, Zou R, Pan S, A R, Gao Y, Yang H, Bai J, Xi S, Wang X, Hong X, Yang W (2021) Hydrogen gas inhalation ameliorates cardiac remodelling and fibrosis by regulating NLRP3 inflammasome in myocardial infarction rats. J Cell Mol Med 25(18):8997–9010. https://doi.org/10.1111/jcmm.16863
25. LeBaron T, Singh RB, Fatima G, Kartikey K, Sharma JP, Ostojic SM, Gvozdjakova A, Kura B, Noda M, Mojto V, Niaz MA, Slezak J (2020) The effects of 24-week, high-concentration hydrogen-rich water on body composition, blood lipid profiles and inflammation biomarkers in men and women with metabolic syndrome: a randomized controlled trial. Diab Metab Syndr Obes Targets Ther 13:889–896
26. Singh RB, Halabi G, Fatima G, Rai RH, Tarnava AT, LeBaron TW. Molecular hydrogen as an adjuvant therapy may be associated with increased oxygen saturation and improved exercise tolerance in a COVID-19 patient. Clin Case Rep 9:e05039. https://doi.org/10.1002/ccr3.5039
27. Akita Y, Higashiyama M, Kurihara C, Ito S, Nishii S, Mizoguchi A, Inaba K, Tanemoto R, Sugihara N, Hanawa Y, Wada A, Horiuchi K, Okada Y, Narimatsu K, Komoto S, Tomita K, Takei F, Satoh Y, Saruta M, Hokari R (2022) Ameliorating role of hydrogen-rich water against NSAID-induced enteropathy via reduction of ROS and production of short-chain fatty acids. Dig Dis Sci 7:1–11. https://doi.org/10.1007/s10620-022-07781-5

28. Koyama Y, Harada S, Sato T, Kobayashi Y, Yanagawa H, Iwahashi T, Tanaka H, Ohata K, Imai T, Ohta Y, Kamakura T, Kobayashi H, Inohara H, Shimada S (2022) Therapeutic strategy for facial paralysis based on the combined application of Si-based agent and methylcobalamin. Biochem Biophys Rep 32:101388. https://doi.org/10.1016/j.bbrep.2022.101388
29. Hong Y, Dong G, Li Q, Wang V, Liu M, Jiang G, Bao D, Zhou J (2022) Effects of pre-exercise H2 inhalation on physical fatigue and related prefrontal cortex activation during and after high-intensity exercise. Front Physiol 13:988028. https://doi.org/10.3389/fphys.2022.988028
30. Eda N, Tsuno S, Nakamura N, Sone R, Akama T, Matsumoto M (2022) Effects of intestinal bacterial hydrogen gas production on muscle recovery following intense exercise in adult men: a pilot study. Nutrients 14:4875. https://doi.org/10.3390/nu14224875
31. Perry RJ, Peng L, Barry NA, Cline GW, Zhang D, Cardone RL, Petersen KF, Kibbey RG, Goodman AL, Shulman GI (2016) Acetate mediates a microbiome–brain–β- cell axis to promote metabolic syndrome. Nature 534:213–217. https://doi.org/10.1038/nature18309
32. De Vadder F, Kovatcheva-Datchary P, Goncalves D, Vinera J, Zitoun C, Duchampt A, Bäckhed F, Mithieux G (2014) Microbiota-generated metabolites promote metabolic benefits via gut-brain neural circuits. Cell 156:84–96. https://doi.org/10.1016/j.cell.2013.12.016
33. Levitt MD (1969) Excretion of hydrogen gas in man. N Engl J Med 281:122–127. https://doi.org/10.1056/NEJM196907172810303
34. Hajishengallis G, Darveau RP, Curtis MA (2012) The keystone-pathogen hypothesis. Nat Rev Microbiol 10(10):717–725. https://doi.org/10.1038/nrmicro2873
35. Lupp C, Robertson ML, Wickham ME, Sekirov I, Champion OL, Gaynor EC, Finlay BB (2007) Host-mediated inflammation disrupts the intestinal microbiota and promotes the overgrowth of Enterobacteriaceae. Cell Host Microbe 2:119–129
36. Donovan SM (2017) Introduction to the special focus issue on the impact of diet on gut microbiota composition and function and future opportunities for nutritional modulation of the gut microbiome to improve human health. Gut Microbes 8(2):75–81
37. Baptista LC, Sun Y, Carter CS, Buford TW (2020) Crosstalk between the gut microbiome and bioactive lipids: therapeutic targets in cognitive frailty. Front Nutr 11(7):17. https://doi.org/10.3389/fnut.2020.00017
38. Christl SU, Murgatroyd PR, Gibson GR, Cummings JH (1992) Production, metabolism, and excretion of hydrogen in the large intestine. Gastroenterology 102(4 Pt 1):1269–1277
39. Suzuki A, Ito M, Hamaguchi T, Mori H, Takeda Y, Baba R, Watanabe T, Kurokawa K, Asakawa S, Hirayama M, Ohno K (2018) Quantification of hydrogen production by intestinal bacteria that are specifically dysregulated in Parkinson's disease. PLoS ONE 13(12):e0208313. https://doi.org/10.1371/journal.pone.0208313
40. Ishibashi T (2019) Therapeutic efficacy of molecular hydrogen: a new mechanistic insight. Curr Pharm Des 25(9):946–955. https://doi.org/10.2174/1381612825666190506123038
41. Dole M, Wilson FR, Fife WP (1975) Hyperbaric hydrogen therapy: a possible treatment for cancer. Science 190:152–154
42. Ohsawa I, Ishikawa M, Takahashi K, Watanabe M, Nishimaki K, Yamagata K, Katsura KI, Katayama Y, Asoh S, Ohta S (2007) Hydrogen acts as a therapeutic antioxidant by selectively reducing cytotoxic oxygen radicals. Nat Med 13:688–694
43. LeBaron TW, Kura B, Kalocayova B, Tribulova N, Slezak J (2019) A new approach for the prevention and treatment of cardiovascular disorders. Molecular hydrogen significantly reduces the effects of oxidative stress. Molecules 24(11):276. https://doi.org/10.3390/molecules24112076
44. LeBaron TW, Sharpe R, Ohno K (2022) Electrolyzed-reduced water: review I. Molecular hydrogen is the exclusive agent responsible for the therapeutic effects. Int J Mol Sci. 23(23):14750. https://doi.org/10.3390/ijms232314750
45. Carr AC, McCall MR, Frei B (2000) Oxidation of LDL by myeloperoxidase and reactive nitrogen species: reaction pathways and antioxidant protection. Arterioscler Thromb Vasc Biol 20:1716–1723
46. Yoshida H, Kisugi R (2010) Mechanisms of LDL oxidation. Clin Chim Acta 411:1875–1882

47. Todorovic N, Fernández, Landa J, Santibañez A, Kura B, Stajer V, Korovljev D, Ostojic SM (2023) The effects of hydrogen-rich water on blood lipid profiles in clinical populations: a systematic review and meta-analysis. Pharmaceuticals 16:142. https://doi.org/10.3390/ph16020142
48. Ge L, Yang M, Yang NN, Yin XX, Song WG (2017) Molecular hydrogen: a preventive and therapeutic medical gas for various diseases. Oncotarget 8(60):102653–102673. https://doi.org/10.18632/oncotarget.21130
49. Nakao A, Toyoda Y, Sharma P, Evans M, Guthrie N (2010) Effectiveness of hydrogen rich water on antioxidant status of subjects with potential metabolic syndrome—an open label pilot study. J Clin Biochem Nutr 46(2):140–149. https://doi.org/10.3164/jcbn.09-100
50. Kajiyama S, Hasegawa G, Asano M, Hosoda H, Fukui M, Nakamura N, Kitawaki J, Imai S, Nakano K, Ohta M, Adachi T, Obayashi H, Yoshikawa T (2008) Supplementation of hydrogen-rich water improves lipid and glucose metabolism in patients with type 2 diabetes or impaired glucose tolerance. Nutr Res 28(3):137–143. https://doi.org/10.1016/j.nutres.2008.01.008
51. Guihua L, Tongsu N (2022) Efficacy and safety of hydrogen inhalation therapy in patients with type 2 diabetes mellitus with poor response to metformin. Adv Clin Med 12(10):9605–9612
52. Zhao Z, Ji H, Zhao Y, Liu Z, Sun R, Li Y and Ni T (2023) Effectiveness and safety of hydrogen inhalation as an adjunct treatment in Chinese type 2 diabetes patients: a retrospective, observational, double-arm, real-life clinical study. Front Endocrinol 13:1114221. https://doi.org/10.3389/fendo.2022.1114221
53. Korovljev D, Trivic T, Drid P, Ostojic SM (2018) Molecular hydrogen affects body composition, metabolic profiles, and mitochondrial function in middle-aged overweight women. Ir J Med Sci 187(1):85–89. https://doi.org/10.1007/s11845-017-1638-4
54. Song G, Li M, Sang H, Zhang L, Li X, Yao S, Yu Y, Zong C, Xue Y, Qin S (2013) Hydrogen-rich water decrease serum low-density lipoprotein cholesterol levels and improves high density lipoprotein function in patients with potential metabolic syndrome. J Lipid Res 54:1884–1893
55. Song G, Lin Q, Zhao H, Liu M, Ye F, Sun Y, Yu Y, Guo S, Jiao P, Wu Y, Ding G, Xiao Q, Qin S (2015) Hydrogen activates ATP-binding cassette transporter A1-dependent efflux ex vivo and improves high-density lipoprotein function in patients with hypercholesterolemia: a double-blinded, randomized, and placebo-controlled trial. J Clin Endocrinol Metab 100(7):2724–2733. https://doi.org/10.1210/jc.2015-1321. Epub 2015 May 15
56. Tao G, Zhang G, Chen W, Yang C, Xue Y, Song G, Qin S (2022) A randomized, placebo-controlled clinical trial of hydrogen/oxygen inhalation for non-alcoholic fatty liver disease. J Cell Mol Med 26(14):4113–4123. https://doi.org/10.1111/jcmm.17456
57. Kura B, Szantova M, LeBaron TW, Mojto V, Barancik M, Szeiffova Bacova B, Kalocayova B, Sykora M, Okruhlicova L, Tribulova N et al (1935) Biological effects of hydrogen water on subjects with NAFLD: a randomized. Placebo-Control Trial Antioxid 2022:11
58. Sobue S, Inoue C, Hori F, Qiao S, Murate T, Ichihara M (2017) Molecular hydrogen modulates gene expression via histone modification and induces the mitochondrial unfolded protein response. Biochem Biophys Res Commun 493(1):318–324. https://doi.org/10.1016/j.bbrc.2017.09.024. Epub 2017 Sep 7
59. Zhang Y, Sun Q, He B, Xiao J, Wang Z, Sun X (2011) Anti-inflammatory effect of hydrogen-rich saline in a rat model of regional myocardial ischemia and reperfusion. Int J Cardiol 148:91–95
60. Turer AT, Hill JA (2010) Pathogenesis of myocardial ischemia-reperfusion injury and rationale for therapy. Am J Cardiol 37:761–771. https://doi.org/10.1016/j.amjcard.2010.03.032
61. Singh RB, Nabavizadeh F, Fedacko J, Pella D, Vanova N, Jakabcin P, Fatima G, Horuichi R, Takahashi T, Mojto V et al (2023) Dietary approaches to stop hypertension via indo-mediterranean foods, may be superior to DASH diet intervention. Nutrients 15:46. https://doi.org/10.3390/nu15010046
62. Nakayama M et al (2018) Novel haemodialysis (HD) treatment employing molecular hydrogen (H2)-enriched dialysis solution improves prognosis of chronic dialysis patients: A prospective observational study. Sci Rep 8:254. https://doi.org/10.1038/s41598-017-18537-x

63. Sugai K, Tamura T, Sano M, Uemura S, Fujisawa M, Katsumata Y, Endo J, Yoshizawa J, Homma K, Suzuki M, Kobayashi E, Sasaki J, Hakamata Y (2020) Daily inhalation of hydrogen gas has a blood pressure-lowering effect in a rat model of hypertension. Sci Rep 10(1):20173. https://doi.org/10.1038/s41598-020-77349-8
64. Chen CH, Manaenko A, Zhan Y, Liu WW, Ostrowki RP, Tang J, Zhang JH (2010) Hydrogen gas reduced acute hyperglycemia-enhanced hemorrhagic transformation in a focal ischemia rat model. Neuroscience 169:402–414
65. Nagatani K, Wada K, Takeuchi S, Kobayashi H, Uozumi Y, Otani N, Fujita M, Tachibana S, Nawashiro H (2012) Effect of hydrogen gas on the survival rate of mice following global cerebral ischemia. Shock (Augusta, Ga) 37:645–652
66. Manaenko A, Lekic T, Ma Q, Ostrowski RP, Zhang JH, Tang J (2011) Hydrogen inhalation is neuroprotective and improves functional outcomes in mice after intracerebral hemorrhage. Acta Neurochir Suppl 111:179–183
67. Ono H, Nishijima Y, Adachi N et al (2012) A basic study on molecular hydrogen (H2) inhalation in acute cerebral ischemia patients for safety check with physiological parameters and measurement of blood H2 level. Med Gas Res 2:21
68. Ono H, Nishijima Y, Ohta S, Sakamoto M, Kinone K, Tohru Horikosi T et al (2017) Hydrogen gas inhalation treatment in acute cerebral infarction: a randomized controlled clinical study on safety and neuroprotection. J Stroke Cerebrovasc Dis 26:2587–2594
69. Yao W, Shen L, Ma Z et al (2022) Efficacy and safety of hydrogen gas versus standard therapy in Chinese patients with cerebral infarction: a pilot study Trop J. Pharm Res 21(3):671
70. Gvozdjáková A, Kucharská J, Kura B, Vančová O, Rausová Z, Sumbalová Z, Uličná O, Slezák J (2020) A new insight into the molecular hydrogen effect on coenzyme Q and mitochondrial function of rats. Can J Physiol Pharmacol 98(1):29–34. https://doi.org/10.1139/cjpp-2019-0281
71. .Kucharská J, Gvozdjáková A, Kura B et al (2018) Effect of molecular hydrogen on coenzyme Q in plasma, myocardial tissue and mitochondria of rats. J Nutr Health Food Eng 8(5):362–364. https://doi.org/10.15406/jnhfe.2018.08.00296
72. Xia Y, Zweier JL (1995) Substrate control of free radical generation from xanthine oxidase in the postischemic heart. J Biol Chem 270:18797–18803. https://doi.org/10.1074/jbc.270.32.18797.[PubMed][CrossRef][GoogleScholar]
73. Zorov DB, Filburn CR, Klotz L-O, Zweier JL, Sollott SJ (2000) Reactive oxygen species (ROS)-induced ROS release: a new phenomenon accompanying induction of the mitochondrial permeability transition in cardiac myocytes. J Exp Med Bull 192:1001–1014. https://doi.org/10.1084/jem.192.7.1001.[PMCfreearticle][PubMed][CrossRef][GoogleScholar]
74. Zhang Y, Tan S, Xu J, Wang T (2018) Hydrogen therapy in cardiovascular and metabolic diseases: from bench to bedside. Cell Physiol Biochem 47:1–10. https://doi.org/10.1159/000489737
75. Ichihara G, Katsumata Y, Moriyama H, Kitakata H, Hirai A, Momoi M, Ko S, Shinya Y, Kinouchi K, Kobayashi E et al (2021) Pharmacokinetics of hydrogen after ingesting a hydrogen-rich solution: a study in pigs. Heliyon 7:e08359
76. Braunwald E, Kloner RA (1985) Myocardial reperfusion: a double-edged sword? J Clin Invest 76:1713–1719
77. Hausenloy DJ, Yellon DM (2007) Preconditioning and postconditioning: united at reperfusion. Pharmacol Ther 116:173–191
78. Ma XL, Weyrich AS, Lefer DJ, Lefer AM (1993) Diminished basal nitric oxide release after myocardial ischemia and reperfusion promotes neutrophil adherence to coronary endothelium. Circ Res 72:403–412
79. Yao L, Chen H, Wu Q, Xie K (2019) Hydrogen-rich saline alleviates inflammation and apoptosis in myocardial I/R injury via PINK-mediated autophagy. Int J Mol Med 44(3):1048–1062
80. Katsumata Y, Fumiya Sano F, Takayuki Abe T et al (2017) The effects of hydrogen gas inhalation on adverse left ventricular remodeling after percutaneous coronary intervention for ST-elevated myocardial infarction—first pilot study in humans. Circ J 81(7):940–947

81. Lin YT, Shi QQ, Zhang L, Yue CP, He ZJ, Li XX et al (2022) Hydrogen-rich water ameliorates neuropathological impairments in a mouse model of Alzheimer's disease through reducing neuroinflammation and modulating intestinal microbiota. Neural Regen Res 17(2):409–417
82. Tan X, Shen F, Dong WL, Yang Y, Chen G (2019) The role of hydrogen in Alzheimer's disease. Med Gas Res 8(4):176–180. https://doi.org/10.4103/2045-9912.248270
83. Wen D, Zhao P, Hui R et al (2017) Hydrogen-rich saline attenuates anxiety-like behaviors in morphine-withdrawn mice. Neuropharmacology 118:199–208
84. Yoshii Y, Inoue T, Uemura Y et al (2017) Complexity of stomach-brain interaction induced by molecular hydrogen in Parkinson's disease model mice. Neurochem Res 42:2658–2665
85. Zanini D, Todorovic N, Korovljev D, Stajer V, Ostojic J, Purac J, Kojic D, Vukasinovic E, Djordjievski S, Sopic M, Guzonjic A, Ninic A, Erceg S, Ostojic SM (2021) The effects of 6-month hydrogen-rich water intake on molecular and phenotypic biomarkers of aging in older adults aged 70 years and over: a randomized controlled pilot trial. Exp Gerontol 155
86. Sun Y, Shuang F, Chen DM, Zhou RB (2013) Treatment of hydrogen molecule abates oxidative stress and alleviates bone loss induced by modeled microgravity in rats. Osteoporos Int 24(3):969–978. https://doi.org/10.1007/s00198-012-2028-4
87. Yang Q, Ji G, Pan R, Zhao Y, Yan P (2017) Protective effect of hydrogen-rich water on liver function of colorectal cancer patients treated with mFOLFOX6 chemotherapy. Mol Clin Oncol 7:891–896
88. Mohd Noor MNZ, Alauddin AS, Wong YH, Looi CY, Wong EH, Madhavan P, Yeong CHA (2023) Systematic review of molecular hydrogen therapy in cancer management. Asian Pacific J Cancer Prev 24(1):37–47. https://doi.org/10.31557/APJCP.2023.24.1.37
89. Hirano SI, Aoki Y, Li XK, Ichimaru N, Takahara S, Takefuji Y (2021) Protective effects of hydrogen gas inhalation on radiation-induced bone marrow damage in cancer patients: a retrospective observational study. Med Gas Res 11:104–109
90. Runtuwene J, Amitani H, Amitani M, Asakawa A, Cheng KC, Inui A (2015) Hydrogen-water enhances 5-fluorouracil-induced inhibition of colon cancer. Peer J. 3:e859
91. Asgharzadeh F, Tarnava A, Mostafapour A, Khazaei M, LeBaron TW (2022) Hydrogen-rich water exerts anti-tumor effects comparable to 5-fluorouracil in a colorectal cancer xenograft model. World J Gastrointest Oncol 14(1):242–252
92. Hasegawa T, Ito M, Hasegawa et al (2022) Molecular hydrogen enhances proliferation of cancer cells that exhibit potent mitochondrial unfolded protein response. Int J Mol Sci 23(5):2888. https://doi.org/10.3390/ijms23052888
93. Wang B, Li Z, Mao L, Zhao M, Yang B, Tao X, Li Y, Yin G (2022) Hydrogen: a novel treatment strategy in kidney disease. Kidney Dis (Basel) 8(2):126–136. https://doi.org/10.1159/000520981
94. Lu Y, Li CF, Ping NN, Sun YY, Wang Z, Zhao GX et al (2020) Hydrogen-rich water alleviates cyclosporine A—induced nephrotoxicity via the Keap1/Nrf2 signaling pathway. J Biochem Mol Toxicol 34(5):e22467
95. Xing Z, Pan W, Zhang J, Xu X, Zhang X, He X et al (2017) Hydrogen rich water attenuates renal injury and fibrosis by regulation transforming growth factor-β induced Sirt1. Biol Pharm Bull 40(5):610–5
96. Chen J, Zhang H, Hu J, Gu Y, Shen Z, Xu L et al (2017) Hydrogen-rich saline alleviates kidney fibrosis following AKI and retains Klotho expression. Front Pharmacol 8:499
97. Li J, Hong Z, Liu H, Zhou J, Cui L, Yuan S et al (2016) Hydrogen-rich saline promotes the recovery of renal function after ischemia/reperfusion injury in rats via anti-apoptosis and anti-inflammation. Front Pharmacol 7:106
98. Xu X, He X, Liu J, Qin J, Ye J, Fan M (2019) Protective effects of hydrogen-rich saline against renal ischemia-reperfusion injury by increased expression of heme oxygenase-1 in aged rats. Int J Clin Exp Pathol 12(4):1488–1496
99. Nishida T, Hayashi T, Inamoto T, Kato R, Ibuki N, Takahara K, et al (2018) Dual gas treatment with hydrogen and carbon monoxide attenuates oxidative stress and protects from renal ischemia-reperfusion injury. Transplant Proc 50(1):250–8

100. Lu H, Ding J, Liu W, Peng Z, Chen W, Sun X et al (2018) UPLC/MS-based metabolomics investigation of the protective effect of hydrogen gas inhalation on mice with calcium oxalate-induced renal injury. Biol Pharm Bull 41(11):1652–1658
101. Zheng Z, Zheng F (2016) Immune cells and inflammation in diabetic nephropathy. J Diabetes Res 2016:1841690
102. Trionfetti F, Marchant V, González-Mateo GT et al (2023) Novel aspects of the immune response involved in the peritoneal damage in chronic kidney disease patients under dialysis. Int J Mol Sci 24(6):5763. https://doi.org/10.3390/ijms24065763
103. Fischer BM, Voynow JA, Ghio AJ (2015) COPD: balancing oxidants and antioxidants. Int J Chron Obstruct Pulmon Dis 10:261–276. https://doi.org/10.2147/COPD.S42414
104. Wang ST, Bao C, He Y, Tian X, Yang Y, Zhang T, Xu KF (2020) Hydrogen gas (XEN) inhalation ameliorates airway inflammation in asthma and COPD patients. QJM 113(12):870–875. https://doi.org/10.1093/qjmed/hcaa164
105. Suzuki Y, Sato T, Sugimoto M, Baskoro H, Karasutani K, Mitsui A, Nurwidya F, Arano N, Kodama Y, Hirano SI, Ishigami A (2017) Hydrogen-rich pure water prevents cigarette smoke-induced pulmonary emphysema in SMP30 knockout mice. Biochem Biophys Res Commun 492(1):74–81
106. Zheng ZG, Sun WZ, Hu JY et al (2021) Hydrogen/oxygen therapy for the treatment of an acute exacerbation of chronic obstructive pulmonary disease: results of a multicenter, randomized, double-blind, parallel-group controlled trial. Respir Res 22:149. https://doi.org/10.1186/s12931-021-01740-w
107. Kohama K, Yamashita H, Aoyama-Ishikawa M, Takahashi T, Billiar TR, Nishimura T, Kotani J, Nakao A (2015) Hydrogen inhalation protects against acute lung injury induced by hemorrhagic shock and resuscitation. Surgery 158:399–407. https://doi.org/10.1016/j.surg.2015.03.038
108. Yang F, Yue R, Luo X, Liu R, Huang X (2020) Hydrogen: a potential new adjuvant therapy for COVID-19 patients. Front Pharmacol 11:543718. https://doi.org/10.3389/fphar.2020.543718
109. Wang M, Peng J, Hui J, Hou D, Li W, Yang J (2021) Hydrogen therapy as an effective and novel adjuvant treatment against COVID-19. QJM 114(1):74–75. https://doi.org/10.1093/qjmed/hcaa301
110. Deryugina AV, Danilova DA, Pichugin VV, Brichkin YD (2023) The effect of molecular hydrogen on functional states of erythrocytes in rats with simulated chronic heart failure. Life 13:418. https://doi.org/10.3390/life13020418
111. Zhang Y, Liu Y, Zhang J (2015) Saturated hydrogen saline attenuates endotoxin-induced lung dysfunction. J Surg Res 198(1):41–49
112. Asada R, Tazawa K, Sato S, Miwa N (2020) Effects of hydrogen-rich water prepared by alternating-current-electrolysis on antioxidant activity, DNA oxidative injuries, and diabetes-related markers. Med Gas Res 10(3):114–121
113. Kamimura N, Ichimiya H, Iuchi K, Ohta S (2016) Molecular hydrogen stimulates the gene expression of transcriptional coactivator PGC-1α to enhance fatty acid metabolism. NPJ Aging Mech Dis 2:16008. https://doi.org/10.1038/npjamd.2016.8
114. Nagatani K, Nawashiro H, Takeuchi S et al (2013) Safety of intravenous administration of hydrogen-enriched fluid in patients with acute cerebral ischemia: initial clinical studies. Med Gas Res 3:13
115. Zanini D, Stajer V, Ostojic SM (2020) Hydrogen vs. caffeine for improved alertness in sleep-deprived humans. Neurophysiology 52:67–72
116. LeBaron WT, Kharman J, McCullough L, Michael (2021) An H2-infused, nitric oxide-producing functional beverage as a neuroprotective agent for TBIs and concussions. J Integr Neurosci 20(3)
117. Takeuchi S, Kumagai K, Toyooka T, Otani N, Wada K, Mori K (2021) Intravenous hydrogen therapy with intracisternal magnesium sulfate infusion in severe aneurysmal subarachnoid hemorrhage. Stroke 52(1):20–27
118. Hirano SI, Aoki Y, Li XK, Ichimaru N, Takahara S, Takefuji Y (2021) Protective effects of hydrogen gas inhalation on radiation-induced bone marrow damage in cancer patients: a

retrospective observational study. Med Gas Res 11(3):104–109. https://doi.org/10.4103/2045-9912.314329
119. Chen JB, Kong XF, Qian W, Mu F, Lu TY, Lu YY, Xu KC (2020) Two weeks of hydrogen inhalation can significantly reverse adaptive and innate immune system senescence patients with advanced non-small cell lung cancer: a self-controlled study. Med Gas Res 10(4):149–154
120. Akagi J, Baba H (2020) Hydrogen gas activates coenzyme Q10 to restore exhausted $CD8^+$ T cells, especially PD-1^+Tim3^+terminal $CD8^+$ T cells, leading to better nivolumab outcomes in patients with lung cancer. Oncol Lett 20(5):258. https://doi.org/10.3892/ol.2020.12121
121. Korovljev D, Trivic T, Stajer V, Drid P, Sato B, Ostojic SM (2019) Short-term H2 inhalation improves cognitive function in older women: a pilot study. Int J Gerontol 14(2):149–150
122. Lv X, Lu Y, Ding G et al (2022) Hydrogen intake relieves alcohol consumption and hangover symptoms in healthy adults: a randomized and placebo-controlled crossover study [published online ahead of print, 2022 Sep 16]. Am J Clin Nutr nqac261. https://doi.org/10.1093/ajcn/nqac261
123. Okada M, Ogawa H, Takagi T, Nishihara E, Yoshida T, Hyodo J, Shinomori Y, Honda N, Fujiwara T, Teraoka M, Yamada H, Hirano S-I, Hato N (2022) A double-blinded, randomized controlled clinical trial of hydrogen inhalation therapy for idiopathic sudden sensorineural hearing loss. Front Neurosci 16:1024634. https://doi.org/10.3389/fnins.2022.1024634
124. Tamura T, Suzuki M, Hayashida K, Kobayashi Y, Yoshizawa J, Shibusawa T, Sano M, Hori S, Sasaki J (2020) Hydrogen gas inhalation alleviates oxidative stress in patients with post-cardiac arrest syndrome. J Clin Biochem Nutr 67(2):214–221
125. Tanaka Y, Kiuchi M, Higashimura Y, Naito Y, Koyama K (2021) The effects of ingestion of hydrogen-dissolved alkaline electrolyzed water on stool consistency and gut microbiota: a double-blind randomized trial. Med Gas Res 2021(11):4
126. Danilova DA, Brichkin YD, Medvedev AP, Pichugin VV, Fedorov SA, Taranov EV, Nazarov EI, Ryazanov MV, Bolshukhin GV, Deryugina AV (2021) Application of molecular hydrogen in heart surgery under cardiopulmonary bypass. Sovrem Tekhnologii Med 13(1):71–75. https://doi.org/10.17691/stm2021.13.1.09
127. Increased sympathetic cardiac activity in healthy females, hydrogen water
128. Li D, Yan M (2016) Clinical assessment of hydrogen-rich water and oxycodone hydrochloride controlled-release tablets in the treatment of malignant neuropathic pain. J Chin Phys 18(9):1329–1331
129. Iketani M, Sekimoto K, Igarashi T, Takahashi M, Komatsu M, Sakane I, Takahashi H, Kawaguchi H, Ohtani-Kaneko R, Ohsawa I (2018) Administration of hydrogen-rich water prevents vascular aging of the aorta in LDL receptor-deficient mice. Sci Rep 8(1):16822. https://doi.org/10.1038/s41598-018-35239-0
130. Kang KM, Kang YN, Choi IB, Gu Y, Kawamura T, Toyoda Y, Nakao A (2011) Effects of drinking hydrogen-rich water on the quality of life of patients treated with radiotherapy for liver tumors. Med Gas Res 1:11
131. Liu B, Jiang X, Xie Y, Jia X, Zhang J, Xue Y, Qin S (2022) The effect of a low dose hydrogen-oxygen mixture inhalation in midlife/older adults with hypertension: A randomized, placebo-controlled trial. Front. Pharmacol 13:1025487. https://doi.org/10.3389/fphar.2022.1025487
132. Guo JD, Li L, Shi YM, Wang HD, Hou SX (2013) Hydrogen water consumption prevents osteopenia in ovariectomized rats. Br J Pharmacol 168(6):1412–1420. https://doi.org/10.1111/bph.12036
133. Singh RB, Gupta AK, Fatima G, Fedacko J et al (2022) Can molecular hydrogen therapy enhance oxygen saturation among patients with chronic lung disease. In: Presented and abstracted in the first conference of the European Academy of molecular hydrogen research in biomedicine. Smolenice Castle, Slovakia, Abstract, p 69
134. Ishibashi T, Kawamoto K, Matsuno K, Ishihara G, Baba T, Komori N (2020) Peripheral endothelial function can be improved by daily consumption of water containing over 7 ppm of dissolved hydrogen: A randomized controlled trial. PLoS ONE 15(5):e0233484
135. Gu Q, Liu B, Xue J, Zhao M, Zhang X, Wang M, Zhang M, Xue Y, Qin S (2020) Effects of drinking hydrogen-rich water in men at risk of peripheral arterial disease: a randomized placebo-controlled trial. https://doi.org/10.21203/rs.3.rs-136097/v1

136. Kubota M, Kawashima M, Inoue S, Imada T, Nakamura S, Kubota S, Watanabe M, Takemura R, Tsubota K (2021) Randomized, crossover clinical efficacy trial in humans and mice on tear secretion promotion and lacrimal gland protection by molecular hydrogen. Sci Rep 11:Article 6434

Chapter 19
Consumption of Hydrogen-Treated Foods Provides Nutritional and Health Benefits

Duried Alwazeer

Abstract Molecular hydrogen (H_2) was proven to be a therapeutic agent for many diseases. In parallel with its biomedical benefits, H_2 has been shown in recent reports to have many applications in horticultural and food industry fields. Many hydrogen-infused products including hydrogen-rich water (HRW), hydrogen-infused beverages, and hydrogen-incorporated atmosphere products have been commercialized or at least studied at the laboratory scale. The hydrogen-treated crops possess preservative properties of phytochemicals e.g. anthocyanins and flavonoids and longer shelf life. The consumption of hydrogen-treated crops and hydrogen-infused beverages can provide many health and nutritional benefits to the consumer. The present report discusses the recent knowledge on the use of molecular hydrogen in crop production and food processing as well as their potential benefits on the health and nutrition of the consumer.

Keywords Molecular hydrogen · Hydrogen-rich water · Foods · Crops · Health · Nutrition

Introduction

Molecular hydrogen (H_2) has recently attracted the attention of many laboratories around the world in different fields such as health, agriculture, and food. Hydrogen known as the smallest molecule and lightest gas in the universe was described by many health-medicine specialists as the "miracle" compound due to its various beneficial health effects [1]. Many research laboratories, especially in Japan, China,

D. Alwazeer (✉)
Department of Nutrition and Dietetics, Faculty of Health Sciences, Iğdır University, 76000 Iğdır, Turkey
e-mail: alwazeerd@gmail.com

Research Center for Redox Applications in Foods (RCRAF), Iğdır University, 76000 Iğdır, Turkey

Application, and Research Center, Innovative Food Technologies Development, Iğdır University, 76000 Iğdır, Turkey

J. Slezak and B. Kura (eds.), *Molecular Hydrogen in Health and Disease*, Advances in Biochemistry in Health and Disease 27, https://doi.org/10.1007/978-3-031-47375-3_19

Korea, and the USA revealed various healing properties of hydrogen against different diseases. The in vitro, in vivo, and animal and human studies linked the beneficial effects of hydrogen with its unique physical, chemical, and biological properties. The specific characteristics of hydrogen that were proved by many studies of different laboratories include, for example, hydrophobic, antioxidant, anti-radical, anti-cancer, anti-inflammatory, anti-apoptosis, and signaling modulator. Another important advantage of hydrogen application in the health field is its non-disturbing of metabolic redox reactions in the cell, non-reduction of the oxidized heme of cytochrome c, and its ability to affect positively some signal transduction pathways as an indirect modulator [2].

The positive effects of hydrogen on health encouraged plant scientists to evaluate its potential effects on the growth of plants and the quality of crops. Besides, other researchers began to study the potential uses of hydrogen in the food industry as a food additive for different purposes. In this report, we will review the recent studies about the use of hydrogen in crop production and food processing, and its potential nutritional and health benefits for the consumer.

Properties of Hydrogen

Hydrogen was first produced by Robert Boyle (1627–1691) but he confused it with ordinary air. The French Henry Cavendish (1731–1810) was the first person to prepare hydrogen by adding acid (hydrochloric acid and sulfuric acid) to various metals (zinc, iron, and tin) [3]. Cavendish in his first paper published in the journal Philosophical Transactions in 1766 described hydrogen as "inflammable air," which we know today as hydrogen.

Molecular hydrogen (H_2) is a colorless, odorless, tasteless gas and the smallest molecule and the oldest element found in the universe. H_2 is electrically neutral and nonpolar with a molecular weight of 2 Da [4]. Due to its small size, high diffusion rate, and nonprotic properties, H_2 can penetrate different biological tissues, barriers, and membranes [5]. The hydrogen molecules are chemically inert and very stable and do not interact with surrounding substances, and do not form hydrogen bonds. These properties facilitate the passage of hydrogen molecules through biopolymers and tissues [6]. The mobility of hydrogen molecules in biological materials and macromolecules like natural rubber and polyethylene was estimated to be 1 mm per min [6].

The hydrogen molecule is found in a gaseous state under ambient conditions and in a liquid state under − 253 °C at 1 atm [7]. The solubility of H_2 in water is 1.6 mg/L at NTP and this value increases with increasing temperature and pressure [8].

Regarding biological properties, hydrogen is an inert and non-toxic gas even at high concentrations with non-effect on the body physiology parameters such as temperature, blood pressure, pH, or pO_2 [9]. Additionally, H_2 is mild enough neither to disturb metabolic redox reactions nor to affect signal-regulating ROS, i.e. $O2^{\cdot -}$, H_2O_2, and $^{\cdot}NO$ [10] (Table 19.1).

Table 19.1 Some biological properties of molecular hydrogen

Property	References
Selective antioxidant against the strong free radicals i.e. hydroxyl radical (·OH) and peroxynitrite ($ONOO^-$) without effect against signal regulating ROS (O2, H_2O2, and NO)	[10]
Non-toxic effects on cell	[4]
Gaseous-signaling molecule	
Regulator of redox and intracellular signaling	
Modulator of cellular antioxidant defenses	
Buffers the oxidative stress	
Downregulates the endoplasmic reticulum	
Radical scavenger	
Modulator of matrix metalloproteinases	
Regulator of intracellular protein signaling	
Modulator of intracellular signaling	
Modulator of gene expression	
Modulator of blood vessel function	
Modulator of myocardial responses to ischemia/reperfusion injury	
Modulator of cellular responses to stress conditions	
Not disturb cellular metabolic redox reactions	
Not disturb intracellular signaling	
Not disturb enzymatic reactions	
Has low reactivity with other gases at therapeutic concentrations and lacks	
Lacks reactivity to nitric oxide (NO·)	
Ables to penetrate biological membranes including the blood–brain barrier and tissues	
Not produced in the body due to the lack of hydrogenase genes	[2]
Anti-allergic function	
Inducer of antioxidant systems (hemeoxygenase-1, SOD, catalase, myeloperoxidase)	
Decreases the expressions of proinflammatory factors (NF-κB, TNF-α, interleukin (IL)-1β, IL-6, IL-10, IL-12, CCL2, and interferon (INF)-γ, ICAM-1, PGE2 and PGE2, high mobility group box 1)	
Stimulator of energy metabolism	
The ideal effective antioxidant against cancer	
Decreases the pro-inflammatory cytokines	
Anti-inflammatory	
Anti-apoptosis	
Protector against LDL oxidation	[11]
Inhibitor of tumor necrosis factor-α (TNF-α)-induced monocyte adhesion to endothelial cells	

(continued)

Table 19.1 (continued)

Property	References
Stimulator of cholesterol efflux from macrophage foam cells	
Protector of endothelial cells from TNF-α-induced apoptosis	
Decreases systolic blood pressure	[12]
Prevents the increase of blood lactate	[13]
Neuroprotector (enhanced production of ghrelin)	[14]

Regarding the safety of the use of hydrogen in industry and laboratories, the flammability of hydrogen ranges from 4 to 75 (vol%) in air and 4.1–94.0 (vol%) in NTP oxygen, and the detonation limits are 18.3–59 (vol%) in NTP air and 15–90 (%vol) in NTP oxygen [15, 16]. The use of hydrogen gas diluted by N_2 at 4/96 (vol%) can give a safe gaseous mix with a reducing property [17]. However, the safest form of hydrogen is its soluble form as various products such as hydrogen-rich water (HRW), hydrogen-rich saline (HRS), and hydrogen-rich eye drops.

Use of Hydrogen in Food Processing

Many studies were conducted to incorporate molecular hydrogen in different foods. The liquids were the first products that were used for this purpose due to the easiness of hydrogen infusion into liquid and the safety concerns. Increasingly, hydrogen-rich water was produced on laboratory and industrial scales and became a kind of commercial "functional water", especially in Japan, China, Korea, and the USA [18].

Hydrogen-Rich Water (HRW)

The history of the HRW started in a small town in Germany called Nordenau where people with different chronic diseases including diabetes, tumors, gastritis, and enteritis were cured after using water from a spring in a superseded mine and call this miracle "Nordenau Phenomenon" [1]. A Japanese doctor, Dr. George Tseng, studied the water and found that it was abundant in molecular hydrogen. The same phenomenon was found in other spring waters around the world e.g. Lourdes (France), Tlacote (Mexico), and Nadana (India). Four folds hydrogen level in the breath of Japanese Centenarians was shown compared to normal health persons [1].

In 1960, an electrolytic device was used in Japan, for preparing HRW and then became a medical device. Nowadays, HRW is available in drug stores, and HRW suppliers are installed at most sports gyms [6].

Hydogen-rich water can be considered functional water due to its various biological properties and different health benefits [19]. Upon 20 min of intaking HRW, about 40% is adsorbed by the body [20].

Although the low solubility of molecular hydrogen in water at normal room conditions, its small size makes low amounts of hydrogen contain huge numbers of hydrogen molecules. For example, the molecular weight of ascorbic acid is 176, vitamin E is 153, beta-carotene is 150, and hydrogen is 2. This small size property of hydrogen gas allows it among other antioxidants to cross the brain-blood barrier [9].

There are many methods of hydrogen-rich water production including hydrogen gas bubbling, water electrolysis, and the reaction of magnesium with water.

Hydrogen-Infused Water

The infusion of hydrogen gas into drinking water is a simple method for preparing HRW. This method depends on bubbling pure hydrogen gas into drinking water for 10 min at least. Although this method is not effective in reaching high dissolved hydrogen levels, it is considered reasonable for obtaining health benefits [21]. There are other methods for preparing supersaturated hydrogen waters. The supersaturation of microbubbles of hydrogen especially in electrolyzed alkaline water showed many physicochemical and biological properties. The mean diameter of hydrogen particles in supersaturated alkaline water (colloidal solution) was estimated to be distributed between 20 and 300 nm and at concentrations higher than that of the saturation level i.e. 0.75 mM [22].

Nanotechnology was also studied for preparing nanomaterials-augmented hydrogen. The advantages of this method are (1) the long-term release of hydrogen, (2) the high concentrations of hydrogen levels, (3) the non-catalyst needed, and (4) the use of biodegradable materials [18]. An ammonia borane-loaded hollow mesoporous silica nanoparticle (AB@hMSN) was prepared to serve as a hydrogen-releasing nanomaterial for sustainable H_2 supply for applications in plants for a long time [23]. The review of Epelle and co-workers discuss the different nanomaterials used for biological hydrogen production [24].

Water Electrolysis

Electrolyzed Reduced Alkaline Water (ERAW) or Electrolyzed Reduced Water (ERW), Alkali-Ionsui Water (AIW), or Electrolyzed Cathodic Water (ECW) are produced by the electrolysis process and are known for their health benefits including gastrointestinal problems, hypertension, diabetes and cancer [25]. In Japan, the ministry of health, labor, and welfare announced (Notification N° 112) that alkaline electrolyzed drinking water is useful for improving gastrointestinal symptoms such as chronic diarrhea, indigestion, abnormal intestinal fermentation activity, and stomach acidity [25].

The hydrogen gas is generated at the negative electrode (cathode) during the electrolysis of water and up to 2 ppm of hydrogen concentration can be obtained. When electrons flow through the electric circuit, protons and oxygen are produced at the anode compartment leading to a decrease in the pH of water up to 4–6 and an increase in the redox potential value up to + 900 mV, with an accumulation of some mineral ions such as HCO_3, Cl^-, HSO4, NO3 [25]. At the cathode, the hydroxyl ions and hydrogen are produced leading to an increase in the pH of water up to 8–10 and a decrease in the redox potential up to − 600 mV with the accumulation of some mineral cations such as Na^+, K^+, Ca^{2+}, Mg^{2+}.

It is important in electrolyzers to avoid contaminating the HRW with electrolytic products such as ozone and hydroxide ions [6]. The regulatory standards of drinking water recommend a pH value between 6.5 and 9, which means that the different HRW types are in agreement with the standards [25].

Magnesium Reaction with Water

This method is based on the insertion of magnesium material (stick or rods) in drinking water producing hydrogen gas and magnesium hydroxide $Mg(OH)_2$. The formed magnesium hydroxide is insoluble and medically classified as a laxative to treat occasional constipation and it also works as an antacid to treat symptoms caused by too much stomach acid [26].

Packaging and Storage of Hydrogen-Infused Water

HRW must be packaged in aluminum bags or cans because these packages can make hydrogen concentration stable for more than 6 months [21]. Whereas, other materials like glass and polyethylene were not appropriate for packaging HRW.

The level of dissolved hydrogen in HRW depends on the gaseous composition of the headspace of the HRW package. In an open package i.e. contact with air, the hydrogen will escape after about 8 h. However, under H_2-included headspace, the hydrogen can be maintained for a longer time i.e. several months. The hydrogen can be kept longer time in the case of Mg-HRW compared to H_2 bubbled water type in an open package [21]. However, it was reported that a hydrogen level as low as 0.8 ppm has similar health benefits to that of a higher concentration i.e. 1.5 ppm [27].

The level of hydrogen in the product at the time of consumption is important. Since hydrogen can easily leak from the product through the package materials due to its capability to penetrate different types of packages including glass and plastic, it is important to choose an appropriate hydrogen infusion process with a hermetic non-hydrogen permeable package (aluminum-included packages are generally recommended).

Properties of Hydrogen-Rich Water

The physicochemical properties of HRW vary according to the method of preparation. Mg-HRW has an alkaline pH value ranging from 8 to 10 due to the formation of magnesium hydroxide. Similar pH values can be seen for electrolyzed reduced alkaline water [25, 28]. Whereas, H_2-bubbled water has a slightly alkaline pH value reaching up to 8 [29] (Table 19.2).

Drinking HRW was shown to be not only safe and well tolerated by the body but also has various beneficial impacts on consumer health [4]. More details are discussed below in the section "nutritional and health benefits of hydrogen-treated foods".

Hydrogen-Rich Beverages

The heat treatment of beverages can induce oxidative degradation of nutrients, e.g. vitamins, pigments, and flavonoids, and sensory quality e.g. color, aroma, taste, and flavor. The infusion of hydrogen into the beverage can alleviate the deteriorative reactions during both heat treatment and the storage period [31–34]. Additionally, the infusion of hydrogen in beverages reduces the dissolved oxygen content in the product leading to better protection of the nutritional and sensorial quality.

The incorporation of beverages with gases was known for a long time, especially in carbonated beverages (CO2-enriched beverages) [35]. However, the infusion of

Table 19.2 Physicochemical properties of HRW [30]

Parameter[a]	Control[b]	HRW[c]
H_2 (mg/L)	0	1.55
ORP (mV)	+ 292	– 453
EC (ds/m)	197	197
TDS (g/L)	197.3	196.7
DO (mg/L)	6.1	5.8
pH	7.6	9.3
Ca^{2+} (ppm)	28	9.1
Mg^{2+} (ppm)	10.2	22.8
Water cluster (Hz)	55.8	54.9

[a] All parameters were measured at 25 °C. EC, electrical conductivity; TDS, total dissolved solids; DO, dissolved oxygen

[b] Control water was filtered from tap water by passage through a calcined ceramic filter, an activated carbon filter, and a magnetized rod

[c] HRW was obtained from the same water apparatus except that the tap water was passed through the first two filters and then reacted with the magnesium–carbon hydrogen storage hybrid materials

Table 19.3 Effect of hydrogen infusion on the stability of fresh (single strength) orange juice and fully pasteurized orange juice (90 °C for 1 min) [31]

Treatment	Ascorbic acid loss (%)	
	Fresh juice	Fully pasteurized juice
H_2/N_2-infusion	7.37 ± 5.40	− 1.70 ± 28.37
N_2-infusion	7.48 ± 20.32	− 13.53 ± 34.74
Control (non-degassing)	22.89 ± 18.23	82.79 ± 24.34

hydrogen into beverages has recently been studied by a few numbers of researchers. In 2003, Alwazeer and co-workers enriched the orange juice with H_2/N_2 (4/96, vol%) before the pasteurization step, and they found that this process allows keeping the ascorbic acid and the color of the product during a 7-weeks storage period [31, 33, 34] (Table 19.3). Additionally, the H_2-infused orange juice i.e. treated with N_2-H_2 gas could restrict the growth of injured cells of spoilage microorganisms. Moreover, the color of H_2-infused juice was better preserved during the storage period.

In another study, a dairy beverage enriched with 2% linseed oil was infused with a gaseous mixture composed of H_2/N_2 (4/96, vol%). The hydrogen-infused dairy beverage reduced oxidative degradation and maintained the color during storage [32]. In another study, an H_2-infused beverage containing nitric-oxide-stimulating citrulline called Hydro Shot produced by H_2BEV company (USA) was shown to have cognitive function improvement [36].

HRW-Washed Butter

The waters used in food processing and preparation have an important role in the safety and quality of the final product. Water forms approximately 95–99% of the total cleaning and sanitizing operations used in food production [37]. Washing is an important step in different food processes aiming to eliminate foreign materials like dust, insects, and dirt, and reduce microbial loads, pesticides, and impurities. The physicochemical and biological quality of the water used in the washing process can affect the quality of the final food product. The water used in the washing process should be of drinking water quality, especially the microbial load and heavy metal and other contaminant contents [37].

Due to the specific properties of HRW, its use in the washing step can bring many advantages for the product as well as the consumer. In a recently published report, our team revealed that the physicochemical and sensory properties of butter were improved when HRW had been used in washing raw cultured butter [38] Additionally, HRW-washed butter restricted the increase in the acid degree value (ADV) and exhibited the highest color notes. This shows that washing raw butter by HRW can improve both the physicochemical and sensory properties of butter.

Heavy metals can attack all the biological components of the body leading to unrepairable modifications like conformational changes of proteins, lipid peroxidation, and DNA damage causing many serious diseases like cancer [39]. Different sources of drinking water including surface water, groundwater, and seawater are likely to be contaminated by heavy metals [40]. In a study performed in our laboratory, washing raw butter with normal tap water showed an increase in the content of arsenic (As), cadmium (Cd), and lead (Pb) [41]. Whereas, the heavy metal contents i.e. As, Cd, Sb, and Hg were significantly reduced when raw cultured butter had been washed with HRW. These results show again the importance of the consideration of HRW in Food preparation. HRW could alleviate the toxic effects of heavy metal-induced oxidative and inflammatory damage in animal models [42, 43].

Biogenic amines (BAs) are bioactive molecules that can exert effects on the central nervous and vascular systems with pathogenic impacts ranging from headaches to death [44]. BAs are found in a variety of foods especially fermented products such as pickles, wine, beer, fish, meat, and dairy [45]. Both the fermentation and spoilage microorganisms can participate in the decarboxylation of amino acids in food to form harmful biogenic amines. In a recently published study, our team proved that washing cultured butter with HRW can restrict significantly the formation of BAs during cold storage [29]. After 90 days of storage, the HRW-washed butter showed the lowest contents of BAs, especially tryptamine, 2-phenylethylamine, spermidine, and spermine when it was compared with butter samples washed with normal tap water. Another advantage seen in this report was the absence of the inhibitive effect of hydrogen on the growth of yogurt starters during the cold storage period.

HRW-Washed Olive Oil

In olive oil manufacturing, the olive fruits are washed before the pressing step. Similar to other food processes, the quality of the water used in the cleaning step (or as an ingredient) can affect the quality and safety of the final product. In our laboratory, our team assayed the use of HRW in the washing of olive fruits during the preparation of olive oil. The results showed that the nutritional quality i.e. phenolics, flavonoids, and antioxidants was better protected in the final olive oil compared with that prepared by tap water washing (non-published data).

Additionally, in another study, we found that using HRW in the washing of crude olive pomace oil (COPO) significantly improved the quality of the oil [46].

Acidity and peroxide values were decreased, while the color, phenolic, and flavonoid contents of oil were increased in the HRW-washed COPO samples compared to those washed with normal water. This extends the potential application of HRW in the washing process to include the vegetable oils sector. This matter brings also additional benefits to the products where the phenolic compounds aiding in the protection of the oil and enhancing its taste, flavor, and aroma were better protected in the case of the HRW washing process. Regarding the consumer, the high contents of nutrients and phytochemicals can better provide the daily nutritional requirements and enhance preventative health.

Hydrogen Atmosphere Packaged Foods

Modified atmosphere packaging is a method widely used for preserving various fresh and processed foods for long shelf life with protecting their sensory and nutritional qualities. Carbon dioxide, oxygen, and nitrogen are the wide gases used in this method [47]. However, recent studies performed in our laboratory revealed that the incorporation of hydrogen gas with other gases can bring many advantages to the product. This new method was called Reducing Atmosphere Packaging (RAP) due to the reducing property of the hydrogen gas included in the atmosphere formulation.

We could find that the use of hydrogen-included gas formulation can maintain the quality of fish i.e. rainbow trout fillets for 15 days at + 2 °C [48]. Furthermore, the formation of biogenic amines in the freshwater and seawater fish, i.e., rainbow trout and horse mackerel, during the cold storage period was restricted in the hydrogen-included samples compared with the control [49]. The formation of different types of BAs i.e. heterocyclic, aromatic, and aliphatic di-amines (histamine, tyramine, putrescine, cadaverine) in the two fish species was inhibited at a potent level in RAP fish-packaged samples rather than modified atmosphere packaging (MAP)—fish packaged samples.

In another study, we could find that the quality notes and the microbial loads of the white cheese packaged under a hydrogen-incorporated atmosphere (RAP) were protected during the storage period [50]. The closest values of color and titratable acidity to the fresh sample were observed for hydrogen-included atmosphere cheese samples compared with MAP ones.

In another study, hydrogen-included atmosphere packaging (RAP) was also tested for extending the shelf life of fresh fruits. The shelf-life of strawberries could be extended up to 8 weeks using hydrogen-included atmosphere packaging (RAP) [51]. The sensory and nutritional qualities of the fruit could be protected during the 12-weeks storage period. The hydrogen-included atmosphere packaging (RAP) showed also the highest total soluble solids (TSS), firmness, color, phenolic and anthocyanin contents, and antioxidant activity compared with MAP and control samples. The hydrogen gas seems to protect the sensible phytochemicals and pigments from oxidative deterioration during the storage period.

Hydrogen-Rich Water-Prepared Pickle

The fermented vegetables, called also pickles, are prepared by using brine that prompts the growth of lactic acid bacteria (LAB) and inhibits spoilage microorganisms. The formation of an acidic medium due to the production of mainly lactic acid by LAB helps to keep low pH conditions that prevent the enzymatic and microbiological deterioration of the product. However, like other fermented products, the pickles are submitted to the risk of biogenic amine formation.

In a recently published study, our team found that the use of HRW in the preparation of brine can limit the formation of BAs in red beetroot pickles during fermentation and storage periods [44]. The levels of BAs i.e. tyramine, 2-phenylethylamine, histamine, tryptamine, and putrescine in HRW-prepared pickles were lower than those prepared by normal water. This study revealed that the conventional method of pickle preparation includes a risk of BA formation, and the use of HRW in the preparation of these products can avoid this safety concern.

Use of Hydrogen in Crop Production and Protection

Due to the multi-biological functions of molecular hydrogen proven in health studies, many laboratories especially in China started to explore the potential benefits of hydrogen in the agriculture and crop production field.

HRW Increases the Phytochemical Concentration of Crops

Much recent research has revealed the advantageous effects of the use of HRW in the cultivation of many crops. HRW can be used by different methods in agriculture including soaking seeds and fruits, spraying leaves, irrigating soil, and incorporating hydroponic solutions. HRW can also increase the anthocyanin and polyphenol content in crops such as radish and flavonoid content in alfalfa [52].

Many reports revealed that HRW can improve the growth of crops. For example, HRW could effectively increase the germination speed, growth speed, and concentrations of bioactive phytochemicals including free vanillic acid, coumaric acid, sinapic acid, conjugated sinapic acid, Ca and Fe, and the hydroxyl radical scavenging rate of blackberry [53]. This increase in the phytochemicals in crops can bring many nutritional benefits to the consumer.

Hydrogen Extends the Shelf Life of Crops

Fresh crops are subject to different spoilage hazards including biological, physical, and chemical factors due to their higher moisture content and nutrient richness. Many processes can be used for protecting fresh crops from deterioration such as refrigeration, freezing, and modified atmosphere packaging (MAP). MAP technology depends on the use of high levels of carbon dioxide (CO_2) and low levels of oxygen for slowing down the metabolism of the crop, reducing enzymatic activities, and inhibiting the growth of spoilage microorganisms [47]. A modified version of MAP based on the incorporation of hydrogen with other gases in the atmosphere of the package including foods has been recently evaluated.

This new method called reducing atmosphere packaging (RAP) is characterized by its ability to preserve the nutritional and sensory properties of fruits such as strawberries [51]. Other studies in our laboratory showed that RAP can protect the quality properties of white fresh cheese [50] and fish [48]. In China, many studies were performed on the use of molecular hydrogen in the form of gas (fumigation) or HRW (immersion) to extend the shelf life of kiwifruit. The hydrogen treatment delayed the riping processes including respiration and ethylene synthesis [54, 55], increased chilling tolerance [56], and maintained the organoleptic of kiwifruits [57, 58].

Health and Nutritional Benefits of Intaking Hydrogen-Treated Foods

Potential Health and Medicinal Benefits of Intaking Hydrogen-Included Product

Regulations Regarding the Use of Hydrogen Gas in Foods and Human Consumption

Hydrogen gas was approved for human consumption by many food and health administrations. In European Community, hydrogen gas is considered a food additive with an E494 number with applications in foods and beverages for infants and young children at *the quantum satis* maximum level [59]. The consumption of HRW was approved to be ad libitum and is more effective than the diluted one [2].

Some Examples of Health Benefits of Intaking HRW

The biomedical benefits of hydrogen were proved by many laboratories around the world. The easiest, simplest, and safe method of hydrogen intake is in its infused form like HRW or H_2-infused beverages.

Example 1

The daily intake of HRW or hydrogen-infused beverages can reduce fatigue, increase force production, and enhance cognitive functions like focus, speed, and plasticity of consumers [60]. HRW improves not only the physical condition but also mental conditions such as mood, anxiety, and autonomic nerve function of consumers [61].

Example 2

The long-term consumption of HRW can stimulate the energy metabolism of the body leading to decrease fat and body weight, levels of plasma glucose, insulin, triglyceride, and total cholesterol and LDL-cholesterol levels, which is beneficial in the prevention of metabolic syndrome [11, 62].

Example 3

The consumption of HRW can lead to an increase in energy production by neutralizing the harmful free radicals as well as reducing post-exercise fatigue development provides additional advantages to consumers generally and to athletes especially [63]. These beneficial effects can be added to the many advantages of hydrogen proved in many in vitro, in vivo, animal, trials, and clinical studies.

Example 4

The consumption of HRW was proven to be a benefit in preventing or impeding type 2 diabetes mellitus and insulin resistance [64]. Drinking HRW by 900 mL a day for 8 weeks reduced LDL levels and normalize glucose tolerance levels indicating that HRW consumption aids lipid and glucose metabolism allowing for preventing or impeding type 2 diabetes mellitus and insulin resistance.

Example 5

An improvement in renal dysfunction was shown in rats after 2-days HRW consumption [65]. An ameliorative effect of HRW against gentamicin-induced nephrotoxicity was obtained.

Example 6

HRW consumption, for a preventive therapeutic purpose, showed its ability to retard the development and progression of Parkinson's disease in rats model [66]. Additionally, a trial study of 48 weeks of HRW consumption showed an improvement in the majority of Parkinson's patients [67].

General Biomedical Properties of Hydrogen

Upon drinking HRW, the maximum level of H_2 in the breath is reached after 10 min following by a decrease to the baseline after 60 min [20]. About 40% of H_2 ingested via HRW is consumed by the body [2], while about 0.1% is released from the skin surface [20].

Molecular hydrogen found in HRW was found to selectively react with the strong free radicals i.e. hydroxyl radical ($\cdot OH$) and peroxynitrite ($ONOO^-$) that can react indiscriminately with many cellular components including membranes, nucleic acids, lipids, and proteins, resulting in DNA damage, lipid peroxidation, and protein deformation [68]. Fortunately, hydrogen does not react with other beneficial free radicals that have essential physiological functions in the body i.e. $O^{\cdot -}$, HO, and $\cdot NO$

(Table 19.1). Additionally, hydrogen showed other positive physiological effects like anti-inflammation, anti-apoptosis, anti-mutation, redox homeostasis agent, and signal modulator (Table 19.1) [68].

Molecular hydrogen plays an important role in preventive therapeutic medicine. As the majority of diseases are occurred due to the high production of free radicals in specific and stressful conditions, the intake of hydrogen can selectively neutralize and alleviate the most aggressive free radicals. The simulation of the enzymatic antioxidant system i.e. dismutase, peroxidase, and catalase, upon the ingestion of hydrogen, can strengthen the body's resistance against oxidative-related diseases. The Ohta report summarizes the different biological and health benefits of hydrogen and HRW [2].

As the crops produced by the HRW method contain higher levels of phytochemicals according to the recently published reports, the consumption of these hydrogen-treated products should effectively provide high amounts of the daily requirements of phytochemicals for the consumer.

Potential Increase of Phytochemical Bioavailability When Intaking HRW

The nutritional benefits of any food product depend not only on its nutrient and non-nutrient contents but also on the bioaccessibility and bioavailability of these compounds. The biological activities of phenolic compounds depend on their bioaccessibility and bioavailability. The interactions between proteins and polyphenolic compounds result in the formation of soluble and non-soluble complexes via different non-covalent bonds including hydrogen bonds, hydrophobic interactions, and Van der Waals forces [69]. These interactions affect the structure of proteins leading to changes in their functionality, digestibility, bioaccessibility, and bioavailability. For example, chlorogenic acid found in coffee and citrus fruits was found to be able to bind to some whey proteins; and the blueberry and cranberry extracts were capable of interacting and binding with peanut proteins [69]. These interactions between proteins and phenolic compounds lead to a decrease in the antioxidant activity of phenolics. It was reported that the intake of jujube juice phenolics dissolved in skimmed milk decrease the plasma concentration of caffeic acid, protocatechuic acid, and vanillic acid as well as the plasma antioxidant capacity compared to the intake of these phenolics in water in Westar rat [70].

Many intrinsic and extrinsic factors can affect the interactions between proteins and phenolic compounds including molecular weight, the type and the position of chemical groups, pH, temperature, and interactions of other compounds found in food [69].

In recently published studies, the infusion of hydrogen in water (HRW) has been found to increase the extraction of phytochemicals including phenolics, flavonoids, anthocyanins, and antioxidant activity [71]. The extraction of some flavonoids and non-flavonoids such as epicatechin, catechin, rutin, gallic acid, chlorogenic acid, p-coumaric acid, and trans-ferulic acid were increased several folds when HRW had been used instead of normal water in the extraction process of phytochemicals from tomato peels, apple peels, lemon peels, carrot, red beetroot, and cabbage [72–75] and olive oil [76]. The use of HRW in rice milk preparation enhanced the contents of some essential minerals and amino acids of product [77].

These recent reports show that the use of HRW in diet may increase the extraction of various phytochemicals including phenolic compounds and antioxidants. This assumes that the intake of HRW with or just upon the meal can enhance the extraction (liberation and release) of some phytochemicals allowing the increase in the bioaccessibility and maybe the bioavailability of these molecules. However, further studies are needed to confirm whether the bioaccessibility and bioavailability of some nutrients and non-nutrient compounds change upon the intake of HRW.

Conclusions

The use of molecular hydrogen in crop production and food processing can bring many nutritional, health, and economical advantages for consumers, producers, and processors. The consumption of hydrogen-treated foods can also provide high amounts of the daily nutrient requirements of phytochemicals, improve performance and mood, enhance the preventive state of the body, and help in healing some oxidative-related diseases of the consumer.

References

1. Zhang J-Y, Liu C, Zhou L, Qu K, Wang R, Tai M, Lei JWL, Wu QF, Wang Z (2012) A review of hydrogen as a new medical therapy. Hepato Gastroenterol Med Surg Trends 59:1026
2. Ohta S (2014) Molecular hydrogen as a preventive and therapeutic medical gas: initiation, development and potential of hydrogen medicine. Pharmacol Ther 144:1–11
3. West JB (2014) Henry Cavendish (1731–1810): hydrogen, carbon dioxide, water, and weighing the world. Am J Physiol Cell Mol Physiol 307:L1–L6
4. Barancik M, Kura B, Lebaron TW, Bolli R, Buday J, Slezak J (2020) Molecular and cellular mechanisms associated with effects of molecular hydrogen in cardiovascular and central nervous systems. Antioxidants 9:1281. https://doi.org/10.3390/antiox9121281
5. Alwazeer D, Liu FF-C, Wu XY, LeBaron WT (2021) Combating oxidative stress and inflammation in COVID-19 by molecular hydrogen therapy: mechanisms and perspectives. Oxid Med Cell Longev. https://doi.org/10.1155/2021/5513868
6. Fukai Y (2020) Molecular hydrogen for medicine: the art of ancient life revived. Springer Nature
7. Aziz M (2021) Liquid hydrogen: a review on liquefaction, storage, transportation, and safety. Energies 14:5917
8. Seo T, Kurokawa R, Sato B (2012) A convenient method for determining the concentration of hydrogen in water: use of methylene blue with colloidal platinum. Med Gas Res 2:1–6

9. Tian Y, Zhang Y, Wang Y, Chen Y, Fan W, Zhou J, Qiao J, Wei Y (2021) Hydrogen, a novel therapeutic molecule, regulates oxidative stress, inflammation, and apoptosis. Front Physiol 2281
10. Ohsawa I, Ishikawa M, Takahashi K, Watanabe M, Nishimaki K, Yamagata K, Katsura K, Katayama Y, Asoh S, Ohta S (2007) Hydrogen acts as a therapeutic antioxidant by selectively reducing cytotoxic oxygen radicals. Nat Med 13:688
11. Song G, Li M, Sang H, Zhang L, Li X, Yao S, Yu Y, Zong C, Xue Y, Qin S (2013) Hydrogen-rich water decreases serum LDL-cholesterol levels and improves HDL function in patients with potential metabolic syndrome. J Lipid Res 54:1884–1893
12. Nakayama M, Nakano H, Hamada H, Itami N, Nakazawa R, Ito S (2010) A novel bioactive haemodialysis system using dissolved dihydrogen (H2) produced by water electrolysis: a clinical trial. Nephrol Dial Transplant 25:3026–3033
13. Aoki K, Nakao A, Adachi T, Matsui Y, Miyakawa S (2012) Pilot study: effects of drinking hydrogen-rich water on muscle fatigue caused by acute exercise in elite athletes. Med Gas Res 2:1–6. https://doi.org/10.1186/2045-9912-2-12
14. Matsumoto A, Yamafuji M, Tachibana T, Nakabeppu Y, Noda M, Nakaya H (2013) Oral 'hydrogen water' induces neuroprotective ghrelin secretion in mice. Sci Rep 3:3273
15. Crowl DA, Jo Y (2007) Do: the hazards and risks of hydrogen. J Loss Prev Process Ind 20:158–164. https://doi.org/10.1016/j.jlp.2007.02.002
16. Najjar YSH (2013) Hydrogen safety: the road toward green technology. http://www.sciencedirect.com/science/article/pii/S036031991301358X
17. Tang C, Huang Z, Jin C, He J, Wang J, Wang X, Miao H (2009) Explosion characteristics of hydrogen-nitrogen-air mixtures at elevated pressures and temperatures. Int J Hydrogen Energy 34:554–561. https://doi.org/10.1016/j.ijhydene.2008.10.028
18. Zhou G, Goshi E, He Q (2019) Micro/Nanomaterials-augmented hydrogen therapy. Adv Healthc Mater 8:1900463
19. Jafta N, Magagula S, Lebelo K, Nkokha D, Mochane MJ (2021) The production and role of hydrogen-rich water in medical applications. Appl Water Sci Fundam Appl 1:273–298. https://doi.org/10.1002/9781119725237.ch10
20. Shimouchi A, Nose K, Shirai M, Kondo T (2012) Estimation of molecular hydrogen consumption in the human whole body after the ingestion of hydrogen-rich water. In: Oxygen transport to tissue XXXIII. Springer, pp 245–250
21. Sun X, Ohta S, Nakao A (2015) Hydrogen molecular biology and medicine. Springer, Berlin
22. Kikuchi K, Takeda H, Rabolt B, Okaya T, Ogumi Z, Saihara Y, Noguchi H (2001) Hydrogen particles and supersaturation in alkaline water from an Alkali–Ion–Water electrolyzer. J Electroanal Chem 506:22–27. https://doi.org/10.1016/S0022-0728(01)00517-4
23. Wang Y, Lv P, Kong L, Shen W, He Q (2021) Nanomaterial-mediated sustainable hydrogen supply induces lateral root formation via nitrate reductase-dependent nitric oxide. Chem Eng J 405:126905. https://doi.org/10.1016/j.cej.2020.126905
24. Epelle EI, Desongu KS, Obande W, Adeleke AA, Ikubanni PP, Okolie JA, Gunes B (2022) A comprehensive review of hydrogen production and storage: a focus on the role of nanomaterials. Int J Hydrogen Energy 47:20398–20431. https://doi.org/10.1016/j.ijhydene.2022.04.227
25. Henry M, Chambron J (2013) Physico-chemical, biological and therapeutic characteristics of electrolyzed reduced alkaline water (ERAW). Water 5:2094–2115
26. WebMD: Magnesium hydroxide suspension—uses, side effects, and more. https://www.webmd.com/drugs/2/drug-8802/magnesium-hydroxide-oral/details
27. Fujita K, Seike T, Yutsudo N, Ohno M, Yamada H, Yamaguchi H, Sakumi K, Yamakawa Y, Kido MA, Takaki A (2009) Hydrogen in drinking water reduces dopaminergic neuronal loss in the 1-methyl-4-phenyl-1, 2, 3, 6-tetrahydropyridine mouse model of Parkinson's disease. PLoS ONE 4:e7247
28. Botek M, Krejčí J, McKune AJ, Sládečková B, Naumovski N (2019) Hydrogen rich water improved ventilatory, perceptual and lactate responses to exercise. Int J Sports Med 40:879–885
29. Bulut M, Çelebi Sezer Y, Ceylan MM, Alwazeer D (2022) Hydrogen-rich water can reduce the formation of biogenic amines in butter. Food Chem. https://doi.org/10.1016/j.foodchem.2022.132613

30. Yao H-T, Yang Y-H, Li M-L (2019) Intake of molecular hydrogen in drinking water increases membrane transporters, p-glycoprotein, and multidrug resistance-associated protein 2 without affecting xenobiotic-metabolizing enzymes in rat liver. Molecules 24:2627
31. Alwazeer D, Delbeau C, Divies C, Cachon R (2003) Use of redox potential modification by gas improves microbial quality, color retention, and ascorbic acid stability of pasteurized orange juice. Int J Food Microbiol 89:21–29. https://doi.org/10.1016/S0168-1605(03)00125-9
32. Giroux HJ, Acteau G, Sabik H, Britten M (2008) Influence of dissolved gases and heat treatments on the oxidative degradation of polyunsaturated fatty acids enriched dairy beverage. J Agric Food Chem 56:5710–5716. https://doi.org/10.1021/jf800516x
33. Alwazeer D (2003) Intérêt de la modification du Eh par des gaz pour la maîtrise des micro-organismes d'altération du jus d'orange et ses implications sur la stabilité de la couleur et de la vitamine C: développement d'un appareil adapté à l'étude des micro-organismes considérés en conditions physico-chimiques définies (Doctoral dissertation, Dijon)
34. Cachon R, Alwazeer D (2019) Quality performance assessment of gas injection during juice processing and conventional preservation technologies. In: Value-added ingredients and enrichments of beverages, Academic Press, pp 465–485
35. Cachon R, Alwazeer D (2019) Quality performance assessment of gas injection during juice processing and conventional preservation technologies
36. LeBaron TW, Kharman J, McCullough ML (2021) An H2-infused, nitric oxide-producing functional beverage as a neuroprotective agent for TBIs and concussions. J Integr Neurosci 20:667–676
37. Bhagwat VR (2019) Safety of water used in food production. In: Food safety and human health. Elsevier, pp 219–247
38. Ceylan MM, Bulut M, Alwazeer D, Koyuncu M (2022) Evaluation of the impact of hydrogen-rich water on the quality attribute notes of butter. J Dairy Res. https://doi.org/10.1017/S0022029922000681
39. Briffa J, Sinagra E, Blundell R (2020) Heavy metal pollution in the environment and their toxicological effects on humans. Heliyon. 6:e04691
40. Chowdhury S, Mazumder MAJ, Al-Attas O, Husain T (2016) Heavy metals in drinking water: occurrences, implications, and future needs in developing countries. Sci Total Environ 569:476–488
41. Alwazeer D, Ceylan MM, Bulut M, Koyuncu M (2022) Evaluation of the impact of hydrogen-rich water on the deaccumulation of heavy metals in butter. J Food Saf. https://doi.org/10.1111/jfs.13005
42. Köktürk M, Atalar MN, Odunkıran A, Bulut M, Alwazeer D (2022) Evaluation of the hydrogen-rich water alleviation potential on mercury toxicity in earthworms using ATR-FTIR and LC–ESI–MS/MS spectroscopy. Environ Sci Pollut Res 29(13):19642–19656. https://doi.org/10.1007/s11356-021-17230-x
43. Köktürk M, Yıldırım S, Eser G, Bulut M, Alwazeer D (2022) Hydrogen-rich water alleviates the nickel-induced toxic responses (inflammatory responses, oxidative stress, DNA damage) and ameliorates cocoon production in earthworm. Biol Trace Elem Res 200(7):3442–3452. https://doi.org/10.1007/s12011-021-02908-7
44. Alwazeer D, Bulut M, Çelebi Y (2022) Hydrogen-rich water can restrict the formation of biogenic amines in red beet pickles. Fermentation 8:741
45. Alwazeer D (2023) Restrictive effect of molecular hydrogen on the formation of biogenic amines: possible mechanisms and perspectives. Process Biochem 132:255–262. https://doi.org/10.1016/j.procbio.2023.07.015
46. Ceylan MM, Silgan M, Elnasanelkasim MA, Alwazeer D (2023) Impact of washing crude olive pomace oil with hydrogen-rich water and incorporating hydrogen into extraction solvents on quality attributes and phytochemical content of oil. J Food Meas Charact. https://doi.org/10.1007/s11694-022-01801-8
47. Alwazeer D (2019) Reducing atmosphere packaging technique for extending the shelf-life of food products. J Inst Sci Technol 9:2117–2123. https://doi.org/10.21597/jist.539744

48. Bulut M, Okutan G, Alwazeer D, Boran G (2023) Hydrogen inclusion in modified atmosphere extends the shelf life of chilled rainbow trout fillets. Turkish J Fish Aquat Sci 23
49. Sezer YÇ, Bulut M, Boran G, Alwazeer D (2022) The effects of hydrogen incorporation in modified atmosphere packaging on the formation of biogenic amines in cold stored rainbow trout and horse mackerel. J Food Compos Anal 112:104688
50. Alwazeer D, Örs B, Tan K (2020) Reducing atmosphere packaging as a novel alternative technique for extending shelf life of fresh cheese. J Food Sci Technol 57:3013–3023. https://doi.org/10.1007/s13197-020-04334-4
51. Alwazeer D, Özkan N (2022) Incorporation of hydrogen into the packaging atmosphere protects the nutritional, textural and sensorial freshness notes of strawberries and extends shelf life. J Food Sci Technol 59:3951–3964. https://doi.org/10.1007/s13197-022-05427-y
52. Li L, Zeng Y, Cheng X, Shen W (2021) The applications of molecular hydrogen in horticulture. Horticulturae 7:513
53. Guan Q, Ding XW, Jiang R, Ouyang PL, Gui J, Feng L, Yang L, Song LH (2019) Effects of hydrogen-rich water on the nutrient composition and antioxidative characteristics of sprouted black barley. Food Chem 299:125095. https://doi.org/10.1016/j.foodchem.2019.125095
54. Hu H, Zhao S, Li P, Shen W (2018) Hydrogen gas prolongs the shelf life of kiwifruit by decreasing ethylene biosynthesis. Postharvest Biol Technol 135:123–130
55. Hu H, Li P, Wang Y, Gu R (2014) Hydrogen-rich water delays postharvest ripening and senescence of kiwifruit. Food Chem 156:100–109. https://doi.org/10.1016/j.foodchem.2014.01.067
56. Liu S, Zha Z, Chen S, Tang R, Zhao Y, Lin Q, Duan Y, Wang K (2022) Hydrogen-rich water alleviates chilling injury-induced lignification of Kiwifruit by inhibiting peroxidase activity and improving antioxidant system. J Sci Food Agric. https://doi.org/10.1002/jsfa.12272
57. Sun Y, Qiu W, Fang X, Zhao X, Xu X, Li W (2023) Hydrogen-rich water treatment of fresh-cut Kiwifruit with slightly acidic electrolytic water: influence on antioxidant metabolism and cell wall stability. Foods 12:426
58. Zhao X, Meng X, Li W, Cheng R, Wu H, Liu P, Ma M (2021) Effect of hydrogen-rich water and slightly acidic electrolyzed water treatments on storage and preservation of fresh-cut kiwifruit. J Food Meas Charact 15:5203–5210
59. European Parliament and the Council of the European Union: Regulation (EC) (2008) No 1333/2008 of the European Parliament and of the Council of 16 December 2008 on food additives. Off J Eur Union. 354:16–33
60. Le Baron TW, Kharman J, McCullough M (2021) Effects of an H_2-infused, nitric oxide-producing functional beverage on exercise and cognitive performance. J Sci Med 3:1–15
61. Mizuno K, Sasaki A, Ebisu K, Tajima K, Kajimoto O, Nojima J, Kuratsune H, Hori H, Watanabe Y (2017) Hydrogen-rich water for improvements of mood, anxiety, and autonomic nerve function in daily life. Med Gas Res 7:247–255. https://doi.org/10.4103/2045-9912.222448
62. Kamimura N, Nishimaki K, Ohsawa I, Ohta S (2011) Molecular hydrogen improves obesity and diabetes by inducing hepatic FGF21 and stimulating energy metabolism in db/db mice. Obesity 19:1396–1403
63. Botek M, Khanna D, Krejčí J, Valenta M, McKune A, Sládečková B, Klimešová I (2022) Molecular hydrogen mitigates performance decrement during repeated sprints in professional soccer players. Nutrients 14:508
64. Dixon BJ, Tang J, Zhang JH (2013) The evolution of molecular hydrogen: a noteworthy potential therapy with clinical significance. Med Gas Res 3:1–12. https://doi.org/10.1186/2045-9912-3-10
65. Matsushita T, Kusakabe Y, Kitamura A, Okada S, Murase K (2011) Protective effect of hydrogen-rich water against gentamicin-induced nephrotoxicity in rats using blood oxygenation level-dependent MR imaging. Magn Reson Med Sci 10:169–176
66. Fu Y, Ito M, Fujita Y, Ito M, Ichihara M, Masuda A, Suzuki Y, Maesawa S, Kajita Y, Hirayama M (2009) Molecular hydrogen is protective against 6-hydroxydopamine-induced nigrostriatal degeneration in a rat model of Parkinson's disease. Neurosci Lett 453:81–85

67. Yoritaka A, Takanashi M, Hirayama M, Nakahara T, Ohta S, Hattori N (2013) Pilot study of H2 therapy in Parkinson's disease: A randomized double-blind placebo-controlled trial. Mov Disord 28:836–839
68. Ge L, Yang M, Yang N-N, Yin X-X, Song W-G (2017) Molecular hydrogen: a preventive and therapeutic medical gas for various diseases. Oncotarget 8:102653
69. Cianciosi D, Forbes-Hernández TY, Regolo L, Alvarez-Suarez JM, Navarro-Hortal MD, Xiao J, Quiles JL, Battino M, Giampieri F (2022) The reciprocal interaction between polyphenols and other dietary compounds: Impact on bioavailability, antioxidant capacity and other physico-chemical and nutritional parameters. Food Chem 375:131904
70. Zhang H, Zheng J, Liu X, Ding Q, Jiang L, Guo H, Ren F (2012) Milk protein and fat play different roles in affecting the bioavailability and the antioxidant activity of jujube juice phenolics in rats. Mol Nutr Food Res 56:1511–1519
71. Alwazeer D et al (2023) Hydrogen-rich solvent method in phytochemical extraction: potential mechanisms and perspectives. Phytochem Anal (Abstract) https://doi.org/10.1002/pca.3304
72. Alwazeer D, Elnasanelkasim MA, Çiçek S, Engin T, Çiğdem A, Karaoğul E (2023) Comparative study of phytochemical extraction using hydrogen-rich water and supercritical fluid extraction methods. Process Biochem
73. Alwazeer D, Elnasanelkasi M (2023) Hydrogen-rich water as a green solvent for the extraction of phytochemicals from agri-food wastes. Sustain Chem Pharm
74. Alwazeer D et al (2023) Hydrogen incorporation into solvents can improve the extraction of phenolics flavonoids anthocyanins and antioxidants: a case-study using red beetroot. Ind Crops Prod 202:117005. https://doi.org/10.1016/j.indcrop.2023.117005
75. Alwazeer D et al (2023) Incorporation of molecular hydrogen into solvents increases the extraction efficiency of phenolics flavonoids anthocyanins and antioxidants: the case of lemon peels. Front Sustain Food Syst 7. https://doi.org/10.3389/fsufs.2023.1223027
76. Alwazeer D, Elnasanelkasim MA, Engin T, Çiğdem A (2023) Use of hydrogen-rich water as a green solvent for the extraction of phytochemicals: case of olive leaves. J Appl Res Med Aromat Plants 100472
77. Zor M, Bulut M, Karagöz SG, Çetintaş Y, Alwazeer D (2024) Use of hydrogen-rich water in rice milk preparation improves the nutritional and sensory properties of product. Food Chem 437:137821. https://doi.org/10.1016/j.foodchem.2023.137821

Chapter 20
Differential Effects of Carbohydrates on the Generation of Hydrogen and Methane in Low- and High-Methane-Producing Rats

Oleg S. Medvedev, Anastasiia Yu. Ivanova, Margarita A. Belousova, Stepan V. Toshchakov, Anastasia S. Krylova, Ivan V. Shirokov, Olga N. Obolenskaya, Tatiana A. Kuropatkina, Grigorii N. Bondarenko, and Ilya B. Gartseev

Abstract The goal of the study was to check the hypothesis that low- and high-methane producers will react differently to the administration of non-digestible carbohydrates. It was discovered in our previous studies that Wistar rats from the Puschino nursery (SPF status) were low methane producers, whereas conventional rats from Stolbovaya nursery were high methane producers. Hydrogen and methane

---

O. S. Medvedev (✉) · A. Yu. Ivanova · M. A. Belousova · I. V. Shirokov · O. N. Obolenskaya · T. A. Kuropatkina
Faculty of Medicine, Lomonosov Moscow State University, Moscow 119991, Russia
e-mail: oleg.omedvedev@gmail.com

A. Yu. Ivanova
e-mail: dnp-reg74@yandex.ru

I. V. Shirokov
e-mail: Iaefree@yandex.ru

O. N. Obolenskaya
e-mail: sliers@mail.ru

O. S. Medvedev · A. Yu. Ivanova
Institute of Experimental Cardiology, National Medical Research Centre of Cardiology, Moscow 121552, Russia

S. V. Toshchakov · A. S. Krylova
Kurchatov Center for Genome Research, National Research Center "Kurchatov Institute", Moscow 123098, Russia

G. N. Bondarenko
Faculty of Chemistry, Lomonosov Moscow State University, Moscow 119991, Russia

I. B. Gartseev
Institute of Artificial Intelligence, Russian Technological University MIREA, Moscow 119454, Russia
e-mail: gartseev@mirea.ru

J. Slezak and B. Kura (eds.), *Molecular Hydrogen in Health and Disease*, Advances in Biochemistry in Health and Disease 27, https://doi.org/10.1007/978-3-031-47375-3_20

breath tests and the taxonomic content of the gut microbiota were evaluated in 25 rats from each nursery. Samples of the exhaled air were taken from awake rats using nose-only apparatus and analyzed by gas chromatography. The taxonomic content of the gut microbiota was evaluated in each rat by the 16S rRNA method. Lactulose, Guar Gum, and inulin were administered by gavage to each rat in 1-week time intervals. Levels of hydrogen and methane in the samples of exhaled air were measured during 8 h after carbohydrate administration. Taxonomic microbiome compositions were quite different between groups. Low methane-producing rats had low alpha and beta diversity, higher abundance of *Christensenellaceae* and *Akkermansia* bacteria, lower abundance of *Helicobacteraceae,* and absence of *Methanobacteriaceae*) that show similarity to the microbiome of the newborns and children. High methane-producing rats (from Stobovaya nursery) had a much higher diversity of microbiota bacteria, a higher abundance of hydrogen-consuming microorganisms, like Helicobacteraceae and Methanobacteriaceae, and close to the microbiota composition in the elderly. The gavage of carbohydrates in low-methane-producing rats was followed just by the increase in hydrogen level in exhaled air, whereas the same carbohydrates evoked an increase in methane level only. We speculate that the administration of the exogenous hydrogen (hydrogen-rich water) will be more efficient in increasing the antioxidant defense in the elderly because taking the food fibers is not followed by the increase of hydrogen level in the blood.

Keywords Hydrogen/methane breath test · Non-absorbable carbohydrates · Taxonomic microbiota composition · Low- and high methane-producing rats

Introduction

In the scientific community it is widely discussed the possibility of gut microbiota involvement not only in the pathogenesis of gastrointestinal diseases—Crohn's disease, irritable bowel syndrome, diarrhea, etc. but also the development of obesity [1], atherosclerosis [2], type 2 diabetes [3], heart failure [4], arterial hypertension [5] and other pathologies. Metagenomic studies allow the detection of a high number of microorganisms in the intestinal microbiome, but they do not allow to evaluate the function or metabolic features of each of the detected microorganisms or their groups and communities. Metabolomic studies, i.e., the study of low-molecular-weight biomarkers produced by microbiota, attract a lot of attention from the scientific community. Such biomarkers include gases, such as hydrogen (H_2), methane (CH_4), hydrogen sulfide (H_2S), and short-chain fatty acids (SCFA) together with other low-molecular-weight volatile substances [6]. It was shown that H_2 was exclusively produced by the gut microbiota, but not by the somatic cells of the host [7, 8]. For a long time, biological CH_4 formation was considered to be produced only by the methanogenic *Archaeae*. However, in recent years it was shown in vitro experiments, that CH_4 could be produced by the human cell lines under hypoxic conditions [9]. Almost 70% of the microbial species listed in the Human Microbiome Project Gastrointestinal Tract (HMP GI) reference genome database encoded hydrogenases which are necessary for demonstrating hydrogenic (production of H_2), or hydrogenotrophic (consuming H_2) activities [10]. One of the first tests to examine H_2 production in the gastrointestinal tract was the lactulose breath test. Lactulose is a carbohydrate that humans do not digest. When administered to humans or rodents, lactulose will traverse the gut until it reaches a depot of intestinal flora capable of digesting the substrate. As a byproduct of fermentation, hydrogen is produced. Hydrogen (as well as other intestinally produced gases) enters the bloodstream, circulates, diffuses into the alveolar space, and is exhaled on the breath [11].

According to a study of normal subjects, the primary constituents of flatus are H_2 (3–20%), CO_2 (9–14%), CH_4 (7.2%), O_2 (2–4%), and H_2S (0.00028%) [12, 13]. *Archaea* of the microbiota metabolize H_2. Two species—*Methanobrevibacter smithii* and *Methanosphaer stadtmanae* are the main methane producers in the human colon. Their number increases all along the colon until reaching a maximal value in the rectum. *Methanobrevibacter smithii* uses hydrogen to reduce CO_2 to methane whereas *Methanosphaera stadtmanae* uses hydrogen for the reduction of methanol to methane. Four hydrogen molecules and one CO_2 molecule are required to synthesize one molecule of methane: $CO_2 + 4H_2 \rightarrow CH_4 + 2H_2O$ [14, 15].

Molecular hydrogen (H_2) up to 2007 was considered a physiologically inert gas. Professor Shigeo Ohta with co-workers has published the keystone article in Nature with proof of the antioxidant activity of H_2 [16]. In the last two decades, accumulating evidence from pre-clinical and clinical studies has indicated that H_2 may act as an antioxidant to exert therapeutic and preventive effects on various disorders, including metabolic diseases [17]. Furthermore, the obtained evidence shows that hydrogen has an impact on signaling pathways whereby information is transmitted across the cell membrane as well as exerts cytoprotection and decreases the synthesis of proinflammatory cytokines and apoptosis. Subsequent clinical trials undertaken to test the antioxidant properties of H_2 largely confirmed the results of previous experimental studies on animals. Thus, researchers corroborated the cardioprotective and neuroprotective effects of H_2 as well as a positive hydrogen effect on endothelial dysfunction [18, 19].

On the one hand, the discovery of the positive effects of molecular hydrogen led to an increase in the number of studies investigating exogenous hydrogen effects, and on the other, revived interest in the analysis of effects of endogenous hydrogen produced by gut microbiota.

Several studies have shown that healthy individuals could be divided into three conditional subgroups according to carbohydrate-evoked hydrogen and methane production. The first group has low levels of hydrogen and methane in the exhaled air after lactulose administration. In the second group, methane-positive, there is an increase in methane and almost no increase in hydrogen. In the third group, an increase in the level of hydrogen only was observed [20, 21]. The use of non-digestible dietary fibers of various structures as prebiotic agents is a safe and effective method to improve the functional state of the microbiota, as they are the main substrate for hydrogen-generating microbes [22–24].

Thus, our study aimed to investigate the possibility of increasing the antioxidant defense of the organism by using different non-digestible carbohydrates possibly via an increase in endogenous H_2 production in low- and high-methanogenic rats.

Materials and Methods

Laboratory Animals

The keeping of animals and all manipulations were carried out following Directive 2010/63/EU of the European Parliament and the Council of the European Union on the protection of animals used for scientific purposes. The experimental protocol was approved by the Ethics Committee of Moscow State University (application number 129-Zh dated May 31, 2021). Animals were obtained from two different nurseries. The first batch, 25 rats—from the nursery of laboratory animals "Puschino" (Institution of Science Institute of Bioorganic Chemistry, Moscow region). Wistar rats had SPF status according to a certificate. The second batch, 25 rats—from the nursery of laboratory animals branch "Stolbovaya" (Federal Medico-Biological Agency of Russia, Moscow district). The rats met the genetic standard, which was confirmed by a quality certificate.

Immediately after arrival, the animals were placed in vivarium rooms and examined for injuries, signs of pain, and/or distress and seated in T3 cages, 2 rats per cage. The animals were kept under standard conditions with 12-h daylight hours and unlimited access to water and food, and regular control of temperature and humidity in the room. The term of adaptation to the conditions of keeping in the vivarium was at least 7 days. For optimal performance of experimental manipulations and to avoid the influence of stress on the results of the study, all animals were subjected to the handling procedure.

Intragastric Administration of Substances

All investigated substances were administered by gavage using a sterile intragastric tube size 16 (for rats 200–300 g). For each group of animals, «Puschino» and «Stolbovaya» the individual intragastric tubes were used to minimize the transfer of microbiota microorganisms between groups.

Dietary Compounds Used for Hydrogen/methane Breath Tests

Lactulose (Fresenius Kabi IPSUM S.r.l., Italy). Lactulose (2 g/kg by gavage) is an artificial disaccharide consisting of galactose and fructose residues.

PHGG (Partially hydrolyzed guar gum, 4 g/kg, Optifiber®, Nestle Health Science, Germany). The main component in its composition is the extract of the fruit of the Cyamopsis tetragonoloba, or guar gum. The chemical structure is 100% guar galactomannans, consisting of mannose and galactose in a ratio of approximately 2:1.

Inulin (Protein company, Russia). Inulin is a polymer of D-fructose, which consists of 30–35 fructose residues in the furanose form. Inulin was administrated at the dosage of 4 g/kg.

Calculation of the weight of the administered substance was made based on the body weight of the animal. All substances were dissolved in 2–3 ml of purified water before administration. Each type of nondigestible carbohydrate was tested on each rat from two groups. The time interval between sequential measurements on each rat was one week to exclude the effect of the previous test.

Experimental Setup for Exhaled Air Samples Collection

The setup used a modification of the life-support system for the quantitative collection of hydrogen gas [25], but instead of placing a rat in the sealed animal chamber we used a version of the Rat Breathalyzer system [26] with the addition of Allay™ restraint collar to facilitate the sampling of the exhaled air for measuring the levels of hydrogen and methane [27] (Fig. 20.1). In the present study, the Allay™ restraint collar was used in conjunction with a ventilated nose-only chamber. The set-up was built up by the one of co-authors (Bondarenko G.).

The experimental setup of a gas installation consists of three key elements—a cylinder case, a life support system, and a circulating ventilation system.

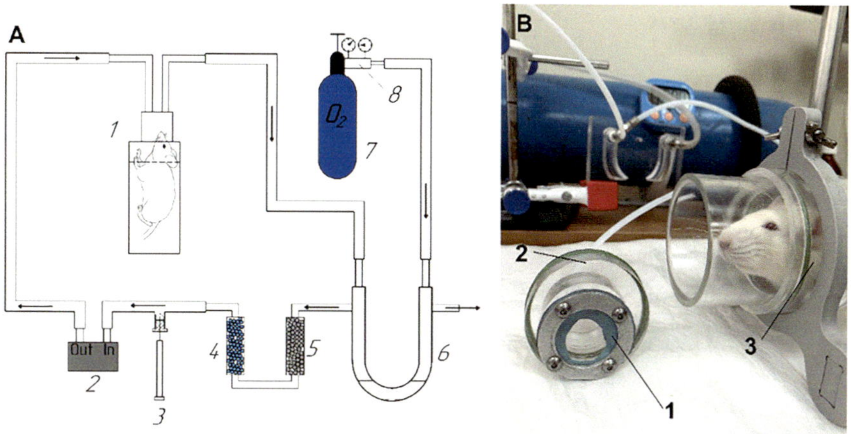

Fig. 20.1 A diagram of the experimental setup for collecting the total air sample in the experiment. 1—cylinder case, 2—air pump, 3—glass syringe, 4—moisture absorbent (hydrogel), 5—carbon dioxide absorbent (soda lime), 6—manometer, 7—oxygen cylinder, 8—gas pressure reducer; **B** photo of an animal cage of an experimental installation: 1—the plane of separation of two chambers with a hole in the rubber membrane, 2—head placement chamber for repeated respiration of the animal, 3—body placement chamber for fixing the animal in the neck area

The animal's head is fixed with a neck plastic clamp, tightly fitting in the neck area only from above and on the sides, which allows the animal to breathe freely. The rat's nose is located in a special head compartment, with a tightly fitting rubber membrane around it. Thus, the tightness of the installation is achieved, and the exhaled air remains circulating inside without leaving the system. Input and output connectors in the head compartment.

The exhaled air recirculated through filters, is cleaned of moisture and CO2. O_2 was maintained at the level of 20–21%. Thus, the accumulation of hydrogen and methane necessary for measuring gas concentrations in the gas installation system occurs. Air intake is carried out through a sealed valve, with a special gas-tight Hamilton syringe for a gas chromatograph. The advantage of our system is in fact that we were able to collect only exhaled air without interferences with the flatus, which make our system similar to the sampling air in the human breath tests.

The first air sample was taken from the animal after 12 h fasting when rats had free access to water. Then the animal was taken out of the cage and a solution of a particular carbohydrate was administered via gavage. In each rat, exhaled air samples were collected five times: baseline, 2, 4, 6, and 8 h after carbohydrate administration. After each sampling, the rat was returned to its cage until the next time point. To rule out crosstalk between different nondigestible carbohydrates in the gut microbiota, rats were kept for several days on a standard diet before measurements were made with another nondigestible carbohydrate.

Sample Analysis

Determination of H_2, CH_4, and CO in exhaled air was carried out on a gas chromatograph (TRILyzer mBA-3000, Japan), after calibration with gases, containing low (H_2, CH_4, and CO—5 ppm each in the air) and high concentrations (H_2, CH_4 and CO—50 ppm each in the air).

The concentration–time curve was plotted for H_2 and CH_4, and the area under the AUC 0–8 h curve (ppm*h) was calculated, which reflects the rate of gas generation during 8 h of the experiment.

Collection of Feces Samples

Fecal samples were collected in individual Eppendorf tubes. The Eppendorf tube was immediately capped, labeled, and frozen at − 80 °C. For each animal, 2 stool samples were taken at a time. The sampling was carried out on the eve of the breath test.

Analysis of Microbial Community Composition

DNA from feces was isolated using QIAampPowerFecal Pro DNA Kit (Qiagen, Hilden, Germany), according to the manufacturer's instructions. Quantity and concentration of DNA were assessed by Qubit™ 4 fluorometer (Thermo Fisher Scientific, MA, USA). 1–2 ng of DNA was used for library preparation. The amplicon libraries of the 16S rRNA V4 hypervariable region were prepared using a two-step PCR with the primers 515F [28] and Pro-mod-805R [29].

PCR of every DNA sample was performed in replicates according to the earlier protocol [30]. Amplification was performed on a Veriti thermal cycler (Applied Biosystems, MA, USA). The resulting libraries were checked on an agarose gel and pooled equimolarly. The final pool was purified using the QIAquick Gel Extraction Kit (Qiagen, Germany) according to the manufacturer's protocols.

Sequencing was performed using the MiSeq™ Personal Sequencing System (Illumina, San Diego, CA, USA) with 2 × 156 bp paired-end.

Statistical Analysis

Statistical analysis of the results and randomization of the groups was performed using GraphPad Prism 8 software. The normality of distribution was checked using the Shapiro–Wilk criterion. A comparison of respiratory test results with a frequency of more than 3 measured points was performed using the area under the curve (AUC) calculation. A one-factor one-way ANOVA analysis of variance was used to compare the mean values of a single index in more than two samples. To determine the simultaneous effect of group and duration of exposure, as well as the interaction between these factors, a two-factor ANOVA analysis of variance was used to identify differences for pairwise comparison of groups, using a paired and unpaired t-test for analysis of dependent and independent samples, respectively. For pairwise comparisons of groups with non-normal distributions, the Wilcoxon test was used for dependent samples and the Mann–Whitney test for independent samples. Correlations were calculated using the Spearman rank correlation coefficient. The exclusion of statistical outliers was performed using ROUT criterion with Q not exceeding 1%. Differences were considered statistically significant at $p < 0.05$. All data are presented as mean ± standard deviation (Mean ± SD). Analysis of microbial community composition, alfa- and beta-diversity metrics, and statistics were performed using R (version 4.2.0). High-quality read pairs were processed with DADA2 pipeline [29], according to the published protocol [31]. The taxonomy of amplified sequence variants (ASVs) was determined with a naive Bayesian classifier using the Silva138 database [32]. For further analysis reads were rarefied using the Phyloseq package [33]. The obtained ASV reference sequences, sample metadata, abundance tables, and taxonomy were imported into the Phyloseq package, and all further operations were performed with the Phyloseq object. Visualization of microbial community composition was performed with a microeco package (multivariate analysis of variance using distance matrices) [34] and vegan package [35].

Results

The Effect of a Single Carbohydrate of Various Structures Load (Lactulose, Inulin, PHGG) on the Content of Gaseous Markers (H_2 and CH_4) in the Rat's Exhaled Air

A significant difference in the level of H_2 and CH_4 within the experimental group "Puschino" was noticed in the case of using lactulose solution and PHGG fibers, $p < 0.0001$ and $p = 0.001$, respectively (Fig. 20.2a). In the inulin group, the levels of hydrogen and methane were approximately the same (about 35–40 ppm*h) and did not differ significantly, $p = 0.6$ (Fig. 20.2a). In a comparison of the H_2 and CH_4 levels within the experimental group "Stolbovaya", a high level of significance, $p < 0.0001$, of differences between the level of gaseous metabolites in all experimental groups, was recorded (Fig. 20.2b). On average, the level of methane in all experimental groups was 12 times higher compared to the level of hydrogen, the average total hydrogen level was 25 ± 8 ppm*h (Lactulose 35 ± 23 ppm*h, PHGG 18 ± 13 ppm*h, Inulin 22 ± 13 ppm*h), the average level of methane 320 ± 38 ppm*h (Lactulose 334 ± 122 ppm*h, PHGG 276 ± 134 ppm*h, Inulin 350 ± 156 ppm*h) (Fig. 20.2b).

When comparing rats with initially different basic levels of hydrogen-producing activity of the microbiota, it was shown that the hydrogen response to some of the carbohydrates in these rats was different. Thus, in the Lactulose and PHGG groups, a significant increase in hydrogen production in response to load was noted in rats

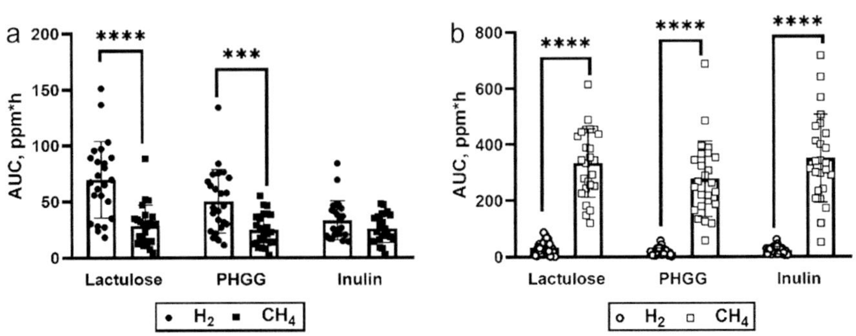

Fig. 20.2 Comparison of the H_2 and CH_4 in the exhaled air after a single gavage of nondigestible carbohydrates of various structures (lactulose, PHGG, and inulin) inside groups of rats from the **a** "Puschino" group and **b** the "Stolbovaya" group

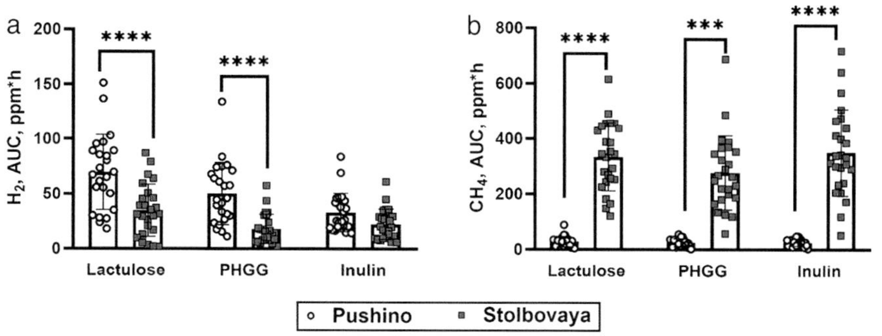

Fig. 20.3 Comparison of the **a** H_2 and **b** CH_4 levels in the exhaled air after a single gavage of different nondigestible carbohydrates among rats from the "Puschino" and "Stolbovaya" groups

from the "Puschino" group compared to rats from the "Stolbovaya" group by 1.9 times and 2.7 times, respectively, $p < 0.0001$ (Fig. 20.3a). The reaction to inulin in rats with an initially high hydrogen potential was 1.4 times higher compared to rats from "Stolbovaya" group, however, this difference was not statistically significant, $p = 0.2$ (Fig. 20.3a).

The methane response to a single gavage of non-digestible carbohydrates in groups of rats with initially different baseline levels of methanogenic activity also differed. In the "Stolbovaya" group the level of methane in response to a single application of nondigestible carbohydrate was 11.5 times or higher than in rats from the "Puschino" group (with the use of Lactulose 12 times higher, PHGG by 11.5 times, Inulin by 14 times), $p < 0.0001$ (Fig. 20.3b).

All rats from the "Puschino" group, which initially had a basic hydrogen potential and did not have a pronounced methanogenic activity, had a hydrogen response to a single application of all types of additional dietary compounds (Fig. 20.4). The increase in hydrogen production increased went as follows: inulin → PHGG → lactulose. The maximum increase in H_2 production was recorded in the case of lactulose injection. All groups differed statistically significantly with a level of significance $p = 0.03$ between the PHGG vs Inulin group, $p = 0.009$ between the Lactulose vs PHGG group, and $p < 0.0001$ between the Lactulose and Inulin groups (Fig. 20.4a). At the same time, when using all nondigestible carbohydrates, neither a significant increase in the level of methane relative to the "fasting" state (zero measurement point) was recorded, nor differences in the level of methane in the exhaled air between the groups, the significance level $p > 0.9$ between all groups (Fig. 20.4b).

In rats from the "Stolbovaya" group, a high methane-producing activity (Fig. 20.4b) and a low hydrogen potential were noted (Fig. 20.4a). When using the same nondigestible carbohydrates as in the experiment with rats from the "Puschino" group, the response of the rats was radically opposite. No increase in hydrogen was recorded in any of the experimental groups, while the baseline level of hydrogen was also very low, $p > 0.9$ (Fig. 20.4a).

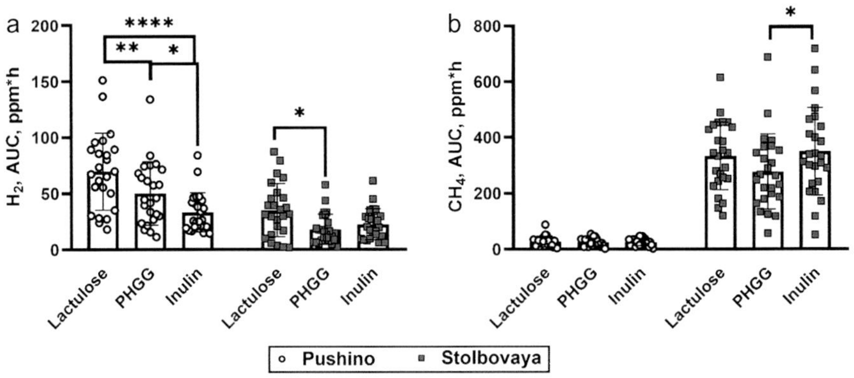

Fig. 20.4 Comparison of the **a** H_2 and **b** CH_4 level in the exhaled air after a single use of different nondigestible carbohydrates inside groups of rats from the "Puschino" and "Stolbovaya" groups

The methanogenic response to fiber applications was high in all three groups and increased consistently in the PHGG → lactulose → inulin series. Statistically significant differences were found only between the PHGG and inulin groups, $p = 0.03$. Differences between the Lactulose vs PHGG and Lactulose vs Inulin groups were $p = 0.1$ and $p = 0.9$, respectively (Fig. 20.4b).

Analysis of the Gut Microbiota Composition

After demultiplexing and read preprocessing, 19,201 read pairs per sample replicate were obtained on average. After DADA2 processing steps (additional filtering, denoising, merge, and chimera removal) the mean read count per replicate was 15,017. Rarefaction analysis showed good saturation for all sequenced samples, supposing the sufficiency of sequencing depth. DADA2 pipeline resulted in the generation of 1854 amplified sequence variants (ASVs).

Microbial Diversity Analysis

Analysis of alpha-diversity metrics performed by the microbiome package showed that the number of observed ASVs, as well as Shannon and inverse Simpson indexes, were significantly higher in the "Stolbovaya" group compared with the "Puschino" group ($p < 0.001$) (Fig. 20.5A). Beta-diversity analysis, performed by adonis2 function of vegan package using Bray–Curtis distances showed that microbial communities of "Puschino" and "Stolbovaya" groups significantly differ from each other (10,000 permutations; F1, 96 = 88.36, $p = 0.001$), which is also clearly supported by PCoA analysis (Fig. 20.5B). Out of 1854 ASVs 717 were specific for "Puschino"

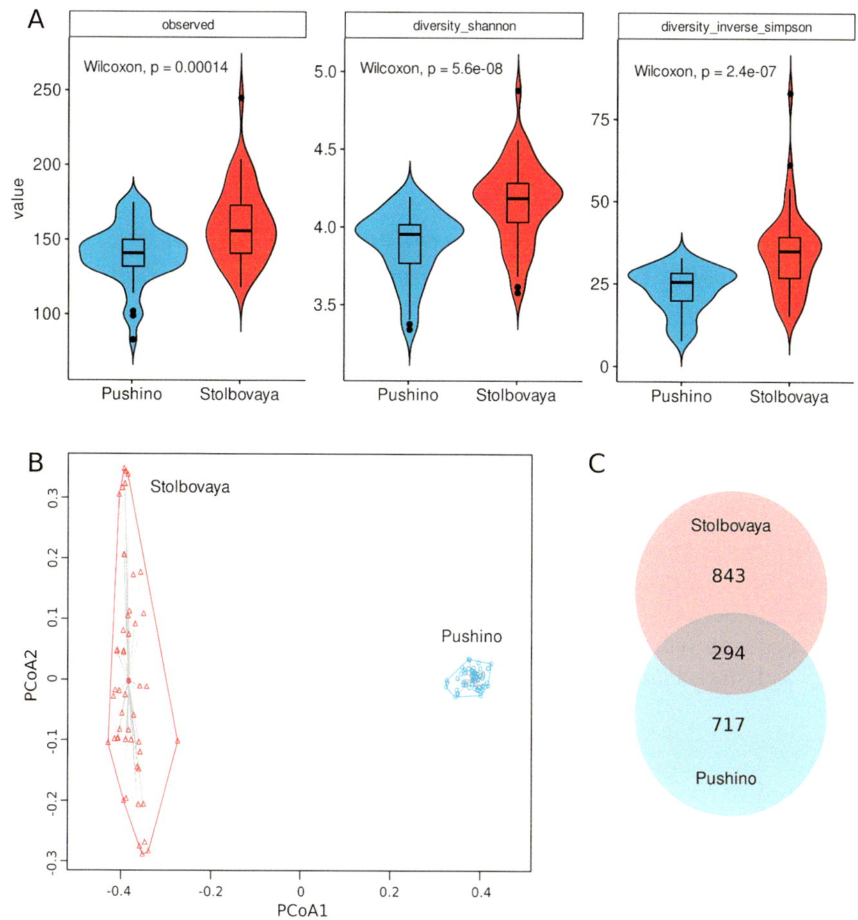

Fig. 20.5 Diversity analysis of experimental groups. **A** visualization of comparative analysis of alpha-diversity metrics; **B** Principal coordinate analysis (PcoA) of bacterial communities in groups of rats from the "Puschino" and "Stolbovaya" groups; **C** Venn diagram of ASVs, detected in experimental groups

group, 843 were specific for "Stolbovaya" group, while 294 ASVs were detected in both experimental groups (Fig. 20.5C).

Microbial Community Composition

The community composition of the gut microbiota in rats was dominated by *Firmicutes* and *Bacteroidota* phyla in all experimental groups. *Campylobacterota* was present in significant numbers in feces samples from the "Stolbovaya" group, while it was practically not detected in the "Puschino" group. The *Bacteroidia* and *Clostridii* classes were prevalent in both groups. On the order level, *Bacteroidales*, *Oscillospirales*, and *Lachnospirales* dominated in both groups. Among other bacterial orders, *Lactobacillales* predominated in the "Stolbovaya" group, *Clostridia UCG-014* prevailed in the "Puschino" group.

Correlation Analysis Between Taxonomic Composition and the Level of Gaseous Metabolites (H_2 and CH_4) of Gut Microbiota

Analysis of correlations between the level of gaseous metabolites (H_2 and CH_4) and microbial community composition was performed by redundancy analysis and analysis of Pearson's correlation in the microeco package. It was shown that the microbial communities in feces samples from the "Stolbovaya" group correlated with the level of methane, while elevated hydrogen levels strongly correlated with the microbial composition of feces samples from the "Puschino" group (Fig. 20.6). Among bacterial taxa the strongest positive correlation with hydrogen was shown for *Bacteroides*, *Alistipes*, and *Enterococcus*, while the level of methane positively correlated with representatives of the genera *Prevotella*, *Prevotellaceae NK3B31-group* and *Rickenellaceae RC9 gut group*.

Discussion

It was shown in our study that rats from two nurseries could be quite different concerning the taxonomic composition of the gut microbiota and the ratio between hydrogen and methane production.

One group of rats was purchased from the Puschino nursery and had an SPF (Specific Pathogen Free) status. These rats were characterized by the low diversity of microbiota organisms, alpha-diversity metrics performed by the microbiome package showed that the Shannon and inverse Simpson indexes were significantly lower in comparison with the other (Stolbovaya) group ($p < 0.001$). Spectrums of the microbiota microorganisms were quite different in the two groups studied. From the 1011 number of amplified sequence variants (ASVs) in the Puschino group only 294 were common with the 1137 ASVs from the Stolbovaya group.

Quite an accidental finding of the robust differences between two groups of rats which could model the existence of high hydrogen- and high methane groups of humans [20, 21, 36] created the basis for testing the hypothesis that endogenous hydrogen availability may depend on the type of taken undigestible carbohydrates as well on the ratio between hydrogenogens and methanogens.

The results of our study show that in the group of hydrogen-producing rats, gavage of lactulose, guar gum, or inulin was followed just by the fiber-dependent increase in

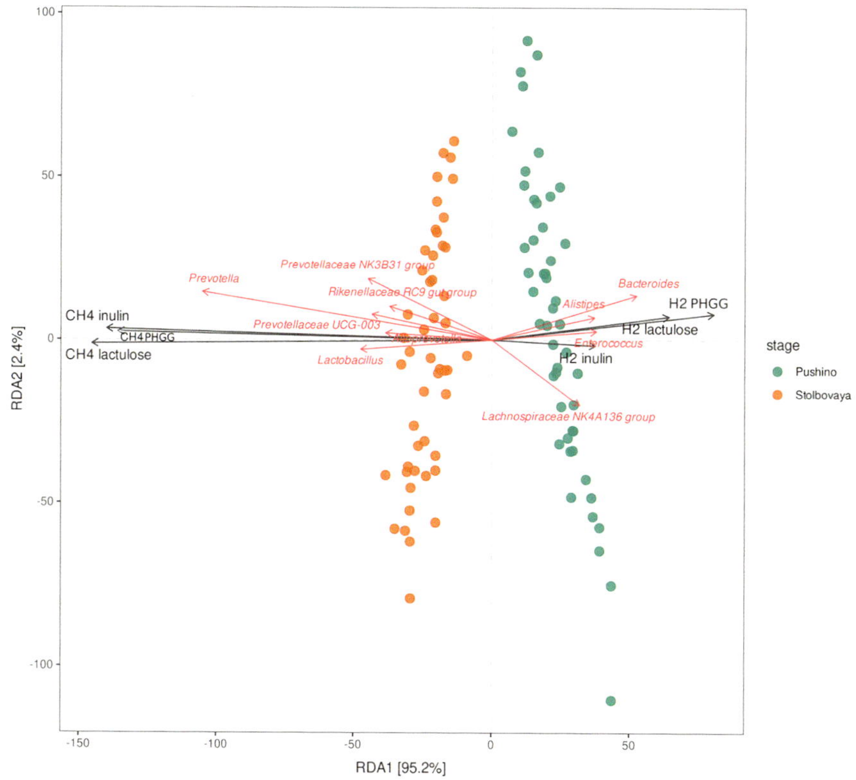

Fig. 20.6 Correlation Analysis. RDA ordination plot of the relationship between the level of gaseous metabolites in exhaled air (H_2 and CH_4) and the relative abundance of bacterial genera in experimental groups

hydrogen release in the exhaled air. On the contrary, gavage of the same carbohydrates to the high methanogenic rats (Stolbovaya nursery) was followed by the increase in methane in the exhaled air, but not in hydrogen.

It is assumed that intestinal H_2 is produced primarily by such bacteria as *Ruminococcus spp.*, *Roseburia spp.*, *Clostridium spp.* belonging to the phylum *Firmicutes*; *Bacteroides spp.* belonging to the phylum *Bacteroidetes* [10, 37]. Furthermore, over 200 pathogenic organisms can produce hydrogen [38]. The inclusion of non-digestible carbohydrates in the diet usually leads to increased hydrogen production by the gut microbiota [7, 10, 23, 37, 39].

The results of rodent studies revealed a significant difference between the amount of hydrogen produced by gut microbiota of experimental rats from different colonies on the background of the equal intake of dietary fiber. Thus, portal blood hydrogen concentration in the first group of rats amounted to 1.54 μmol/l versus 17.4 μmol/l in the second group.

Per oral transplantation of rat colonic microbiota with high H_2 production to the low H_2-generating animals of the first group led to an increase of the H_2 concentration in the portal vein from 3.07 to 9.95 μmol/l as well as a growing number of bacteria belonging to the genera *Bifidobacterium* (phylum *Actinobacteria*), *Allobaculum* (phylum *Firmicutes*) and *Parabacteroides* (phylum *Bacteroidetes*). At the same time, there was a decrease in the levels of *Bacteroides* (phylum *Bacteroidetes*), *Ruminococcus* (phylum *Firmicutes*), and *Escherichia* (phylum *Proteobacteria*) [40].

Comparison of the taxonomic differences between H_2-producing and CH_4-producing groups of rats shows, that the microbiota of H_2-producers is close to the composition of microbiota of new-borns and young children. H_2-producers have several times higher abundance of *Christensenellaceae* bacteria which is a well-known leader in producing H_2.

Age was negatively correlated with the abundances of *Christensenellaceae* and *Christensenella* in humans, indicating that younger subjects carry greater relative abundances of *Christensenellaceae* and *Christensenella* [41]. Goodrich et al. analyzed the gut microbiota of 416 pairs of twins and found that the abundance of *Christensenellaceae* was correlated with low BMI and that the transplantation of *Christensenella minuta* into germ-free mice reduces weight gain [42]. *Akkermansia* genes were found in H_2-producing rats but not in methane-producing. *Verrucomicrobia* abundance at taxonomic phylum and *Akkermansia* abundance at taxonomic genus *Akkermansiaceae muciniphila* is found in about 90% of healthy humans, makes up about 1% to 3% of the fecal microbiota, and colonizes the gut during the first year of life. Its prevalence can decrease with age or in disease states [43].

The genome of *Akkermansia* contains plentiful hydrogenases, such as HypE, HypD, HypA, HypB, HypF, HypC, HybG, and HupF and thus *Akkermansia* is able to catalytically decompose and utilize H_2. Therefore, it is possible that a sustained and abundant H_2 supply propagates this bacterium as a nutrient [44]. Everard et al. showed that the relative abundance of *A. muciniphila*, which has been linked to a favorable effect on glucose metabolism, is reduced in obese and diabetic mice and humans [45].

It was shown in our experiments that H_2-producing rats had four times less abundance of *Helicobacteraceae* bacteria, than methanogenic rats. *Helicobacteraceae* (*Helicobacter pylor*i and *Helicobacter ssp.*) oxidize H_2, which leads to a decrease in the intestinal pool of hydrogen [38]. We could not detect the abundance of *Methanobacteriaceae* in the H_2-producing group of rats (might be their low level was below the detection limit of our method), but these *Archaeae* were discovered in the methane-producing group.

So, taking into account the following observations (low alpha and beta diversity, higher abundance of *Christensenellaceae* and *Akkermansia* bacteria, lower abundance of *Helicobacteraceae,* and absence of *Methanobacteriaceae*) allow us to assume that a group of high H_2-producing rats could serve as a model of the young human microbiota.

Experiments with germ-free rats created the basis for the long-lasting view that methane, like hydrogen, is exclusively produced by the gut microbiota [46]. Several studies concluded that methane could be produced by human cells as well, but all the pieces of evidence in favor of such a conclusion were received in vitro and up to now were not confirmed by the in vivo experiments. Based on these data, it was suggested that next to microbial origin there might be other, as yet unidentified sources for endogenous CH_4 production [47]. *Methanobrevibacter smithii* and *Methanosphaera stadtmanae* are the main methane producers in human colon. Their number increases all along the colon until reaching a maximal value in the rectum [48]. *Methanobrevibacter smithii* uses hydrogen to reduce CO_2 to methane whereas *Methanosphaera stadtmanae* uses hydrogen for the reduction of methanol to methane [49, 50]. Four hydrogen molecules and one CO_2 molecule are required to synthesize one molecule of methane: $CO_2 + 4H_2 \rightarrow CH_4 + 2H_2O$. The number of methanogens in the gut microbiota changes with age. Typically, the gut microbiota of newborns does not contain methanegens [46], as well as children up to three years [51]. The percentage of humans who produced methane increases with age, reaching 40–77% in the age group 80–90 years [52, 53]. We speculate that high methane-producing rats could serve as a model of elderly people's microbiome.

Breath tests with the non-digestible carbohydrates lactulose, guar gum (the chemical structure is 100% guar galactomannans, consisting of mannose and galactose in a ratio of approximately 2:1) and inulin (polymer of D-fructose) loads evoked hydrogen increase of different amounts only in H_2-producing rats. The maximal increase in H_2 response was after gavage of lactulose, whereas amplitudes of H_2 responses to the guar gum and inulin were lower despite the fact, that doses of food fibers (guar gum and inulin) were two times higher than of the lactulose. Close results were shown in the experiments of Bond and Levitt on humans. Such differences in the H_2-producing effects of individual carbohydrates were shown both in vitro [24, 54] and in vivo experiments [22] and explained by the difference in taxonomic composition of the host microbiota. For example, galactomannans are degraded mainly by *B. ovatus*, whereas inulin—is by *Bacteroides ovatus* and *Bacteroides caccae* [55]. The maximal H_2-stimulating effect of lactulose may be explained by its simple structure (artificial disaccharide, containing fructose and galactose) and by the fact, that 35 species were screened in human feces with the ability to utilize lactulose [56].

Unexpected effects of all investigated carbohydrates were observed in the methane-producing group of rats. Gavage of any of the three carbohydrates studied was followed by the increase in the production of methane only. Similar results were demonstrated in human studies, as well [21, 23]. Hydrogen, produced by the gut microbiota could be consumed by several hydrogenotrophs, like methanogenic *Archaeae*, sulfate-reducing bacteria producing H_2S, and acetogens [57]. Different species of methanogenerating *Archaeae* are using H_2 in pair with CO_2, or methanol or trimethylamine (TMA) for the synthesis of methane [10, 58]. Since the availability of H_2 is necessary for the synthesis of CH_4 it is incorrect to say that methanogenic humans or animals do not produce H_2. We hypothesize that the results of our experiments with carbohydrates loads, that lead to the exclusive production of methane could be explained by the existence of highly active hydrogenotrophs (methanogens in our case), who consume any amount of hydrogen available. The strong support for our assumptions was the results of the elegant study of Ruaud et al. who were able to cultivate such anaerobes as *Christensenella* (producer of H_2) and *Methanobrevibacter* (producer of CH_4). During separate cultivation, *Christensenella* bacteria generated H_2, whereas *Methanobrevibacter Archaea*—generated CH_4. During the cultivation of both microorganisms in the same tube they formed tight contacts with each called flocks and the whole H_2 produced by *Christensenella* was consumed by the *Methanobrevibacter* [41]. A similar effect was shown by the pair of H_2 producer (*Bacteroides thetaiotaomicron*) and hydrogenotroph (*Desulfovibrio vulgaris*), producing H_2S [59].

We suppose that the best way to evaluate the total fermentative capacity of the gut microbiome is to assess the original, primary amount of H_2, which later could be exhaled via the lungs, to be consumed by the different hydrogenotrophs. The amount of the primary produced H_2 could be calculated as the sum of the exhaled H_2 (in ppm) and the exhaled CH_4, (in ppm) multiplied by the factor of 4. Such calculations were used in the previous works [60]. We assume that the result of this study is important for the recommendation of healthy nutrition and increases the antioxidant capacity of the organism. Advise to increase the consumption of food fibers is well known but it will be ineffective in case of the high methanogenic activity of the gut microbiome. We suppose that drinking hydrogen-rich water will be followed by the absorption of H_2 in the upper part of the gastrointestinal tract, diffusion to the blood, and reaching the liver via the portal system and will act as an antioxidant.

The short note concerning the methodology of microbiome fermentative activity study. A hydrogen breath test is dominating in clinical studies for diagnosing SIBO, malabsorption and tolerance of different carbohydrates, orocaecal transit time, and some other conditions. In the majority of experimental studies on rodents (mice and rats), animals are placed in a hermetically sealed glass bottle or metabolic cage [25, 61–64].

Such a method is not similar to the breath test in humans because, during occasional gas passes (flatus) in the sealed volume, the portion of intestinal gas with high concentration (in % but not in ppm like in exhaled air) produces a peak in concentrations of the gases studied that increases the dispersion of the results. In this work, we used a modified nose-only system combined with *the Allay*™ restraint collar [26, 27] to facilitate the breath test procedure and to collect just the H_2 in exhaled air in rats, which makes the method close to the breath test in humans and facilitate the comparison of the experimental results with the clinical ones.

Conclusion

The effectiveness of different food fibers in the increase of endogenous hydrogen release depends on the abundance ratio between hydrogenogens and methanogens in the microbiota. The higher presentation of the methanogens the lower will be hydrogen stimulating effects of the food fibers. Due to the higher percentage of methane producers with age, we speculate to expect a lower positive response to the administered food fibers. We suppose that the recommendation to drink hydrogen-rich water or inhalation of hydrogen will be more efficient in the elderly group of humans to increase the antioxidant protection of the organism.

References

1. DiBaise JK, Zhang H, Crowell MD et al (2008) Gut microbiota and its possible relationship with obesity. Mayo Clin Proc 83(4):460–469
2. Komaroff AL (2018) The Microbiome and Risk for Atherosclerosis. JAMA 319(23):2381–2382
3. Zhao L, Zhang F, Ding X et al (2018) Gut bacteria selectively promoted by dietary fibers alleviate type 2 diabetes. Science 359(6380):1151–1156. https://doi.org/10.1126/science.aao5774
4. Branchereau M, Burcelin R, Heymes C (2019) The gut microbiome and heart failure: a better gut for a better heart. Rev Endocr Metab Disord 20(4):407–414
5. Wilck N, Matus MG, Kearney SM et al (2017) Salt-responsive gut commensal modulates T_H17 axis and disease. Nature 551(7682):585–589
6. Drabińska N, Flynn C, Ratcliffe N et al (2021) A literature survey of all volatiles from healthy human breath and bodily fluids: the human volatilome. J Breath Res 15(3)
7. Levitt MD (1969) Production and excretion of hydrogen gas in man. N Engl J Med 281(3):122–127. https://doi.org/10.1056/NEJM196907172810303
8. Hartmann L, Taras D, Kamlage B (2000) A new technique to determine hydrogen excreted by gnotobiotic rats. Lab Anim 34(2):162–170. https://doi.org/10.1258/002367700780457617
9. Keppler F, Ernst L, Polag D et al (2022) ROS-driven cellular methane formation: Potential implications for health sciences. Clin Transl Med 12(7):e905. https://doi.org/10.1002/ctm2.905
10. Wolf PG, Biswas A, Morales SE (2016) H2 metabolism is widespread and diverse among human colonic microbes. Gut Microbes 7(3):235–245. https://doi.org/10.1080/19490976.2016.1182288
11. Pimentel M, Mathur R, Chang C et al (2013) Gas and the microbiome. Curr Gastroenterol Rep 15(12):356. https://doi.org/10.1007/s11894-013-0356-y
12. Sahakian AB, Jee SR, Pimentel M et al (2010) Methane and the gastrointestinal tract. Dig Dis Sci 55(8):2135–2143. https://doi.org/10.1007/s10620-009-1012-0
13. Modesto A, Cameron NR, Varghese C et al (2022) Meta-analysis of the composition of human intestinal gases. Dig Dis Sci 67(8):3842–3859. https://doi.org/10.1007/s10620-021-07254-1
14. Gaci N, Borrel G, Tottey W et al (2014) Archaea and the human gut: new beginning of an old story. World J Gastroenterol 20(43):16062–16078. https://doi.org/10.3748/wjg.v20.i43.16062
15. Ivashkin VT, Medvedev OS, Poluektova EA et al (2022) Direct and indirect methods for studying human gut microbiota. Russ J Gastroenterol Hepatol Coloproctol 32(2):19–34. https://doi.org/10.22416/1382-4376-2022-32-2-19-34
16. Ohsawa I, Ishikawa M, Takahashi K et al (2007) Hydrogen acts as a therapeutic antioxidant by selectively reducing cytotoxic oxygen radicals. Nat Med 13(6):688–694. https://doi.org/10.1038/nm1577
17. Slezak J, Kura B, LeBaron TW et al (2021) Oxidative stress and pathways of molecular hydrogen effects in medicine. Curr Pharm 27(5):610–625. https://doi.org/10.2174/1381612826666200821
18. Ohta S (2021) Direct targets and subsequent pathways for molecular hydrogen to exert multiple functions: focusing on interventions in radical reactions. Curr Pharm 27(5):595–609. https://doi.org/10.2174/1381612826666200806101137
19. Xie F, Song Y, Yi Y et al (2023) Therapeutic potential of molecular hydrogen in metabolic diseases from bench to bedside. Pharm Basel 16(4):541. https://doi.org/10.3390/ph16040541

20. Gottlieb K, Le C, Wacher V et al (2017) Selection of a cut-off for high- and low-methane producers using a spot-methane breath test: results from a large North American dataset of hydrogen, methane and carbon dioxide measurements in breath. Gastroenterol Rep Oxf. 5(3):193–199. https://doi.org/10.1093/gastro/gow048
21. Pimentel M, Saad RJ, Long MD et al (2020) Rao SSC. ACG clinical guideline: small intestinal bacterial overgrowth. Am J Gastroenterol 115(2):165–178
22. Bond JH, Levitt MD (1978) Effect of dietary fiber on intestinal gas production and small bowel transit time in man. Am J Clin Nutr 31(10 Suppl):S169–S174
23. Kalantar-Zadeh K, Berean KJ, Burgell RE et al (2019) Intestinal gases: influence on gut disorders and the role of dietary manipulations. Nat Rev Gastroenterol Hepatol 16(12):733–747
24. Yu X, Gurry T, Nguyen LTT et al (2020) Prebiotics and Community Composition Influence Gas Production of the Human Gut Microbiota. mBio 11(5):e00217–20
25. Gumbmann MR, Williams SN, Booth AN (1971) The quantitative collection and determination of hydrogen gas from the rat and factors affecting its production. Proc Soc Exp Biol Med 137(4):1171–1251
26. Javors MA, Ginsburg BC, Friesenhahn G et al (2005) Rat breathalyzer. Alcohol Clin Exp Res 29(1853–1857):2
27. House A, Shafer R, Shemesh Y et al (2014) Use of the Allay™ restraint collar to facilitate the measurement of ventilation in conscious rats. J Pharmacol Toxicol Method 70(3):348. https://doi.org/10.1016/j.vascn.2014.03.135
28. Hugerth LW (2014) DegePrime, a program for degenerate primer design for broad-taxonomic-range PCR in microbial ecology studies. Appl Environ Microbiol 80(16):5116
29. Merkel AY (2019) Analysis of 16S rRNA primer systems for profiling of thermophilic microbial communities. Microbiol Russian Fed 88(6):671–680
30. Toshchakov SV, Izotova AO, Vinogradova EN et al (2021) Culture-independent survey of thermophilic microbial communities of the north caucasus biology 10(6):1352
31. Callahan BJ, McMurdie PJ, Rose MJ et al (2016) DADA2: high-resolution sample inference from illumina amplicon data. Nat Methods 13:581–583. https://doi.org/10.1038/nmeth.3869
32. Quast C, Pruesse E, Yilmaz P et al (2013) The SILVA ribosomal RNA gene database project: improved data processing and web-based tools. Nucleic Acids Res 41:590–596. https://doi.org/10.1093/nar/gks1219.8
33. McMurdie PJ, Holmes S (2013) Phyloseq: an R package for reproducible interactive analysis and graphics of microbiome census data. PLoS ONE 8:e61217. https://doi.org/10.1371/journal.pone.0061217.9
34. Liu C, Cui Y, Li X et al (2021) Microeco: an R package for data mining in microbial community ecology. FEMS Microbiol Ecol 97(fiaa255):10
35. Dixon P (2003) VEGAN, a package of R functions for community ecology. J Veg Sci 14:927–930
36. Chaudhary PP, Conway PL, Schlundt J (2018) Methanogens in humans: potentially beneficial or harmful for health. Appl Microbiol Biotechnol 102(7):3095–3104. https://doi.org/10.1007/s00253-018-8871-2
37. Carbonero F, Benefiel AC, Gaskins HR (2012) Contributions of the microbial hydrogen economy to colonic homeostasis. Nat Rev Gastroenterol Hepatol 9(9):504–518. https://doi.org/10.1038/nrgastro.2012.85
38. Benoit SL, Maier RJ, Sawers RG et al (2020) Molecular hydrogen metabolism: a widespread trait of pathogenic bacteria and protists. Microbiol Mol Biol Rev 84(1):e00092-e119. https://doi.org/10.1128/MMBR.00092-19
39. Flint HJ, Scott KP, Duncan SH et al (2012) Microbial degradation of complex carbohydrates in the gut. Gut Microbes 3(4):289–306. https://doi.org/10.4161/gmic.19897
40. Nishimura N, Tanabe H, Komori E et al (2018) Transplantation of high hydrogen-producing microbiota leads to generation of large amounts of colonic hydrogen in recipient rats fed high amylose maize starch. Nutrients 10(2):144. https://doi.org/10.3390/nu10020144
41. Ruaud A, Esquivel-Elizondo S, de la uesta-Zuluaga J et al (2020) Syntrophy via interspecies H_2 transfer between christensenella and methanobrevibacter underlies their global cooccurrence in the human gut. mBio 11:e03235–19. https://doi.org/10.1128/mBio.03235-19

42. Goodrich JK, Waters JL, Poole AC et al (2014) Human genetics shape the gut microbiome. Cell 159 789–799. https://doi.org/10.1016/j.cell.2014.09.053
43. Iwaza R, Wasfy R, Dubourg G et al (2022) Akkermansia muciniphila: the state of the art, 18 years after its first discovery. Front Gastroenterol https://doi.org/10.3389/fgstr.2022.1024393
44. Jin Z, Sun Y, Yang T et al (2021) Nanocapsule-mediated sustained H_2 release in the gut ameliorates metabolic dysfunction-associated fatty liver disease. Biomaterials 276:121030. https://doi.org/10.1016/j.biomaterials.2021.121030
45. Everard A, Belzer C, Geurts L et al (2013) Cross-talk between Akkermansia muciniphila and intestinal epithelium controls diet-induced obesity. PNAS 110:9066–9071. https://doi.org/10.1073/pnas.1219451110
46. Bond JH Jr, Engel RR, Levitt MD (1971) Factors influencing pulmonary methane excretion in man. An indirect method of studying the in situ metabolism of the methane-producing colonic bacteria. J Exp Med 133(3):572–88. https://doi.org/10.1084/jem.133.3.572
47. Keppler F, Boros M, Frankenberg C et al (2009) Methane formation in aerobic environments. Environ Chem 6:459–465. https://doi.org/10.1071/EN09137
48. Nava GM, Carbonero F, Croix JA et al (2012) Abundance and diversity of mucosa-associated hydrogenotrophic microbes in the healthy human colon. ISME J 6(1):57–70. https://doi.org/10.1038/ismej.2011.90
49. Miller TL, Wolin MJ (1985) Methanosphaera stadtmaniae gen. nov., sp. nov.: a species that forms methane by reducing methanol with hydrogen. Arch Microbiol 141(2):116–22. https://doi.org/10.1007/BF00423270
50. Gaci N, Borrel G, Tottey W et al (2014) Archaea and the human gut: new beginning of an old story. World J Gastroenterol 20(43):16062–78. https://doi.org/10.3748/wjg.v20.i43.16062
51. Peled Y, Gilat T, Liberman E et al (1985) The development of methane production in childhood and adolescence. J Pediatr Gastroenterol Nutr 4:575–579
52. Polag D, Leiß O, Keppler F (2014) Age dependent breath methane in the German population. Sci Total Environ 481:582–587. https://doi.org/10.1016/j.scitotenv.02.086
53. Takakura W, Oh SJ, Singer-Englar T et al (2020) Comparing the rates of methane production in patients with and without appendectomy: results from a large-scale cohort. Sci Rep 10(1):867. https://doi.org/10.1038/s41598-020-57662-y
54. Karppinen S, irpa, Liukkonen K, Aura A-M, Forssell P, Poutanen K (2000) In vitro fermentation of polysaccharides of rye, wheat and oat brans and inulin by human faecal bacteria. J Sci Food Agric 80:1469–1476
55. Koropatkin NM, Cameron EA, Martens EC (2012) How glycan metabolism shapes the human gut microbiota. Nat Rev Microbiol 10(5):323–35. https://doi.org/10.1038/nrmicro2746
56. Mao B, Li D, Zhao J et al (2014) In vitro fermentation of lactulose by human gut bacteria. J Agric Food Chem 62(45):10970–7. https://doi.org/10.1021/jf503484d
57. Krajmalnik-Brown R, Ilhan ZE, Kang DW et al (2012) Effects of gut microbes on nutrient absorption and energy regulation. Nutr Clin Pract 27(2):201–14. https://doi.org/10.1177/0884533611436116
58. Mahnert A, Blohs M, Pausan MR et al (2018) The human archaeome: methodological pitfalls and knowledge gaps. Emerg Top Life Sci 2(4):469–482. https://doi.org/10.1042/ETLS2018003
59. Birg A, Ritz N, Barton LL et al (2023) Hydrogen availability is dependent on the actions of both hydrogen-producing and hydrogen-consuming microbes. Dig Dis Sci 68(4):1253–1259. https://doi.org/10.1007/s10620-022-07743-x
60. Christl SU, Murgatroyd PR, Gibson GR et al (1992) Production, metabolism, and excretion of hydrogen in the large intestine. Gastroenterology. 102(4 Pt1):1269–77
61. Wagner JR, Carson JF, Becker R et al (1977) Comparative flatulence activity of beans and bean fractions for man and the rat. J Nutr 107(4):680–9. https://doi.org/10.1093/jn/107.4.680
62. Chen X, Zhai X, Shi J et al (2013) Lactulose mediates suppression of dextran sodium sulfate-induced colon inflammation by increasing hydrogen production. Dig Dis Sci 58(6):1560–1568. https://doi.org/10.1007/s10620-013-2563-7
63. Zhao L, Wang Y, Zhang G et al (2019) L-Arabinose elicits gut-derived hydrogen production and ameliorates metabolic syndrome in C57BL/6J mice on high-fat-diet. Nutrients 11(12):3054. https://doi.org/10.3390/nu11123054

64. Ivanova AY, Shirokov IV, Toshchakov SV et al (2023) Effects of Coenzyme Q10 on the Biomarkers (Hydrogen, Methane, SCFA and TMA) and composition of the gut microbiome in rats. Pharmaceuticals 16:686. https://doi.org/10.3390/ph16050686

Chapter 21
Natural Biomolecules, Plant Extracts and Molecular Hydrogen—New Antioxidant Alternatives in the Management of Male Infertility

Eva Tvrdá, Michal Ďuračka, and Eva Ivanišová

Abstract Oxidative stress resulting from an imbalance between the levels of reactive oxygen species (ROS) and antioxidants, plays a pivotal role in the pathophysiology of male sub- or infertility. Substantial evidence highlights the importance of oral antioxidant therapy in the management of male reproductive dysfunction. Nevertheless, studies focused on traditional antioxidant supplement have often come to inconclusive or contradictory results. In the meantime, scientific focus has shifted to alternative remedies such as plant extracts, plant-based biomolecules as well as molecular hydrogen which present with numerous health benefits and powerful antioxidant properties. As such, the focus of this chapter is to present recent evidence assessing the in vivo effects of alternative antioxidant remedies on the structural, functional, and oxidative indicators of male reproductive function with a special emphasis on rats as attractive animal models.

Keywords Male fertility · Oxidative stress · Oral antioxidants · Biomolecules · Plant extracts · Molecular hydrogen · Rats

E. Tvrdá (✉)
Faculty of Biotechnology and Food Sciences, Institute of Biotechnology, Slovak University of Agriculture, Tr. A. Hlinku 2, 949 76 Nitra, Slovakia
e-mail: eva.tvrda@uniag.sk

M. Ďuračka
AgroBioTech Research Centre, Slovak University of Agriculture, Tr. A. Hlinku 2, 949 76 Nitra, Slovakia
e-mail: michal.duracka@uniag.sk

E. Ivanišová
Faculty of Biotechnology and Food Sciences, Institute of Food Sciences, Slovak University of Agriculture, Tr. A. Hlinku 2, 949 76 Nitra, Slovakia
e-mail: eva.ivanisova@uniag.sk

J. Slezak and B. Kura (eds.), *Molecular Hydrogen in Health and Disease*, Advances in Biochemistry in Health and Disease 27, https://doi.org/10.1007/978-3-031-47375-3_21

Introduction

General medical and scientific consensus defines infertility as a disease of the male or female reproductive system leading to the inability to accomplish a pregnancy following at least 12 months of regular unprotected sexual intercourse [1]. Current data published by the World Health Organization (WHO) estimate that around 17.5% of the adult population—roughly 1 in 6 worldwide—experience infertility, showing the urgent need to increase access to affordable, high-quality fertility care for those in need [1, 2]. According to Leslie et al. [3] males are solely responsible for approximately 20% and contribute to another 30–40% of all infertility cases. Overall, male factor infertility is a substantial contributor to around 50% of all cases of reproductive failure in couples [1, 3].

Male sub- or infertility is a complex and multifactorial condition that may be caused by genetic factors [4], endocrine imbalance [5], primary pathologies of the male urogenital tract or systemic diseases [6], lifestyle [7], infections [8] or environmental exposure [9]. No matter the primary cause of male reproductive dysfunction, oxidative stress, resulting from an imbalance between reactive oxygen species (ROS) production and inherent antioxidant defense mechanisms of the organism has been unanimously recognized as a contributing factor to male reproductive dysfunction [recently reviewed by 10].

A continuously ongoing process of spermatogenesis presents with high energy requirements and an exceptional metabolic activity of male germ cells, which on the other hand renders male reproductive structures to be major ROS producers and at the same time targets of potential oxidative insults [11]. Reactive oxygen and nitrogen species (ROS and RNS) encompass a large family of highly reactive chemicals such as superoxide anion ($O_2^{\bullet}$), hydrogen peroxide (H_2O_2), hydroxyl radical ($^{\bullet}OH$), nitric oxide (NO) or peroxynitrite ($ONOO^-$) most of which are known to act as signaling molecules and maintain the cellular oxidative balance under physiological conditions. Moderate ROS and RNS levels are furthermore required for the regulation of normal sperm function such as sperm production and maturation, capacitation, hyperactivation, acrosomal reaction and fertilization. These beneficial effects are however overturned into pathological ones if excessive and uncontrolled ROS production occurs [10, 11]. Decades of research have identified three molecular mechanisms which are considered to be key in ROS-associated male sub- or infertility: (a) lipid peroxidation (LPO) leading to the loss of membrane permeability and fluidity and permeability with a negative impact on the sperm motion and ability to interact with the oocyte; (b) protein alterations causing a reduced sperm metabolism and ATP synthesis; and (c) an increased damage to sperm DNA resulting in poor conception rates [12].

Treatment of testicular and/or seminal oxidative stress should first identify and treat the cause for increased ROS levels in the male reproductive system. As a complementary therapy, oral antioxidants are used based the premise that oxidative imbalance occurs in part due to a deficiency in antioxidants present in the testicular tissue and/or semen [13]. Ideally, oral antioxidants should reach levels high enough to support the endogenous antioxidant system, primarily represented by the enzymes superoxide dismutase (SOD), catalase (GPx), glutathione peroxidase (GPx) and glutathione reductase (GR) as well as the non-enzymatic antioxidant glutathione (GSH) whilst repleting a deficiency of vital elements important for spermatogenesis. The supplement should fortify the scavenging capacity and reduce high ROS concentrations [14], while at the same time these should not be entirely suppressed as this may impair normal sperm production and physiological functions which depend on low ROS levels [15].

Since the incorporation of oral antioxidants infertility has become a routine step in the management of male sub- or infertility, the search for an optimal remedy that would support an adequate oxidative balance of the male reproductive system while at the same time supporting a proper spermatogenesis, steroidogenesis and sperm function has become the center of widespread research attempts. Many promising compounds have been tested for their antioxidant properties in vitro [16], nevertheless several limitations must be kept in mind. In vitro tests do not account for issues related to the absorption, distribution, or excretion of the studied compound. Their failure to capture the inherent complexity of organ systems. At the same time, in vitro models may not account for interactions between cells and biochemical processes that occur during turnover and metabolism. As a result, in vitro studies have developed a reputation for being "less translatable" to humans. The use of animals in in vivo studies addresses many of the shortcomings of in vitro experiments, since the safety, toxicity, and efficacy of a drug candidate can be better evaluated in a complex model [17].

As such, this chapter aimed to review the effects of traditional as well as alternative antioxidant supplements on the structural integrity and functional activity of the male reproductive system by gathering evidence from in vivo studies primarily using the rat as the most common and popular laboratory animal in biomedical research.

Alternative Plant-Derived Bioactive Molecules

Various conventional biocompounds have been extensively studied as potential oral remedies to improve the reproductive potential and alleviate seminal oxidative stress in subjects diagnosed with sub- or infertility caused by various etiologies. While supplementation of ascorbic acid (vitamin C) and tocopherol (vitamin E), L-carnitine, coenzyme Q10, zinc or selenium considered as primary or secondary antioxidants have yielded positive outcomes in several studies, other research suggests little to no effects of the above-mentioned molecules or trace elements on the testicular function or sperm quality in subjects suffering from sub- or infertility [13, 18]. Moreover, conservative therapy based on hormonal stimulation with gonadotropins and androgen agonists or antagonists may not be necessarily effective for subjects a substandard semen quality and does not present with antioxidant benefits [19, 20]. Infertility management in these cases is still unpredictable and challenging, which is why more than one third of men with fertility issues seek alternative or complementary therapy options such as dietary supplements, nutraceuticals or functional foods fortified with plant extracts stemming from ethnopharmacological phytomedicine [21–23].

Natural biomolecules, such as flavonoids, phenolic acids, lignans or tannins have recently attracted substantial scientific as well as public attention. This notoriety may be primarily driven by ever-increasing evidence accentuating health-promoting effects of plant-derived biomolecules, and thus their potential implementation in prevention and treatment strategies of a broad array of diseases including cardiovascular, autoimmune, metabolic, malignant, and neurodegenerative patophysiologies [24, 25]. As opposed to conventional medication, natural biomolecules present with the ability to target and modulate multiple molecular pathways at once, an intrinsic biologic activity, low toxicity, multivariable pharmacological properties, and a formidable chemical diversity [26]. As such, these have become the focal point of extensive research for their potential to be used as alternative supplements for the management of male infertility.

Resveratrol

Resveratrol (3,5,40-trihydroxistilbene; RES) is a polyphenol that acts as a phytoalexin and a phytoestrogen able to modulate estrogen-sensitive systems [27]. For several decades now, RES has attracted a widespread scientific attention because of the "French paradox" hypothesis [27, 28], and its potential in the prevention or management of numerous ailments for its anti-inflammatory, antimicrobial and antioxidant effects [24] as well as its ability to modulate the most-conserved mammalian NAD^+-dependent protein sirtuin 1 [29].

Numerous studies agree that RES presents with a complex nature and affects numerous molecular and cellular targets within the male reproductive system. Properly selected RES doses have been shown to stimulate the hypothalamic-pituitary-gonadal axis, testosterone production and penile erection, leading to an increased sperm count and motility [28, 30, 31]. These phenomena may be primarily attributed to the ability of RES to interact with molecular pathways critical for glycolysis, mitochondrial metabolism and respiratory balance which are essential for a proper testicular function. At the same time, it was revealed that RES stabilizes the pronuclear and mitochondrial genome, and aids to prevent DNA defects, which are commonly observed in subjects suffering from sub- or infertility [12]. Novel proteomic studies have also unraveled that RES was able to affect the expression patterns of key spermatogenic regulatory proteins such as heat shock proteins, clusterin, centrin 1 and zona pellucida binding protein, indicating that the molecule could interact with cellular processes critical for a proper sperm production and function [32].

The ability of RES as an effective ROS-quencher and stabilizer of the testicular antioxidant milieu is considered as a hallmark property of the molecule [12]. According to several studies the exceptional antioxidant ability of RES may be attributed to four principal mechanisms of action, namely (1) a direct prevention of ROS over generation by the mitochondrial respiratory chain, (2) the ability to scavenge and neutralize superoxide, hydroxyl radical, and radicals created by metal-catalyzed oxidation, (3) inhibition of lipid peroxidation, and (4) regulation of endogenous antioxidants [27, 31, 33].

RES has been shown to act as a notable agent to stabilize the testicular antioxidant balance and sperm quality besides modulating the testicular polyADP-ribose polymerase (PARP) signaling pathway in experimental models of metabolic diseases such as hyperthyroidism [34], Type 1 or Type 2 diabetes mellitus [35, 36]. On the other hand, studies on animals subjected to exercise-induced stress revealed potent anti-inflammatory properties of RES through decreasing the testicular levels of pro-inflammatory interleukin 6 (IL-6) and tumor necrosis factor α (TNF-α) [32, 37].

A significant impact of RES on the molecular machinery of the sperm production, maturation and functionality has been suggested to be mediated through the modulation of the pro-apoptotic cleaved caspase-3, p53, calpain-1/cleaved caspase-12, or p-ERK1/2 and anti-apoptotic p-Akt/p-Bad pathways [38]. In the meantime, RES has been shown to modulate the phosphorylation and thus activity of AMP-activated protein kinase (AMPK) through increased AMP/ATP ratio, inhibition of mitochondrial ATP synthase or ROS overproduction [39, 40]. Finally, it was suggested that RES may be able to affect key sperm paternal transcripts serving as potential markers for male fertility (protamine 1 and 2) and pregnancy success (adducin 1 alpha), primarily through AMPK stimulation and improved interactions amongst sperm mRNAs, and thus an improved sperm resilience towards stress [41]. The most important findings from studies addressing the in vivo effects of oral RES administration on male reproductive parameters in rat models are provided in Table 21.1.

Quercetin

Quercetin (3,30,40,5,7-pentahydroxylflavone; QUE) is a flavonol-type flavonoid which has been reported to exhibit antidiabetic, anti-inflammatory, antibacterial and anticarcinogenic effects (Ulusoy and Sanlier 2020). Within the family of flavonoids, QUE is suggested to be the most powerful scavenger of ROS and NO thanks to its direct free radical-quenching and metal chelating properties, the ability to interact with and penetrate through lipid bilayers, modulate the activity of antioxidant enzymes or induce cellular repair [47].

QUE proved to be a highly effective male fertility-promoting biomolecule. Currently available studies speculate that beneficial effects of QUE on male reproduction may stem from its ability to stimulate the hypothalamic-pituitary-testicular axis as well as accessory glands responsible for the creation of a proper milieu of male reproductive fluids [48]. Furthermore, it was reported that QUE administration may contribute to an increased testicular and epididymal weight, stabilization of the testicular tubular architecture and prevention of excessive germ cell apoptosis, which may in turn lead to a rise in the sperm motion behavior, membrane, and acrosome integrity as well as mitochondrial activity [49].

Table 21.1 Summary of most relevant studies addressing the effects of oral resveratrol (RES) administration on reproductive parameters in rat models

Rat model	Administration	Effects	Reference
Healthy Sprague–Dawley rats	20 mg RES/kg BW, oral gavage, 90 days	↑ tubular density ↑ sperm count ↑ follicle stimulating hormone (FSH), luteinizing hormone (LH) and testosterone levels	[30]
Sprague Dawley rats subjected to high-intensity exercise	50 mg RES/kg BW/day, oral gavage, 9 weeks	↑ FSH and testosterone levels ↑ sperm density ↓ testicular tumor necrosis factor-alpha (TNF-α) and interleukin-6 (IL-6) levels ↓ testicular lesions	[32]
Wistar rats treated with 400 mg valproic acid/kg BW/day	10 mg RES/kg BW/day, oral gavage, 28 days	↑ sperm motility ↑ testicular and epididymal total antioxidant capacity (TAC) and glutathione (GSH) levels ↓ oxidative damage to sperm proteins and lipids	[34]
Wistar rats exposed to forced swimming exercise	50 mg RES/kg BW/day, oral gavage, 8 weeks	↑ FSH, LH and testosterone levels ↑ sperm count, motility, and morphology ↑ testicular SOD activity ↑ Johnsen's testicular score ↑ Number of Sertoli and Leydig cells ↓ testicular LPO, ornithine decarboxylase (ODC), putrescine, spermidine and spermine levels ↓ testicular TNF-α and IL-6 levels ↓ testicular lesions and germ cell apoptosis	[37]
Wistar rats fed a diet rich in lipids and simple carbohydrates	30 mg RES/kg body weight (BW)/day, oral gavage, 2 months	↑ sperm motility and viability ↑ germ cell proliferation ↓ germ cell apoptosis	[42]
Wistar rats treated with 5 mg finasteride/kg BW/day	20 mg RES/kg BW/day, oral gavage, 8 weeks	↑ testosterone levels ↑ testicular superoxide dismutase (SOD), catalase (CAT) and glutathione peroxidase (GPx) activity ↑ seminiferous tubular diameter and epithelial height ↓ testicular lipid peroxidation (LPO) ↓ testicular lesions and germ cell apoptosis	[43]

(continued)

Table 21.1 (continued)

Rat model	Administration	Effects	Reference
Wistar rats, varicocele induced by a standard surgical procedure	300 mg RES/kg BW/day, oral gavage, 58 days	↑ sperm vitality ↑ testicular and epididymal CAT activity ↑ sperm DNA integrity and chromatin protamination ↑ gestational index and development of preimplantation embryos ↓ testicular and sperm LPO ↓ sperm reactive oxygen species (ROS) production	[44]
Wistar rats treated with 200 mg vancomycin	20 mg RES/kg BW/day, oral gavage, 7 days	↑ testosterone levels ↑ sperm count and motility ↓ FSH and LH levels ↓ testicular lesions	[45]
Wistar rats, diabetes induced with 65 mg streptozotocin (STZ)/kg BW	150 mg RES/kg BW/day, oral gavage, 21 days	↑ FSH, LH and testosterone levels ↑ sperm count, motility and morphology ↑ testicular SOD, CAT and GPx activity ↓ testicular LPO ↓ testicular TNF-α and IL-6 levels ↓ testicular lesions	[46]

Metabolic studies have unraveled that QUE administration may lead to increased testicular mRNA expression levels of major antioxidant enzymes increased while reducing the occurrence of oxidative insults to the proteins and lipids and the levels of prime pro-inflammatory cytokines, suggesting strong antioxidant and anti-inflammatory properties of QUE in counteracting reproductive complications resulting from diabetes [50, 51]. According to numerous reports, QUE is a potent molecule to ameliorate structural or functional damage to the male reproductive system exposed to a wide array of toxicants. QUE treatment has resulted in an improved testicular structure and qualitative sperm parameters which was accompanied by a stabilization of the levels of male reproductive hormones as well as a notable improvement in the endogenous antioxidant system of animals exposed to heavy metals [52–54], herbicides [55], insecticides [56] or plasticizers [57].

In the meantime, a variety of reports hypothesize that besides testicular structures QUE also exhibits its effects directly on male gametes. It was previously suggested that QUE may affect the activity of CatSper channels which play a critical role in the sperm capacitation, hyperactivation and acrosome reaction [58]. Stimulating effects of QUE on the sperm motility may also be carried out through its interaction with intracellular cyclic adenosine monophosphate (cAMP) and subsequently Ca^{2+}-ATPase which are intimately involved in supporting sperm motility [59].

Within spermatozoa, QUE seems to be effective in preventing mitochondrial dysfunction by its ability to accumulate within the mitochondrial membranes and to counteract any eventual ROS overproduction through its antioxidant activity [60] and ability to inhibit NADPH oxidase and NADH-dependent oxidoreductase and/or mimic SOD [60, 61]. Accordingly, cytochrome B, NADH 5 and succinate dehydrogenase levels were found to be significantly increased in the presence of QUE [62], indicating a protective role of the flavonoid on components crucial for the mitochondrial respiratory chain, particularly during the initiation of oxidative chain reactions, keeping superoxide in physiological levels. This phenomenon also enables QUE to prevent excessive hydrogen peroxide (H_2O_2) production, which could result in LPO. Correspondingly, lower malondialdehyde (MDA) levels in spermatozoa is a frequently reported observation in QUE-centered studies [50, 63]. The most important findings from studies addressing the in vivo effects of oral QUE administration on male reproductive parameters in rat models are displayed in Table 21.2.

Lycopene

Lycopene (ψ,ψ-Carotene) (LYC) is a natural carotenoid that has only recently emerged as a subject of scientific interest for its multivariable biological activity and protective effects against cardiovascular, malignant or neurodegenerative diseases [65]. At the same time, LYC acts as a very powerful antioxidant, which has been reported to neutralize singlet oxygen twice as fast in comparison to β-carotene and ten times more effectively as opposed to α-tocopherol [65, 66].

Pivotal studies have unraveled that LYC concentrations in the testes are higher when compared to other tissues suggesting that the carotenoid may play important roles in the process of sperm production [67]. Subsequent research [68, 69] has revealed that the amount of LYC in the seminal plasma increases following its oral supplementation. As such, it may be hypothesized that LYC intake may provide an extra layer of protection against testicular oxidative stress, that may be followed by reproductive dysfunction.

While several mechanisms of action of LYC have been suggested, the principal one seems to be mediated through its antioxidant potential. Besides acting as a singlet oxygen, hydroxyl radical and nitrogen dioxide scavenger [70], LYC is lipophilic and easily incorporated into the cell membranes [65, 66] and readily available to protect critical sperm structures against oxidative damage. Additional hypotheses include the ability of LYC to increase the amount and/or activity of antioxidant enzymes [13] and reduce the expression patterns of proinflammatory cytokines [67, 71].

Currently available literature indicates that LYC supplementation to animal models exposed to drugs [72–74], PCBs [75], mycotoxins [76, 77] or bacterial toxins [78] significantly increased the sperm concentration, motility and viability whilst preventing ROS over generation and a subsequent lipid peroxidation. Furthermore, all studies agree on a notable antioxidant activity of LYC that has provided additional support to testicular antioxidant capacity of rats treated with cisplatin.

Table 21.2 Summary of most relevant studies addressing the effects of oral quercetin (QUE) administration on reproductive parameters in rat models

Rat model	Administration	Effects	Reference
Sprague–Dawley rats, diabetes induced with 55 mg STZ/kg BW	10, 25 or 50 mg QUE/kg BW/day, orally, 28 days	↑ sperm count, motility, viability and morphology ↑ sperm SOD, CAT and GPx levels and mRNA expression of the corresponding genes ↓ sperm DNA fragmentation and LPO ↓ sperm NF-κβ-p65 and TNF-α expression levels	[50]
Zucker diabetic fatty (ZDF) rats spontaneously developing Type 2 diabetes	20 mg QUE/kg BW/day, orally, 6 weeks	↑ testicular TAC, SOD, CAT, GPx and GSH levels ↓ testicular ROS production, LPO and protein oxidation ↓ testicular TNF-α, interleukin 1 (IL-1), IL-6 and interleukin 18 (IL-18) levels ↓ testicular BAX, p53 and caspase-3 expression ↓ testicular lesions	[51]
Sprague–Dawley rats exposed to 50 ppm sodium arsenite	50 mg QUE/kg BW, 49 days	↑ number of spermatogonia, primary and secondary spermatocytes, spermatids and Leydig cells ↑ testosterone levels ↑ testicular SOD, peroxidase (POD), CAT and GSH levels ↓ testicular LPO ↓ testicular abnormalities	[52]
Albino rats exposed to 5 mg cadmium chloride ($CdCl_2$)/kg BW	50 mg QUE/kg BW, oral gavage, 4 weeks	↑ FSH, LH and testosterone levels ↑ sperm count, motility and morphology ↑ Johnsen's testicular score	[53]
Sprague–Dawley rats administered with 2 mg $CdCl_2$/kg BW	50 mg QUE/kg BW/day, oral gavage, 4 weeks	↑ testicular SOD, CAT, GPx and GSH levels ↓ testicular LPO ↓ testicular abnormalities ↓ testicular expression of autophagy-related proteins P62 and LC3B	[54]

(continued)

Table 21.2 (continued)

Rat model	Administration	Effects	Reference
Wistar rats supplemented with 120 mg atrazine/kg BW	10 mg QUE/kg BW, orally, 16 days	↑ testicular SOD levels ↑ 3 -hydroxysteroid dehydrogenase (3β-HSD), 17β -hydroxysteroid dehydrogenase (17β-HSD) levels ↑ testicular and epididymal sperm number ↑ daily sperm production ↑ sperm motility, viability and morphology ↓ lactate dehydrogenase (LDH) levels ↓ testicular LPO	[55]
Sprague–Dawley rats administered with 900 mg/kg BW/day of a mixture of dibutyl phthalate, di(2-ethyhexyl) phthalate and butyl benzyl phthalate	10, 30 or 90 mg QUE/kg BW/day, orally, 30 days	↑ FSH, LH, testosterone, and estradiol levels ↑ expression of PWIL 1 and 2, ↓ expression of StAR, P450scc, CYP17A1, 17β-HSD, P450arom ↓ testicular abnormalities	[57]
Wistar rats, hypertension induced with 40 mg N-nitro-l-arginine methyl ester/kg BW/day	50 mg QUE/kg BW, oral gavage, 4 weeks	↑ FSH, LH and testosterone levels ↑ sperm progressive motility and viability ↑ testicular and epididymal NO levels ↓ testicular and epididymal arginase levels ↓ testicular and epididymal ROS levels ↑ testicular and epididymal GSH and thiol levels ↓ testicular and epididymal LPO ↓ testicular abnormalities	[63]
Albino rats treated with 15 mg monosodium glutamate/kg BW	14 mg QUE/kg BW, oral gavage, 30 days	↑ testicular SOD, GPx and GSH levels ↑ testosterone levels ↓ testicular LPO ↓ testicular structural and ultrastructural abnormalities	[64]

According to several authors [74, 75], LYC attenuated the mitochondrial damage thanks to its ability to accumulate in the membranous structures, and subsequently stabilize the mitochondrial metabolism [67]. Correspondingly, structurally, and functionally stabilized mitochondria will be less prone to initiate the apoptotic machinery leading to cell death as unraveled by a significant reduction of the expression profiles of pro-apoptotic proteins caspase-3 and caspase-9, Bax, Bad and Bid in the germ cells in the presence of LYC [74, 75, 79]. Protective effects of LYC on the male reproductive structures may be accompanied by its anti-inflammatory actions as revealed by a significant reduction of prominent pro-inflammatory cytokines such as testicular interleukin 1alpha (IL-1α), interleukin 1 beta (IL-1β), interleukin 6 (IL-6) or tumor necrosis factor-alpha (TNF-α) [78]. Finally, several studies imply that LYC may prevent sperm DNA damage, thus increasing the success of fertilization and embryogenesis. This protection may be offered on a direct level of the DNA molecule as well as on a more complex level in terms of a proper chromatin structure and packaging [80, 81].

Summarizing all above mentioned research outputs, two hypotheses may explain the roles LYC on either improving or attenuating male fertility. The first one hypothesizes that LYC as a lipophilic substance easily passes through the membranous structures and swiftly enters the intracellular environment where it stabilizes and fortifies the inherent antioxidant network. The second one suggests that LYC binds to the membranes where it contributes to the protection of lipoproteins against possible oxidative damage [82]. The most important findings from studies addressing the in vivo effects of oral LYC administration on male reproductive parameters in rat models are shown in Table 21.3.

Curcumin

Curcumin [1,7-bis(4-hydroxy-3-methoxyphenyl)-1,6-heptadiene-3,5-dione] (CUR) is a curcuminoid with antitumor, anti-inflammatory, radioprotective and neuroprotective properties [85]. CUR is an effective scavenger of superoxide, hydroxyl radical, nitrogen dioxide as well as a strong inhibitor of LPO [85, 86]. All these attributes suggest the molecule could be a suitable preventive pharmacological intervention for male reproductive dysfunction.

One of the first studies to observe the effects of CUR on male fertility has reported that sperm concentration, motility and morphology were improved correspondingly to the increasing CUR doses (50, 100 or 150 mg/kg BW) in male Wistar rats, with a concomitant rise in the testicular spermatogenic activity of the testicular tissue [87]. Furthermore, higher Johnsen's score accompanied by improved morphometrical indices of the testicular tissue and qualitative sperm parameters (count, motility, viability and/or morphology) were observed following CUR administration to experimental animals with induced health ailments [88–90] or exposed to heavy metals [91–93], insecticides [94], medicaments [95, 96], radiation [97], indicating a broad versatility of the molecule under different endogenous or exogenous pathologies.

Table 21.3 Summary of most relevant studies addressing the effects of oral lycopene (LYC) administration on reproductive parameters in rat models

Rat model	Administration	Effects	Reference
Wistar rats treated with 100 mg gentamycin/kg BW/day	4 mg LYC/kg BW/day, oral gavage, 15 days	↑ sperm count, motility and viability ↑ testosterone levels ↑ LDH and glucose-6-phosphate dehydrogenase (G6PDH) levels ↑ testicular SOD, CAT, GPx, GR and GSH levels ↓ testicular LPO and H_2O_2 levels ↓ testicular caspase-3 and caspase-9 levels	[74]
Wistar rats exposed to 2 mg Aroclor 1254/kg BW/day	4 mg LYC/kg BW/day, oral gavage, 30 days	↑ sperm count and motility ↓ H_2O_2, •OH, and LPO in Sertoli cells ↓ caspase-8, Bax, Bad and Bid expression levels in Sertoli cells	[75]
Wistar-Albino rats exposed to 2.5 mg aflatoxin B1/kg BW	10 mg LYC/kg BW/day, oral gavage, 15 days	↑ sperm motility and morphology ↓ testicular LPO ↓ testicular lesions	[76]
Sprague Dawley rats exposed to 0.5 mg ochratoxin A/kg BW/day	5 mg LYC/kg BW/day, oral gavage, 7 or 14 days	↑ testicular selenium (Se) and zinc (Zn) levels	[77]
Sprague–Dawley rats administered with 5 mg lipopolysaccharide/kg BW	5 mg LYC/kg BW, oral gavage, 4 weeks	↑ testicular TAC, SOD and CAT levels ↓ testicular IL-1, IL-6, TNF-α, monocyte chemotactic protein 1 (MCP-1) levels ↓ testicular LPO ↓ testicular lesions	[78]
Wistar rats, varicocele induced by a standard surgical procedure	4 or 10 mg LYC/kg BW/day, oral gavage, 2 months	↑ sperm concentration, motility, viability and membrane integrity ↓ testicular lesions ↓ testicular expression of Bax, hypoxia-inducible factor 1α and heat-shock protein A2 genes	[79]
Wistar albino rats with surgically induced ischemia/reperfusion	4 mg LYC/kg BW, oral gavage, 30 days	↑ sperm concentration, motility and morphology ↑ testicular CAT and GPx levels ↓ testicular lesions	[83]
Wistar albino rats, diabetes induced with 55 mg STZ/kg BW	4 mg LYC/kg BW, oral gavage, 28 days	↓ testicular LPO ↓ testicular lesions	[84]

This bioactivity of CUR has been attributed to its ability to competitively inhibit the cytochrome P450 isoenzymes [98] as well as to suppress NF-κB DNA-binding activity [99], which may subsequently lead to a reduced expression of pro-apoptotic and autophagic proteins such as Bax, caspases or calpain-1 [89]. Besides it was observed that CUR possesses chromatin-stabilizing properties stemming from its protective effects on the chromatin condensation, lysine-rich histones, cytosine-rich and guanosine sequences which will sustain the degree of sperm chromatin protamination [100].

Most experimental studies agree on significant antioxidant properties of CUR through its ability to either quench ROS directly, or to stimulate the internal testicular ROS-scavenging system of male reproductive cells and tissues under oxidative stress [89, 95, 101]. The molecule is able to prevent superoxide, hydroxyl radical and nitric oxide overproduction by inhibiting the Fenton reaction, xanthine oxidase and nitric oxidase [88, 102]. Finally, it has been repeatedly suggested that CUR prevents oxidative insults to the membranous structures of male reproductive cells most likely through its ability to trap and neutralize lipid peroxyls before these can interact with intact membrane lipids [89, 90, 93, 101].

Despite promising benefits stemming from a broad biological activity of CUR, its bioavailability is relatively low because of its reduced aqueous solubility, light sensitivity and instability during extended time periods [86]. In this sense, CUR encapsulation and/or liposomal nanoparticles represent a strategy that may enhance its bioavailability alongside its therapeutic potential. Correspondingly, pivotal studies have observed that encapsulated or liposomal CUR may be equally effective as pure CUR in preventing germ cell apoptosis and excessive oxidative damage to the testicular tissue while at the same time preserving a desirable sperm concentration, motility and morphology [89, 103]. The most important findings from studies addressing the in vivo effects of oral CUR administration on male reproductive parameters in rat models are provided in Table 21.4.

Plant Extracts

Complementary therapies for the prevention or management of male sub- or infertility have recently received growing attention, giving rise to novel nutritional supplements, and medicinal plants-derived remedies have been proposed as alternative strategies to treat reproductive dysfunction in men) [104]. This approach stems from experience gathered from developing countries, where infertile couples take advantage of both contemporary therapies as well as herbal medicines for treatment. Moreover, because of the limited access to conservative medicine, traditional remedies are often the first available treatment option of infertile couples in these countries [105]. Consequently, a complementary approach to infertility treatment based on herbal remedies has gained in popularity in the western civilization [106] as confirmed by the European Association of Urology which has recently defined ethnopharmacology as a multidimensional integrative approach to male infertility management

Table 21.4 Summary of most relevant studies addressing the effects of oral curcumin (CUR) administration on reproductive parameters in rat models

Rat model	Administration	Effects	Reference
Wistar rats, varicocele induced by a standard surgical procedure	100 mg CUR/kg BW, oral gavage, 8 weeks	↑ sperm count, motility, viability and morphology ↓ testicular NO levels	[88]
Wistar rats, diabetes induced with 50 mg STZ/kg BW	100 mg CUR/kg BW, oral gavage, 8 weeks	↑ testosterone, 3β-HSD and 17β-HSD levels ↑ sperm count, motility and morphology ↑ testicular SOD, CAT, GR, glutathione-S-transferase (GST) and GPx activity ↓ testicular ROS production, LPO and protein carbonylation ↓ testicular TNF-α and IL-1 level ↓ testicular expression of NFκB, Lamin B1, IκBα, COX-2, phospho-PI3K, phospho-Akt, Nrf-2, TNF-R1, phospho-p38, phospho-JNK, Bad, calpain-1, caspase-12, caspase-3, caspase-9, Bax, PARP, caspase-8 and Bid	[89]
Wistar rats, diabetes induced with 40 mg STZ/kg BW and high-fat diet	100 or 200 mg CUR/kg BW/day, oral gavage, 16 weeks	↑ testosterone levels ↑ testicular SOD and GPx levels ↓ testicular lesions ↓ testicular TUNEL positivity and BAX expression ↓ testicular ROS production and LPO ↓ testicular ratio of $NADP^+$/NADPH and the expression levels of $p47^{phox}$, $p67^{phox}$, and $gp91^{phox}$ ↓ testicular expression of MAPK family members, including JNK, p38MAPK, and ERK	[90]
Wistar rats exposed to 1 mg cadmium/kg BW	100 mg CUR/kg BW/day, oral gavage, 4 weeks	↑ testosterone levels ↑ mean seminiferous tubule diameter ↑ Johnsen's testicular score ↓ germ cell and Leydig cell apoptosis	[92]
Wistar rats administered with 50 mg lead acetate/kg BW	100, 200 or 400 mg CUR/kg BW, oral gavage, 35 days	↑ sperm count, motility and viability ↑ testicular SOD and GPx levels ↓ testicular lesions ↓ testicular LPO	[93]

(continued)

Table 21.4 (continued)

Rat model	Administration	Effects	Reference
Wistar albino rats treated with 7 mg cisplatin/kg BW	200 mg CUR/kg BW/day, orally, 10 days	↑ mean seminiferous tubule diameter and germ cell layer ↑ testosterone levels ↑ testicular GPx and GSH levels ↓ testicular lesions ↓ testicular LPO and NO levels	[95]
Wistar rats exposed to 100 mg copper(II) sulfate ($CuSO_4$)/kg BW	80 mg pure CUR or liposomal nano-CUR/kg BW, orally, 7 days	↑ FSH, LH and testosterone levels ↑ testicular mRNA expression StAR, 3β-HSD, CYP17A1, and 17β-HSD ↑ testicular protein expression of Nrf2, HO-1, Bcl-2 and PCNA ↑ testicular SOD, CAT and GSH levels ↓ testicular lesions ↓ testicular LPO and NO levels ↓ testicular protein expression of NF-κB p65, iNOS, and TNF-α, Bax and caspase-3	[101]

[107]. Accordingly, WHO advocates for the integration of medicinal plants into the health care system [108]. Besides, a substantial interest in herbal medicine stems from scientists searching for natural biomolecules to replace synthetic drugs at least partially [109].

Medicinal plants that are currently known to assist in the management of male fertility issues are empirically used as oils, extracts, herbal infusions or decoctions or partially purified compounds. Current evidence suggests that plants are rich in novel chemical entities such as polyphenols or flavonoids that could enhance male reproductive health which has been demonstrated in experimental animal models on numerous occasions [110]. The outcomes depend on the plant extract, dose and/or treatment duration, any may vary from an improved libido, erectile function and ejaculatory function to a higher semen quality and hormonal profile. Numerous studies have revealed that herbal remedies may increase the number of testicular vessels, protect the structural integrity and functional activity of germ cells, and thus increase the lifespan and number of spermatozoa produced [111–114]. Different plant extracts may also enhance the activity of the hypothalamic-pituitary–gonadal axis on different levels, hence affecting the secretion of pituitary hormones and testosterone [112, 115, 116]. Most scientists agree that medicinal plants exhibit their fertility-enhancing by their antioxidant activity, through the prevention of ROS formation and LPO, leading to the reduction of oxidative damage to the sperm cells [115, 117–119].

Despite the wide popularity herbal remedies have gained over the past years, scientific knowledge on their properties and roles in male reproduction is still scarce. As such, it is essential to undertake more complex research on the chemical composition, bioavailability, effectiveness, safety and mechanism of herbal treatments may exhibit on male reproductive cells and tissues [110]. The most important findings from studies addressing the in vivo effects of most relevant plant extract administration on male reproductive parameters in rat models are summarized in Table 21.5.

Table 21.5 Summary of most relevant studies addressing the effects of oral plant extract administration on reproductive parameters in rat models

Rat model	Administration	Effects	Reference
Wistar rats exposed to 2.5 mg flutamide/kg BW in utero	*Basella alba*, leaves, 1 mg/kg BW, orally, 45 or 70 days	↑ fecundity ↑ testosterone levels ↓ testicular lesions	[105]
Holtzman rats	*Lepidium meyenii* (Maca), dried hypocotyls, water extract, 2 g/kg BW, orally, 1, 3, 5, 7 or 12 days	↑ daily sperm production and epididymal sperm count	[111]
Wistar rats treated with 0.1% lead acetate	*Fumaria parviflora*, leaves, ethanolic extract, 200 mg/kg BW/day, oral gavage, 70 days	↑ tubular diameter ↑ sperm concentration, motility and morphology ↑ testicular SOD and GPx levels ↑ FSH, LH and testosterone levels ↓ testicular LPO	[112]
Healthy Wistar rats	*Apium graveolens*, leaves, hydro- alcoholic extract, 100 or 200 mg/kg BW/every two days, orally, 60 days	↑ number of Sertoli cells, and primary spermatocytes ↑ sperm count ↑ thickness of germinal epithelium	[113]
Wistar rats exposed to 200 mg para-nonylphenol/kg BW/day	Green tea, leaves, water extract, 200 mg/kg/day, oral gavage, 8 weeks	↑ sperm number, motility, viability, morphology, and chromatin integrity ↓ testicular LPO ↓ testicular lesions	[114]
Wistar rats, diabetes induced with 55 mg STZ/kg BW	*Zingiber officinale*, roots powdered and dissolved in 2 cc distilled water, 100 mg/kg BW/day, oral gavage, 8 weeks	↑ sperm concentration, motility, viability, and morphology ↑ FSH, LH and testosterone levels ↑ testicular TAC, SOD, CAT and GPx levels ↓ testicular LPO	[115]
Wistar rats, diabetes induced with 55 mg STZ/kg BW	*Cinnamon zeylanicum*, powdered and dissolved in 2 cc distilled water, 75 mg/kg BW/day, oral gavage, 8 weeks	↑ FSH and testosterone levels ↓ testicular LPO	[115]

(continued)

Table 21.5 (continued)

Rat model	Administration	Effects	Reference
Wistar rats, infertility induced by 100 mg sulphasalazine/kg BW	*Pedalium murex*, fruit, methanolic extract, 50 mg/kg BW, orally, 61 days	↑ FSH, LH and testosterone levels ↑ sperm density and fertility ↑ number of Sertoli cells, spermatogonia, primary and secondary spermatocytes and round spermatids ↓ testicular abnormalities	[116]
Healthy Wistar rats	*Garcinia kola*, seed powdered and dissolved in corn oil, 50, 500 or 1 000 mg/kg BW, once daily, orally, 6 weeks	↑ testosterone levels ↑ testicular SOD, CAT, GPx and GST levels ↑ sperm CAT and GST levels ↓ testicular and sperm LPO	[117]
Sprague Dawley rats exposed to 500 mg sodium valproate/kg BW	*Tribulus terrestris*, fruits, methanolic extract, 2.5, 5 and 10 mg/kg, orally, 60 days	↑ FSH, LH and testosterone levels ↑ sperm number, motility, viability and morphology ↑ testicular SOD, CAT and GPx levels ↓ testicular lesions	[118]
Sprague–Dawley rats, varicocele induced by a standard surgical procedure	*Schisandra chinensis* Baillon, fruits, ethanolic extract, 200 mg/kg BW, orally, 28 days	↑ testosterone levels ↑ sperm count and motility ↑ Johnsen's testicular score ↓ testicular TUNEL positivity and apoptotic index ↓ testicular LPO, ROS and RNS production ↑ testicular SOD, CAT and GPx levels ↓ testicular TNF-α and IL-6 levels ↓ testicular lesions	[119]

Molecular Hydrogen

Unlike originally thought to be a biologically inactive gas in mammalian cells, molecular hydrogen (H_2) has become a major focus point for current research since the discovery of its abilities to effectively alleviate oxidative damage by selectively scavenging hydroxyl radicals and nitro peroxide anions [120]. Subsequent studies have unraveled versatile benefits of H_2 including anti-apoptotic, anti-inflammatory, and anti-allergic effects in over 200 pathophysiologies [121]. In addition to its antioxidant effects, H_2 is able to interact with various signal transduction pathways and modulate the expression profiles of genes responsible for energy metabolism or ROS-dependent peroxidation via the Ca^{2+} signal transduction pathway [122].

Despite its undeniable benefits in the prevention or management of ailments, very little research has been done in terms of its effects on reproductive dysfunction. Whilst a few in vitro studies have suggested that H_2 supplementation may increase the intracellular ATP content in spermatozoa, only one published report has revealed its protective effects on damage inflicted to the reproductive system of Sprague–Dawley rats exposed to 1 000 kV/m electromagnetic pulse [123]. Administration of hydrogen-rich water has led to a lower occurrence of testicular lesions, germ cell apoptosis and lipid peroxidation to the testicular structures, which in turn resulted in a higher sperm motility and viability. Furthermore, higher levels of antioxidant enzymes including SOD and GPx were observed in H_2-treated animals. Similarly, our preliminary data suggest that H_2 administered to Wistar rats exposed to radiation of 10 Gy has led to an improvement in the sperm count, motility, viability, and DNA integrity while preventing testicular ROS-overgeneration followed by oxidative damage to the proteins and lipids (data not shown).

The proposed mechanism of action may lie in the ability of H_2 to penetrate the cellular membranous structures and rapidly diffuse into the intracellular milieu without affecting the signal transduction process. Once H_2 enters the subcellular structures, it may rapidly trap and neutralize excessive ROS and RNA overproduced under stress and thus exert a protective role on a proper structure and function of cellular compartments [124]. By doing so, H_2 may also alleviate and restore oxidative balance that has been previously disrupted by a stressful input. Moreover, H_2 has been shown to exert its protective mechanisms on regulatory proteins involved in the behavior of intracellular ribosomes, cell adhesion events, cellular senescence, phagocytosis, and intracellular metabolism [123]. Besides, it is ought that H_2 may play a role in modulating signals involved in the process of steroidogenesis and testosterone production to improve male reproductive potential compromised by redox imbalance [125]. This versatility allows H_2 to be considered as an appealing alternative strategy for the prevention and/or treatment of male sub- or infertility. Nevertheless, more complex, and detail-oriented research is necessary to open a new horizon for molecular hydrogen in the field of reproductive medicine and biology.

Conclusions

This chapter aimed to summarize currently available information on the effects of alternative compounds with antioxidant properties on the structural integrity, functional activity and oxidative balance of male reproductive cells and tissues, with a particular emphasis on in vivo studies utilizing rats as animal models. By and large, all discussed substances exhibited protective and beneficial roles on male fertility, primarily through their ROS-quenching and neutralizing activity, antiapoptotic and

DNA protecting properties, their ability to stimulate mitochondrial metabolism and hence sperm motility and fertilizing potential. New discoveries in the field of ethnopharmacology and chemical engineering open new horizons in the search for an optimal oral antioxidant that could serve as a remedy in the management of male sub- or infertility. Nevertheless, the exact mechanisms underlying the behavior of bioactive compounds has only recently started to be elucidated more thoroughly, a critical approach is advised before drawing definite conclusions on potential benefits of oral antioxidant supplementation on the male reproductive system. Since a compound may exhibit a specific biological effect in animal models that may or may not be the same in human subjects, further studies are necessary to translate the activity of adjacent metabolites into human physiology. At the same time, it must be remembered that in vivo, most molecules may act in synergy or antagony depending on the conditions, which needs to be considered in the interpretation of exact behavior. Finally, establishment of a normal physiological range of particularly novel biomolecules in reproductive fluids and tissues is a critical prerequisite that may assist to determine if the effects achieved from a certain dose are physiologically relevant, since a proportion of scientific studies may employ doses that exceed normal physiologic levels. Accordingly, all these aspects should be taken into consideration in future studies providing more clarification to the positive or negative effects of conventional or alternative oral antioxidants and/or their combination on male fertility.

Competing Interest This publication was supported by the projects APVV-21-0095, KEGA 008SPU-4/2021 and by the Operational program Integrated Infrastructure within the project Demand-driven research for sustainable and innovative food, Drive4SIFood 313011V336, co-financed by the European Regional Development Fund. We also wish to thank the CeRA Team of Excellence for their support. Authors declare no conflict of interests.

References

1. World Health Organization (2023) Infertility. https://www.who.int/news-room/fact-sheets/detail/infertility. Accessed 30 May 2023
2. The Lancet Global Health (2022) Infertility-why the silence? Lancet Glob Health 10(6):e773. https://doi.org/10.1016/S2214-109X(22)00215-7
3. Leslie SW, Soon-Sutton TL, Khan MAB (2023) Male infertility. [Updated 2023 Mar 3]. In: StatPearls [Internet]. StatPearls Publishing, Treasure Island (FL). Available from: https://www.ncbi.nlm.nih.gov/books/NBK562258/
4. Shamsi MB, Kumar K, Dada R (2011) Genetic and epigenetic factors: role in male infertility. Indian J Urol 27(1):110–120. https://doi.org/10.4103/0970-1591.78436
5. Kathrins M, Niederberger C (2016) Diagnosis and treatment of infertility-related male hormonal dysfunction. Nat Rev Urol 13(6):309–323. https://doi.org/10.1038/nrurol.2016.62
6. Leaver RB (2016) Male infertility: an overview of causes and treatment options. Br J Nurs 25(18):S35–S40. https://doi.org/10.12968/bjon.2016.25.18.S35
7. Durairajanayagam D (2018) Lifestyle causes of male infertility. Arab J Urol 16(1):10–20. https://doi.org/10.1016/j.aju.2017.12.004
8. Schuppe HC, Pilatz A, Hossain H (2017) Urogenital infection as a risk factor for male infertility. Dtsch Arztebl Int 114(19):339–346. https://doi.org/10.3238/arztebl.2017.0339

9. Selvaraju V, Baskaran S, Agarwal A (2021) Environmental contaminants and male infertility: effects and mechanisms. Andrologia 53(1):e13646. https://doi.org/10.1111/and.13646
10. Mannucci A, Argento FR, Fini E et al (2022) The impact of oxidative stress in male infertility. Front Mol Biosci 8:799294. https://doi.org/10.3389/fmolb.2021.799294
11. Baskaran S, Finelli R, Agarwal A (2021) Reactive oxygen species in male reproduction: a boon or a bane? Andrologia 53(1):e13577. https://doi.org/10.1111/and.13577
12. Illiano E, Trama F, Zucchi A et al (2020) Resveratrol-based multivitamin supplement increases sperm concentration and motility in idiopathic male infertility: a pilot clinical study. J Clin Med 9(12):4017. https://doi.org/10.3390/jcm9124017
13. Zini A, Al-Hathal N (2011) Antioxidant therapy in male infertility: fact or fiction? Asian J Androl 13(3):374–381. https://doi.org/10.1038/aja.2010.182
14. Zini A, San Gabriel M, Baazeem A (2009) Antioxidants and sperm DNA damage: a clinical perspective. J Assist Reprod Genet 26(8):427–432. https://doi.org/10.1007/s10815-009-9343-5
15. Du Plessis SS, Agarwal A, Halabi J et al (2015) Contemporary evidence on the physiological role of reactive oxygen species in human sperm function. J Assist Reprod Genet 32(4):509–520. https://doi.org/10.1007/s10815-014-0425-7
16. Alam MN, Bristi NJ, Rafiquzzaman M (2013) Review on in vivo and in vitro methods evaluation of antioxidant activity. Saudi Pharm J 21(2):143–152. https://doi.org/10.1016/j.jsps.2012.05.002
17. Hermans N, Cos P, De Meyer GR et al (2007) Study of potential systemic oxidative stress animal models for the evaluation of antioxidant activity: status of lipid peroxidation and fat-soluble antioxidants. J Pharm Pharmacol 59(1):131–136. https://doi.org/10.1211/jpp.59.1.0017
18. Dimitriadis F, Borgmann H, Struck JP et al (2023) Antioxidant supplementation on male fertility—a systematic review. Antioxidants 12:836. https://doi.org/10.3390/antiox12040836
19. Madhukar D, Rajender S (2009) Hormonal treatment of male infertility: promises and pitfalls. J Androl 30(2):95–112. https://doi.org/10.2164/jandrol.108.005694
20. Patel AS, Leong JY, Ramos L et al (2019) Testosterone is a contraceptive and should not be used in men who desire fertility. World J Mens Health 37(1):45–54. https://doi.org/10.5534/wjmh.180036
21. Ly C, Yockell-Lelièvre J, Ferraro ZM et al (2015) The effects of dietary polyphenols on reproductive health and early development. Hum Reprod Update 21(2):228–248. https://doi.org/10.1093/humupd/dmu058
22. Martins RVL, Silva AMS, Duarte AP (2021) Natural products as protective agents for male fertility. BioChem 1:122–147. https://doi.org/10.3390/biochem1030011
23. Ferramosca A, Zara V (2022) Diet and male fertility: the impact of nutrients and antioxidants on sperm energetic metabolism. Int J Mol Sci 23:2542. https://doi.org/10.3390/ijms23052542
24. Rana A, Samtiya M, Dhewa T et al (2022) Health benefits of polyphenols: a concise review. J Food Biochem 46(10):e14264. https://doi.org/10.1111/jfbc.14264
25. Ullah A, Munir S, Badshah SL et al (2022) Important flavonoids and their role as a therapeutic agent. Molecules 25(22):5243. https://doi.org/10.3390/molecules25225243
26. Alarcón de la Lastra C (2008) Curcumin: a promising spice for therapeutics. Mol Nutr Food Res 52(9):985. https://doi.org/10.1002/mnfr.200890036
27. Gambini J, Inglés M, Olaso G et al (2015) Properties of resveratrol: in vitro and in vivo studies about metabolism, bioavailability, and biological effects in animal models and humans. Oxid Med Cell Longev 2015:837042. https://doi.org/10.1155/2015/837042
28. Mongioì LM, Perelli S, Condorelli RA et al (2021) The role of resveratrol in human male fertility. Molecules 26(9):2495. https://doi.org/10.3390/molecules26092495
29. Mohar DS, Malik S (2012) The sirtuin system: the holy grail of resveratrol? J Clin Exp Cardiolog 3(11):216. https://doi.org/10.4172/2155-9880.1000216
30. Juan ME, González-Pons E, Munuera T et al (2005) Trans-resveratrol, a natural antioxidant from grapes, increases sperm output in healthy rats. J Nutr 135(4):757–760. https://doi.org/10.1093/jn/135.4.757

31. Pasquariello R, Verdile N, Brevini TAL et al (2020) The role of resveratrol in mammalian reproduction. Molecules 25(19):4554. https://doi.org/10.3390/molecules25194554
32. Guo Y, Wang A, Liu X et al (2019) Effects of resveratrol on reducing spermatogenic dysfunction caused by high-intensity exercise. Reprod Biol Endocrinol 17:42. https://doi.org/10.1186/s12958-019-0486-7
33. Pervaiz S, Holme AL (2009) Resveratrol: its biologic targets and functional activity. Antioxid Redox Signal 11(11):2851–2897. https://doi.org/10.1089/ars.2008.2412
34. Ourique GM, Finamor IA, Saccol EM et al (2013) Resveratrol improves sperm motility, prevents lipid peroxidation and enhances antioxidant defences in the testes of hyperthyroid rats. Reprod Toxicol 37:31–39. https://doi.org/10.1016/j.reprotox.2013.01.006
35. Abdelali A, Al-Bader M, Kilarkaje N (2016) Effects of trans-resveratrol on hyperglycemia-induced abnormal spermatogenesis, DNA damage and alterations in poly (ADP-ribose) polymerase signaling in rat testis. Toxicol Appl Pharmacol 311:61–73. https://doi.org/10.1016/j.taap.2016.09.023
36. Bahmanzadeh M, Goodarzi MT, Rezaei Farimani A et al (2019) Resveratrol supplementation improves DNA integrity and sperm parameters in streptozotocin-nicotinamide-induced type 2 diabetic rats. Andrologia 51(8):e13313. https://doi.org/10.1111/and.13313
37. Khalil SS, Abdelghany EMA, Abd El-Fatah SS et al (2023) Protective effect of resveratrol against testicular dysfunction induced by forced swimming exercise in adult rats. Zagazig Med J 29(1):266–276. https://doi.org/10.21608/zumj.2022.152238.2607
38. Shati AA (2019) Resveratrol improves sperm parameter and testicular apoptosis in cisplatin-treated rats: effects on ERK1/2, JNK, and Akt pathways. Syst Biol Reprod Med 65(3):236–249. https://doi.org/10.1080/19396368.2018.1541114
39. Emerling BM, Weinberg F, Snyder C et al (2009) Hypoxic activation of AMPK is dependent on mitochondrial ROS but independent of an increase in AMP/ATP ratio. Free Radic Biol Med 46(10):1386–1391. https://doi.org/10.1016/j.freeradbiomed.2009.02.019
40. Zheng J, Ramirez VD (2000) Inhibition of mitochondrial proton F0F1-ATPase/ATP synthase by polyphenolic phytochemicals. Br J Pharmacol 130(5):1115–1123. https://doi.org/10.1038/sj.bjp.0703397
41. Shabani Nashtaei M, Nekoonam S, Naji M et al (2018) Cryoprotective effect of resveratrol on DNA damage and crucial human sperm messenger RNAs, possibly through 5' AMP-activated protein kinase activation. Cell Tissue Bank 19(1):87–95. https://doi.org/10.1007/s10561-017-9642-5
42. de Oliveira FA, Costa WS, Sampaio FJB et al (2019) Resveratrol attenuates metabolic, sperm, and testicular changes in adult Wistar rats fed a diet rich in lipids and simple carbohydrates. Asian J Androl 21(2):201–207. https://doi.org/10.4103/aja.aja_67_18
43. Shalaby AM, Alabiad MA, El Shaer DF (2020) Resveratrol ameliorates the seminiferous tubules damages induced by finasteride in adult male rats. Microsc Microanal 26(6):1176–1186. https://doi.org/10.1017/S1431927620024514
44. Mendes TB, Simas JN, Fischer LW et al (2022) Resveratrol benefits on sperm DNA, chromatin structure and reproductive outcomes of varicocelized rats. Andrologia 54(6):e14417. https://doi.org/10.1111/and.14417
45. Alshehri FS (2023) Resveratrol ameliorates vancomycin-induced testicular dysfunction in male rats. Medicina 59(3):486. https://doi.org/10.3390/medicina59030486
46. Elsawy H, Famurewa AC, Sedky A (2023) Resveratrol mitigates diabetic testicular dysfunction, endocrine deficits, and insulin resistance via suppression of sperm-endocrine aberrations and oxidative inflammation in rats. Andrologia 2023:6385767. https://doi.org/10.1155/2023/6385767
47. Xu D, Hu MJ, Wang YQ (2008) Antioxidant activities of quercetin and its complexes for medicinal application. Molecules 24(6):1123. https://doi.org/10.3390/molecules24061123
48. Taepongsorat L, Tangpraprutgul P, Kitana N et al (2008) Stimulating effects of quercetin on sperm quality and reproductive organs in adult male rats. Asian J Androl 10(2):249–258. https://doi.org/10.1111/j.1745-7262.2008.00306.x

49. Naseer Z, Ahmad E, Aksoy M et al (2020) Impact of quercetin supplementation on testicular functions in summer heat-stressed rabbits. World Rabbit Sci 28:19–27. https://doi.org/10.4995/wrs.2020.12420
50. Yelumalai S, Giribabu N, Karim K (2019) In vivo administration of quercetin ameliorates sperm oxidative stress, inflammation, preserves sperm morphology and functions in streptozotocin-nicotinamide induced adult male diabetic rats. Arch Med Sci 15(1):240–249. https://doi.org/10.5114/aoms.2018.81038
51. Tvrdá E, Kováč J, Ferenczyová K et al (2022) Quercetin ameliorates testicular damage in Zucker diabetic fatty rats through its antioxidant, anti-inflammatory and anti-apoptotic properties. Int J Mol Sci 23(24):16056. https://doi.org/10.3390/ijms232416056
52. Jahan S, Iftikhar N, Ullah H (2015) Alleviative effect of quercetin on rat testis against arsenic: a histological and biochemical study. Syst Biol Reprod Med 61(2):89–95. https://doi.org/10.3109/19396368.2014.998350
53. Badr GM, Elsawy H, Sedky A et al (2019) Protective effects of quercetin supplementation against short-term toxicity of cadmium-induced hematological impairment, hypothyroidism, and testicular disturbances in albino rats. Environ Sci Pollut Res 26:8202–8211. https://doi.org/10.1007/s11356-019-04276-1
54. Wang J, Zhu H, Wang K et al (2020) Protective effect of quercetin on rat testes against cadmium toxicity by alleviating oxidative stress and autophagy. Environ Sci Pollut Res 27:25278–25286. https://doi.org/10.1007/s11356-020-08947-2
55. Abarikwu SO, Farombi EO (2016) Quercetin ameliorates atrazine-induced changes in the testicular function of rats. Toxicol Ind Health 32(7):1278–1285. https://doi.org/10.1177/0748233714555389
56. Ben Abdallah F, Fetoui H, Zribi N et al (2013) Quercetin attenuates lambda cyhalothrin-induced reproductive toxicity in male rats. Environ Toxicol 28(12):673–680. https://doi.org/10.1002/tox.20762
57. Xia LZ, Jiang MZ, Liu LL et al (2023) Quercetin inhibits testicular toxicity induced by the mixture of three commonly used phthalates in rats. J Sci Food Agric 103(3):1541–1549. https://doi.org/10.1002/jsfa.12251
58. Askari Jahromi M, Movahedin M, Mazaheri Z et al (2014) Evaluating the effects of Escanbil (Calligonum) extract on the expression level of Catsper gene variants and sperm motility in aging male mice. Iran J Reprod Med 12(7):459–466
59. Williams KM, Ford WC (2003) Effects of Ca-ATPase inhibitors on the intracellular calcium activity and motility of human spermatozoa. Int J Androl 26(6):366–375. https://doi.org/10.1111/j.1365-2605.2003.00438.x
60. Ulusoy HG, Sanlier N (2020) A minireview of quercetin: from its metabolism to possible mechanisms of its biological activities. Crit Rev Food Sci Nutr 60(19):3290–3303. https://doi.org/10.1080/10408398.2019.1683810
61. Tvrdá E, Benko F, Slanina T et al (2021) The role of selected natural biomolecules in sperm production and functionality. Molecules 26(17):5196. https://doi.org/10.3390/molecules26175196
62. Diao R, Gan H, Tian F et al (2019) In vitro antioxidation effect of Quercetin on sperm function from the infertile patients with leukocytospermia. Am J Reprod Immunol 82(3):e13155. https://doi.org/10.1111/aji.13155
63. Al-Fattah AA, El-Raouf YMA (2019) Dietary supplementation of quercetin and the reproductive functions in hypertensive male rats. Med J Cairo Univ 87(1):535–545
64. Abdul-Hamid M, Galaly SR, Ahmed RR et al (2021) Histopathological and biochemical effect of quercetin on monosodium glutamate supplementation-induced testicular toxicity. Beni-Suef Univ J Basic Appl Sci 10:73. https://doi.org/10.1186/s43088-021-00167-y
65. Przybylska S (2020) Lycopene—a bioactive carotenoid offering multiple health benefits: a review. Int J Food Sci Technol 55:11–32. https://doi.org/10.1111/ijfs.14260
66. Arballo J, Amengual J, Erdman JW Jr (2021) Lycopene: a critical review of digestion, absorption, metabolism, and excretion. Antioxidants 10(3):342. https://doi.org/10.3390/antiox10030342

67. Durairajanayagam D, Agarwal A, Ong C, Prashast P (2014) Lycopene and male infertility. Asian J Androl 16(3):420–425. https://doi.org/10.4103/1008-682X.126384
68. Goyal A, Chopra M, Lwaleed BA (2007) The effects of dietary lycopene supplementation on human seminal plasma. BJU Int 99(6):1456–1460. https://doi.org/10.1111/j.1464-410X.2007.06804.x
69. Yamamoto Y, Aizawa K, Mieno M et al (2017) The effects of tomato juice on male infertility. Asia Pac J Clin Nutr 26(1):65–71. https://doi.org/10.6133/apjcn.102015.17
70. Pirayesh Islamian J, Mehrali H (2015) Lycopene as a carotenoid provides radioprotectant and antioxidant effects by quenching radiation-induced free radical singlet oxygen: an overview. Cell J 16(4):386–391. https://doi.org/10.22074/cellj.2015.485
71. Oborna I, Malickova K, Fingerova H et al (2011) A randomized controlled trial of lycopene treatment on soluble receptor for advanced glycation end products in seminal and blood plasma of normospermic men. Am J Reprod Immunol 66(3):179–184. https://doi.org/10.1111/j.1600-0897.2011.00984.x
72. Türk G, Ateşşahin A, Sönmez M (2007) Lycopene protects against cyclosporine A-induced testicular toxicity in rats. Theriogenology 67(4):778–785. https://doi.org/10.1016/j.theriogenology.2006.10.013
73. Salem EA, Salem NA, Maarouf AM (2012) Selenium and lycopene attenuate cisplatin-induced testicular toxicity associated with oxidative stress in Wistar rats. Urology 79(5):1184.e1-1184.e11846. https://doi.org/10.1016/j.urology.2011.12.006
74. Aly H (2019) Testicular toxicity of gentamicin in adult rats: ameliorative effect of lycopene. Hum Exp Toxicol 38(11):1302–1313. https://doi.org/10.1177/0960327119864160
75. Krishnamoorthy G, Selvakumar K, Venkataraman P et al (2013) Lycopene supplementation prevents reactive oxygen species mediated apoptosis in Sertoli cells of adult albino rats exposed to polychlorinated biphenyls. Interdiscip Toxicol 6(2):83–92. https://doi.org/10.2478/intox-2013-0015
76. Taş M, Saruhan GB, Kurt D et al (2010) Protective role of lycopene on aflatoxin B1 induced changes sperm characteristics and testicular damages in rats. Kafkas Univ Vet Fak Derg 16:597–604
77. Palabiyik SS, Erkekoglu P, Kızılgun M et al (2017) Lycopene restores trace element levels in ochratoxin A-treated rats. Arh Hig Rada Toksikol 68(2):135–141. https://doi.org/10.1515/aiht-2017-68-2905
78. Li Y, Zhan M, Li J (2023) Lycopene alleviates lipopolysaccharide-induced testicular injury in rats by activating the PPAR signaling pathway to integrate lipid metabolism and the inflammatory response. Transl Androl Urol 12(2):271–285. https://doi.org/10.21037/tau-22-864
79. Babaei A, Asadpour R, Mansouri K et al (2022) Lycopene improves testicular damage and sperm quality in experimentally induced varicocele: relationship with apoptosis, hypoxia, and hyperthermia. Food Sci Nutr 10(5):1469–1480. https://doi.org/10.1002/fsn3.2762
80. Zini A, San Gabriel M, Libman J (2010) Lycopene supplementation in vitro can protect human sperm deoxyribonucleic acid from oxidative damage. Fertil Steril 94(3):1033–1036. https://doi.org/10.1016/j.fertnstert.2009.04.004
81. Tvrda E, Mackovich A, Greifova H et al (2017) Antioxidant effects of lycopene on bovine sperm survival and oxidative profile following cryopreservation. Vet Med 62:429–436. https://doi.org/10.17221/86/2017-VETMED
82. Paiva SA, Russell RM (1999) Beta-carotene and other carotenoids as antioxidants. J Am Coll Nutr 18(5):426–433. https://doi.org/10.1080/07315724.1999.10718880
83. Hekimoglu A, Kurcer Z, Aral F (2009) Lycopene, an antioxidant carotenoid, attenuates testicular injury caused by ischemia/reperfusion in rats. Tohoku J Exp Med 218(2):141–147. https://doi.org/10.1620/tjem.218.141
84. Kara Ö, Baş H, Pandır D (2016) Furan toxicity on testes and protective role of lycopene in diabetic rats. J Turk Ger Gynecol Assoc 17(4):191–196. https://doi.org/10.5152/jtgga.2016.16144

85. Amalraj A, Pius A, Gopi S et al (2016) Biological activities of curcuminoids, other biomolecules from turmeric and their derivatives—a review. J Tradit Complement Med 7(2):205–233. https://doi.org/10.1016/j.jtcme.2016.05.005
86. Gupta SC, Patchva S, Agarwal BB (2013) Therapeutic roles of curcumin: lessons learned from clinical trials. AAPS J 15(1):195–218. https://doi.org/10.1208/s12248-012-9432-8
87. Iranloye B, Uweru O (2016) Fertility effects of Curcumin-a kitchen spice on male Wistar rat. Endocrine Abstr 44:P215
88. Izadpanah M, Alizadeh R, Bagher Minaee M et al (2015) The effects of curcumin on sperm parameters and nitric oxide production in varicocelized rats. Int J Morphol 33(4):1530–1535
89. Rashid K, Sil PC (2015) Curcumin ameliorates testicular damage in diabetic rats by suppressing cellular stress-mediated mitochondria and endoplasmic reticulum-dependent apoptotic death. Biochim Biophys Acta 1852(1):70–82. https://doi.org/10.1016/j.bbadis.2014.11.007
90. Zha W, Bai Y, Xu L et al (2018) Curcumin attenuates testicular injury in rats with streptozotocin-induced diabetes. Biomed Res Int 2018:7468019. https://doi.org/10.1155/2018/7468019
91. Chandra AK, Chatterjee A, Ghosh R et al (2007) Effect of curcumin on chromium-induced oxidative damage in male reproductive system. Environ Toxicol Pharmacol 24(2):160–166. https://doi.org/10.1016/j.etap.2007.04.009
92. Aktas C, Kanter M, Erboga M et al (2012) Anti-apoptotic effects of curcumin on cadmium-induced apoptosis in rat testes. Toxicol Ind Health 28(2):122–130. https://doi.org/10.1177/0748233711407242
93. Sudjarwo SA, Sudjarwo GW, Koerniasari (2017) Protective effect of curcumin on lead acetate-induced testicular toxicity in Wistar rats. Res Pharm Sci 12(5):381–390. https://doi.org/10.4103/1735-5362.213983
94. Sharma P, Singh R (2010) Protective role of curcumin on lindane induced reproductive toxicity in male Wistar rats. Bull Environ Contam Toxicol 84(4):378–384. https://doi.org/10.1007/s00128-010-9942-y
95. Ilbey YO, Ozbek E, Cekmen M et al (2009) Protective effect of curcumin in cisplatin-induced oxidative injury in rat testis: mitogen-activated protein kinase and nuclear factor-kappa B signaling pathways. Hum Reprod 24(7):1717–1725. https://doi.org/10.1093/humrep/dep058
96. Roshankhah SH, Salahshoor MR, Aryanfar S et al (2017) Effects of curcumin on sperm parameters abnormalities induced by morphine in rat. J Med Biomed Sci 6:1–10
97. Khalaji N, Namyari M, Rasmi Y et al (2018) Protective effect of curcumin on fertility of rats after exposure to compact fluorescent lamps: an experimental study. Int J Reprod Biomed 16(7):447–454
98. Mathuria N, Verma RJ (2008) Curcumin ameliorates aflatoxin-induced toxicity in mice spermatozoa. Fertil Steril 90(3):775–780. https://doi.org/10.1016/j.fertnstert.2007.07.1300
99. Singh S, Aggarwal BB (1995) Activation of transcription factor NF-kappa B is suppressed by curcumin (diferuloylmethane). J Biol Chem 270(42):24995–25000. https://doi.org/10.1074/jbc.270.42.24995
100. Shahedi A, Talebi AR, Mirjalili A et al (2021) Protective effects of curcumin on chromatin quality, sperm parameters, and apoptosis following testicular torsion-detorsion in mice. Clin Exp Reprod Med 48(1):27–33. https://doi.org/10.5653/cerm.2020.03853
101. Sarawi WS, Alhusaini AM, Fadda LM et al (2022) Nano-curcumin prevents copper reproductive toxicity by attenuating oxidative stress and inflammation and improving Nrf2/HO-1 signaling and pituitary-gonadal axis in male rats. Toxics 10(7):356. https://doi.org/10.3390/toxics10070356
102. Wei SM, Yan ZZ, Zhou J (2009) Curcumin attenuates ischemia-reperfusion injury in rat testis. Fertil Steril 91(1):271–277. https://doi.org/10.1016/j.fertnstert.2007.10.082
103. Ahmed-Farid OAH, Nasr M, Ahmed RF et al (2017) Beneficial effects of curcumin nano-emulsion on spermatogenesis and reproductive performance in male rats under protein deficient diet model: enhancement of sperm motility, conservancy of testicular tissue integrity, cell energy and seminal plasma amino acids content. J Biomed Sci 24(1):66. https://doi.org/10.1186/s12929-017-0373-5

104. Abdi F, Roozbeh N, Mortazavian AM (2017) Effects of date palm pollen on fertility: research proposal for a systematic review. BMC Res Notes 10(1):363. https://doi.org/10.1186/s13104-017-2697-3
105. Nantia EA, Manfo PF, Beboy NE et al (2012) Effect of methanol extract of Basella alba L. (Basellaceae) on the fecundity and testosterone level in male rats exposed to flutamide in utero. Andrologia 44(1):38–45. https://doi.org/10.1111/j.1439-0272.2010.01104.x
106. Yao DF, Mills JN (2016) Male infertility: lifestyle factors and holistic, complementary, and alternative therapies. Asian J Androl 18(3):410–418. https://doi.org/10.4103/1008-682X.175779
107. Madersbacher S, Berger I, Ponholzer A (2008) Plant extracts: sense or nonsense? Curr Opin Urol 18(1):16–20. https://doi.org/10.1097/MOU.0b013e3282f0d5c8
108. Boroujeni SN, Malamiri FA, Bossaghzadeh F et al (2022) The most important medicinal plants affecting sperm and testosterone production: a systematic review. JBRA Assist Reprod 26(3):522–530. https://doi.org/10.5935/1518-0557.20210108
109. Hasani-Ranjbar S, Larijani B, Abdollahi M (2009) A systematic review of the potential herbal sources of future drugs effective in oxidant-related diseases. Inflamm Allergy Drug Targets 8:2–10. https://doi.org/10.2174/187152809787582561
110. Abarikwu SO, Onuah CL, Singh SK (2020) Plants in the management of male infertility. Andrologia 52(3):e13509. https://doi.org/10.1111/and.13509
111. Gonzales GF, Nieto J, Rubio J et al (2006) Effect of Black maca (Lepidium meyenii) on one spermatogenic cycle in rats. Andrologia 38(5):166–172. https://doi.org/10.1111/j.1439-0272.2006.00733.x
112. Dorostghoal M, Seyyednejad SM, Jabari A (2014) Protective effects of Fumaria parviflora L. on lead-induced testicular toxicity in male rats. Andrologia 46:437–446. https://doi.org/10.1111/and.12100
113. Kooti W, Mansouri E, Ghasemiboroon M et al (2014) The effects of hydroalcoholic extract of Apium graveolens leaf on the number of sexual cells and testicular structure in rat. Jundishapur J Nat Pharm Prod 9:e17532. https://doi.org/10.17795/jjnpp-17532
114. Azizi M, Soleimani Mehranjani DM (2019) The effect of green tea extract on the sperm parameters and histological changes of testis in rats exposed to para-nonylphenol. Int J Reprod Biomed 17(10):717–726. https://doi.org/10.18502/ijrm.v17i10.5290
115. Khaki A, Khaki AA, Hajhosseini L (2014) The anti-oxidant effects of ginger and cinnamon on spermatogenesis dys-function of diabetes rats. Afr J Tradit Complement Altern Med 11:1–8. https://doi.org/10.4314/ajtcam.v11i4.1
116. Sharma M, Arya D, Bhagour K et al (2019) Modulatory effects of methanolic fruit fraction of Pedalium murex on sulphasalazine-induced male reproductive disruption. Andrologia 51(2):e13190. https://doi.org/10.1111/and.13190
117. Farombi EO, Adedara IA, Oyenihi AB (2013) Hepatic, testicular and spermatozoa antioxidant status in rats chronically treated with Garcinia kolaseed. J Ethnopharmacol 146(2):536–542. https://doi.org/10.1016/j.jep.2013.01.018
118. Shalaby MA, Hammouda AA (2014) Assessment of protective and anti-oxidant properties of Tribulus terrestris fruits against testicular toxicity in rats. J Intercult Ethnopharmacol 3(3):113–118. https://doi.org/10.5455/jice.20140627123443
119. Karna KK, Choi BR, Kim MJ et al (2019) The effect of schisandra chinensis baillon on cross-talk between oxidative stress, endoplasmic reticulum stress, and mitochondrial signaling pathway in testes of varicocele-induced SD rat. Int J Mol Sci 20(22):5785. https://doi.org/10.3390/ijms20225785
120. Ohsawa I, Ishikawa M, Takahashi K et al (2007) Hydrogen acts as a therapeutic antioxidant by selectively reducing cytotoxic oxygen radicals. Nat Med 13(6):688–694. https://doi.org/10.1038/nm1577
121. Ichihara M, Sobue S, Ito M (2015) Beneficial biological effects and the underlying mechanisms of molecular hydrogen - comprehensive review of 321 original articles. Med Gas Res 5:12. https://doi.org/10.1186/s13618-015-0035-1

122. Iuchi K, Imoto A, Kamimura N et al (2016) Molecular hydrogen regulates gene expression by modifying the free radical chain reaction-dependent generation of oxidized phospholipid mediators. Sci Rep 6:18971. https://doi.org/10.1038/srep18971
123. Ma L, Hao W, Feng WB et al (2022) Molecular hydrogen reduces electromagnetic pulse-induced male rat reproductive system damage in a rodent model. Oxid Med Cell Longev 2022:3469474. https://doi.org/10.1155/2022/3469474
124. Santini SJ, Cordone V, Falone S et al (2020) Role of mitochondria in the oxidative stress induced by electromagnetic fields: focus on reproductive systems. Oxid Med Cell Longev 2020:5203105. https://doi.org/10.1155/2020/5203105
125. Begum R, Bajgai J, Fadriquela A et al (2018) Molecular hydrogen may enhance the production of testosterone hormone in male infertility through hormone signal modulation and redox balance. Med Hypotheses 121:6–9. https://doi.org/10.1016/j.mehy.2018.09.001

Chapter 22
Comparison of Free-Radical Scavenging Activity of Various Sources of Molecular Hydrogen

Katarína Valachová, Branislav Kura, Ján Slezák, Mojmír Mach, and Ladislav Šoltés

Abstract Molecular hydrogen is well-known for its antioxidative and antiinflammatory properties in numerous in vitro and in vivo experiments. High-molar-mass hyaluronan (HMM HA), a natural polysaccharide present in tissues of all vertebrates, was used as a marker of the polysaccharide degradation by reactive oxygen species. The radical scavenging capacity of various sources of molecular hydrogen dissolved in water was assessed by the DPPH assay. In experiments using rotational viscometry, HMM HA was oxidatively degraded by cupric ions (1 μM) and ascorbic acid (100 μM). Further, the effects of molecular hydrogen from various sources were assessed before beginning $^{\bullet}$OH radical-induced HA degradation or one hour later, when alkyloxy- and alkylperoxy-type radical-induced HA degradation prevailed. The results of the DPPH assay showed that of the examined samples, only H_2 tablet dissolved in both distilled and drinkable water mildly scavenged DPP$^{\bullet}$. In contrast, experiments from rotational viscometry showed that the inhibition of reactive oxygen species induced-hyaluronan degradation was observed when examining molecular hydrogen saturated in both distilled and drinkable water.

Keywords Free radical scavenging capacity · Hyaluronan · Reactive oxygen species

---

K. Valachová (✉) · B. Kura · J. Slezák · M. Mach · L. Šoltés
Centre of Experimental Medicine, Slovak Academy of Sciences, Dúbravská Cesta 9, 84104 Bratislava, Slovakia
e-mail: katarina.valachova@savba.sk

B. Kura
e-mail: branislav.kura@savba.sk

J. Slezák
e-mail: jan.slezak@savba.sk

M. Mach
e-mail: mojmir.mach@savba.sk

L. Šoltés
e-mail: ladislav.soltes@savba.sk

J. Slezak and B. Kura (eds.), *Molecular Hydrogen in Health and Disease*, Advances in Biochemistry in Health and Disease 27, https://doi.org/10.1007/978-3-031-47375-3_22

Introduction

Molecular hydrogen is an odorless, tasteless, and non-toxic gas even when applied at high concentrations. It rapidly crosses different tissue barriers and penetrates various organelles such as mitochondria and nucleus to regulate transcription [1–5]. In animal models H_2 suppressed inflammation, excessive oxidative stress, and cell death thereby protected the liver from various acute and chronic injuries [1]. H_2 can stimulate a specific nucleotide-bearing domain and inhibit NLRP3 in macrophages and inflammatory reactions [5]. First therapeutic effects of H_2 were reported in 1975 by Dole et al. [6] in a skin squamous carcinoma mouse model.

Recent studies have shown that molecular hydrogen has a significant protective effect on multiple organs and physiological barriers in septic animal models [3]. Sim et al. [7] performed a randomized, double-blind, placebo-controlled study. The healthy adults (20–59 y) were administered either 1.5 L/d of hydrogen water (HW, $n = 20$) or plain water (PW, $n = 18$) for 4 weeks. In those aged ≥ 30 y, biological antioxidant potential was greater in the HW group compared to the PW group. Apoptosis of peripheral blood mononuclear cells was significantly less in the HW group. Flow cytometry analysis of $CD4^+$, $CD8^+$, $CD14^+$, $CD20^+$ and $CD11b^+$ cells showed that the frequency of $CD14^+$ cells decreased in the HW group.

One of medicinal approaches that uses hydrogen is the breath hydrogen test, which is performed by measuring the amount of hydrogen generated by intestinal bacteria that permanently synthesize hydrogen owing to the fermentation of unabsorbed carbohydrates. This test is also used as a biomarker in clinical and scientific research including biochemistry, dentistry, and physiology [2]. In vivo experiments and clinical trials have reported that H_2 had a radioprotective effect [8]. H_2 is also effective in the treatment of temporary and chronic forms of oxidative stress-associated fatigue [9]. During hemodialysis using H_2-enriched dialysis solution the levels of plasma monocyte chemoattractant protein 1 and myeloperoxidase were markedly decreased [10]. One of the earliest direct uses of H_2 was in deep sea diving. Hydroliox was used as a mixture of hydrogen, helium, and oxygen to protect divers being at great depths from the development of decompression syndrome [11]. The first mode of the application of hydrogen was a hyperbaric chamber with an atmosphere rich in hydrogen gas [5]. There are several ways how to apply H_2, which involve inhalation, oral administration of hydrogen-rich water (HRW) or hydrogen tablets, and hydrogen-saturated saline injections [1, 9, 10].

A simple method how to administer H_2 therapeutically is by inhalation using a facemask, ventilator circuit, or nasal cannula. The effect of inhaled H_2 is rapid and this method can be used to treat acute oxidative stress. An experiment in rats showed that inhalation of H_2 mixed with nitrous oxide, O_2, and N_2 dose dependently increased levels of H_2 dissolved in arterial blood in higher concentrations than in venous blood, demonstrating that administered H_2 was incorporated into tissues. No adverse effects of inhaling H_2 were demonstrated. Despite the fact that the inhalation of H_2 is fast, this method is not appropriate for daily preventive usage. For this reason, H_2 concentrations and dosages must be thoroughly controlled [4, 5, 12].

The beneficial effects of H_2 inhalation in chronic obstructive pulmonary disease and severe bronchial asthma have also been reported. Cole et al. [13] found out that in healthy animals, a continuous inhalation of a gas mixture containing 2.4% hydrogen for 72 h did not result in any changes in physiological parameters. In 2001, Gharib et al. [14] showed that inhalation of H_2 enhanced the catalytic properties of superoxide dismutase. Inhalation of hydrogen gas (1% H_2 in air) or drinking a saturated level of H_2 water attenuated mortality and body-weight loss caused by cisplatin (chemotherapeutic agent), and alleviated nephrotoxicity [10].

Unlike gaseous H_2, H_2-dissolved water or HRW is safe, portable, and readily administered. H_2 can be dissolved in water up to 0.8 mM (1.6 mg/L) at room temperature under atmospheric pressure without changing pH. H_2 at this concentration had a positive effect on obesity in mice model. Some studies found that drinking HRW had a positive effect in disease models, such as Parkinson's disease, radiation-induced oxidative injuries, oral palatal wounds, depression, periodontal tissue aging [5, 10, 12, 15, 16]. Clinical tests have revealed that drinking H_2-water ameliorated markers of oxidative stress in patients with high levels of glucose and cholesterol [10]. Kang et al. [17] reported that HRW was beneficial for patients suffering from liver cancer receiving radiotherapy. In Japan hydrogen is licensed as a food additive, and HRW is already being sold as a safe drinking product. Magnesium sticks and HRW made by electrolysis are also being examined for acute/sub-acute toxicity and, mutagenicity [10]. Hydrogen rich water can be readily generated by exposing water to magnesium by dissolving electrolyzed hydrogen into water, or by dissolving molecular hydrogen into water under high pressures [2, 4, 10]. A clinical study of HRW intake found that it decreased oxidative stress and improved lipid and glucose metabolism in patients with metabolic syndrome, type 2 diabetes, and impaired glucose tolerance. Animal studies have shown that blood concentrations of orally ingested H_2 reached the peak at approximately 15 min and returned to baseline levels at 20–30 min after administration. In humans drinking hydrogen-enriched water, H_2 concentrations in exhaled air reached a peak at approximately 10 min and returned to baseline levels at 1 h after administration [18]. Ostojic et al. [19] showed that the oral and topical administration of hydrogen-rich water for two weeks attenuated the inflammation, serum C-reactive protein, pain intensity, degree of swelling, enhanced recovery and functional abilities in male and female athletes after acute soft-tissue injury.

Although oral administration is safe and comfortable, hydrogen in water tends to be eliminated over time and some hydrogen penetrates into the stomach or intestine, thus it is difficult to control the concentration of ingested hydrogen. Administration of hydrogen via an injectable hydrogen saline vehicle may allow the delivery of more accurate concentrations of hydrogen [10]. The paper published by Ohsawa et al. [20] more markedly allevated the interest of researchers into the biological and medical effects of hydrogen. The paper relates to the results of a successful treatment using inhalation of H_2 to prevent the damage induced by ischemia–reperfusion after ischemic stroke in a rat model. When using an injection of hydrogen solutions, H_2 enters blood, where it modulates oxidative metabolism by capturing of free radicals and rejuvenating of endogenous antioxidants [5].

Another way is the administration of H_2 in eye drops, which attenuated ischemia–reperfusion injury of the retina in rats. After their continuous administration, H_2 concentration was increased in the vitreous body, and $^{\bullet}OH$ level decreased during retinal ischemia–reperfusion (treatment of glaucoma). Hydrogen readily penetrates the skin and distributes throughout the body via blood flow. Thus, taking a warm water bath dissolving H_2 is a mode of incorporating H_2 into the body daily, especially in Japan [10]. Alternative ways how to administer H_2 are: H_2-rich nanocomposites/microbubbles, orally taken solid-state H_2 sustained-release agents, stimulation of intestinal microbes to produce H_2, hyperbaric H_2 chambers, tube feeding of H_2-rich solutions, rinsing cavities H_2-rich water/solutions [5].

Unlike most antioxidant compounds, H_2 is able to pass through the blood brain barrier [10, 12]. H_2 inhibits oxidative stress-induced inflammatory tissue injury via downregulation of pro-inflammatory and inflammatory cytokines, such as IL-1β, IL-6, intercellular cell adhesion molecule-1, TNF-α, NF-κB and prostaglandin E2 [1, 5, 10, 12]. H_2 exerts anti-apoptotic effects by up- or downregulating apoptosis-related factors. For example, H_2 inhibits expression of the pro-apoptotic factors, caspase-3, caspase-8, and caspase-12, B-cell lymphoma-2-associated X-protein and upregulates the anti-apoptotic factors, B-cell lymphoma-2 and B-cell lymphoma [1, 12]. H_2 administration induces expression of diverse genes, including NF-κB, N-terminal kinase, c-Jun proliferation cell nuclear antigen, vascular endothelial growth factor, glial fibrillary acidic protein, and creatine kinase [12]. Hydrogen appears to act via hormesis in which it activates the Nrf2 pathway and subsequent induction of phase 2 antioxidant enzymes. Additionally, H_2 modulates signal transduction, protein phosphorylation cascades, which allows modulation of inflammatory pathways (NFAT, NF-κB, TLRs), increases PGC-1α (marker of mitochondrial biogenesis), alleviates resilience to stress, maintains mitochondrial membrane potential and ATP production and influence other signaling molecules (e.g., Nrf2, Nox1, STAT3) [21].

H_2 is a specific scavenger of $^{\bullet}OH$ and $ONOO^-$, which are very strong oxidants that react indiscriminately with nucleic acids, lipids, and proteins followed by DNA fragmentation, lipid peroxidation, and protein inactivation. However, H_2 is not able to react with signaling oxidants such as $O_2^{\bullet-}$, $NO^{\bullet}$, H_2O_2. H_2 administration attenuated the expression of various oxidative stress markers, such as myeloperoxidase, malondialdehyde, 8-hydroxy-desoxyguanosine, 8-iso-prostaglandin F2a, and thiobarbituric acid reactive substances in all human diseases and rodent models. Recent reports also revealed that H_2-selective anti-oxidation mitigates certain pathological processes in plants and retains freshness in fruits [2, 10, 12, 21].

We selected hyaluronan (HA) as a marker of oxidative degradation, which is a naturally occurring non-sulfated glycosaminoglycan (GAG). It is composed of repeating units of the disaccharide [-D-glucuronic acid-β-1,3-*N*-acetyl-D-glucosamine-β-1,4-]$_n$ biopolymer, and is the critical component of extracellular matrix. This GAG is widely distributed in vertebrate tissues and fluids. It has remarkable physicochemical and biological properties. The occurrence of HA in tissues varies, and the highest concentration of HA are in human joint synovial fluid, vitreous body, human skin and brain. In healthy individuals, HA is composed of 10,000 disaccharide units, which correspond to up to 40,000 kDa.

The HA chains tend to entangle at low concentrations, showing viscoelastic behavior, i.e., while they act as a viscous liquid at low frequencies, they show elastic behavior at a higher frequency. In vivo, degradation of HA may arise from hydroxyl radicals, peroxynitrite, and hypochlorite anion and is likely to be enhanced in inflammation. The hydroxyl radical can abstract hydrogen atoms at all ring C-H bonds except C-2 of *N*-acetyl hexosamine. The abstraction of hydrogen atoms generates carbon-centered radicals. The radicals at carbons, which form glycosidic bonds, undergo a β-scission reaction resulting in the breakdown of the HA chains [22, 23].

The aim of the study was to assess free radical scavenging properties of molecular hydrogen saturated in water and hydrogen-rich water commercial preparations.

Materials and Methods

Materials

Sodium hyaluronate (M_w 1.67 MDa) was purchased from Lifecore Biomedical, Chaska, MN, USA. Ascorbic acid, 2,2-diphenyl-1-picrylhydrazyl and methanol were purchased from Merck KGaA, Darmstadt, Germany. NaCl and $CuCl_2 \cdot 2H_2O$ p.a. were purchased from Slavus Ltd., Bratislava, Slovakia. Methylene blue was from Mikrochem, Pezinok, Slovakia. H_2-producing tablets (Drink HRW) were obtained from HRW Natural Health Products Inc. New Westminster BC, Canada. Ultra-high concentrated hydrogen-infused reverse osmosis (RO) water was obtained from ProLife Group, Ltd., Bratislava, Slovakia, drinkable and distilled water saturated with molecular hydrogen. Deionised high-purity grade water, with conductivity of $\leq$ 0.055 μS/cm, was produced by using the TKA water purification system (Water Purification Sys-tems GmbH, Niederelbert, Germany).

DPPH Assay

$DPP^{\bullet}$ was prepared by dissolving 2,2-diphenyl-1-picrylhydrazyl (1.1 mg) in 50 mL of methanol. Various sources of H_2 water (1 mL) were added to $DPP^{\bullet}$ (2 mL) and kinetic measurements were performed for 10 or 15 min at the wavelength 517 nm.

Rotational Viscometry

Sodium hyaluronate (1.75 mg/mL) was dissolved overnight in 0.15 M aqueous NaCl in two steps: At first, 4.0 mL of the solvent was added to 14 mg HA. After 6 h 3.9 or 2.9 mL of 0.15 M NaCl was added. The stock solutions of ascorbic acid (16 mM) and cupric chloride (160 μM) were also prepared in 0.15 M aqueous NaCl. H_2-producing tablets were dissolved in drinkable or distilled water (330 mL).

Degradation of HA was initiated by $CuCl_2$ (1 μM) and ascorbic acid (100 μM), the source of reactive oxygen species, in the following way: $CuCl_2$ stock solution (50 μL) was added to the HA solution (7.9 mL), stirred for 0.5 min. After not stirring the solution for 7.5 min, the stock solution of ascorbic acid (50 μL) was added to the HA solution and stirred again for 0.5 min. The final solution was then immediately transferred into the viscometer Teflon® vessel.

The procedures to examine the effect of molecular hydrogen water as an inhibitor of HA degradation were as followed:

(a) The stock solution of $CuCl_2$ (50 μL) was added to the HA solution (6.9 mL), stirred for 0.5 min. The solution was not stirred for another 7.5 min. Then, molecular hydrogen water (1 mL) was added, and the solution was stirred again for 0.5 min. Finally, the stock ascorbic acid solution (50 μL) was added, and the final solution was stirred for 0.5 min. The solution mixture was then immediately transferred into the viscometer Teflon® vessel.
(b) The stock solution of $CuCl_2$ (50 μL) was added to the HA solution (6.9 mL), stirred for 0.5 min. The solution was not stirred for another 7.5 min. Further, the stock ascorbic acid solution (50 μL) was added, and the solution was stirred for 1 h. In the end, molecular hydrogen water (1 mL) was added and stirred again for 0.5 min.

Dynamic viscosity of the HA solutions was measured for 5 h by reporting the data in 3-min intervals using the program Rheocalc of the rotational viscometer (Brookfield Engineering Labs, Inc., Middleboro, MA, USA). Other parameters were: temperature 25 °C, 180 rpm of the viscometer Teflon® spindle and shear rate equaling to 237 s^{-1}.

Results and Discussion

There are several methods of evaluating hydrogen atom transfer properties. One of them is the DPPH assay (Fig. 22.1).

$$DPP^{\bullet} + H \rightarrow DPPH$$

The disadvantage of this method is the solubility of DPPH in ethanol or methanol and the fact that $DPP^{\bullet}$ radical is not present in human individuals.

As shown in Fig. 22.2, either the addition of H_2 dissolved in RO water (1 mL, 80 μM, triangle), or molecular hydrogen saturated in drinkable or distilled water (1 mL, 190 μM, grey and white circles, respectively) were not effective in scavenging $DPP^{\bullet}$.

Results in Fig. 22.3, left panel show that molecular hydrogen tablet dissolved in drinkable water was effective in part in scavenging of $DPP^{\bullet}$ (black circle). The

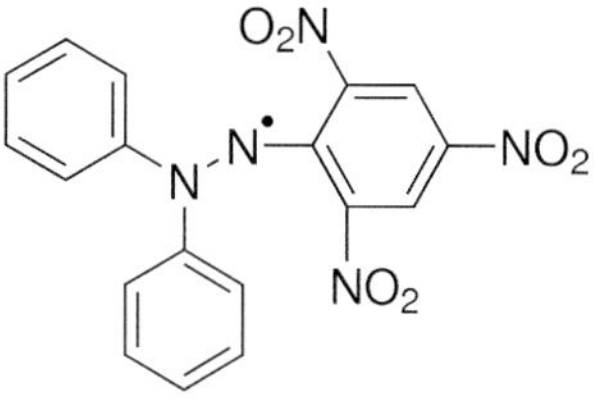

Fig. 22.1 Structure of DPP•

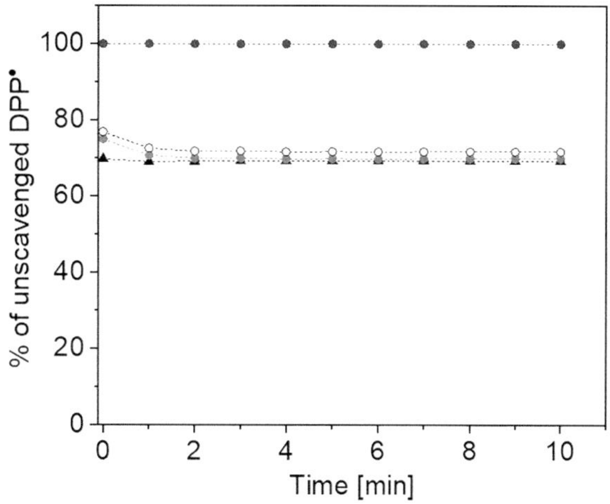

Fig. 22.2 Percentage of unscavenged DPP• (black circle) by H_2 dissolved in RO water (80 μM, triangle), H_2 saturated in drinkable water (190 μM, grey circle), H_2 saturated in distilled water (190 μM, white circle)

percentage of unscavenged DPP• decreased from 78% in time 0 min to 38% after 15 min (white triangle). Drinkable water itself (black triangle), bubbled out H_2 from the tablet (black square), bubbled out H_2 from the tablet and resaturated with H_2 (white circle) were not able to scavenge DPP•. Similar results were observed in right panel. A positive effect was observed only when adding H_2 tablet dissolved in distilled water, where the percentage of unscavenged DPP• decreased from 74% in time 0 min to 35.8% after 15 min.

Fan et al. [24] confirmed our results that HRW at volume 4 mL was poor in scavenging DPP• (4 mL). The efficacy of HRW was 6%.

We have been applying the oxidative system composed of cupric ions with ascorbate, which is a source of reactive oxygen species such as •OH, alkyloxy, alkylperoxy-type radicals, which deteriorate all molecules, including a polysaccharide hyaluronan. Hyaluronan is readily susceptible to oxidative degradation and protective effects of numerous antioxidants or drugs were shown, which had H atom donor properties [25–27]. This oxidative system represents pathophysiological conditions and occurs in human individuals. Of methods we used rotational viscometry, which monitors values of dynamic viscosity of hyaluronan solutions versus time.

Results in Fig. 22.4, left panel show that hyaluronan subjected to the exposition of reactive oxygen species degraded by 9.1 mPa·s (black curve). As shown, H_2

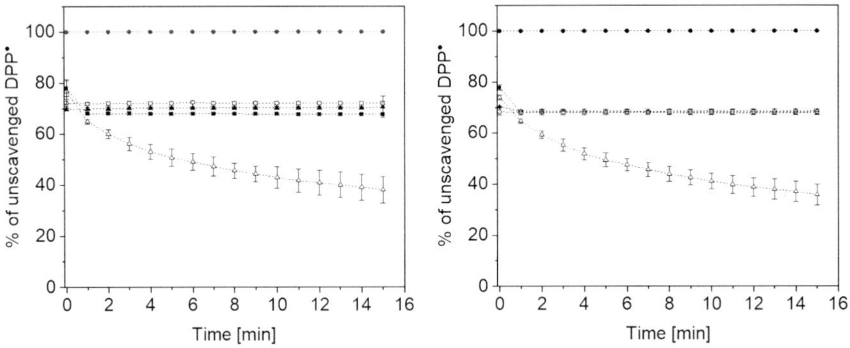

Fig. 22.3 Percentage of unscavenged DPP• (●) by H_2 tablet dissolved in water (190 μM, Δ), water (▲), bubbled out H_2 from the tablet (■), bubbled out H_2 from the tablet and resaturated with H_2 (130 μM, ○). Left panel: H_2 was dissolved in drinkable water, right panel: H_2 was dissolved in distilled water. Experiments were performed in triplicate

dissolved in drinkable water at the concentration 75 μM (grey curve) protected in part hyaluronan molecules from •OH-induced degradation. The dynamic viscosity of the hyaluronan solution decreased by 3.0 mPa·s within 5 h. A new finding of our group is the ability of H_2 to scavenge moderately also alkyloxy- and alkylperoxy-type radicals (right panel). Hyaluronan was shown to degrade mildly, whereas the dynamic viscosity of the hyaluronan solution decreased by 4.15 mPa·s within 4 h.

As obvious from results in Fig. 22.5, H_2 dissolved in distilled water (75 μM, grey curve) was less effective in inhibiting •OH and alkyloxy and alkylperoxy-type radical-induced hyaluronan degradation compared to the results in Fig. 22.4. The dynamic viscosity values decreased by 6.47 (left panel) and 4.95 mPa·s (right panel).

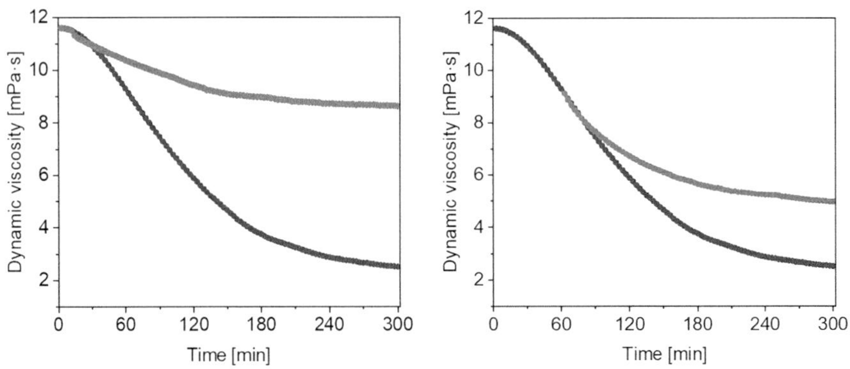

Fig. 22.4 Time-dependent changes in dynamic viscosity of hyaluronan solution exposed to 1 μM Cu(II) and 100 μM ascorbate (black curve). The addition of H_2 dissolved in drinkable water (75 μM, grey curve) before beginning the experiment (left panel, adapted from Kura et al. [28] and 1 h later (right panel)

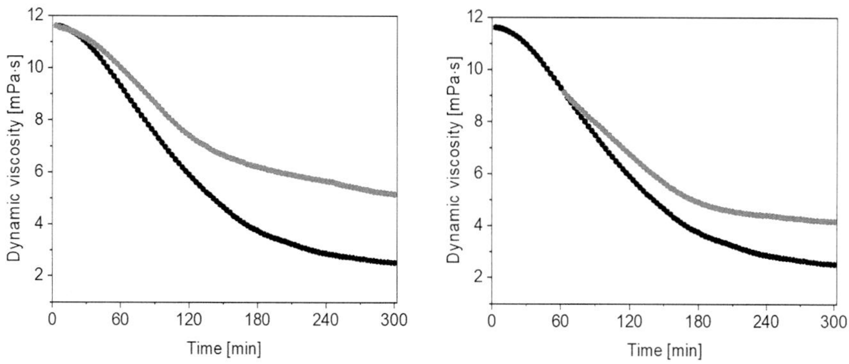

Fig. 22.5 Time-dependent changes in dynamic viscosity of hyaluronan solution exposed to 1 μM Cu(II) and 100 μM ascorbate (black curve). The addition of H_2 dissolved in distilled water (75 μM, grey curve) before beginning the experiment (left panel) and 1 h later (right panel)

Results in Fig. 22.6 show that commercially used ultra-high concentrated hydrogen-infused RO water at concentration 80 μM slightly scavenged $^{\bullet}OH$ (left panel). However, its protective effect against alkyloxy and alkylperoxy-type radicals was not shown (right panel).

Further, we examined another commercial preparation Drink HRW, where molecular hydrogen from the tablet is produced by the active ingredient metallic magnesium (80 mg), which reacts with water to produce H_2 gas and magnesium hydroxide according to the reaction: $Mg + 2H_2O \rightarrow H_2(g) + Mg(OH)_2$. The tablets also contain organic acids such as malic acid and tartaric acid, which neutralize $Mg(OH)_2$ and catalyze the reaction rate [21].

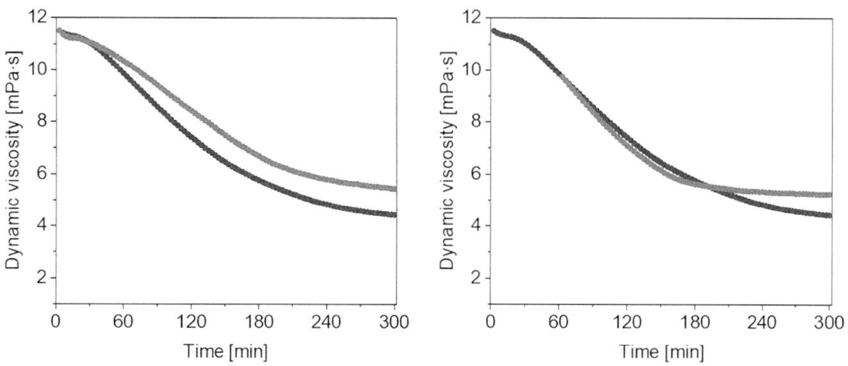

Fig. 22.6 Time-dependent changes in dynamic viscosity of hyaluronan solution exposed to 1 μM Cu(II) and 100 μM ascorbate (black curve). The addition of ultra-high concentrated hydrogen-infused RO water (80 μM, grey curve) before beginning the experiment (left panel) and 1 h later (right panel)

As shown in Fig. 22.7, left panel at first hyaluronan solution was exposed to 1 μM Cu(II) ions and 100 μM ascorbate as a source of reactive oxygen species. Within 5 h hyaluronan degraded by 7.07 mPa·s (black circle). The addition of 1 mL of H_2 tablet dissolved in drinkable water (190 μM) resulted in a significant $^{\bullet}OH$ radical-induced degradation of hyaluronan, which was the most obvious within 2 h. The decrease in dynamic viscosity of the hyaluronan solution was 8.3 mPa·s after 5 h of the measurement (black square). This massive hyaluronan degradation is for the presence of malic and tartaric acids. Ye at. [29] by using the EPR method detected $O_2^{\bullet -}$, carbon-centered radical ($^{\bullet}R$) and $H^{\bullet}$ in the Cu(II)/tartaric acid/H_2O_2 system, which contributed to the promotion of the Cu(II)/Cu(I) redox cycle. Cabelli and Bielski [30] carried out the reactions between $^{\bullet}OH$ and malic acid using pulse radiolysis technique and they showed that when malic acid interacts with $^{\bullet}OH$, α-malate radical is formed. The addition of H_2 tablet dissolved in drinkable water with bubbled out H_2 resulted in a bit faster hyaluronan degradation (white triangle) compared with the reference. However, the resaturation of this water with H_2 even at concentration 220 μM had only a little impact on the inhibition of hyaluronan degradation (black triangle). For comparison, we examined also the effect of drinkable water itself, where the most rapid degradation within 1 h, followed by a significant retardation of the hyaluronan degradation (white circle).

Similar results were observed when adding H_2 tablet dissolved in drinkable water (190 μM) 1 h after initiating the oxidative degradation of hyaluronan (right panel). Again a massive decrease in dynamic viscosity of the hyaluronan solution 8.0 mPa·s was recorded (black square). After the addition of H_2 tablet dissolved in drinkable water with bubbled out H_2 and only distilled water decreased the rate of hyaluronan degradation in a similar way (white triangle and white circle, respectively). However, the resaturation of this water with H_2 (160 μM) positively affected alkyloxy and

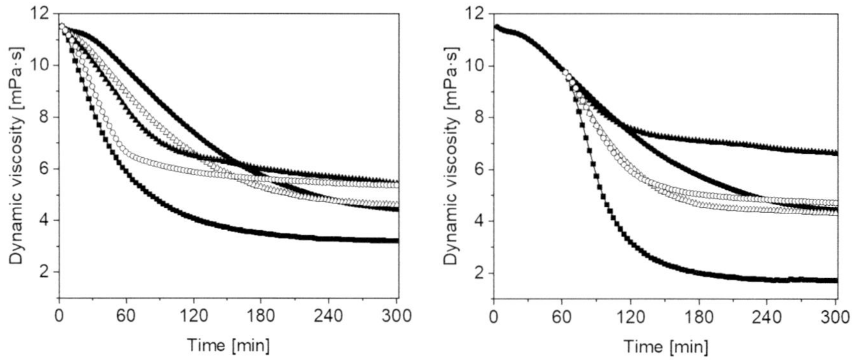

Fig. 22.7 Time-dependent changes in dynamic viscosity of hyaluronan solutions subjected to oxidative degradation by 1 μM Cu(II) ions and 100 μM ascorbate (●). The addition of H_2 tablet (1 mL) dissolved in drinkable water (■), bubbled out H_2 from the tablet (Δ), bubbled out H_2 from the tablet and resaturated with H_2 (▲), drinkable water (○). The sample was added at the beginning of the experiment (left panel) or 1 h later (right panel)

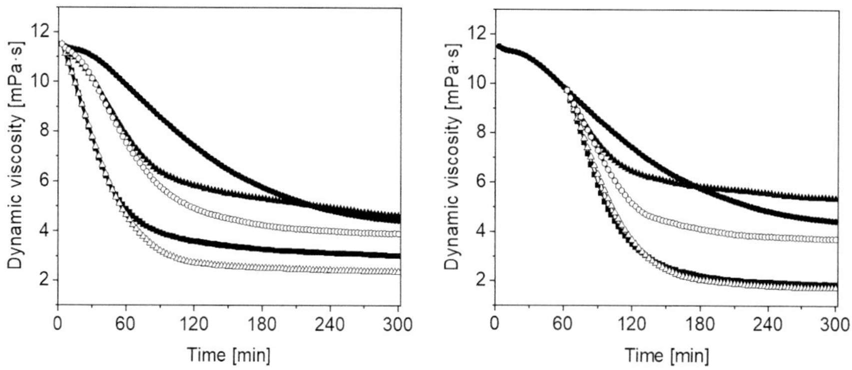

Fig. 22.8 Time-dependent changes in dynamic viscosity of hyaluronan solutions subjected to oxidative degradation by 1 μM Cu(II) ions and 100 μM ascorbate (●). The addition of H_2 tablet (1 mL) dissolved in distilled water (■), bubbled out H_2 from the tablet (△), bubbled out H_2 from the tablet and resaturated with H_2 (▲), distilled water (◯). The sample was added at the beginning of the experiment (left panel) or 1 h later (right panel)

alkylperoxy-type radical-induced hyaluronan degradation. The viscosity was lower by 3.31 mPa·s (black triangle).

Results in Fig. 22.8 show that the addition of 1 mL of H_2 from the tablet (125 μM) dissolved in distilled water resulted in a more rapid $^{\bullet}OH$ induced degradation of hyaluronan (black circle) compared to results with drinkable water. The decrease in dynamic viscosity of the hyaluronan solution was 8.48 mPa·s after 5 h of the measurement (black square). The addition of H_2 tablet dissolved in distilled water with removed of H_2 mildly promoted hyaluronan degradation (white triangle) compared with the curve (black square). On the other hand, the addition of this water resaturated with H_2 (125 μM) attenuated hyaluronan degradation. The decrease in dynamic viscosity of the hyaluronan solution was 6.88 mPa·s (black triangle). For comparison, we examined also the effect of distilled water itself, which effect was a bit weaker (white circle) than in black triangle curve.

More significant pro-oxidative effects of H_2 tablet dissolved in distilled water (compared to results in Fig. 22.7) were shown also 1 h after initiating the degradation of hyaluronan (right panel). Very similar results were shown when examining H_2 from the tablet (125 μM) dissolved in distilled water and the water after bubbling out of H_2 (black square and white triangle curves, respectively). However, the resaturation of this water with H_2 (125 μM) had the least deteriorating effect on alkyloxy and alkylperoxy-type radical-induced hyaluronan degradation. The viscosity was lower by 4.4 mPa·s (black triangle).

Different results with Drink HRW tablets obtained Kura et al. [31], who administered HRW to 30 individuals suffering from non-alcoholic fatty liver disease in a randomized, double-blinded, placebo-controlled manner for 8 weeks. Phenotypically, they observed a decreased weight (≈1 kg) and body mass index in the HRW group. HRW was well-tolerated, with no significant changes in liver enzymes, improved lipid profile and reduced lactate dehydrogenase levels. HRW non-significantly decreased levels of nuclear factor kappa B, heat shock protein 70 and matrix metalloproteinase-9. However, mild, increased levels of 8-hydroxy-2'-deoxyguanosine and malondialdehyde in the HRW group were observed.

LeBaron et al. [21] examined the effect of acute supplementation with HRW (13 ppm) on exercise performance in 19 individuals as measured by respiratory exchange ratio heart rate, and respiratory rate. The authors showed that the administration with H_2 did not facilitate maximal exercise performance in young healthy individuals; yet, it significantly diminished exercising heart rate at submaximal intensities.

In ancient times, the atmosphere of the Earth contained, among others gaseous components, hydrogen molecules [32]. Cosmic radiation, γ-radiation and the other ones that fell on the upper layers of the atmosphere provided enough energy for the reaction of dissociation of hydrogen molecules into two atoms of hydrogen. Hydrogen atom is a reactive radical element, which interacts intensively with numerous compounds.

However, under common biogenic conditions, dissociation of H_2 requires as much energy that the formation of two atoms of hydrogen from its molecule is almost impossible. In organic chemistry if dissociation of hydrogen in chemical reactions is ever possible, at the same time it is supposed the presence of some of the biogenic transition metals, such as Fe(II)/Fe(III), Cu(II)/Cu(I), Co(II)/Co(III), Ni(II)/Ni(III) and Mn(II)/Mn(III). The presence of these transition metal ions is unamnigously responsible for catalytic activity of numerous redox active enzymes.

The starting point for considering the homogenous dissociation of the hydrogen molecule into H plus H in the presence of a transition metal ion in a higher oxidation state is the following hypothetical reduction reaction

$$Me^{n+1} + H^{\bullet} \rightarrow Me^{n} + H^{+}$$

However, this reaction might result into an irreversible reduction of metal ion in enzyme, which is considered minor.

In this book entitled "Molecular Hydrogen in Health and Disease", particularly in the chapter "The emergence, development, and future mission of hydrogen medicine and biology" by Ohta, the author, based on his numerous scientific papers related to the scavenging effect of molecular hydrogen in living organisms, selected as the most plausible biogenic reagent an oxidative form of porphyrin, i.e. PrP-Fe(III)-OH as well as its reduced functional form, i.e. PrP-Fe(II). It is obvious, that the selection of metalloprotein porphyrin is the most plausible for the suggestion of the below denoted reaction scheme.

In the primary step the oxidative form of the porphyrin participates in the catalytic dissociation of molecular hydrogen to form a hydride form of the porphyrin and molecules of water.

$$\text{PrP-Fe(III)-OH} + H_2 \rightarrow \text{PrP-Fe(III)-H} + H_2O$$

However, it is just the formed product, i.e., the hydride form of porphyrin, which enters the reaction with a free radical $-R^{\bullet}$, which is sufficiently reactive for the abstraction of hydrogen atom $H^{\bullet}$ to form a molecule −R-H [33].

Since highly reactive free radicals include radicals with high positive standard electric potential $E^{o\prime}$, we can consider that in a biological system $-R^{\bullet}$ will be, e.g. $^{\bullet}OH$, alkyloxy, alkylperoxy, whose $E^{o\prime}$ are 2310, 1600 and 1000 eV, respectively.

Ohta in its paper "The emergence, development, and future mission of hydrogen medicine and biology" mentioned: "In pathological conditions, porphyrins are oxidatively damaged and H_2 can activate different types of porphyrin-enzymes by repairing porphyrins. Thus, H_2 can express many functions. Importantly, H_2 exerts its function only on the porphyrins damaged by oxidation, and H_2 does not act on functional porphyrins. In other words, H_2 exerts its function only in pathological conditions [33]".

As we showed in our experiments in Figs. 22.4, 22.5 and 22.6, molecular hydrogen rich water inhibited free-radical degradation of biogenic high-molar-mass hyaluronan, i.e. scavenged $^{\bullet}OH$, alkyloxy, alkylperoxy radicals in the reaction mixture [28]. An informed reader could ask a question, why results in Figs. 22.7 and 22.8 showed not antioxidative but prooxidative effect of H_2 supplied by the tablet. However, after dissolving the H_2 tablet there is evident some surplus of free organic acids such as malic and tartaric acids. Despite these acids do not interact with any target in an organism, the reaction of hyaluronan under acid conditions resulted in the promoted degradation of its macromolecules.

Conclusion

In this chapter the authors would like to point out the fact that the above-mentioned forms of the porphyrin are plausible in acute and chronic inflammation, yet we used the Weissberger biogenic oxidative system (WBOS), which is a similar redox milieu in reaction of molecular hydrogen inhibited hyaluronan degradation (Scheme 22.1).

$$H_2O_2 + [\text{Cu(I)---komplex}] \rightarrow^{\bullet} OH + HO^{-} + [\text{Cu(II)---komplex}]$$

In WBOS as a transition metal ion is Cu(II)/Cu(I), which instead of a complexing porphyrin under in vivo conditions, in our in vitro conditions is excellently fulfilled by copper ions complexed with ascorbate (cf. Scheme 22.1).

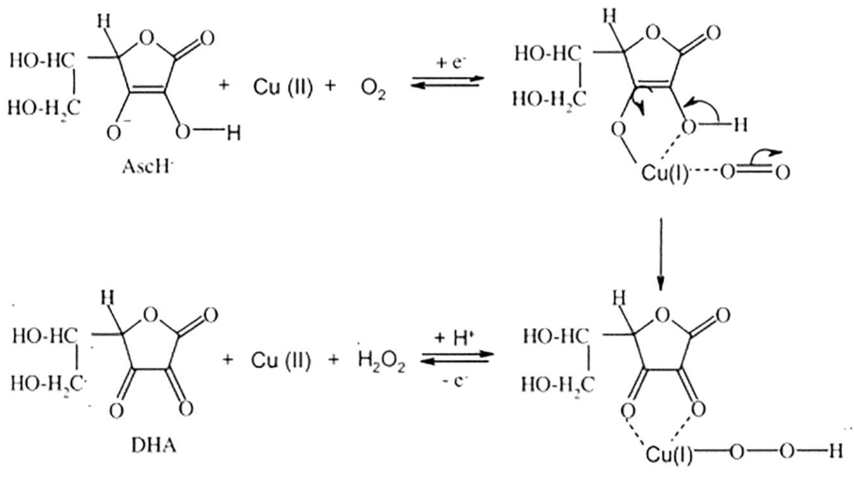

$$H_2O_2 + [Cu(I) \text{ --- complex}] \longrightarrow {}^{\bullet}OH + HO^{-} + [Cu(II) \text{ --- complex}]$$

Scheme 22.1 Chemistry of the Weissberger biogenic oxidative system

Competing Interest The study was supported by the grant VEGA 2/0008/23, VEGA 2/0092/22, APVV-19–0317. The authors declare no conflict of interest.

References

1. Adzavon YM, Xie F, Yi Y, Jiang X, Zhang X, He J, Zhao P, Liu M, Ma S, Ma X (2022) Long-term and daily use of molecular hydrogen induces reprogramming of liver metabolism in rats by modulating NADP/NADPH redox pathways. Sci Rep 12:390
2. Dixon BJ, Tang J, Zhang JH (2013) The evolution of molecular hydrogen: a noteworthy potential therapy with clinical significance. Med Gas Res 3:10
3. Qiu P, Liu Y, Zhang J (2019) Recent advances in studies of molecular hydrogen against sepsis. Int J Biol Sci 15:1261–1275
4. Tian Y, Zhang Y, Wang Y, Chen Y, Fan W, Zhou J, Qiao J, Wei Y (2021) Hydrogen, a novel therapeutic molecule, regulates oxidative stress, inflammation, and apoptosis. Front Physiol 12:789507
5. Artamonov MY, Martusevich AK, Pyatakovich FA, Minenko IA, Dlin SV, LeBaron TW (2023) Molecular hydrogen: from molecular effects to stem cells management and tissue regeneration. Antioxidants 12:636
6. Dole M, Wilson FR, Fife WP (1975) Hyperbaric hydrogen therapy: a possible treatment for cancer. Science 190:152–154
7. Sim M, Kim CS, Shon WJ, Lee YK, Choi EY, Shin DM (2020) Hydrogen-rich water reduces inflammatory responses and prevents apoptosis of peripheral blood cells in healthy adults: a randomized, double-blind, controlled trial. Sci Rep 10:12130
8. Hirano SI, Ichikawa Y, Sato B, Yamamoto H, Takefuji Y, Satoh F (2021) Molecular hydrogen as a potential clinically applicable radioprotective agent. Int J Mol Sci 22:4566

9. Lucas K, Rosch M, Langguth P (2021) Molecular hydrogen (H_2) as a potential treatment for acute and chronic fatigue. Arch Pharm 354:e2000378
10. Ohta S (2012) Molecular hydrogen is a novel antioxidant to efficiently reduce oxidative stress with potential for the improvement of mitochondrial diseases. Biochim Biophys Acta 1820:586–594
11. Lanphier EH (1972) Human respiration under increased pressures. Symp Soc Exp Biol 26:379–394
12. Ge L, Yang M, Yang NN, Yin XX, Song WG (2017) Molecular hydrogen: a preventive and therapeutic medical gas for various diseases. Oncotarget 8(60):102653−102673
13. Cole AR, Sperotto F, DiNardo JA, Carlisle S, Rivkin MJ, Sleeper LA, Kheir JN (2021) Safety of prolonged inhalation of hydrogen gas in air in healthy adults. Crit Care Explor 3:e543
14. Gharib B, Hanna S, Abdallahi OM, Lepidi H, Gardette B, De Reggi M (2001) Anti-inflammatory properties of molecular hydrogen: Investigation on parasite-induced liver inflammation. Comptes Rendus Acad Sci III 324:719–724
15. Iida A, Nosaka N, Yumoto T, Knaup E, Naito H, Nishiyama C, Yamakawa Y, Tsukahara K, Terado M, Sato K, Ugawa T, Nakao A (2016) The clinical application of hydrogen as a medical treatment. Acta Med Okayama 70(5):331–337
16. Wu C, Zou P, Feng S, Zhu L, Li F, Liu TCY, Duan R, Yang L (2023) Molecular hydrogen: an emerging therapeutic medical gas for brain disorders. Mol Neurobiol 60:1749–1765
17. Kang KM, Kang YN, Choi IB, Gu Y, Kawamura T, Toyoda Y, Nakao A (2011) Effects of drinking hydrogen-rich water on the quality of life of patients treated with radiotherapy for liver tumors. Med Gas Res 1:11
18. Hara F, Tatebe J, Watanabe I, Yamazaki J, Ikeda T, Morita T (2016) Molecular hydrogen alleviates cellular senescence in endothelial cells. Circ J 80:2037–2046
19. Ostojic SM (2012) The effects of hydrogen-rich formulation for treatment of sport-related soft tissue injuries. Clin Trials. http://clinicaltrials.gov/ct2/show/NCT01759498
20. Ohsawa I, Ishikawa M, Takahashi K, Watanabe M, Nishimaki K, Yamagata K, Ken-Katsura I, Katayama Y, Asoh S, Ohta S (2007) Hydrogen acts as a therapeutic antioxidant by selectively reducing cytotoxic oxygen radicals. Nat Med 13(6):688–694
21. LeBaron TW, Larson AJ, Ohta S, Mikami T, Barlow J, Bulloch J, DeBeliso M (2019) Acute supplementation with molecular hydrogen benefits submaximal exercise indices. Randomized, double-blinded, placebo-controlled crossover pilot study. J Lifestyle Med 9(1):36–43
22. Stern R (2008) Hyaluronan in cancer biology. Semin Cancer Biol 18(4):237
23. Berdiaki A, Neagu M, Spyridaki I, Kuskov A, Perez S, Nikitovic D (2023) Hyaluronan and reactive oxygen species signalling-novel cues from the matrix? Antioxidants 12:824
24. Fan L, Chen H, Liang J, Chen D, Huang Y (2021) Controllable synthesis of hydrogen bubbles via aeration method for efcient antioxidant process. App Nanosci 11:833–840
25. Valachová K, Mendichi R, Šoltés L (2010) Effect of L-glutathione on high-molar-mass hyaluronan degradation by oxidative system Cu(II) plus ascorbate. In Pethrick RA, Petkov P, Zlatarov A, Zaikov GE, Rakovsky SK (eds) Monomers, oligomers, polymers, composites, and nanocomposites. Nova Science Publishers, New York, pp 101–111
26. Valachová K, Hrabárová E, Priesolová E, Nagy M, Baňasová M, Juránek I, Šoltés L (2011) Free-radical degradation of high-molecular-weight hyaluronan induced by ascorbate plus cupric ions. Testing of bucillamine and its SA981-metabolite as antioxidants. J Pharm Biomed Anal 56:664–670
27. Valachová K, Vargová A, Rapta P, Hrabárová E, Dráfi F, Bauerová K, Juránek I, Šoltés L (2011) Aurothiomalate in function of preventive and chain-breaking antioxidant at radical degradation of high-molar-mass hyaluronan. Chem Biodivers 8:1274–1283
28. Kura B, Bagchi AK, Singal PK, Barancik M, LeBaron TW, Valachova K, Šoltés L, Slezák J (2019) Molecular hydrogen: potential in mitigating oxidative-stress induced radiation injury. Can J Physiol Pharmacol 97:287–292
29. Ye Q, Xu H, Wang Q, Huo X, Wang Y, Huang X, Zhou G, Lu J, Zhang J (2021) New insights into the mechanisms of tartaric acid enhancing homogeneous and heterogeneous copper-catalyzed Fenton-like systems. J Hazard Mater 407:124351

30. Cabelli DE, Bielski BHJ (1985) A pulse radiolysis study of some dicarboxylic acids of the citric acid cycle. The kinetics and spectral properties of the free radicals formed by reaction with the OH radical. Z Naturforsch 40b:1731–1737
31. Kura B, Szantova M, LeBaron TW, Mojto V, Barancik M, Szeiffova Bacova B, Kalocayova B, Sykora M, Okruhlicova L, Tribulova N, Gvozdjakova A, Sumbalova Z, Kucharska J, Faktorova X, Jakabovicova M, Durkovicová Z, Macutek J, Koscová M, Jan Slezak J (2022) Biological effects of hydrogen water on subjects with NAFLD: a randomized, placebo-controlled trial. Antioxidants 11:1935
32. Zahnle K, Schaefer L, Fegley B (2010) Earth's earliest atmospheres. Cold Spring Harb Perspect Biol 2(10):a004895
33. Ohta S. The emergence, development, and future mission of hydrogen medicine and biology. In: Molecular hydrogen in health and disease. Springer Nature, in press

Chapter 23
Development of a Preclinical Tool for Measuring Percutaneous Transfer of Dihydrogen, with a View to Optimizing Medical Devices Adapted to Focal Therapies in Dermatology

C. Salomez-Ihl, S. Tanguy, F. Boucher, V. Pascal Mousselard, P. Bedouch, A. Stephanou, J. P. Alcaraz, and P. Cinquin

Abstract Numerous studies have demonstrated the efficacy of dihydrogen (H_2) in dermatological pathologies, with no attributed adverse effects. However, there is no formal proof of percutaneous transfer of H_2 through the skin. The aim of the present study is to demonstrate this transfer and to characterize the H_2 diffusion coefficient in the skin. Rat abdominal skin is affixed to the center of a diffusion cell. In a first

---

C. Salomez-Ihl · S. Tanguy · F. Boucher · V. Pascal Mousselard · P. Bedouch · A. Stephanou · J. P. Alcaraz · P. Cinquin (✉)
University of Grenoble Alpes, CNRS, UMR 5525, VetAgro Sup, Grenoble INP, TIMC, 38000 Grenoble, France
e-mail: Philippe.Cinquin@univ-grenoble-alpes.fr

C. Salomez-Ihl
e-mail: cordelia.ihl@univ-grenoble-alpes.fr

S. Tanguy
e-mail: stephane.tanguy@univ-grenoble-alpes.fr

F. Boucher
e-mail: Francois.Boucher@univ-grenoble-alpes.fr

P. Bedouch
e-mail: PBedouch@chu-grenoble.fr

A. Stephanou
e-mail: angelique.stephanou@univ-grenoble-alpes.fr

J. P. Alcaraz
e-mail: jean-pierre.alcaraz@univ-grenoble-alpes.fr

C. Salomez-Ihl · P. Bedouch
Department of Pharmacy, Grenoble University Hospital, 38000 Grenoble, France

P. Cinquin
Centre for Clinical Investigation Technological Innovation, INSERM CIC803, Grenoble University Hospital, 38000 Grenoble, France

J. Slezak and B. Kura (eds.), *Molecular Hydrogen in Health and Disease*, Advances in Biochemistry in Health and Disease 27, https://doi.org/10.1007/978-3-031-47375-3_23

compartment, a gas mixture containing different proportions of H_2 combined with air is injected. A second compartment contains either air or physiological medium + air. A H_2 sensor is connected to the second compartment and records H_2 concentration. A bi-exponential model of H_2 concentration is then fitted. The mean value of the flow is 0.40 $nmol.s^{-1}.cm^{-2}$, and the mean value of the H_2 diffusion coefficient is 4.2 10^{-5} $cm^2.s^{-1}$, which is very close to H_2 diffusion coefficient in water (D = 4.58 10^{-5} $cm^2.s^{-1}$). Our model confirms the major interest of designing new approaches to deliver H_2 for dermatology applications, because the local concentration of H_2 in the target zones will be by several orders of magnitude higher than what can be achieved with oral administration.

Keywords Hydrogen · Dermatology · Percutaneous Administration · Pharmacokinetics · Diffusion chamber

Introduction

Dihydrogen (H_2) is a molecule of potential therapeutic interest in numerous pathologies, particularly those with an inflammatory or oxidative component. In the 1970s, the scientific community began to take an interest in the health properties of H_2. In 1975, the team of Dole [1] was the first to demonstrate, on a mouse model with UV-induced squamous cell carcinoma, that exposure to a mixture of 2.5% oxygen and 97.5% H_2 at 8 atmospheres for up to 2 weeks induced a marked regression of the tumors compared to placebo mice exposed to 2.5% oxygen and 97.5% helium under the same conditions. Real renewed interest in the medical applications of H_2 came in 2007 with the publication of a leading paper by Ohsawa et al. in Nature Medicine in 2007. This study, carried out on a rat model of cerebral infarction, showed that H_2 therapy drastically reduced the size of cerebral necroses [2]. Since then, studies on H_2 as a drug candidate have been steadily increasing on a global scale in many medical specialties as diverse as neurology, virology, oncology, cardiology, pneumology, immunology and dermatology. In the latter context of dermatological pathologies, seven clinical trials have been conducted with very positive conclusions. Used in baths, hydrogen-water was shown to improve psoriasis and parapsoriasis symptoms. 24.4% of patients receiving hydrogen-water bathing achieved at least 75% improvement in Psoriasis Area Severity Index (PASI) score compared with 2.9% of patients of the control group ($p = 0.022$) [3]. Two studies investigated the impact of topical application of hydrogenated water produced by hydrolysis on acne-prone skin and seem to have demonstrated H_2's interesting properties in this indication, despite methodological weaknesses [4, 5]. Addition of oral or topical H_2 to conventional treatment protocols was found potentially effective in the treatment of sport-related soft tissue injuries in male professional athletes [6]. The use of H_2-releasing mask packs was shown to significantly improve skin density of the participants compared to a standard commercial cosmetic sample [7]. In a prospective open-label phase 2 study, patients received hydrogen-rich water (HRW) orally with a great efficacy on chronic graft-versus-host-disease affecting different organs. In this study, HWR ingestion reduced by 66.7% the severity score of the disease in the skin compared to standard of care alone [8]. Finally, HRW administered via tube-feeding has been shown to reduce pressure sores size and promote healing in a cohort of 22 elderly hospitalized patients [9]. These clinical trials demonstrate the therapeutic impact of H_2 on widespread human skin diseases with major public health implications, with no adverse effects attributable to H_2. Five clinical reports involving a total of 14 patients have also studied H_2 in various ways of administration (infusion, inhalation, HRW ingestion, local application and HRW bath) in psoriasis [10], pemphigus vulgaris [11], chronic wound [12], acute erythematous diseases [13], skin blotch and visceral fat [14]. The conclusions of these clinical cases were always the same: beneficial effects of H_2 and absence of attributable adverse effects. Pre-clinical studies have also demonstrated interesting properties of H_2 in various animal models such as mice, dogs, and rats, with different ways of administration (HRW, inhalation, intravenous or intraperitoneal injection of hydrogen rich saline (HRS) and H_2-generating silicon agent). The experimental pathologies studied in this context were also heterogeneous: atopic dermatitis [15–17], cutaneous wounds [18], diabetic wounds [19], cutaneous lesions linked to ischemia reperfusion [20–24], skin aging [25], cutaneous lesions generated by radiotherapy [26, 27], inflammatory and oxidative skin conditions [28] and UVB-related skin lesions [29]. Finally, in vitro studies have also demonstrated the efficacy of H_2 (dissolved in DMEM, incorporated in titanium oxide nanorods, released from silica material) in different experimental pathological conditions (oxidative stress lesions caused by UVB exposure [30], high glucose exposure [31] and skin aging related to UVA lesions [32]). If H_2 is effective in all the pathologies studied despite their apparent diversity, it is because this molecule possesses both anti-inflammatory and antioxidant properties that are sufficiently general to find numerous applications in dermatology. Despite this apparent evidence of efficacy, there is still no formal proof of the transcutaneous passage of H_2. In addition, there is no consensus about how to deliver H_2 to achieve optimal therapeutic effect, either in terms of mode of administration, dose, or administration schedule.

The aim of the present study is to characterize the coefficient of diffusion of H_2 in the skin and its flow through the skin. This represents the first step towards a pharmacokinetic model governing transcutaneous administration of H_2. The final goal will be to develop a model for the evaluation of new formulations in the context of hydrogen-therapy.

Methods

Skin Sampling

The skin samples used in this study were taken from 11 rats (2 Fischer rats and 9 Wistar rats, from 200 to 410 g body weight). These rats had been included in other validated experimental protocols conducted in our work environment. They were euthanized by sodium pentobarbital injection (200 $mg.kg^{-1}$ *body weight*, intra peritoneal) prior skin sampling. The abdominal skin was shaved with particular attention to avoid cuts and wounds. A circular abdominal skin sample, 3 cm in diameter, was delicately cut. The absence of skin injuries was guaranteed by visual inspection by two independent experimenters. Samples were immediately stored in modified Tyrode buffer (310 mOsm, HEPES 10 mM, pH = 7.4) at room temperature and used within a maximum of five minutes after collection.

H_2 Preparation

H_2 gas was generated by the reaction between calcium hydride (Sigma Aldrich ref 208027) and water according to the chemical reaction shown in Fig. 23.1 (Panel a). CaH_2 was finely ground extemporaneously in a mortar. 30 to 100 mg of the obtained powder was placed in a 10 mL syringe and 1 mL of water was placed in a second 10 mL syringe. The 2 syringes were then connected to a three-way medical tap, along with an empty 50 mL syringe (Fig. 23.1; panel a). The three-way tap was opened to mix CaH_2 with water and form pure H_2, which is collected in the 50 mL syringe. In order to produce 10% (fraction f = 0.1) and 50% (fraction f = 0.5) H_2 mixtures, we mix air and this pure H_2 using another 50 mL syringe.

Experimental Set-Up

The diffusion experiments were conducted in a Side-bi-Side diffusion cell apparatus (PermeGear) at room temperature (Fig. 23.1; panel a). This device has two compartments. $Compartment_1$ refers to the compartment where the gas mixture containing H_2 was introduced at atmospheric pressure. $Compartment_2$ is the compartment in which H_2 concentration was monitored in *ppm (parts per millions)* every 5 s using a SKY2000 gas detector (ATO, USA) placed in a bypass (bypass flow rate: 0.375 L/min) connected to $Compartment_2$ (Fig. 23.1; Panel b). The SKY2000 sensor was calibrated daily using a Quintron BreathTracker H_2Plus®, following manufacturer's recommendations.

The skin sample was sandwiched between the two compartments, avoiding folds and leaks, with the outer face of the skin facing $Compartment_1$. The system

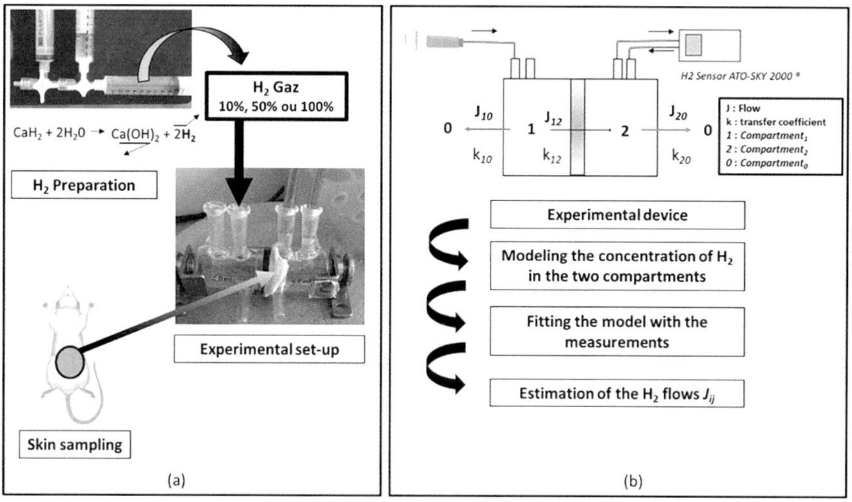

Fig. 23.1 Methodology of the study

was secured thanks to a plastic fastener adapted to the external diameter of the compartments.

A first series of experiments with skin samples from two Fisher rats (E_g^n, where $n \in \{1, 2\}$ represents the skin sample number) was conducted with pure H_2 in *Compartment*$_1$ and ambient air in *Compartment*$_2$. The total volume of the analyzed air, including the volume of *Compartment*$_2$ and that of the bypass, was estimated at 30 mL in this series of experiments (E_g^n).

The second series of experiments (E_b^n) was conducted with 10%, 50% or 100% of H_2 in *Compartment*$_1$ and 5 mL of stirred physiological buffer (modified Tyrode) in *Compartment*$_2$. The total volume of the analyzed air for E_b^n was estimated at 25 mL. Skin samples of 9 Wistar rats were utilized for E_b^n experiments. Gas mixtures with H_2 concentrations of 10, 50 or 100% were then slowly injected in *Compartment*$_1$. H_2 measurements in *Compartment*$_2$ started at the end of the injection. The total duration of the experiment ranged from 30 to 90 min, depending on the H_2 concentration of the injected mixture.

Modeling H2 Concentrations in the Two Compartments in E_g^n Experiments (Gas/Skin/Air)

We note with subscript $i \in \{1, 2, O\}$ and $j \in \{1, 2, O\}$ the variables corresponding to each compartment (where subscript O denotes the *outside* of the compartments, thus *Compartment*$_O$ denotes ambient air).

Let:

$$\boldsymbol{t} = \boldsymbol{time}\ (\text{seconds}),$$

$$\boldsymbol{T} = \boldsymbol{room\,temperature}\ (^{\circ}\text{C}),$$

$$\boldsymbol{V_m} = \boldsymbol{molar\ volume}\ \left(\text{L.mol}^{-1}\right),$$

$$\boldsymbol{S} = \boldsymbol{surface\,of\,the\,rat\,skin\,through\,which\,H_2\,flows}\ \left(\text{m}^2\right),$$

$$\boldsymbol{e} = \boldsymbol{thickness\,of\,the\,rat\,skin\,through\,which\,H_2\,flows}\ (\text{m}),$$

$$\boldsymbol{V_i} = \boldsymbol{volume\,of\,Compartment_i}\ (\text{L}),$$

$$\boldsymbol{m_i} = \boldsymbol{quantity\,of\,H_2\,in\,Compartment_i}\ (\text{mol}),$$

$$\boldsymbol{c_i} = \frac{\boldsymbol{m_i}}{\boldsymbol{V_i}} = \boldsymbol{molar\,concentration\,of\,H_2\,in\,Compartment_i}\ \left(\text{mol.L}^{-1}\right),$$

$$\boldsymbol{f_i} = \boldsymbol{c_i V_m} = \boldsymbol{fraction\,of\,H_2\,in\,Compartment_i},$$

$$\boldsymbol{p_i} = \boldsymbol{f_i}.10^6 = \boldsymbol{c_i V_m}.10^6 = \boldsymbol{H_2\,concentration}\ (\text{ppm}),$$

$$\boldsymbol{J_{ij}} = \boldsymbol{flow\,of\,H_2\,from\,Compartment_i\,to\,Compartment_j}\left(\text{mol.s}^{-1}\right),$$

$$\boldsymbol{k_{ij}} = \frac{\boldsymbol{J_{ij}}}{\left(\boldsymbol{c_i} - \boldsymbol{c_j}\right)} = \boldsymbol{transfer\,coefficient}\ \left(\text{L.s}^{-1}\right),$$

$$\boldsymbol{L_{12}} = \frac{\boldsymbol{J_{12}}}{\boldsymbol{S}} = \boldsymbol{specific\,flow\,from\,Compartment_1\,to\,Compartment_2}\left(\text{mol.s}^{-1}.\text{cm}^{-2}\right),$$

$$\boldsymbol{q} = \boldsymbol{H_2 solubility\,at\,room\,temperature}\ \left(\text{mol.L}^{-1}\right).$$

The following equations model the exchanges of H_2 between compartments. H_2 concentration in the ambient air is considered as ~550 ppb (*parts per billions*) [33], therefore it can be assumed that $c_O(t) = 0$. On this basis, we can assume that H_2 leaks from $Compartment_i$ are proportional to c_i. Fick's law states that H_2 flow J_{ij} across the skin is driven by the gradient of concentrations of H_2 between $Compartment_i$

and $Compartment_j$, so that $J_{ij} = k_{ij}\left(c_i - c_j\right)$, where k_{ij} is a transfer coefficient depending on the thickness of the skin and on H_2 coefficient of diffusion in the skin.

Therefore

$$\frac{dm_1}{dt} = -J_{1O} - J_{12} = -k_{1O}c_1 - k_{12}(c_1 - c_2) \tag{23.1}$$

It is reasonable to assume that c_1 can be neglected with respect to c_2, since $c_2(0) = 0$, and since, as will be described in the results section, the duration of the experiments does not allow p_2 to rise above 400 ppm (corresponding to $c_2 = 17.10^{-6}$ mol.L^{-1}), while p_1 remains greater than 10,000 ppm (corresponding to $c_1 = 420.10^{-6}$ mol.L^{-1}) so that $\frac{c_1}{c_2} > 25$.

Thus

$$\frac{dm_1}{dt} = -k_{1O}c_1 - k_{12}c_1 = -(k_{1O} + k_{12})c_1 \tag{23.2}$$

$$V_1\frac{dc_1}{dt} = -(k_{1O} + k_{12})c_1 \tag{23.3}$$

$$\frac{dc_1}{dt} = -\frac{(k_{1O} + k_{12})}{V_1}c_1 \tag{23.4}$$

$$c_1(t) = c_1(0)e^{-\frac{(k_{1O}+k_{12})}{V_1}t} \tag{23.5}$$

H_2 concentrations measured in $Compartment_2$ are the result of H_2 flow from $Compartment_1$, through the skin sample, and leakage from $Compartment_2$.

Therefore

$$\frac{dm_2}{dt} = J_{12} - J_{2O} = k_{12}(c_1 - c_2) - k_{2O}c_2 \tag{23.6}$$

$$V_2\frac{dc_2}{dt} = k_{12}c_1 - (k_{12} + k_{2O})c_2 \tag{23.7}$$

$$\frac{dc_2}{dt} = -\frac{(k_{12} + k_{2O})}{V_2}c_2 + k_{12}\frac{c_1(0)}{V_2}e^{-\frac{(k_{1O}+k_{12})}{V_1}t} \tag{23.8}$$

Since $c_2(0) = 0$, the solution of the differential Eq. (23.8) is

$$c_2(t) = a\left(e^{-bt} - e^{-ct}\right) \tag{23.9}$$

where

$$a = \frac{k_{12}}{\frac{k_{12}+k_{2O}}{V_2} - \frac{k_{1O}+k_{12}}{V_1}}.\frac{c_1(0)}{V_2} \tag{23.10}$$

$$b = \frac{k_{1O} + k_{12}}{V_1} \tag{23.11}$$

$$c = \frac{k_{12} + k_{2O}}{V_2} \tag{23.12}$$

Fitting the Model with the Measurements in E_g^n Experiments (Gas/Skin/Air)

The SKY2000 sensor measures $p_2 = c_2 V_m 10^6$.

Therefore

$$p_2(t) = d\left(e^{-bt} - e^{-ct}\right) \tag{23.13}$$

with

$$d = aV_m 10^6 = \frac{k_{12}}{\frac{k_{12}+k_{2O}}{V_2} - \frac{k_{1O}+k_{12}}{V_1}} . \frac{c_1(0)}{V_2} . V_m 10^6 \tag{23.14}$$

Leaks from $Compartment_2$ are significantly more important than leaks from $Compartment_1$, because the measurements of H_2 with the SKY2000 sensor imply the use of tubes to circulate the gas from the inside of $Compartment_2$ to the measuring chamber of the sensor. Therefore, $k_{2O} > k_{1O}$, so that $c > b$ (since V_1 and V_2 are not very different). As a consequence, at some point, $\exists t_m, t > t_m \Rightarrow e^{-bt} \gg e^{-ct}$, so that the latter can be neglected:

$$\exists t_m, t > t_m \Rightarrow p_2(t) \sim de^{-bt} \tag{23.15}$$

$$\exists t_m, t > t_m \Rightarrow Ln[p_2(t)] \sim Ln(d) - bt \tag{23.16}$$

t_m is defined by visual observation of the graph of $Ln[p_2(t)]$, as the value of t where it can be accepted that this graph becomes a straight line (which is verified by computation of the coefficient correlation of a linear model to the values of $Ln[p_2(t_i)]$ for all points $t_i > t_m$).

This linear fit of the final part of the graph of $Ln[p_2(t)]$ thus yields coefficients d and b. We can now estimate coefficient c, since from (23.13) we draw

$$de^{-bt} - p_2(t) = de^{-ct} \tag{23.17}$$

$$Ln\left[de^{-bt} - p_2(t)\right] = Ln(d) - ct \tag{23.18}$$

To estimate coefficient c, we therefore fit a linear model to the values of $Ln\left[de^{-bt_i} - p_2(t_i)\right]$ and we check the fitness of this model by computing the corresponding correlation coefficient.

Finally, we validate the fitting of the complete model by computing the correlation coefficient between the sets $\{p_2(t_i)\}$ and $\left\{d\left(e^{-bt_i} - e^{-ct_i}\right)\right\}$.

Estimation of the H_2 Flows J_{ij} in E_g^n Experiments (Gas/Skin/Air)

Now that we have estimated coefficients b, c and d, we can solve the linear system of the 3 Eqs. (23.10)–(23.12) with unknowns k_{12}, k_{1O}, k_{2O} (using Eq. (23.14) to derive a from d). We obtain

$$\begin{cases} \boldsymbol{k}_{12} = \frac{c-b}{c_1(0)} a \frac{V_2}{V_m} 10^{-6} \\ \boldsymbol{k}_{1O} = b\boldsymbol{V}_1 - \boldsymbol{k}_{12} \\ \boldsymbol{k}_{2O} = c\boldsymbol{V}_2 - \boldsymbol{k}_{12} \end{cases} \tag{23.19}$$

This yields the values of the flows of H_2 between compartments:

$$\boldsymbol{J}_{ij} = \boldsymbol{k}_{ij}\left(\boldsymbol{c}_i - \boldsymbol{c}_j\right) \tag{23.20}$$

Estimation of Flow J12 in the E_b^n Experiments (Gas/Skin/Buffer/Air)

The model we previously described does no longer work, since now in $Compartment_2$ we have two phases (modified Tyrod liquid and air). The presence of a liquid phase generates a delay t_m before H_2 can be measured in the gaseous phase, because H_2 first needs to go through the liquid phase, then needs to equilibrate its liquid concentration with the gaseous concentration. Besides, for sufficiently small values of t ($t < t_M$), $c_1(t)$ can be considered almost constant, since in this limited time, very limited amounts of H_2 cross the skin or leak outside (so that we can assume that for $t < t_M$, $c_1(t) \sim c_1(0)$). Finally, $c_2(0) = 0$ and t_M can be chosen so that $c_2(t)$ remains neglectable for $t < t_M$. Thus, for $t_m < t < t_M$,, Eq. (23.6) can be simplified into

$$\frac{\boldsymbol{dm}_2}{\boldsymbol{dt}} = \boldsymbol{k}_{12}(\boldsymbol{c}_1 - \boldsymbol{c}_2) - \boldsymbol{k}_{2O}\boldsymbol{c}_2 = \boldsymbol{k}_{12}\boldsymbol{c}_1(0) = \boldsymbol{k} \tag{23.21}$$

where k is a constant. This implies

Table 23.1 Application of our model yields estimations of flow and specific flow for each E_g^n experiment (gas/skin/air)

Parameters of the model	E_g^n Experiment (gas/skin/air)	
	E_g^1	E_g^2
Coefficient of correlation (r)	0.995	0.999
Flow (J_{12})	0.84 nmol.s^{-1}	0.56 nmol.s^{-1}
Specific flow ($L_{12}^n = \frac{J_{12}^n}{S}$)	0.48 nmol.s^{-1}.cm^{-2}	0.32 nmol.s^{-1}.cm^{-2}

The surface area S of the skin between the two compartments in both experiments is set by the diffusion device at 1.77 cm^2

$$m_2(t) = kt \tag{23.22}$$

To estimate t_m and t_M, we visually observe the graph of $p_2(t)$ and define t_m and t_M as the limit values where the assumption of linearity of the graph holds (this assumption is verified by computation of the correlation coefficient). It should be noted that these approximations provide an underestimation of the real value of J_{12}, since we do not take into account the leaks of H_2.

Results

Estimation of the Flow Across the Skin with E_g^n Experiments (Gas/Skin/Air)

The previously described bi-exponential model was applied on two independent experiments of the E_g^n protocol (Gas/Skin/Air). Table 23.1 shows H_2 estimated transcutaneous flows. The two estimated specific flows are $L_{12}^1 = 0.48$ nmol.s^{-1}.cm^{-2} and $L_{12}^2 = 0.32$ nmol.s^{-1}.cm^{-2}. Figure 23.2 illustrates the fitting between experimental data and the theoretical model on E_g^1 experiment, with a very good correlation coefficient $r = 0.995$ (for E_g^2 experiment, we obtained $r = 0.999$).

Estimation of the Flow Across the Skin with E_b^n Experiments (Gas/Skin/Buffer/Air)

This experiment was run on 9 skin samples for values of $f_1(0)$: $\{1; 0.5; 0.1\}$. Raw data of the E_b^4 experiment is shown in Fig. 23.3 for illustrative purposes.

For each of the 9 skin samples, according to the previously described method, a window where the graph could be visually considered linear was selected, and a

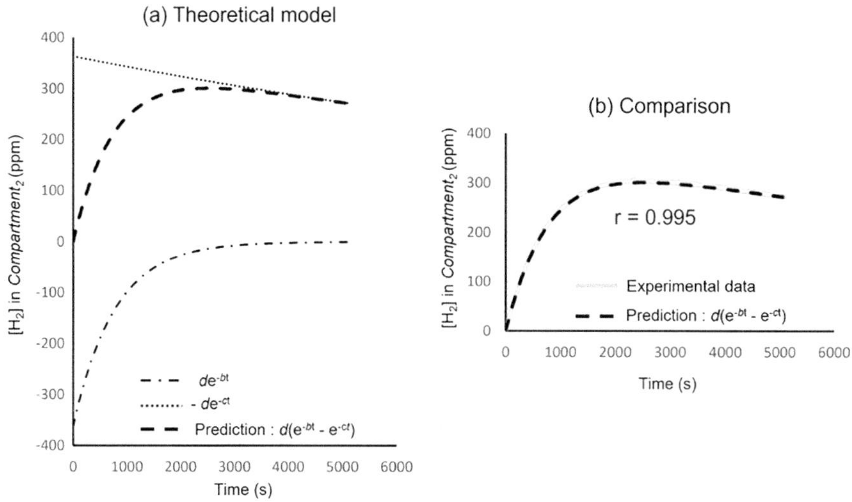

Fig. 23.2 Modelling of H_2 concentration in *Compartment*$_2$ in E_g^1 experiment (Gas/Skin/Air) (a) Theoretical model. (b) Fitting between theoretical model and experimental data of rat 1. t = time; *b*, *c* and *d* are defined in methods. Experimental data were measured by the SKY2000 sensor (initial H2 concentration in *Compartment*$_1$: $f_1(0) = 1$)

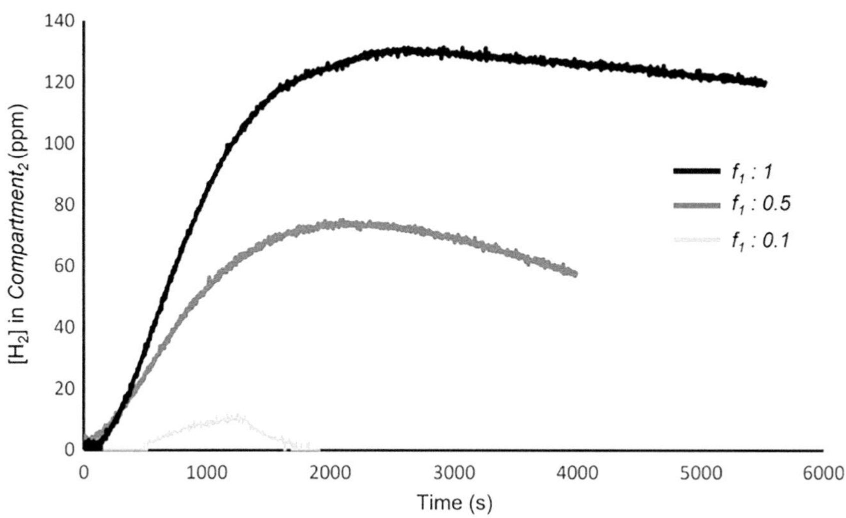

Fig. 23.3 Kinetics of H_2 transfer across the skin sample (between *Compartment*$_1$ and *Compartment*$_2$) with $f_1(0)$: {1; 0.5; 0.1}. Experimental data were measured in experiment E_b^4 (gas/skin/buffer/air) by the SKY2000 sensor

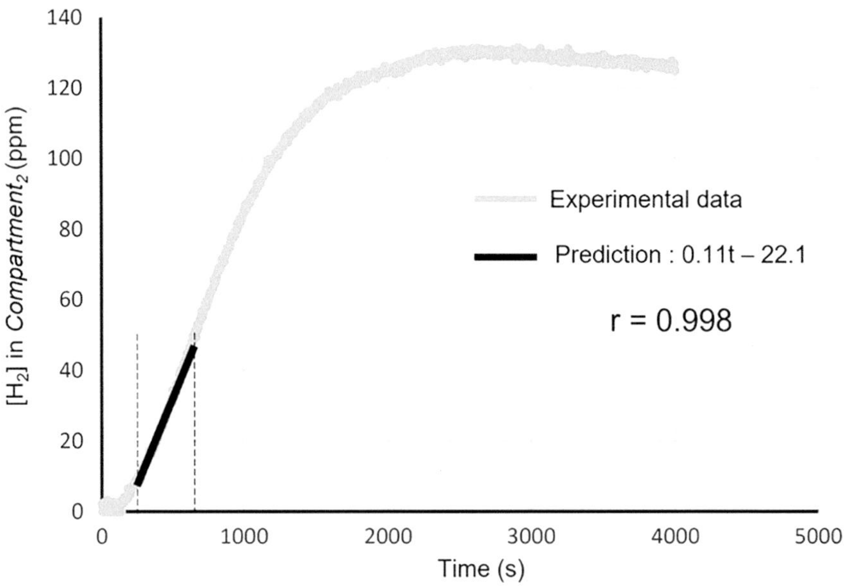

Fig. 23.4 Example of linear model application to E_b^4 A window (between dashed bars) where the graph can be visually considered linear is selected, and a linear model is fitted (black line). t = time. Experimental data were measured in experiment E_b^4 (gas/skin/buffer/air) by the SKY2000 sensor (initial H_2 fraction in *Compartment*$_1$: $f_1(0) = 1$)

linear model was fitted. The linear model applied to the E_b^4 window is shown in Fig. 23.4 as an example.

At room temperature $T = 20\ °\text{C}$, $V_m = 24\ \text{L.mol}^{-1}$. Let k^n be the slope of the simplified linear model we use in skin sample n of the E_b^n set of experiments. Since $p_2 = f_2.10^6 = c_2 V_m.10^6 = m_2 \frac{V_m}{V_2}.10^6$,

$$\boldsymbol{J}_{12}^{\boldsymbol{n}} = \boldsymbol{k}^{\boldsymbol{n}}.\frac{\boldsymbol{V}_2}{\boldsymbol{V}_{\boldsymbol{m}}} 10^{-6}\ \text{moles.s}^{-1} = \boldsymbol{k}^{\boldsymbol{n}}.1.25\ \text{nanomoles.s}^{-1}$$

Since the surface of the skin of the rat through which H_2 flows is $S = 1.77\ \text{cm}^2$, we can estimate the specific flow

$$\boldsymbol{L}_{12}^{\boldsymbol{n}} = \frac{\boldsymbol{J}_{12}^{\boldsymbol{n}}}{\boldsymbol{S}} = \boldsymbol{k}^{\boldsymbol{n}}.\frac{1.25}{1.77}\ \text{nmol.s}^{-1}.\text{cm}^{-2} = \boldsymbol{k}^{\boldsymbol{n}}.0.71\ \text{nmol.s}^{-1}.\text{cm}^{-2}$$

The specific flows L_{12}^n of H_2 estimated in the 9 skins (E_b^1 to E_b^9) are represented in Table 23.2.

Table 23.2 H_2 specific flows (L_{12}^n) in nmol.s^{-1}.cm^{-2} with respect to initial concentration of H_2 in *Compartment*$_1$ for each skin used in E_b^n experiments (Gas/Skin/Buffer/Air)

		L_{12}^1	L_{12}^2	L_{12}^3	L_{12}^4	L_{12}^5	L_{12}^6	L_{12}^7	L_{12}^8	L_{12}^9
H_2 fractions in Compartment$_1$	0.1	0.006	0.010	0.008	0.009		0.006			
	0.5	0.030	0.046	0.032	0.025	0.039		0.043	0.042	0.030
	1	0.053	0.038	0.050	0.071		0.052	0.079	0.078	0.044

Discussion

The results of the second set of experiments E_b^n (Gas/Skin/Buffer/Air) confirm that H_2 crosses the skin. As can be seen in Fig. 23.5, the hypothesis of a linear relationship between $f_1(0)$, the initial concentration of H_2 in *Compartment*$_1$, and the specific flows, can be accepted ($r = 0.985$). This is coherent with what Fick's law teaches (the flow is proportional to the difference of concentrations in the two compartments).

It should be noted that the estimated specific flows of E_b^n are about 7 times lower than the estimations obtained in the first set of experiments E_g^n (gas/skin/air). Indeed, the mean value of the estimations of specific flows in experiments E_b^n, with $f_1(0) = 1$, is 0.061 nmol.s^{-1}.cm^{-2}, compared to the two estimations of specific flows in experiments E_g^1: 0.48 nmol.s^{-1}.cm^{-2} and E_g^2: 0.32 nmol.s^{-1}.cm^{-2}. This is the consequence of the already mentioned fact that the method used to estimate the flow of H_2 in E_b^n does not take the leaks into account, so that the transcutaneous flows are underestimated. The model used to interpret experiments E_g^n enable us to take into account the leaks, and to estimate more accurately the transcutaneous flows. The coefficient of diffusion of H_2 through the skin can then be estimated in the following

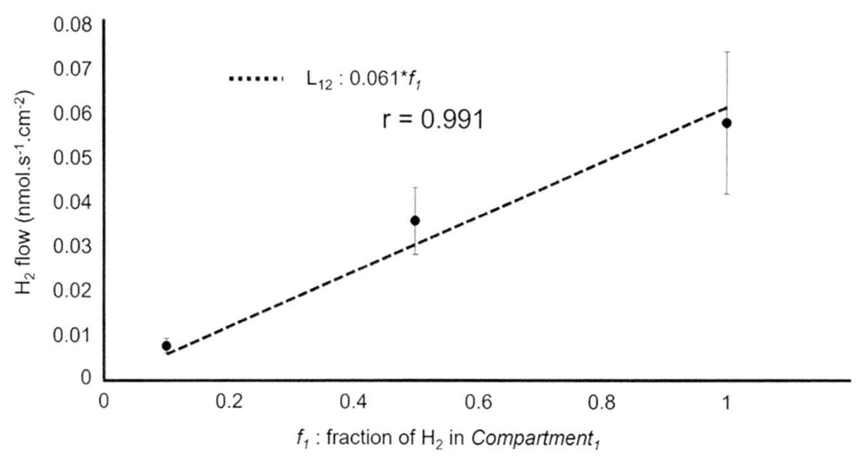

Fig. 23.5 Linear relationship between H_2 fraction in *Compartment*$_1$ and H_2 transcutaneous flow. H_2 flows were calculated from experiments E_b^n (Gas/Skin/Buffer/Air) by the SKY2000 sensor. 5 to 8 skin samples were used for each fraction f_1 of H_2 in *Compartment*$_1$, see Table 23.2

way. In the present study, the thickness of the skin was estimated by light microscopy from 0.40 to 0.95 mm, with a mean value of $e = 0.8$ mm, which is in accordance with the literature [34, 35]. Let d_1 (resp. d_2) be the concentration of H_2 in the skin, at the interface between skin and *Compartment*$_1$ (respectively *Compartment*$_2$). Since skin's salinity is about 9 g.L^{-1} and room temperature is $T = 20$ °C, solubility of H_2 in the skin can be estimated at 770 μmol.L^{-1} [36]. In E_g^n experiments, $f_1(0) = 1$, and $f_2(0) = 0$. Therefore, $d_1 = 770$ μmol.L^{-1} and $d_2 = 0$ μmol.L^{-1}. Let D_s^n be the estimation of the coefficient of diffusion of H_2 through the skin in rat n of E_g^n. Fick's law states that

$$\boldsymbol{D_s^n} = \boldsymbol{L_{12}^{\mathrm{n}}}.\frac{\boldsymbol{e}}{(\boldsymbol{d_1} - \boldsymbol{d_2})}$$

Taking into account the obtained estimations of L_{12}^n, we get $D_s^1 = 5 \times 10^{-5}$ cm^2.s^{-1}, and $D_s^2 = 3.3$ cm^2.s^{-1}. These estimations are very close to H_2 diffusion coefficient in water ($D = 4.58 \times 10^{-5}$ cm^2.s^{-1}, [37]). We can therefore conclude that skin is no more a barrier to H_2 than water. These findings confirm that transcutaneous administration of H_2 can be a very efficient way of treatment of cutaneous or subcutaneous pathologies, because local concentrations of H_2 will be very significantly higher than with ingestion of Hydrogen Rich Water (HRW). Let us assume that we target a cutaneous lesion of 1 cm^2 in surface and 1 cm in depth, and that the surface of this lesion is exposed during 1 h to a liquid saturated with H_2, (this can be achieved for instance with baths of HRW such as those used in [3, 14]). It should be noted that using a bath with water saturated in H_2 is equivalent to exposing the skin to a gas with 100% H_2, as we did in experiments E_g^n. Let us assume that the flow of H_2 through the lesion surface has the smallest value we estimated in these experiments, $L_{12}^2 = 0.32$ nmol.s^{-1}.cm^{-2}. In 1 h, the 1 cm^3 volume of the lesion will have received 1.2 μmol. This has to be compared with what the same volume would have received with other means of H_2 administration. Ito's team reports that after ingestion of 200 mL H_2-saturated water, the H_2 concentration in exhaled air returns to baseline in about 45 min [38]. The maximum peak H_2 concentration in exhaled air is about 25 ppm. With a solubility of H_2 at 37 °C around 730 μmol.L^{-1}, this corresponds to a H_2 concentration in the venous blood of about 18 nmol.L^{-1}. For an adult human being of 70 kg, let us assume a water volume of 45 L and a cardiac flow of 5 L.min^{-1}. Each liter of the water volume is irrigated by a cardiac flow of 0.11 L.min^{-1}. If we consider that H_2 venous concentration remains at 18 nmol.L^{-1} during 45 min, this implies that a target with a volume of 1 cm^3 receives in 45 min 0.08 nmol, which is 4 orders of magnitude less than what is delivered in a 1 cm^3 cutaneous lesion by local administration of water saturated in H_2. This accounts for the remarkable efficacy results of HRW baths on cutaneous pathologies, for instance of psoriasis [3].

Conclusion

We have demonstrated that H_2 crosses the skin according to Fick's law and obtained a coefficient of diffusion of H_2 through the skin very close to the coefficient of diffusion of H_2 in water. These findings confirm the major interest in designing new approaches to deliver H_2 for dermatology applications, because the local concentration of H_2 in the target zones will be by several orders of magnitude higher than what can be achieved with oral administration.

References

1. Dole M et al (1975) Hyperbaric hydrogen therapy: a possible treatment for cancer. Science 190(4210):152–154
2. Ohsawa I et al (2007) Hydrogen acts as a therapeutic antioxidant by selectively reducing cytotoxic oxygen radicals. Nat Med 13(6):688–694
3. Zhu Q et al (2018) Positive effects of hydrogen-water bathing in patients of psoriasis and parapsoriasis en plaques. Sci Rep 8(1):1–8
4. Chilicka K et al (2021) Effects of topical hydrogen purification on skin parameters and acne vulgaris in adult women. Healthc Multidiscipl Digital Publish Inst 9(2):144
5. Chilicka K et al (2022) Efficacy of hydrogen purification and cosmetic acids in the treatment of acne vulgaris: a preliminary report. J Clin Med 11(21):6269
6. Ostojic SM et al (2014) Effectiveness of oral and topical hydrogen for sports-related soft tissue injuries. Postgrad Med 126(5):188–196
7. Kwon HJ et al (2020) Antioxidant activity of hydrogen water mask pack composed of gel-type emulsion and hydrogen generation powder. Int J Mol Sci 21(24):9731
8. Qian L et al (2020) Hydrogen in patients with corticosteroid-refractory/dependent chronic graft-versus-host-disease: a single-arm, multicenter, open-label, phase 2 trial. Front Immunol 11:598359
9. Li Q et al (2013) Hydrogen water intake via tube-feeding for patients with pressure ulcer and its reconstructive effects on normal human skin cells in vitro. Med Gas Res 3:1–17
10. Ishibashi T et al (2015) Improvement of psoriasis-associated arthritis and skin lesions by treatment with molecular hydrogen: a report of three cases. Mol Med Rep 12(2):2757–2764
11. Yang F et al (2019) Skin ulcers infected with conditional pathogenic strains treated with local hydrogen water packing in two pemphigus vulgaris patients: case reports with follow-up for 2 months. Dermatol Ther 32(5):e13027
12. Zhao PX et al (2022) Effect of hydrogen intervention on refractory wounds after radiotherapy: a case report. World J Clin Cases 10(21):7545
13. Ono H et al (2012) Hydrogen (H_2) treatment for acute erythematous skin diseases. A report of 4 patients with safety data and a non-controlled feasibility study with H_2 concentration measurement on two volunteers. Med Gas Res 2(1):1–9
14. Asada R et al (2019) Effects of hydrogen-rich water bath on visceral fat and skin blotch, with boiling-resistant hydrogen bubbles. Med Gas Res 9(2):68
15. Ignacio RMC et al (2013) The drinking effect of hydrogen water on atopic dermatitis induced by Dermatophagoides farinae allergen in NC/Nga mice. Evid Based Complementary Alternat Med. https://doi.org/10.1155/2013/538673
16. Yoon YS et al (2014) Positive Effects of hydrogen water on 2, 4-dinitrochlorobenzene-induced atopic dermatitis in NC/Nga mice. Biol Pharm Bull 37(9):1480–1485
17. Kajisa T et al (2017) Hydrogen water ameliorates the severity of atopic dermatitis-like lesions and decreases interleukin-1β, interleukin-33, and mast cell infiltration in NC/Nga mice. Saudi Med J 38(9):928

18. Qi DD et al (2021) The therapeutic effects of oral intake of hydrogen rich water on cutaneous wound healing in dogs. Vet Sci 8(11):264
19. Chen S et al (2022) Photocatalytic glucose depletion and hydrogen generation for diabetic wound healing. Nat Commun 13(1):5684
20. Otani N et al (2022) Hydrogen-generating Si-based agent protects against skin flap ischemia–reperfusion injury in rats. Sci Rep 12(1):1–8
21. Liu YQ et al (2015) Hydrogen-rich saline attenuates skin ischemia/reperfusion induced apoptosis via regulating Bax/Bcl-2 ratio and ASK-1/JNK pathway. J Plast Reconstr Aesthet Surg 68(7):e147–e156
22. Fang W et al (2018) Hydrogen gas inhalation protects against cutaneous ischaemia/reperfusion injury in a mouse model of pressure ulcer. J Cell Mol Med 22(9):4243–4252
23. Dong XH et al (2019) Postconditioning with inhaled hydrogen attenuates skin ischemia/reperfusion injury through the RIP-MLKL-PGAM5/Drp1 necrotic pathway. Am J Transl Res 11(1):499
24. Zhao L et al (2013) Protective effect of hydrogen-rich saline on ischemia/reperfusion injury in rat skin flap. J Zhejiang Univ Sci B 14:382–391
25. Kiyoi T et al (2023) Intermittent environmental exposure to hydrogen prevents skin photoaging through reduction of oxidative stress. Geriatr Gerontol Int 23(4):304–312
26. Zhou P et al (2019) The healing effect of hydrogen-rich water on acute radiation-induced skin injury in rats. J Radiat Res 60(1):17–22
27. Watanabe S et al (2014) Protective effect of inhalation of hydrogen gas on radiation-induced dermatitis and skin injury in rats. J Radiat Res 55(6):1107–1113
28. Sahin AE et al (2021) Hydrogen-rich saline reduces tissue injury and improves skin flap survival on a rat hindlimb degloving injury model. J Plast Reconstr Aesthet Surg 74(9):2095–2103
29. Guo Z et al (2012) Hydrogen–rich saline protects against ultraviolet B radiation injury in rats. J Biomed Res 26(5):365–371
30. Zhang B et al (2018) Hydrogen ameliorates oxidative stress via PI3K-Akt signaling pathway in UVB-induced HaCaT cells. Int J Mol Med 41(6):3653–3661
31. Yu P et al (2011) Hydrogen-rich medium protects human skin fibroblasts from high glucose or mannitol induced oxidative damage. Biochem Biophys Res Commun 409(2):350–355
32. Xiao L et al (2021) Hydrogen-generating silica material prevents UVA-ray-induced cellular oxidative stress, cell death, collagen loss and melanogenesis in human cells and 3D skin equivalents. Antioxidants 10(1):76
33. Patterson JD et al (2021) H_2 in Antarctic firn air: atmospheric reconstructions and implications for anthropogenic emissions. Proc Natl Acad Sci 118(36):e2103335118
34. Ngawhirunpat T et al (2002) Changes in electrophysiological properties of rat skin with age. Biol Pharm Bull 25(9):1192–1196
35. Zaki SM et al (2015) Characteristics of the skin of the female albino rats in different ages: histological, morphometric and electron microscopic study. J Cytol Histol 3(3):1
36. Wiesenburg DA et al (1979) Equilibrium solubilities of methane, carbon monoxide, and hydrogen in water and sea water. J Chem Eng Data 24(4):356–360
37. Ferrell RT et al (1967) Diffusion coefficients of hydrogen and helium in water. Chem Eng Sci 13(4):702–708
38. Ito M et al (2012) Drinking hydrogen water and intermittent hydrogen gas exposure, but not lactulose or continuous hydrogen gas exposure, prevent 6-hydorxydopamine-induced Parkinson's disease in rats. Med Gas Res 2(1):1–7

Chapter 24
Intraosseous Administration of Molecular Hydrogen: A Novel Technique—From Molecular Effects to Tissue Regeneration

Mikhail Yu. Artamonov, Tyler W. LeBaron, Evgeniy L. Sokov, Lyudmila E. Kornilova, Felix A. Pyatakovich, and Inessa A. Minenko

Abstract In recent decades, molecular hydrogen has been shown to have diverse biological effects. By the end of 2022, more than 2000 articles have been published in the field of hydrogen medicine, many of which are original studies. There is some preliminary evidence on the regenerative effect of molecular hydrogen when administered by inhalation or via drinking hydrogen-rich water. Here we propose a novel method of administering hydrogen via intraosseous route, which potentially enhances the tissue regenerative effect. The purpose of this review was to systematize ideas about the nature, characteristics, and mechanisms of the influence of molecular hydrogen on various types of cells, including stem cells, as well as to introduce the hypothesis of the potential advantages of intraosseous administration route. The molecular, cellular, tissue and systemic effects of hydrogen are also reviewed. The existing literature indicates that the molecular and cellular effects of hydrogen qualify it to be a potentially effective agent in regenerative medicine. Administration of

M. Yu. Artamonov (✉) · F. A. Pyatakovich · I. A. Minenko
MJA Research and Development Inc., East Stroudsburg, PA 18301, USA
e-mail: contact@mjahealthcare.com

F. A. Pyatakovich
e-mail: piatakovich@gmail.com

I. A. Minenko
e-mail: minenko_i_a@staff.sechenov.ru

T. W. LeBaron
Department of Kinesiology and Outdoor Recreation, Southern Utah University, Cedar City, UT 84720, USA

Molecular Hydrogen Institute, Enoch, UT, USA

T. W. LeBaron
e-mail: tylerlebaron@suu.edu

E. L. Sokov · L. E. Kornilova
Department of Algology and Rehabilitation, Peoples' Friendship University, Moscow, Russia
e-mail: elsokov@mail.ru

J. Slezak and B. Kura (eds.), *Molecular Hydrogen in Health and Disease*, Advances in Biochemistry in Health and Disease 27, https://doi.org/10.1007/978-3-031-47375-3_24

molecular hydrogen via intraosseous route may serve as an optimal way to stimulate a bone marrow stem cell pool.

Keywords Molecular hydrogen · Intraosseous therapy · Hydrogen-rich saline · Regenerative medicine · Oxidative stress · Mesenchymal stem cells · Tissue regeneration

Introduction: Features of Bone Tissue Metabolism and Its Role in the Pathogenesis of Disease

In phylogenetic aspect, the bone is the youngest tissue. It is still in a period of adaptation to existence in gravity, upright posture, and in recent centuries to a sitting position. In morphofunctional terms, bone is one of the most complex and biologically active tissues [1, 2]. In many respects, it surpasses other body systems as the most massive and multifunctional, with a high metabolic and reparative activity. The structure of the bone marrow stroma includes undifferentiated stem mesenchymal cells—reticular, connective tissue, endosteal fibroblast-like, endothelial cells, adipocytes, differentiated bone cells (osteoblasts, osteoclasts, osteocytes), intercellular substance, endosteal and periosteal cells, bone marrow, vascular, lymphatic and nerve formations intimately associated with surrounding soft tissues [1, 2].

In the bone tissue, two oppositely directed processes constantly occur—bone resorption (bone breakdown) and bone deposition (formation of new bone). The ratio of these processes depends on various factors, including physical stress on the bone, hormonal status, and age. Osteogenesis is thought to occur at the expense of endosteal, periosteal, and bone marrow cells. The regulation of osteogenesis has three levels: local, systemic, and genetic. Local regulation is carried out by the microenvironment through various cytokines, many growth factors, a number of polypeptides, enzymes, and intercellular contacts. Systemic neuroendocrine regulation is carried out by hormones and hormone-like substances [3–5].

The morphofunctional relationship between osteogenesis and blood circulation is carried out not only anatomically, but also closely functionally. This has been confirmed by numerous studies of the connections between the intraosseous and extraosseous systems of the arterial, venous [6], lymphatic channels [7], and nervous regulation with osteoreception [1]. The red bone marrow is a blood depot, a hematopoietic organ, a highly sensitive reflexogenic zone, the central link of the immune system [8]. The red bone marrow is a source of an almost inexhaustible pool of mesenchymal stem fibroblast-like cells—precursors of osteoblasts, capable of not only potentiating osteogenesis, but also building a hematopoietic microenvironment and regulating hematopoiesis itself [9].

Reparative regeneration is the restoration of tissue after damage. The mechanisms of physiological and reparative regeneration of bone tissue are qualitatively the same and are carried out on the basis of common patterns. Reparative regeneration—there is to some extent enhanced and accelerated physiological regeneration [10]. One of the most natural, significant and strong inducers of reparative regeneration and metabolism of bone tissue is its traumatic injury [11–14]. Many researchers point to the possibility of local excitation to induce the reparative regeneration of bone tissue, thereby changing its metabolism, by the method of osteotomy, trepanation, tunneling or perforation in the necessary areas of the bone. Localized and dosed alteration of bone tissue is used as a means of therapeutic action and leads to the relief of degenerative-dystrophic disorders. The therapeutic effect is manifested locally in the stimulation zone and regionally in segmental areas due to the intensification of hemocirculation [11–14].

During the histological examination of bone tissue, it's been shown intraosseous administration of various liquids or blood within a day results in the destruction of bone trabeculae, damage to the stroma and parenchyma of the bone marrow tissue, and impaired blood circulation at the site of administration. These effects are attributed to the mechanical damage caused during needle insertion into the bone. By day 60, bone and bone marrow tissues completely restored their cellular structure [15].

The bones are supplied with blood from nearby arteries, which form plexuses and networks with a large number of anastomoses in the periosteal region. The blood supply of the thoracic and lumbar spine is provided by the branches of the aorta, the cervical spine—by the vertebral artery. The presence of vessels penetrating the bone was proven histologically [15]. Through small openings, arterioles penetrate the bone, branch dichotomously, and form a branched closed system of hexagonal sinuses, anastomosing with each other. The capacity of the intramedullary venous plexus exceeds the arterial bed by several tens of times. Due to its large cross-sectional area, blood flow in the spongy bone is exceptionally slow, and in certain sinuses it stops for 2–3 min. Venules form plexuses upon exiting the sinuses and exit the bone through small holes. The intraosseous administration method is the only way to effectively deliver drugs to the vascular bed of the bone [16].

The studies using venospondylography showed that when a contrast agent is injected into the spinous processes of the vertebrae, the veins of the spongy substance of the spinous processes and vertebrae are evenly filled 6–8 segments above and 3–4 segments below the injection site, venous vessels of the periosteum, internal, and then the external vertebral plexus, veins of the epidural space, veins of the dura mater, venous plexuses of the spinal nodes and nerves. The dye penetrates the spongy tissue of the spinous processes and vertebrae, which indicates the absence of valves in the veins of the vertebral plexuses [17].

The system of blood inflow and outflow in the bones is functionally balanced and regulated by the nervous system. Under the influence of osteoclastic and osteoblastic processes, bone tissue is constantly and actively renewed. The blood flow in the trabeculae of the bone, is associated, among other things, with the physical impact on the spine [18]. When a compression load occurs on the vertebral bodies, elastic deformation of the bone trabeculae and an increase in pressure in the cavities filled with red bone marrow occur. Considering the converging direction of the nuclear-articular axes, for example, when walking, an increase in pressure alternately occurs in the anterior right half of the vertebra (decrease in the anterior left), and then in the anterior (decrease in the anterior right). The red bone marrow shifts alternately from a zone of higher pressure to a zone of lower pressure. This allows us to consider the vertebral bodies as a kind of biological hydraulic shock absorbers. At the same time, pressure fluctuations in the cavities of the spongy substance of the vertebral bodies contribute to the penetration of young blood cells into the sinus capillaries and the outflow of venous blood from the spongy substance to the internal vertebral plexus [18].

The disruption of intraosseous blood flow results in increased intraosseous pressure. Prolonged elevation of this pressure leads to specific structural changes in the bone, including resorption of intraosseous trabeculae and sclerosis of the cortical layer in the spongy tissue of the vertebral body's endplates. These changes can eventually lead to the formation of cysts and necrosis [19].

Some researchers are already talking about "intraosseous nociceptors", which are activated under the influence of various stimuli, especially under the influence of increased intraosseous pressure and bone marrow edema, which is the most important factor in the formation of skeletal pain [3, 20–22].

Characteristics of Molecular and Cellular Aspects of the Biological Action of H_2

In recent decades, medical gases have deservedly attracted much attention from specialists in the field of biomedicine. At the same time, the spectrum of gases for which biological and beneficial effects have been discovered and observed is quite wide and continues to increase [23–25].

Recently, molecular hydrogen, which has several unique characteristics, has occupied a special place in medical gas therapy. This gas, which has no specific color or smell, created from the lightest chemical element, is ubiquitous, and, due to its size and minimal molecular weight, can penetrate through any biological barrier [26–30]. The high bioavailability of H_2 satisfies the first requirement of any pharmacological agent to have a biological effect.

The body of knowledge in the field of H_2 biomedicine is being actively updated, including over 100 clinical trials and more than 2000 articles that have been published by the end of 2022. Various aspects of the issue are considered in detail, but the focus of research is largely shifted towards the cardiac [29, 31], neurological [32, 33], and radioprotective effects [34, 35] of hydrogen. On the other hand, cellular effects, which are fundamentally significant for regenerative medicine, are revealed only indirectly.

The combination of routes of introducing H_2 into the body is necessary for the realization of numerous biological and beneficial effects of this gaseous molecule. These include antioxidant, anti-apoptotic, anti-inflammatory activity, the regulation of gene expression, etc. However, diverse molecular mechanisms are primarily involved that mediate the unique effects of molecular hydrogen (Fig. 24.1).

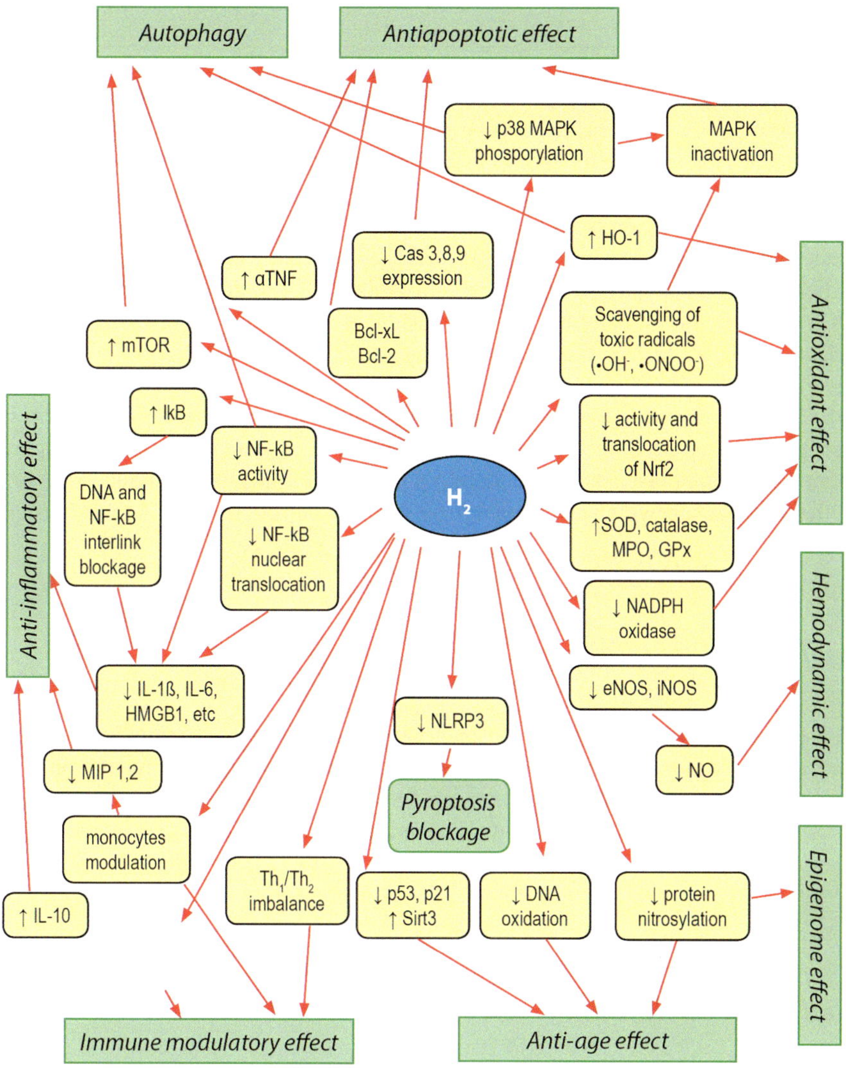

Fig. 24.1 Molecular effects of hydrogen in living organisms. The figure shows some of the proposed mechanisms by which the main effects of hydrogen are mediated on the molecular and cellular level. The brown links represent shifts in regulatory molecules that lead to the development of specific cellular effects (indicated by green blocks). Abbreviations: MAPK–mitogen-activated protein kinase, HO–heme oxygenase, TNF–tumor necrosis factor, SOD–superoxide dismutase, MPO–myeloperoxidase, NOS–nitric oxide synthase (inducible–iNOS; endothelial–eNOS), GPx–glutathione peroxidase, Cas–caspase, HMGB1-high-mobility group protein B1, NLRP–Nucleotide-binding oligomerization domain, Th–T-cytotoxic lymphocyte. Reprinted with permission from Molecular Hydrogen: From Molecular Effects to Stem Cells Management, by Mikhail Yu. Artamonov et al., Antioxidants 2023, 12, 636. https://doi.org/10.3390/antiox12030636

Antioxidant Effects

Currently, many studies have assumed that the effect of H_2 on the state of oxidative metabolism consists of the direct capture of free radicals and a regulatory effect on antioxidant systems. The first component of the antioxidant effect of the gas in question is realized by the direct interaction of hydrogen with toxic oxidants, primarily hydroxyl (•OH) and peroxynitrite (ONOO−) [36]. This is of fundamental importance, since it is these oxidant molecules that have the maximum oxidative activity, which is shown using a specific fluorescent probe (2', 7'-dichlorodihydrofluorescin; DCF) [37]. It should be emphasized that H_2 is a new type of antioxidant, since it selectively disposes of $^{\bullet}OH$ and $ONOO^-$ without reacting with hydrogen peroxide or the superoxide radical, and thus does not violate the mechanisms of cellular signaling [38, 39]. It is important to emphasize that the neutralization of radicals can occur in the extracellular space (including in biological fluids), as well as in any cell compartments, including plasma and mitochondrial membranes, due to the unique small size of the H_2 molecule, which gives it a high permeability through any biological barriers [39, 40].

Anti-inflammatory Effects

The anti-inflammatory effects of H_2 are closely associated with its antioxidant effects and involve similar or identical initial mechanisms in their implementation. For example, molecular hydrogen influences ASK1- and p38 MAPK-dependent regulatory pathways, which are also involved in inflammation. Excessive generation of bioradicals can stimulate an inflammatory response due to activation of NF-kB, the p53 gene, hypoxia-inducible factor-1α, matrix metalloproteinases, etc. [41, 42].

Anti-apoptotic Effects

It is shown that H_2 can have multifactorial effects on the process of cell death including autophagy and apoptosis [43]. In general, this gas inhibits apoptosis by affecting the signaling pathways regulating apoptosis and its associated proteins, including phosphatidylinositol-3-kinase (PI3K), protein kinase B (Akt), and glycogen synthase kinase-3ß (GSK3ß) [44]. The use of hydrogen affects the degree of activation of the cascades, including ASK1/JNK [45], ERK 1/2, and MEK 1/2, and inhibits of the activity of caspases 3, 8, and 9, as well as the Bcl/Bax system [45, 46]. In addition, the anti-apoptotic effect of H_2 is closely associated with the two effects described above, since reducing the severity of inflammatory reactions and relieving oxidative stress reduces the need for cell removal by apoptosis [47].

The regulation of autophagy serves to reduce the intensity of apoptosis, forming an important balance between these processes.

Modulation of Autophagy

Autophagy, also called partial macro-autophagy, is a catabolic process that supports cellular homeostasis and is implemented with the participation of lysosomes and the ubiquitin–proteasome system [48]. This process, which initially has exclusively physiological significance, under certain conditions (in particular, under the action of stressors of abnormal intensity and/or duration) acquires a disadaptive character [49]. It has been shown that H_2 can have a dualistic effect on the autophagy process. On the one hand, hydrogen can stimulate a specific nucleotide-bearing domain and inhibit NLRP3 in macrophages and limit inflammatory reactions [50, 51]. This makes it possible to achieve a protective effect by stimulating autophagy and providing an "intracellular renewal" procedure.

Regulation of Pyroptosis

Pyroptosis is a relatively new form of programmed cell death, fundamentally different from apoptosis in that it induces an inflammatory reaction. Inflammation in this case is provoked by the activation of pattern-recognizing receptors in special structures called inflammasomes [52]. It is believed that this mechanism has a protective quality, but excessive stimulation of this process contributes to the development or progression of pathology. The factors contributing to the induction of pyroptosis include reactive oxygen species, casapase-1, and the inflammasomes [53]. According to the data presented above, molecular hydrogen has an anti-inflammatory and antioxidant effect, which makes it possible to block all these factors, preventing excessive activation of pyroptosis.

Routes of Introducing Molecular Hydrogen into the Body

Currently, the range of routes of introducing H_2 into the body is extremely wide. First, it is important to emphasize that these methods differ not only in the convenience of application for a specific pathology (e.g., in the case of the treatment of dermatological diseases, hydrogen baths may be the preferred option), but also in the pharmacokinetics of the molecule, which alters its pharmacological activity [38].

The most common options for molecular hydrogen therapy are the inhalation of H_2-containing gas mixtures of various compositions, the use of hydrogen-saturated water, and the infusion/injection of a sodium chloride solution saturated with H_2 [30].

Each of these pathways has its own characteristics, advantages, and disadvantages, as well as possible different molecular mechanisms of action.

Most studies aimed at evaluating the effectiveness of the use of H_2 have used hydrogen-saturated solutions (i.e., hydrogen water, hydrogen saline, etc.). However, studies on the inhalation of hydrogen, especially in clinical use, are increasing [29, 35, 36, 54]. Inhalation of H_2 is a fairly simple way of exposure for both laboratory animals and humans.

Indeed, this was the method employed by Ohsawa et al. [36] using the rat model of ischemia–reperfusion. In addition, an important advantage of the technology is the possibility of the strict dosing of hydrogen by regulating the exposure time and concentration of H_2 in the gas mixture [38, 55]. The effectiveness of H_2 inhalation in chronic obstructive pulmonary disease [56, 57] and severe bronchial asthma [58] has also been reported. The expediency of such an approach is associated with the variable dose dependence of the antioxidant and anti-inflammatory properties of H_2 [38, 43].

The most convenient way of introducing molecular hydrogen in clinical practice is drinking water saturated with molecular hydrogen. This option eliminates the explosion and fire hazard of the therapy and ensures its portability, opening up the possibility for the widespread use of H_2-containing water. However, this path also has disadvantages associated with low gas solubility [38]. It is known that the saturation of dissolved hydrogen is 0.78 mM (1.57 mg/L) at normal atmospheric pressure and room temperature [59]. This circumstance may be significant since it does not always allow achieving the necessary dose of the molecule to ensure a full clinical effect. In addition, when using this route, it should be considered that the prepared hydrogen water should be applied immediately, since it has a very short period of maintaining the concentration of H_2. Moreover, upon ingestion of hydrogen water, a significant amount (>90%) is lost via normal expiration [60]. This at least indicates the high uptake of H_2 to pass through the gastrointestinal tract and into the venous system where it reaches the lungs and is exhaled. At the same time, the distribution of hydrogen in various tissues and organs after drinking hydrogen water is not the same. In particular, the penetration of H_2 into brain cells when using the route of administration under consideration is minimal [61], which may be of fundamental importance for determining indications for its clinical use.

The third main route of introducing H_2 into the body is the use of injections and infusions of an H_2-saturated isotonic saline solution [35, 38, 55]. The specified path also has advantages and disadvantages. For one, this path allows the dosing of the injected amount of hydrogen with high accuracy, the use of different concentrations, an increase in the bioavailability of the agent to the target organ, and, if necessary, a carrying out of the topical effects on strictly defined areas of tissues (e.g., surface localization or areas of catheterization and injection). At the same time, injections of hydrogen solutions pose a certain risk of invasiveness and, consequently, infection, while also requiring the involvement of experienced medical personnel for manipulation. The administration of these solutions is mainly performed intravenously (in patients) or intraperitoneally (in experimental studies using laboratory animals) [35].

In our opinion, the clinical potential of this route of administration is not fully developed, as evidenced the rich experience of intravenous ozone therapy, also based on the effects of a medical gas [62, 63].

Thus, at present, there is a wide range of routes of introducing molecular hydrogen into the body, differing not only in the topical and physico–chemical parameters, but also in the pharmacokinetics of the action of the molecule.

The Effect of Molecular Hydrogen on Various Cell Pools and Regeneration Processes

As already mentioned, the multifactorial biological activity of H_2 creates prerequisites for its modulating effect on the processes of cell formation, starting with early precursors—stem cells. This hypothesis is confirmed in the available experimental data obtained in vitro and in vivo and testifying to the positive effect of H_2 on all stages of the formation of differentiated cells (Fig. 24.2).

Special attention should be paid to the effect of H_2 on mtUPR as a recently identified mediator involved in mitochondrial function [64] and cell survival in adverse conditions [65]. This effect may be associated with the stimulation of heat-shock protein (HSP) production [66]. The stimulating effect of hydrogen on mtUPR is manifested as a change in the level of phosphorylation of eIF2a [66] and shifts in the expression of ATF4 [67] and ATF5 [68]. Such a cascade response provides activation of the protein folding processes. Additionally, this response is facilitated by the induction of HSP60 [69] leading to an increase in collagen synthesis, which is necessary for both cell growth and the formation of intercellular matter. Another factor by which hydrogen may increase its cellular proliferation is its ability to activate GFAP (glial fibrillary acidic protein), a marker of differentiation in glioblastoma cells [70].

An important aspect ensuring the activation of the processes of proliferation, differentiation, and growth of emerging cells is the development of the microenvironment of stem cells. This is facilitated by an increase in the number of colony-forming factors, as well as the regulating effect of cytokines. In particular, the use of H_2 induces the activation of CCL-2, which leads to a decrease in the level of proinflammatory cytokines (TNFα, IL-6, IFN-y) [67]. The second mechanism determining the decrease in the concentration of these cytokines is the inhibition of NF-kB, which can also be modulated by H_2 [71]. In addition, another antioxidation/detoxification enzyme induced by H_2 is hemoxygenase-1. This enzyme acts as a strong antioxidant [72] and promotes the stimulation of the synthesis of the anti-inflammatory cytokine IL-10 and the formation of differentiation markers on the cell surface [71]. Although several studies have used various types of cancer cells, the collective results indicate that H_2 may provide conditions for the accelerated proliferation, differentiation, and growth of stem cells.

This property of H_2 is of great importance for regenerative medicine since its principal task is to develop the most sparing technologies for stimulating tissue

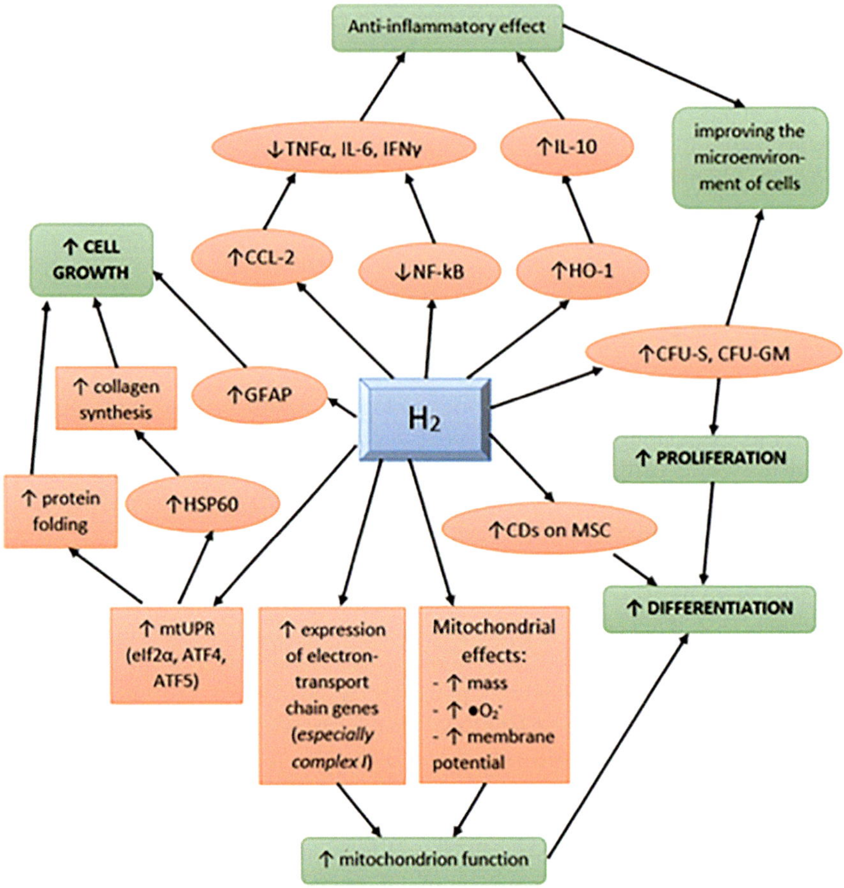

Fig. 24.2 The influence of molecular hydrogen on the state of stem cells. This diagram shows the molecular regulators (indicated by red clouds) and the cellular effects caused by them (green blocks) that are significant for the functioning of stem cells. Abbreviations: HSP–heat shock protein, CFU–colony-forming unit, HO–heme oxygenase, TNF–tumor necrosis factor, IFN–interferon, CD–cluster differentiation marker, MSC–mesenchymal stem cell, GFAP—glial fibrillary acidic protein, IL–interleukin. Reprinted with permission from Molecular Hydrogen: From Molecular Effects to Stem Cells Management, by Mikhail Yu. Artamonov et al., Antioxidants 2023, 12, 636. https://doi.org/10.3390/antiox12030636

regeneration processes. The combination of molecular, cellular and tissue effects of H_2 suggest its multifaceted, pro-regenerative activity, some aspects of which are presented in Fig. 24.3. In particular, the ability of H_2 to reduce free radical damage is its fundamental value. Relieving pronounced oxidative stress that inevitably occurs in the tissues of a wound or other tissue defect is key for cellular regeneration [73, 74]. The creation of favorable conditions for the restoration of the cellular composition of the tissue is also attained by an anti-inflammatory effect [74–76] as well as the formation of intercellular substance components (in particular, collagen). Direct replacement of cellular deficiency occurs due to activation of mesenchymal stem cells, and a stimulation of their proliferation and differentiation which occurs under the influence of molecular hydrogen. This process is additionally induced and regulated by a set of pro-regenerative cytokines and cell migration into a definitive niche, which is facilitated by activation of the expression of cell adhesion molecules [77]. In general, the effect of H_2 on the state of mesenchymal stem cells and on tissue regeneration processes should be considered favorable.

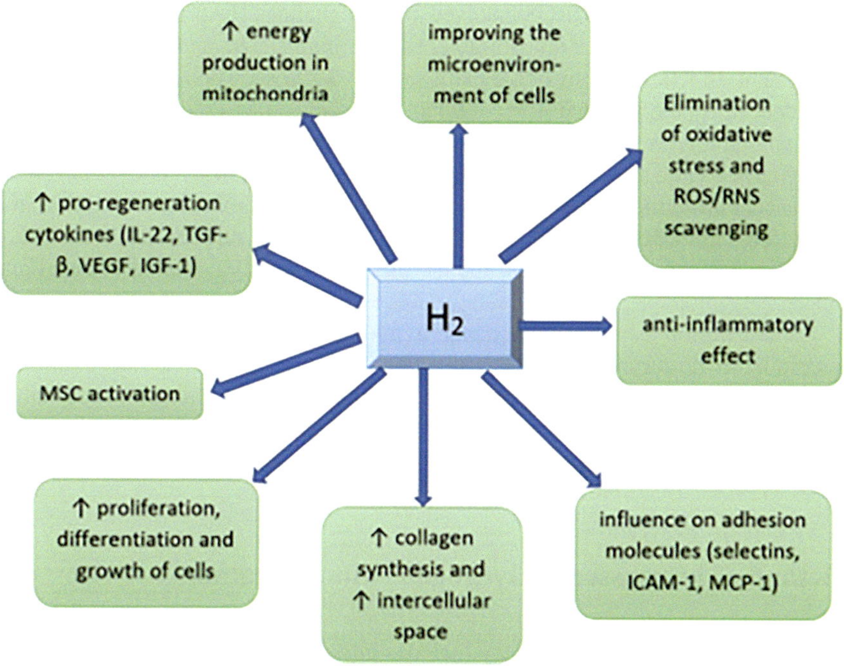

Fig. 24.3 The influence of molecular hydrogen on tissue regeneration. The figure illustrates the effects of molecular hydrogen, potentially significant for stimulating the regeneration and differentiation of stem cells. Abbreviations: IL–interleukin, TGF–tumor growth factor, ROS–reactive oxygen species, RNS–reactive nitrogen species, VEGF–vasculo-endothelial growth factor, IGF—insulin-like growth factor, ICAM-intercellular adhesion molecule, MCP-monocyte chemotactic protein. Reprinted with permission from Molecular Hydrogen: From Molecular Effects to Stem Cells Management, by Mikhail Yu. Artamonov et al., Antioxidants 2023, 12, 636. https://doi.org/10.3390/antiox12030636

Hypothesis: Intraosseous Administration of Molecular Hydrogen

Our hypothesis was that hydrogen-rich saline may be a promising, effective, and safe approach to stimulate the pool of autologous stem cells in the bone marrow, based on the following. Hydrogen, a new medical gas, could potentially selectively reduce $^{\bullet}OH$ and $ONOO^{-}$, exerting organ-protective effects through regulating oxidative stress and inflammation. It is so mild that it does not disturb metabolic oxidation–reduction reactions or disrupt the ROS involved in cell signaling. The primary advantage of hydrogen-rich saline is that it is portable, easily administered, and safe, with similar antioxidant effects. It is physiologically safe at any dose. The safety of hydrogen is evidenced by the fact that it is continuously produced by colonic bacteria in the body and circulates normally in the bloodstream. Also, the tissue compatibility of hydrogen is greater than that of many other antioxidants because it is an endogenous substance. Furthermore, hydrogen can also penetrate biomembranes and diffuse into the cytosol, mitochondria, and nuclei, thereby protecting nuclear DNA and mitochondria, suggesting that it could reduce the risk of lifestyle-related diseases and cancer.

Administration can be done using a hydrogen-rich normal saline solution. According to the current data, essentially all medications that can be safely administered through intravenous route can be used in intraosseous approach. As described above, the intraosseus approach has been established as safe and effective in reaching the bone marrow.

Method of Intraosseous Hydrogen Administration

The method of intraosseous therapy includes intraosseous blockade, the essence of which is that a local anesthetic approved for intravenous administration (usually low concentrations of cocaine or Novocaine) is injected into the spongy bone tissue in order to minimize irritation of intraosseous slow-conducting receptors and reduce or relieve pain.

The method of intraosseous therapy also includes intraosseous infusions (bolus or drip) of various drugs for the purpose of drug effects on the biochemical mechanisms of the local pathological process.

The method of intraosseous therapy also includes the intraosseous administration of biological solutions as well as the intraosseous administration of gas solutions or blood that has been exposed to physical factors, such as, for example, laser irradiation.

Intraosseous therapy is a type of intraosseous administration of drugs. Intraosseous administration of drugs is mainly used for par enteral administration, as an analogue of intravenous access. For par enteral administration of drugs, they can be injected into any bone formations: the internal condyle of the tibia, the head of the humerus, the calcareous, the sternum, etc. The main thing is that the drug is in the vascular bed of the body.

In intraosseous therapy, the place of intraosseous administration of the drug is fundamental. We perform intraosseous blockades in the following bone protrusions: spinous processes of the cervical, thoracic, lumbar, and sacral vertebrae, posterior and anterior iliac wing spines, scapular spine, sternum, acromion, head of the humerus, diaphysis of the radius and ulna, zygomatic bone, lower jaw, trochanter of the femur, condyle of the tibia, head of the fibula, outer and inner malleolus, calcaneus.

Since 1980, the authors have performed more than 40,000 intraosseous therapy procedures and have not noted a single case of a serious complication. After determining the site of intraosseous entry determined by clinical situation, we infiltrate the skin and soft tissues up to the periosteum with a 0.5% lidocaine solution. Then, through the anesthetized soft tissues, we pass an intraosseous or spinal needle to the periosteum; insert the needle into the bone to a depth of 5–10 mm. We remove the stylet, attach a syringe containing 5.0 ml of a 0.5% lidocaine solution, and conduct an aspiration test, drawing 2–4 ml of blood into the syringe. A positive aspiration test—free drawing of blood with small droplets of fat into the syringe indicates that the needle is inserted correctly and the blood comes from the spongy bone tissue. Then, without removing the needle, the contents of the syringe are mixed and the resulting mixture is injected into the spongy tissue.

After the confirmation of the proper intraosseous access, intraosseous drip administration of drugs approved for intravenous use can be carried out. In our case, the intraosseous drip of hydrogen-rich saline is administered.

Conclusion

The presented evidence suggests that hydrogen-rich saline can be a suitable candidate to provide protective and therapeutic effects via intraosseous route by stimulating and mobilizing the pool of autologous stem cells from the bone marrow. In addition, hydrogen is regarded mainly as a preventive measure, and its therapeutic effect in established disease needs further study. We also believe hydrogen-rich saline could be used as an adjuvant therapy for any disease in conjunction with other methods, as it could delay the progress of the disease, reduce its severity, and increase the treatment effect, thereby improving patients' quality of life. Before hydrogen-rich saline can really become a promising novel therapy, a great many experiments both in vitro and in vivo need to be carried out to determine its effectiveness and feasibility. We also plan to perform an animal experiment to test this hypothesis. After effective animal experiments, multicenter clinical studies are also essential, and because the existing clinical research is limited, the optimal strategy for using intraosseous administration of hydrogen-rich saline must also be explored. Therefore, conducting targeted research in this area can open new horizons of regenerative medicine and create an innovative technology for accelerated tissue repair.

References

1. Yankovsky GA (1982) Osteoreception. Riga, p 346
2. Shevelev OA, Sokov EL, Khodorovich NA (1995) Role of intraosseous receptors in afferent and motor reaction modulation. Bull Exp Biol Med 120:685–687
3. Ishida T, Tanaka S, Sekiguchi T, Sugiyama D, Kawamata M (2016) Spinal nociceptive transmission by mechanical stimulation of bone marrow. Mol Pain 12:1–15
4. Stewart K, Walsh S, Screen J et al (1999) Further characterization of cells expressing STRO-1 in cultures of adult human bone marrow stromal cells. J Bone Miner Res 14:1345–1356
5. Zhogolev KD, Nikitin YuV, Gypsy EN, Knyazkin IV, Pakhomov EYu (2003) Regulation osteogenesis and reparative process. C. Pb, p 168
6. Shih C, Bernard GW (1997) Neurogenic substance P stimulates osteogenesis in vitro. Peptides 18(2):323–326
7. Shimizu T, Mehdi R, Yoshimura Y et al (1998) Sequential expression of bone morphogenetic protein, tumor necrosis factor, and their receptors in bone-forming reaction after mouse femoral marrow ablation. Bone 23(2):127–133
8. Marie PJ (1999) Cellular and molecular alterations of osteoblasts in human disorders of bone formation. Histol Histopathol 2:525–538
9. Lane JM, Yasko AW, Tomin E et al (1999) Bone marrow and recombinant human bone morphogenetic protein-2 in osseous repair. Clin Orthop 361:216–227
10. Susuki A, Kaneko E, Maeda I et al (1997) Mesoderm induction by BMP-4 and -7-heterodimers. Biochem Biophys Res Commun 232(1):153–156
11. Bunov BS, Schevzov VI (1999) Effect of arterial insufficiency of the lower extremities on intraosseous pressure. Orthopedichesky Geniy 2:51–56
12. Zusmanovich FN (1996) Revascularized bone trepanation in treatment of peripheral vascular disease. Kurgan, p 92
13. Shevzov VI, Schurov VI (1999) Ultrasound Doppler sonography of extracranial arteries. Orthopedichesky Geniy 3:56–59
14. Larionov AA, Rechkin My, Assonova SN (1999) Bone trepanation effect on bone regeneration and regional blood flow. Orthopedichesky Geniy 2:80–85
15. Atyasov NI (1965) Intraosseus infusions in dermatological and orthopedia. Diss. d-ra. med.nauk, Gorky
16. Protasov VYa (1970) Vertebral venous system as a central collector of the human body. Diss. kand. med. nauk. Blagoveschensk
17. Shabanov VV (1992) Administration of antibiotics in pancreatitis. Diss. kand. med.nauk, Санкт- Петербург
18. Yudelson YB, Nechaev VI (2000) Morphofunctional aspects of vertebroneurology. Nevrologichesky Zhurnal 6:33–35
19. Arnoldi CC (1989) Intraosseus engorgement—pain syndromes. The pathomechanism of pain. In: Bone circulation and bone necrosis. Tulusa, pp 253–259
20. Sokov EL (1994) Intraosseous Blockades – pathogenetic treatment of lumbar osteochondrosis. Terapevtichesky Archiv 3:84–88
21. Sokov EL (1996) Osteogenic afferent reaction to clinical pathogenesis of osteochondrosis and mechanisms. Diss. d-ra. med.nauk, Moscow, p 177
22. Sokov LP, Sokov EL, Sokov SL (2002) Rukovodstvo by neuroorthopedics. RUDN, p 541
23. Zafonte RS, Wang L, Arbelaez CA, Dennison R, Teng YD (2022) Medical gas therapy for tissue, organ, and CNS protection: a systematic review of effects, mechanisms, and challenges. Adv Sci 9:2104136
24. Teng YD, Wang L, Kabatas S, Ulrich H, Zafonte RD (2018) Cancer stem cells or tumor survival cells? Stem Cells Dev 27:1466–1478
25. Nazarov EI, Khlusov IA, Noda M (2021) Homeostatic and endocrine responses as the basis for systemic therapy with medical gases: Ozone, xenon and molecular hydrogen. Med Gas Res 11:174–186

26. Martusevich A, Surovegina A, Popovicheva A, Didenko N, Artamonov M, Nazarov V (2022) Some Beneficial effects of inert gases on blood oxidative metabolism: in vivo study. BioMed Res Int 2022:5857979
27. Ichihara M, Sobue S, Ito M, Ito M, Hirayama M, Ohno K (2015) Beneficial biological effects and the underlying mechanisms of molecular hydrogen–comprehensive review of 321 original articles. Med Gas Res 5
28. Shen M, Zhang H, Yu C, Wang F, Sun X (2014) A review of experimental studies of hydrogen as a new therapeutic agent in emergency and critical care medicine. Med Gas Res 4:17
29. Hirano SI, Ichikawa Y, Kurokawa R, Takefuji Y, Satoh F (2020) A "philosophical molecule", hydrogen may overcome senescence and intractable diseases. Med Gas Res 10:47–49
30. Tao G, Song G, Qin S (2019) Molecular hydrogen: current knowledge on mechanism in alleviating free radical damage and diseases. Acta Biochim Biophys Sin 51:1189–1197
31. Wang L, Zhao C, Wu S, Xiao G, Zhuge X, Lei P, Xie K (2018) Hydrogen gas treatment improves the neurological outcome after traumatic brain injury via increasing miR-21 expression. Shock 50:308–315
32. Wu J, Wang R, Yang D, Tang W, Chen Z, Sun Q, Liu L, Zang R (2018) Hydrogen postconditioning promotes survival of rat retinal ganglion cells against ischemia/reperfusion injury through the PI3K. Akt pathway. Biochem Biophys Res Commun 495:2462–2468
33. Hirano S-I, Ichikawa Y, Sato B, Yamamoto H, Takefuji Y, Satoh F (2021) Molecular hydrogen as a potential clinically applicable radioprotective agent. Int J Mol Sci 22:4566. https://doi.org/10.3390/ijms22094566
34. Hu Q, Zhou Y, Wu S, Wu W, Deng Y, Shao A (2020) Molecular hydrogen: a potential radioprotective agent. Biomed Pharmacother 130:110589
35. Runtuwene J, Amitani H, Amitani M, Asakawa A, Cheng KC, Inui K (2015) Hydrogen-water enhances 5-fluorouracil-induced inhibition of colon cancer. PeerJ 3:e859
36. Guan WJ, Chen RC, Zhong NS (2020) Strategies for the prevention and management of coronavirus disease 2019. Eur Respir J 55:2000597
37. Ohta S (2014) Molecular hydrogen as a preventive and therapeutic medical gas: initiation, development and potential of hydrogen medicine. Pharmacol Ther 144:1–11
38. Fu Z, Zhang J (2022) Molecular hydrogen is a promising therapeutic agent for pulmonary disease. J Zhejiang Univ Sci B 23:102–122
39. Ohta S (2015) Molecular hydrogen as a novel antioxidant: overview of the advantages of hydrogen for medical applications. Methods Enzymol 555:289–317
40. Gharib B, Hanna S, Abdallahi OM, Lepidi H, Gardette B, De Reggi M (2001) Anti-inflammatory properties of molecular hydrogen: Investigation on parasite-induced liver inflammation. C R Acad Sci III(324):719–724
41. Rimessi A, Previati M, Nigro F, Wieckowski MR, Pinton P (2016) Mitochondrial reactive oxygen species and inflammation: molecular mechanisms, diseases and promising therapies. Int J Biochem Cell Biol 81(Pt B):281–293
42. Chen M, Zhang J, Chen Y, Qiu Y, Luo Z, Zhao S, Du L, Tian D (2018) Hydrogen protects lung from hypoxia/re-oxygenation injury by reducing hydroxyl radical production and inhibiting inflammatory responses. Sci Rep 8:8004
43. Cole AR, Sperotto F, DiNardo JA, Carlisle S, Rivkin MJ, Sleeper LA, Kheir JN (2021) Safety of prolonged inhalation of hydrogen gas in air in healthy adults. Crit Care Explor 3:e543
44. Chen K, Wang N, Diao Y, Dong W, Sun Y, Liu L, Wu X (2017) Hydrogen-rich saline attenuates brain injury induced by cardiopulmonary bypass and inhibits microvascular endothelial cell apoptosis via the PI3K/Akt/GSK3β signaling pathway in rats. Cell Physiol Biochem 43:1634–1647
45. Liu YQ, Liu YF, Ma XM, Xiao YD, Wang YB, Zhang MZ, Cheng AX, Wang TT, Li JL, Zhao PX et al (2015) Hydrogen-rich saline attenuates skin ischemia/reperfusion induced apoptosis via regulating Bax/Bcl-2 ratio and ASK-1/JNK pathway. J Plast Reconstr Aesthetic Surg 68:147–156
46. Mo XY, Li XM, She CS, Lu XQ, Xiao CG, Wang SH, Huang GQ (2019) Hydrogen-rich saline protects rat from oxygen glucose deprivation and reperfusion-induced apoptosis through VDAC1 via Bcl-2. Brain Res 1706:110–115

47. Li J, Hong ZJ, Liu H, Zhou J, Cui L, Yuan S, Chu X, Yu P (2016) Hydrogen-rich saline promotes the recovery of renal function after ischemia/reperfusion injury in rats via anti-apoptosis and anti-inflammation. Front Pharmacol 7:106
48. Yang ZF, Klionsky DJ (2010) Mammalian autophagy: Core molecular machinery and signaling regulation. Curr Opin Cell Biol 22:124–131
49. Maiuri MC, Zalckvar E, Kimchi A, Kroemer G (2007) Self-eating and self-killing: crosstalk between autophagy and apoptosis. Nat Rev Mol Cell Biol 8:741–752
50. Zhang G, Li Z, Meng C, Kang J, Zhang M, Ma L, Zhou H (2018) The anti-inflammatory effect of hydrogen on lung transplantation model of pulmonary microvascular endothelial cells during cold storage period. Transplantation 102:1253–1261
51. Chen H, Mao X, Meng X, Li Y, Feng J, Zhang L, Zhang Y, Wang Y, Yu Y, Xie K (2019) Hydrogen alleviates mitochondrial dysfunction and organ damage via autophagy-mediated NLRP3 inflammasome inactivation in sepsis. Int J Mol Med 44:1309–1324
52. Shi JJ, Gao WQ, Shao F (2017) Pyroptosis: Gasdermin-mediated programmed necrotic cell death. Trends Biochem Sci 42:245–254
53. Zha QB, Wei HX, Li CG, Liang YD, Xu LH, Bai WJ, Pan H, He XH, Ouyang DY (2016) ATP-induced inflammasome activation and pyroptosis is regulated by AMP-activated protein kinase in macrophages. Front Immunol 7:597
54. Tian Y, Zhang Y, Wang Y, Chen Y, Fan W, Zhou J, Qiao J (2021) Hydrogen, a novel therapeutic molecule, regulates oxidative stress, inflammation, and apoptosis. Front Physiol 12:789507
55. Guan WJ, Wei CH, Chen AL, Sun XC, Guo GY, Zou X, Shi JD, Lai PZ, Zheng ZG, Zhong NS (2020) Hydrogen/oxygen mixed gas inhalation improves disease severity and dyspnea in patients with Coronavirus disease 2019 in a recent multicenter, open-label clinical trial. J Thorac Dis 12:3448–3452
56. Zheng ZG, Sun WZ, Hu JY, Jie ZJ, Xu JF, Cao J, Song YL, Wang CH, Wang J, Zhao H et al (2021) Hydrogen/oxygen therapy for the treatment of an acute exacerbation of chronic obstructive pulmonary disease: Results of a multicenter, randomized, double-blind, parallel-group controlled trial. Respir Res 22:149
57. Huang P, Wei S, Huang W, Wu P, Chen S, Tao A, Wang H, Liang Z, Chen R, Yan J et al (2019) Hydrogen gas inhalation enhances alveolar macrophage phagocytosis in an ovalbumin-induced asthma model. Int Immunopharmacol 74:105646
58. LeBaron TW, Kura B, Kalocayova B, Tribulova N, Slezak J (2019) A new approach for the prevention and treatment of cardiovascular disorders. Molecular hydrogen significantly reduces the effects of oxidative stress. Molecules 24:2076
59. Shimouchi A, Nose K, Shirai M, Kondo T (2012) Estimation of molecular hydrogen consumption in the human whole body after the ingestion of hydrogen-rich water. Adv Exp Med Biol 737:245–250
60. Liu C, Kurokawa R, Fujino M, Hirano S, Sato B, Li XK (2014) Estimation of the hydrogen concentration in rat tissue using an airtight tube following the administration of hydrogen via various routes. Sci Rep 4:5485
61. Kawamura M, Imamura R, Kobayashi Y, Taniguchi A, Nakazawa S, Kato T, Namba-Hamano T, Abe T, Uemura M, Kobayashi H et al (2020) Oral administration of Si-based agent attenuates oxidative stress and ischemia-reperfusion injury in a rat model: a novel hydrogen administration method. Front Med 7:95
62. Martínez-Sánchez G, Schwartz A, Di Donna V (2020) Potential cytoprotective activity of ozone therapy in SARS-CoV-2/COVID-19. Antioxidants 9:389
63. Zhu Q, Wu Y, Li Y, Chen Z, Wang L, Xiong H, Dai E, Wu J, Fan B, Ping L et al (2018) Positive effects of hydrogen-water bathing in patients of psoriasis and parapsoriasis en plaques. Sci Rep 8:8051
64. Fiorese CJ, Schulz AM, Lin YF, Rosin N, Pellegrino MW, Haynes CM (2016) The transcription factor ATF5 mediates a mammalian mitochondrial UPR. Curr Biol 26:2037–2043
65. Wu Z, Senchuk MM, Dues DJ, Johnson BK, Cooper JF, Lew L, Machiela E, Schaar CE, DeJonge H, Blackwell TK et al (2018) Mitochondrial unfolded protein response transcription factor ATFS-1 promotes longevity in a long-lived mitochondrial mutant through activation of stress response pathways. BMC Biol 16:147

66. Lin YF, Haynes CM (2016) Metabolism and the UPR(mt). Mol Cell 61:677–682
67. Zhao YS, An JR, Yang S, Guan P, Yu FY, Li W, Li JR, Guo Y, Sun ZM, Ji ES (2019) Hydrogen and Oxygen mixture to improve cardiac dysfunction and myocardial pathological changes induced by intermittent hypoxia in rats. Oxid Med Cell Longev 2019:7415212
68. Wang YT, Lim Y, McCall MN, Huang KT, Haynes CM, Nehrke K, Brookes PS (2019) Cardio-protection by the mitochondrial unfolded protein response requires ATF. Am J Physiol Heart Circ Physiol 317:H472–H478
69. Berger E, Rath E, Yuan D, Waldschmitt N, Khaloian S, Allgauer M, Staszewski O, Lobner EM, Schottl T, Giesbertz P et al (2016) Mitochondrial function controls intestinal epithelial stemness and proliferation. Nat Commun 7:13171
70. Liu M-Y, Xie F, Zhang Y, Wang T-T, Ma S-N, Zhao P-X, Zhang X, LeBaron TW, Yan X-L, Ma X-M (2019) Molecular hydrogen suppresses glioblastoma growth via inducing the glioma stem-like cell differentiation. Stem Cell Res Ther 10:145
71. Fang W, Wang G, Tang L, Su H, Chen H, Liao W, Xu J (2018) Hydrogen gas inhalation protects against cutaneous ischaemia/reperfusion injury in a mouse model of pressure ulcer. J Cell Mol Med 22:4243–4252
72. Buchholz BM, Masutani K, Kawamura T, Peng X, Toyoda Y, Billiar TR, Bauer AJ, Nakao A (2011) Hydrogen-enriched preservation protects the isogeneic intestinal graft and amends recipient gastric function during transplantation. Transplantation 92:985–992
73. Deng L, Du C, Song P, Chen T, Rui S, Armstrong DG, Deng W (2021) The role of oxidative stress and antioxidants in diabetic wound healing. Oxidative Med Cell Longev 2021:8852759
74. Lin T-K, Zhong L, Santiago JL (2017) Anti-inflammatory and skin barrier repair effects of topical application of some plant oils. Int J Mol Sci 19:70
75. Litwiniuk M, Krejner A, Speyrer MS, Gauto AR, Grzela T (2016) Hyaluronic acid in inflammation and tissue regeneration. Wounds 28:78–88
76. Werner S, Grose R (2003) Regulation of wound healing by growth factors and cytokines. Physiol Rev 83:835–870
77. Dohi K, Kraemer BC, Erickson MA, McMillan PJ, Kovac A, Flachbartova Z, Hansen KM, Shah GN, Sheibani N, Salameh T, Banks WA (2014) Molecular hydrogen in drinking water protects against neurodegenerative changes induced by traumatic brain injury. PLoS ONE 9:e108034

Chapter 25
Perspective of Nanomaterials and Nanomedicine Procedures in Molecular Hydrogen Therapy

Štefan Luby

Abstract In parallel with the development of nanoscience, nanomedicine and its branches were established. Its remarkable development is evidenced by milestones, awarded several times with Nobel prizes. Milestones also include the renaissance of molecular hydrogen therapy from 2007, which uses H_2 as an effective antioxidant in preventive and therapeutic applications. In this chapter, in connection with the developed targeted delivery of drugs, we are dedicated to its modification in the form of local generation and release of hydrogen from nanocarriers such as nanoparticles, nanorods etc. The results are already reported in Alzheimer's disease and cancer treatment, wound healing, etc. Hydrogen therapy requires support and service which includes safe solid state storage of hydrogen to avoid potential explosions, where, in addition to hydrides such as MgH_2, functionalized nanomaterials from the carbon family—as graphene or carbon nanotubes are applied. Attention must also be paid to the leakage of hydrogen and its presence in the environment, where nanosensors based on metal oxides and especially graphene-based sensors are the promising solution. Considering the numerous clinical trials and studies, H_2 therapy can today be situated at the rise part of the Gartner's cycle heading for the production plateau.

Keywords Molecular hydrogen therapy · Nanomaterials · Nanomedicine · Graphene · Carbon · Sensors

Introduction

The term nanotechnology was used first time by Taniguchi in 1974 [1]. Advances in microelectronics and nanoelectronics are at the vanguard of the development of nanotechnology. However, the trend of miniaturization as a source of new discoveries and inventions in those areas postulated by Feynman in 1959 [2] penetrated quickly into medicine as well. It was reflected in the establishment of n*anomedicine*

Š. Luby (✉)
Institute of Physics, Slovak Academy of Sciences, Dubravska cesta 9, 84511 Bratislava, Slovakia
e-mail: stefan.luby@savba.sk

J. Slezak and B. Kura (eds.), *Molecular Hydrogen in Health and Disease*, Advances in Biochemistry in Health and Disease 27, https://doi.org/10.1007/978-3-031-47375-3_25

and the development of corresponding branches, materials and products with the prefix nano-, such as nanotoxicology, nanodentistry, nanopharmacy, nanosurgery, nanooncology, nanotherapy&nanodiagnostics (theranostics), nanoflake, nanocarrier, nanoplatform, nanocatalyst or nanosensor. The first branches of nanomedicine were dentistry, traditionally based on fine mechanics, and medical cosmetics, where physiological barriers were lower as elsewhere. Today nanomedicine quickly penetrates into newly created operational areas.

The size of 2-dimensional (2D), 1-dimensional (1D) and 0-dimensional (0D) nanostructures used in the inorganic world is postulated between 1 and 100 nm [3]. Due to the unique effects based on the nanostructures in medicine, the upper limit of at least one of their dimensions is increased to about 300 nm [4]. Milestones in the development of nanoscience have been identified by several works, e.g. [5, 6]. Based on them, we selected as nanomedicine milestones some important discoveries, inventions or breakthrough experiments, especially those that have been awarded Nobel Prizes (NoP):

- M. Knoll, E. Ruska, Invention of electron microscope (1933), NoP in physics for Ruska 1986.
- J. Watson, F. Crick, Double-helix structure of DNA (1953), NoP in physiology or medicine 1962.
- N. Seeman, Development of the concept of DNA nanotechnology (1982).
- G. Binnig, Ch. Gerber, C. F. Quate, Invention of atomic force microscope (1986).
- A. Geim, K. Novoselov, Groundbreaking experiments regarding 2D material graphene (2004), NoP in physics 2010.
- J. P. Sauvage, J. F. Stoddart, B. L. Feringa, Design and synthesis of molecular machines (80s and 90s), NoP in chemistry 2016.
- E. Charpentier, J. A. Doudna, Development of a method of genome editing (2009–2012), NoP in chemistry 2020.

These examples show that nanomedicine acts synergistically with advances in nanoscience and nanotechnology corresponding to physics, chemistry, measurement and imaging techniques. Graphene, as the most studied nanomaterial in the last decade (European Union flagship project 2013–2023) with excellent properties such as high electrical and thermal conductivity, large specific surface area, impermeability to gases, etc. [7] finds application in all areas of nanoscience, including nanomedicine [8], especially for tissue engineering.

Regarding the content of this book, we will mention two more milestones from the field of molecular hydrogen (H_2) therapy:

- M. Dole, R. F. Wilson, W. P. Fife, Treatment for cancer by hyperbaric hydrogen therapy [9].
- S. Ohta et al., Hydrogen as therapeutic antioxidant [10].

Molecular hydrogen has potential as an antioxidant in preventive and therapeutic applications. An example is given in Fig. 25.1. The idea found many followers and in the course of the last years great progress was achieved, summarized e.g. in [11, 12]. The method is also applied in sports [13] and agriculture [14]. After the paper

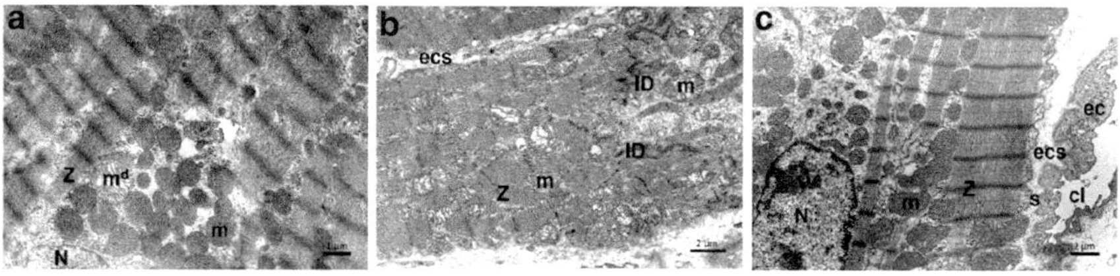

Fig. 25.1 Electron-microscopic examination of the left ventricle ultrastructure of the transplanted heart after the hydrogen application: **a** control myocardium, **b** transplanted myocardium—numerous mitochondrial changes, damage of capillary endothelium, etc. **c** improved ultrastructure after hydrogen treatment (Courtesy J. Slezák, Centre of Experimental Medicine SAS, Bratislava). In 2022, the European Academy for Molecular Hydrogen Research in Biomedicine was founded, which had its first conference "Hydrogen in Biomedicine" in Smolenice, Slovakia, October 2022. The pioneers of this therapy in our country are Slezák et al. [16]

[9] and a few more in 70s and 80s the turning point was the publication [10] by Ohta et al. Since then, a total of 1226 papers on hydrogen therapy were retrieved from the PubMed database until July 30, 2021 [15].

Next, we will focus on three topics. Today, the most frequent area of research in nanomedicine is targeted delivery of drug directly into the selected organ, which possibly overcomes also the suppressive/protective factors in the tissue. It is also referred to as the Trajan Horse approach [17]. In the following part we will show that an analogous approach has its advantages in hydrogen therapy, where it involves localized generation and release of hydrogen. Then we will deal with safe solid state storage of H_2 for therapeutic applications and finally with monitoring hydrogen in the environment using nanosensors, which is also an important task from the point of view of safety and prevention of potential explosions.

Therapy with Local Generation and Controlled Release of Hydrogen

Targeted drug delivery (TDD) was mentioned as early as 1913 by P. Ehrlich. He reasoned that if a compound could be made that will selectively target a diseased organ, then drug could be delivered along with that agent. His term for therapeutic agent was *magic bullet* (*Zauberkugel*). Today TDD research has reached the point where treatment can be tailored to the individual patient, thus fulfilling the concept of precision medicine. The release and absorption of the drugs, their protection against degradation, carriers (drug vectors), such as colloids, polymers, monoclonal antibodies, nanoparticles (NPs), graphene, etc. are studied. According to the type of carrier, we recognize physical, chemical and biological targeting [18, 19]. TTD include cancer therapy, brain and blood pressure treatment, DNA delivery, etc. The number of Scopus publications using the search strings nanoparticle & drug & delivery in title, abstract and keywords reached 10,800 by 2022.

H_2 has strong tissue penetration ability, therefore its delivery to certain locations and organs is not so topical as in the case of standard drugs. On the other hand, it is more difficult to maintain H_2 at the site of action. Therefore, it is timely to develop methods of local generation and controlled release of hydrogen in sufficient quantity and for a sufficiently long time in order to increase therapeutic efficiency. At the same time, we need to detect its presence in various places in the body. Several approaches such as isotope tagging or gas chromatography have been developed, but tissues have to be extracted. Let us mention at least activity-based sensing approach for in vivo and in vitro H_2 detection [20]: Pd NPs encapsulated in mesoporous silica capture and catalyze H_2 into active atoms for hydrogenation of NDI/N_3 into NDI/NH_2, enabling fluorescence emission around 540 nm. An in vivo experiment with the mouse brain showed that after 5 min of inhalation hydrogen molecules crossed the blood–brain barrier. Now, we will present several examples of local generation and release of hydrogen that move hydrogen therapy to the position of specialized nanomedicine. They are too summarized in Table 25.1.

- In vivo mice experiment documented the diabetic wound healing by sustainable anti-inflamatory H_2 release and local glucose depletion. Hydrogen incorporated TiO_2 nanorods as visible light-sensitive photocatalyst ensured H_2 generation [21].

Table 25.1 Basic characteristics of the mentioned examples of local generation and release of hydrogen

Aim	Role of H_2	Carrier	H_2 generation	Reference
Diabetic wound healing	Anti-inflammatory agent	H_2 in TiO_2 nanorods	Visible induced irradiation	[21]
Treatment of Alzheimer's disease	Antioxidant, scavenger of ROS	PdH nanoparticles	Catalytic hydrogenation effect of Pd	[22]
Treatment of breast cancer	Decrease of intracellular ROS levels	$PdH_{0.2}$ nanocrystals	NIR induced H_2 release and heating	[23]
Cervical and breast cancer treatment	Hydrogen/hole therapy destroys cellular ADS	$SnS_{1.68}WO_{2.41}$ nanocatalyst	NIR irradiation	[24]
Overcoming of cancer drug resistance	Eradication of toxic side-effects of therapy	Porphyrin-Fe MOF nanocrystals	Acid-responsive degradation	[25]
Radioprotective effect on male fertility	Free radical scavenger	MgH_2 nanoparticles	Release from carriers in aqueous phase	[26]
Colon carcinoma treatment	ROS scavenging and expression of P53 protein	Mg wires or sheets	H_2 release by degradation of Mg in situ	[27]

ADS antioxidant defence system, *MOF* metal–organic framework

Effective healing of bacterial infection by PdH-nanohydride released hydrogen reported Yu et al. [28].

- Hydrogen was used to treat Alzheimer's disease (AD) [22]. By virtue of the catalytic hydrogenation of Pd hydrogen was relased in AD brain from PdH NPs.
- In vivo hydrogenothermal treatment of mice exhibited a synergistic effect against breast cancer [23]. Owing to the targeted delivery of $PdH_{0.2}$ nanpocrystals that release hydrogen under near-infrared (NIR) irradiation together with the irradiation caused increase of local temperature tumor inhibition was observed.
- Concept of toxic drug-free therapy of cervical and breast cancer was elaborated by Zhan [24]. NIR activated nanocatalyst $SnS_{1.68}WO_{2.41}$ producing oxidative holes and hydrogen molecules have synergetically induced tumor cell apoptosis.
- Hydrogen may help to overcome multidrug resistance by codelivery of drug (doxorubicin hydrochloride) loaded in pore channels of carrier generating H_2 by acid-response degradation in situ [25]. Efficiency of combined nanomedicine was studied by tumor-bearing mouse models.
- Radiation induced damage to the male fertility was alleviated by suppressing oxidative stress via H_2 free radical scavenging [26]. Improvement of sperm motility and the maintenance of spermatogenesis after exposure to ionizing radiation was confirmed.
- The apoptosis of colon carcinoma cells was induced by in situ released H_2 produced by Mg degradation in aqueous environment [27]. Reactive oxygen species (ROS) scavenging and expression of tumor suppressor protein P53 was studied by inserting Mg wire into the subcutaneous tumor in a mouse.

Other examples of therapy with sustained delivery of hydrogen, applied nanocarriers and triggering of H_2 release are given by Zhou et al. [29].

Safe Solid-State Storage of Hydrogen

Hydrogen is considered as an energy carrier and the fuel of the future. It is noteworthy that in the work that deals with the future applications of hydrogen, its pharmaceutical applications are mentioned, including molecular hydrogen therapy [30]. Common forms of hydrogen storage for the above purposes are (i) cryogenic liquid, (ii) pressurized gas and (iii) solid sources such as metal hydrides or carbon materials. The advantage of the last method is the safety [31]. History remembers great disasters caused by the explosion of hydrogen, such as the destruction of the airship Hindenburg, the space shuttle Challenger or nuclear power plant Fukushima. In the case of widely decentralized hydrogen therapy, often even in the home environment, safety is paramount. In the following, we will focus on solid-state storage and release of hydrogen.

Almost all elements form binary hydrides with either ionic, covalent or metallic bonding. Storage capacity (SC) is measured in wt% of hydrogen or its volume density kg/m^3 in the respective compound. Here we will start from the goals of 6.5 wt% and

62 kg/m^3 recommended by the US Department of Energy (DOE) [31]. It is natural that light metals such as Li, Be, Na, Mg, Al are prefered for storage. Palladium is not mentioned because it belongs to heavier metals. However, it absorbs up to 900 times its own volume of H_2 at room temperature (RT), the limit of SC at normal pressure is $PdH_{0.7}$. Pd is therefore advantageous in microscopic applications, e.g. as Pd nanocarriers in targeted delivery of H_2 or in hydrogen sensors. Next, we present examples of hydrides with storage capacity SC > 6.5 wt% [31]:

- MgH_2, MgH_2:1 at% Al and MgH_2:5 at% Ge provide the capacity of 7.00, 7.30 and 7.60 wt%, respectively. However, desorption temperatures (T_{des}) are 350, 340 and 50–150 °C, respectively. Absorption temperatures (T_{abs}) are about 200–300 °C. A similar picture can be found in the case of Mg-Ni based hydrides.
- In the group of sodium alanates, both T_{ads} and T_{des} are in the range of 100–200 °C, but SC drops below 5 wt%. An example is $NaAlH_4$ with both temperatures between 80 and 180 °C and SC = 5 wt %.
- The properties of RT hydrides at acceptable pressure conditions are discussed in [32] based on the data in [31]. Adsorption and desorption temperatures around (1–2) RT can be reached in the group of intermetallic compounds, but SC drops to 1–3 wt%. Examples are compounds FeTi, $LaNi_5$, $TiCr_{10}Mn_{15}V_{32}$. Of course, the weight of the storage media is higher than in the previous groups.

We can conclude that further work needs to be carried out in optimizing these materials and their storage properties. Nevertheless, in the case of H_2 therapy, the demands for SC are not as high as in energy sector, so safe storage hydride-based media could already be used here today. Apart from that, we will also mention intensively studied storage media based on a groundbreaking material from the carbon family—graphene.

Carbon based systems are promising for high gravimetric SC due to carbon small atomic mass. Therefore, attention was paid to graphene (G), carbon nanotubes, graphene oxide, etc. Proper calculations with the contribution of quantum effects showed that the DOE specifications can be approached in graphene nanostructures [33]. However, H_2 binding energy in pure carbon nanostructures is only about 0.10 eV/H_2 and hydrogen desorbs well below the RT. Therefore, carbon-based nanomaterials are functionalized with metals and metal hydrides [34]. Many first-principal calculations were performed on this topic, of which 5 cases are summarized in Table 25.2. The binding energy around 0.26 eV/H_2 shows that desorption would take place at ambient conditions.

Efficiency of hydrogen storage can be improved also by creating defects in graphene, e.g. by irradiation with Ar ions (100 eV). The hydrogen uptake was thus increased up to 12.8 wt% [39]. Finally, it should be mentioned that the use of graphene power source has already been demonstrated in the successful operation of hydrogen fuel cell [40].

Table 25.2 Functionalized graphene-based hydrogen storage media

Functionalization of graphene	Coverage	H_2 binding energy	Graphene SC	References
		eV/H_2	wt%	
Li in double vacancy	On both sides	0.26	7.26	[34]
Al—adsorbed G	On both sides	0.19	13.79	[35]
Ca—adsorbed G	On zigzag edges	0.20	~ 5	[36]
Ta embedded in double vacancy	On both sides	0.31	6.3	[37]
Zr attached on G	On both sides	0.34*	11	[38]

* T_{des} is proportional to the magnetic moment of the system

Monitoring Hydrogen in the Environment Using Nanosensors

Hydrogen is a colorless, odorless, tasteless and non-toxic gas that cannot be detected using human senses. It has a low explosion limit ~ 4% and a wide explosion range (4–75%) in air. It may also lead to hypoxic asphyxia with the accumulation of H_2 in air [41]. It is therefore obvious that in the environment in which molecular hydrogen therapy takes place, the presence of hydrogen either due to leakage or improper handling of hydrogen sources must be consistently monitored. The relevant range of H_2 concentration in air should be from 10 to 100 ppm up to 4–5%, i.e. up to three orders of magnitude. Due to the extended therapy in various environments, often in a self-service conditions, laboratory equipment is not feasible, but personal portable hydrogen sensors are a suitable choice. The requirements are (i) low price, (ii) low power consumption, supply from battery and, therefore, near RT operation, important also as explosion prevention [42], (iii) monitoring interval indicated above, (iv) selectivity to hydrogen and (v) the flexibility of the sensor [43], which can be attached to the clothes. Solution is offered by nanosensors based on nanostructures, e.g. 0D nanoparticles, nanocrystals, 1D nanorods, nanotubes, nanowires, nanobelts, nanoribbons and 2D nanoflakes, nanosheets.

Nanostructured materials applied in gas sensors include metal oxides, carbon family materials, chalcogenides, metal–organic frameworks, perovskites, etc. [44]. The principle of different devices corresponds to the output quantity such as resistance, thermal conductivity, acoustic, mechanic, optical or electrochemical properties, work function, etc. [45]. Due to the limited scope of this work, we will focus on the semiconductor metal-oxide sensors that have been most frequently used so far and on the graphene sensors, which have been studied intensely during the last two decades, both of them with electrical registration. The first of them were invented by Seiyama et al. in 1962 [46], the second were created as a part of graphene research after 2004 (cf. the milestone in the introduction). (The sensors on carbon nanotubes invented by Iijima [47] in 1991 also belong to the carbon family.)

Metal Oxides-Based Hydrogen Sensors

In the category of oxides N-type γ-Fe_2O_3, SnO_2, TiO_2, WO_3, ZnO, MoO_3 Nb_2O_5 and P-type CuO, NiO are used. Sensors are produced by many companies, the leading countries are China, USA, Germany, India Japan, UK, Korea, Italy, France, Russia. Metal-oxides have a large energy gap of 2–3.5 eV and therefore the working temperature of sensors is usually up to 500 °C, only exceptionally around RT. High temperature is necessary for the fast surface reaction of the semiconductor surface with oxidizing gases such as CO_2, NO_2, O_2, which capture electrons from the semiconductor, or reducing gases such as CH_4, CO, H_2, which, on the contrary, release electrons. As a result, the electrical conduction of the oxide changes. Here we have presented only the simplest detection mechanism, which in practice complicates due to the structure of the material, the interaction between the present gases, etc.

From the overview of H_2 monitoring using metal oxide nanosensors [41] it follows that from 45 listed devices with N and P type oxides and 0D, 1D and 2D nanostructures, only 10 worked around RT. Here belong the sensors based on MoO_3 nanoribbons with a measurement range of 0.5–1000 ppm H_2 in air or devices with a special, e.g. porous structure. Appropriate, at least in part of the required detection interval, was Nb_2O_5 nanorods sensor that measured H_2 at RT in the interval 1000–6000 ppm. The highest measured concentration of hydrogen in air was 1% with a MoO_3 nanoflake sensor at 200 °C. The situation has not fundamentally changed even after decoration of metal oxides with noble metals such as Pd, Pt or non-noble metals such as Co, Ni.

In view of the aforementioned findings, we continued to concentrate on sensors from 1D nanostructures. Arafat et al. summarized 46 of them based on ZnO, SnO_2 and TiO_2 oxides [48]. Unfortunately, even among these sensors, no one was found that would monitor wide range of concentration at RT. Higher concentrations of H_2 (in one case of O_2) in the region of several % are recorded by devices with large interface area of the sensing material on the substrate and, correspondingly, large volume of detection material with resulting high adsorption capacity of gas. Nevertheless, at the highest measured concentrations of hydrogen in the air, saturation of adsorption and also of the output signal vs concentration manifests itself. In the next we present some examples of these sensors with characteristics: *the measured range, area on the substrate, nanostructure, working temperature, reference*:

- 100 ppm—4% H_2 in air, 20 mm^2, TiO_2 nanotubes, 180–400 °C [49].
- 0.5%–2% H_2 in air, 120 mm^2, TiO_2–SnO_2 nanofibers, 300–600 °C [50].
- 200 ppm—20% O_2 in nitrogen, 30 mm^2, TiO_2 nanotubes, 50–100 °C [51].

Let's add that 1D nanostructures form interwoven clusters are more stable than nanoparticle arrays bound only by weak van der Waals force.

On the contrary, RT sensors used small quantities of sensing material, in extreme cases they contain only single nanorod or nanowire. We believe that here the exchange of electrons between the nanostructure and the gas is facilitated by the strongly curved surface, edges or ends of the structures. But the measured concentration ranges are smaller. Also, here we give a couple of examples including the dimensions of single nanostructure:

- 1–1000 ppm H_2 in air, 1 μm × 10 μm, ZnO nanorod, RT [52].
- 1–1000 ppm H_2 in air, (10–30) nm × (50–250) nm, ZnO nanowire, RT [53].
- 100–1000 ppm H_2 in air, (20–80) nm × (100–800) nm, TiO_2 nanowire, RT [53].

Obviously, if we use a combination of sensors from metal oxides, we can cover the entire required measurement range and at the same time get closer to the RT, which is a matter of the sensing material, while the range is also a matter of the construction of the device.

Graphene-Based Hydrogen Sensors

Breakthrough experiments with graphene by Geim, Novoselov et al. carried out and published starting in 2004 (e.g. [54]) showed that this semimetal is among other applications suitable for gas and vapor sensors due to its combination of properties. Graphene has a large specific surface area—theoretically 2630 m^2/g, low noise, high thermal conductivity, strength and flexibility. Thanks to high conductivity and ballistic transport, graphene chemiresistors, unlike metal oxides-based devices, work around room temperature. On the other hand, graphene has not dangling bonds on the surface which are necessary for the adsorption of gases and vapors. Therefore, it is often functionalized by Pd, Pt or other materials [55–57]. Graphene is also a platform to carry other components, and thanks to the planar structure, graphene sensors can be integrated into Si electronic circuits [58]. Extreme sensitivity of graphene micrometer-size sensor to detect single gas molecules reported Schedin et al. [59].

Graphene is prepared by exfoliation from graphite, epitaxial growth, vapor deposition, and chemical reduction. The consequence is a great variability of graphene gas/hydrogen sensors. It follows from review papers [55, 60] that among the included 32 graphene and 17 graphene oxide (GO) H_2 sensors, 41 of them are operated at RT. For others, the temperatures are 50–200 °C. The reason is the increase of both the output signal and reaction rates, i.e. the reduction of response and recovery times, which in some cases can take ten minutes or more. Nevertheless, it can be concluded that from the point of view of safe operation at RT, graphene and GO hydrogen sensors are a satisfactory solution. Considering the measured concentration range, several sensors presented in [60] cover the interval of three orders of magnitude but starting from 1 ppm. A suitable sensor provides a range from 25 ppm to 1% H_2 at RT [61]. The graphene was decorated with Pd and the sensor area was quite large—4 × 3 mm^2, which is consistent with our conclusions in the previous section.

Similar sensors were prepared in our laboratory [56]. Graphene nanoflakes 5 nm thick with lateral dimensions around 300 nm were prepared by liquid exfoliation from expanded graphite. The flakes were deposited onto the substrate from oxidized Si by Langmuir- Schaefer technique. The deposit was decorated by colloid Pd NPs using one or more spin-coating cycles. After heat treatment the effective sensing area 20

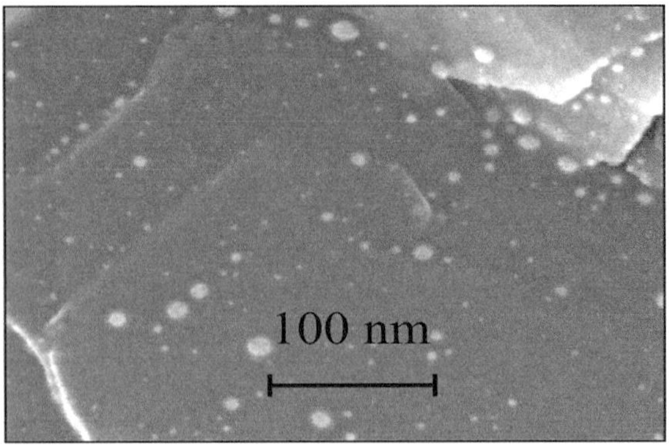

Fig. 25.2 SEM picture of graphene decorated by Pd nanoparticles

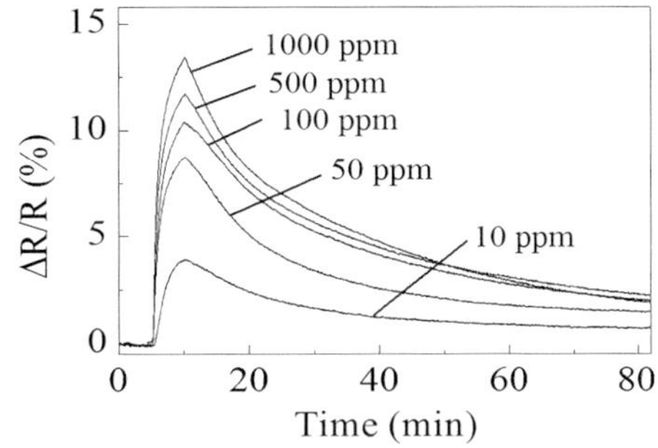

Fig. 25.3 Time response of Pd-decorated graphene sensor at 10–1000 ppm of H_2 in air

mm^2 was defined by silver paste contacts. The scanning electron microscope (SEM) examination confirmed quite uniform distribution of NPs over the nanofilm surface with preferential location at the edges of flakes or holes which are the chemically "hot" sites with dangling bonds (Fig. 25.2). The response of the sensor within the range of 10–1000 ppm of H_2 in air is given in Fig. 25.3. Nevertheless, the sensor responded also to higher concentration around 1% of H_2, although with the saturation of the output signal.

Cross Sensitivity

Crosssensitivity is a general problem of solid state sensors including metal oxide and graphene-based devices. They often indicate insufficient selectivity, because the reactions of the sensors are analogous for entire groups, e.g. oxidizing or reducing gases. The outcome lies in the further research of multicomponent or hybrid nanostructures. The selectivity can also be modified by the right choice of decorating metals. An example is the decoration of graphene with palladium [62], which increased the sensitivity to hydrogen 2.5 times and at the same time suppressed the sensitivity to NO_2 more than 15 times.

History and Perspectives

A suitable entry to the topic of H_2 therapy might be the paper [63]. It points to the repeating periods of growth and decline of interest in hydrogen medicine applications. If we skip the first reports from antiquity, these fluctuations date back to 1793, when T. Beddoes drew attention to the importance of hydrogen in the treatment of pulmonary tuberculosis. In 1888, hydrogen was used to locate penetrating wounds in the gastrointestinal tract by rectal insufflation. In 1941 H_2 was used in diving to prevent decompression sickness. During the twentieth century, ERW (electrolyzed reduced water) was studied mainly in Japan (it is obtained during the production of hydrogen by electrolysis at the cathode and it contains molecular hydrogen). Many studies confirmed its antioxidant and anti-inflammatory effects [64]. In 1965 ERW was officially approved in Japan. We may continue with the two milestones mentioned in the Introduction—hyperbaric hydrogen therapy for cancer in 1975 and the invention by Ohta et al. [10] who showed that hydrogen acts as a therapeutic antioxidant. Since then, the medical use of hydrogen has become established in research laboratories around the world, the number of publications is constantly growing, but this method is not yet firmly established in today's medicine.

A comprehensive bibliometric analysis of medical research on molecular hydrogen was published in 2023 [15]. It analyzes 1226 publications retrieved from PubMed database until July 30, 2021 (in [63] it is stated that 2000 articles have been published on this topic in the last 12 years). According to [15], publications are growing linearly from 10 to 125 per year since 2007 until today. The main keywords by which the works were searched are *molecular hydrogen, hydrogen-rich water, oxidative stress, hydrogen gas, and inflammation.* Later, in the years 2019–2021, the keywords *gut microbiota, pyroptosis and COVID-19* were in the first three places. The most actively publishing journals are *Antioxidants* MDPI, *Medical Gas Research, Scientific Reports, Shock, Journal of Surgical Research and Current Pharmaceutical Design.* Top 5 authors who published the most articles were *X.-J. Sun, K.-L. Xie, Y.-H. Yu, S. Ohta and H.-G. Chen.* Among the 10 first authors we find 7 Chinese and three Japanese names. According to the SCOPUS database, they have an affiliation either in China or in Japan (Shanghai, Tianjin, Tokyo) with the exception of John H. Zhang from Loma Linda Univ. USA. Of course, co-authors from other western laboratories participate in the work. However, principal investigators who created the main scientific schools and publication clusters are often engaged also in other research topics. The first three most cited papers at the time of analysis were [10], 1233 citations, [65], 203 citations and [66], 172 citations.

When considering the reason for the focus of this research in East Asia, the influence of traditional Chinese and Japanese medicine might be suggested. Chinese medicine recognizes water (shui) as one among five basic "elements". From there, short cut leads to ancient "curative waters", to ERW and today's hydrogen-rich water. In this context it is symptomatic that in 2015 Tu Youyou from China Academy of Traditional Chinese Medicine won the first Chinese Nobel Prize in physiology and medicine for the discovery of a new malaria therapy.

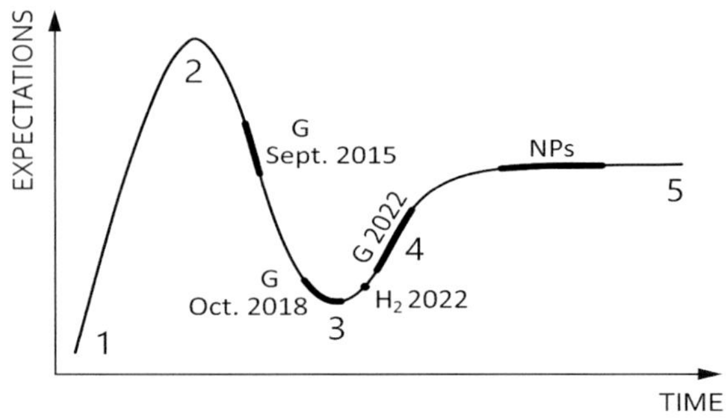

Fig. 25.4 Gartner cycle of graphene with the superimposed position of molecular H_2 therapy

Considering the methods of hydrogen administration, among which inhalation, bathing and drinking of H_2-rich water, hyperbaric chamber or rectal insufflation of the gas dominated until now, hydrogen therapy resembles a spa rather than a clinical treatment. In work [64], however, other methods of administration are already mentioned, such as H_2 rich dialysis solution for hemodialysis or H_2 production by intestinal bacteria. This suggests that hydrogen therapy is knocking on the doors of clinics. Zhou et al. [29] mentioned more than 50 clinical trials in many oxidative stress-/inflammation-related diseases which have been confirmed. Li et al. [15] cite pilot studies related to liver, pulmonary or Parkinson's diseases, muscular damage, etc. Hydrogen treatment has also been used in clinical trials for COVID-19 in certain regions with positive effects. Japanese government approved H_2 inhalation as "advanced medicine" [63] and virtual *Molecular Hydrogen Institute* started in 2013. Some soluble precursors for the preparation of hydrogen water were approved by national food and drug administrations.

The well-known Gartner's cycles illustrate the time evolution of any technology from the start (1) through the peak of expectations (2) and through the valley of rationalization (3) to the rising curve (4) and the production plateau (5). In Fig. 25.4 we show our gradually updated cycle of graphene to which, based on the stated level of knowledge and clinical ambitions, we drew the position of H_2 therapy. It is on the rising curve just below present graphene position. This is also in line with the fact that the breaking papers on graphene and H_2 therapy appeared close to each other in 2004 and 2007, respectively.

An important role of hydrogen therapy now lies in the detailed verification of clinical effects with the further research that will advance it into the field of nanomedicine.

Conclusion

The advantages and promising perspective of hydrogen therapy result from the low price of the input materials and the technologies used so far. Due to the molecular size of the new medical gas—hydrogen of 0.12 nm, H_2 therapy tends towards the field of nano- or even "picomedicine". The enormous hype, which was fully manifested in such areas as TDD [67] avoided hydrogen therapy so far. However, it doesn't have sufficient support yet in the development and production of medical devices. That is why we have included sections about the safe storage of hydrogen and monitoring its presence in the environment. Possible compromise by hydrogen explosions, which would harm the new therapy, should be avoided. The aim of the section on local generation and controlled release of hydrogen is to point out the expected shift of hydrogen therapy to the most sophisticated areas of medicine, which will enhance its influence and possibilities.

Competing Interest Scientific Grant Agency VEGA, Bratislava, grant 2/0142/23 is gratefully acknowledged. Author declares no conflict of interests.

References

1. Taniguchi N, Arakawa C, Kobayashi T (1974) On the basic concept of nano-technology. In: Proceedings of the international conference on production engineering, Japan Society of Precision Engineering, Tokyo, 26–29 Aug 1974
2. Feynman R (1960) There's plenty of room at the bottom. Eng Sci 23:22–36
3. Boholm M, Arvidsson R (2016) A definition framework for the terms nanomaterial and nanoparticle. NanoEthics 10:25–40
4. Etheridge ML, Campbell SA, Erdman AG et al (2013) The big picture of nanomedicine: the state of investigational and approved nanomedicine product. Nanomedicine 9(1):1–14
5. Bayda S, Muhammad A, Tuccinardi T et al (2020) The history of nanoscience and nanotechnology: from chemical-physical applications to nanomedicine. Molecules 25:112(1–15)
6. Luby S (2022) Milestones in the development of nanoscience and nanotechnology (1850–2022). In: Feiler M, Ziman M, Kovacova S et al (eds) (2022) Proceedings of ADEPT international conference, Slovak Physical Society, 20–24 June 2022
7. Mertens R (2023) The graphene handbook. MetalGrass, Kfar Saba, Israel
8. Kumar S, Chatterjee K (2016) Comprehensive review on the use of graphene-based substrates for regenerative medicine and biomedical devices. ACS Appl Mater Interfaces 8(40):26431–26457
9. Dole M, Wilson FR, Fife WP (1975) Hyperbaric hydrogen therapy: a possible treatment for cancer. Science 190(4210):152–154
10. Ohsawa I, Ishikawa M, Takahashi K et al (2007) Hydrogen acts as a therapeutic antioxidant by selectively reducing cytotoxic radicals. Nat Med 13:688–694
11. Ge L, Yang M, Yang NN et al (2017) Molecular hydrogen: a preventive and therapeutic medical gas for various diseases. Oncotarget 8(60):102653–102673
12. Noor MMZM, Alauddin AS, Wong YH et al (2023) A systematic review of molecular hydrogen therapy in cancer management. Asian Pac J Cancer Prev 24(1):37–47
13. Kawamura T, Higashida K, Muraoka I (2020) Application of molecular hydrogen as a novel antioxidant in sports science. Oxid Med Cell Longevity 2020:2328768
14. Wang YQ, Liu YH, Wang S (2020) Hydrogen agronomy: research progress and prospects. J Zhejiang Univ: Sci B 21(11):841–855
15. Li H, Ma HY, Hua WL et al (2023) Trend of research on the medical use of molecular hydrogen: a bibliometric analysis. Med Gas Res 13(4):212–218
16. Barancik M, Kura B, LeBaron TW et al (2020) Molecular and cellular mechanisms associated with effects of molecular hydrogen in cardiovascular and central nervous systems. Antioxidants 9:1281

17. Nelson D, Fisher S, Robinson B (2014) The "Trajan horse" approach to tumor immunotherapy: targeting the tumor microenvironment. J Immun Res 2014:789069
18. Manzari MT, Shamay Y, Kiguchi H et al (2021) Targeted drug delivery strategies for precision medicine. Nat Rev Mater 6:351–370
19. Tewabe A, Abate A, Tamrie M et al (2021) Targeted drug delivery—from magic bullet to nanomedicine: principles, challenges, and future perspectives. J Multidiscip Healthc 14:1711–1724
20. Gong W, Jiang L, Zhu Y et al (2022) An activity-based ratiometric fluorescent probe for in vivo real-time imaging of hydrogen molecules. Angew Chem 134:202114594
21. Chen S, Zhu Y, Xu Q et al (2022) Photocatalytic glucose depletion and hydrogen generation for diabetic wound healing. Nat Comm 13:5684
22. Zhang L, Zhao P, Yue C et al (2019) Sustained release of bioactive hydrogen by Pd hydride nanoparticles overcomes Alzheimer's disease. Biomaterials 197:39–404
23. Zhao P, Jin Z, Chen Q et al (2018) Local generation of hydrogen for enhanced photothermal therapy. Nature Comm 9:4241
24. Zhao B, Wang Y, Yao X et al (2021) Photocatalysis-mediated drug-free sustainable cancer therapy using nanocatalyst. Nature Comm 12:1345
25. Yao X, Chen D, Zhao B et al (2022) Acid-degradable hydrogen-generating metal-organic framework for overcoming cancer resistance/metastasis and off-target side effects. Adv Sci 9(10):2101965
26. Ma J, Dong S, Lu H et al (2022) The hydrogen storage nanomaterial MgH_2 improves irradiation-induced male fertility impairment by suppressing oxidative stress. Biomater Res 26:20
27. Yan R, Wang H, Cai W et al (2022) Controlled release of hydrogen by implantation of magnesium induces P53-mediated tumor cells apoptosis. Bioactive Mater 9:385–396
28. Yu S, Li G, Zhao P et al (2019) NIR-laser-controlled releasing PdH nanohydride for synergic hydrogen-photothermal antibacterial and wound-healing therapies. Adv Funct Mater 29:1905697
29. Zhou G, Goshi E, He Q (2019) Micro/nanomaterials-augmented hydrogen therapy. Adv Healthc Mater 8(16):1900463
30. Okolie JA, Patra BR, Mukherjee A et al (2021) Futuristic applications of hydrogen in energy, biorefining, aerospace, pharmaceuticals and metallurgy. Int J of Hydrogen Energy 46:888–8905
31. Sakintuna B, Lamari-Darkrim F, Hirscher M (2007) Metal hydride materials for solid hydrogen storage: a review. Int J Hydrogen Energy 32:112–1140
32. Modi P, Aguey-Zinsou KF (2021) Room temperature metal hydrides for stationary and heat storage applications: a review. Front Energy Res 9:616115
33. Patchovskii S, Tse JS, Yurchenko SN et al (2005) Graphene nanostructures as tunable storage media for molecular hydrogen. Proc NAS 102:10439–10444
34. Seenithurai S, Pandyan RK, Kumar SV et al (2014) Li-decorated double vacancy graphene for hydrogen storage applications: a first principle study. Int J Hydrogen Energy 39:11016–11026
35. Ao ZM, Peeters FM (2010) High-capacity hydrogen storage in Al-adsorbed graphene. Phys Rev B 81:205406
36. Lee H, Ihm J, Cohen ML et al (2010) Calcium-decorated graphene-based nanostructures for hydrogen storage. Nano Lett 10(3):793–798
37. Chu S, Hu L, Hu X et al (2011) Titanium-embedded graphene as high-capacity hydrogen-storage media. Int J Hydrogen Energy 36:1232–1238
38. Yadav A, Chalkraborty B, Gangan A et al (2017) Magnetic moment controlling desorption temperature in hydrogen storage: a case of zirconium-doped graphene as a high capacity storage medium. J Phys Chem C 121(31):16721–16730
39. Sunnardianto GK, Bokas G, Hussein A et al (2021) Efficient hydrogen storage in defective graphene and its mechanical stability: a combined density functional theory and molecular dynamics simulation study. Int J Hydrogen Energy 46:5485–5494
40. Morse JR, Zugell DA, Matis BR et al (2020) Macroscale evaluation and testing of chemically hydrogenated graphene for hydrogen storage applications. Int J Hydrogen Energy 45(3):2135–2144

41. Shi Y, Xu H, Liu T et al (2021) Advanced development of metal oxide nanomaterials for H_2 gas sensing applications. Mater Adv 2:1530–1569
42. Zhang XY, Ma RH, Li LS et al (2021) A room-temperature ultrasonic hydrogen sensor based on a sensitive layer of reduced graphene oxide. Sci Rep 11:2404
43. Chung MG, Kim DH, Seo DK et al (2012) Flexible hydrogen sensor using graphene with palladium nanoparticle decoration. Sens Actuators B 169:387–392
44. Sonker KS, Singh K, Sonkawade R (eds) (2022) Smart nanostructure materials and sensor technology. Springer Nature, Singapore
45. Pour GB, Aval LF, Sarvi MN et al (2019) Hydrogen sensors: palladium-based electrode. J Mater Sci 30:8145–8153
46. Seiyama T, Kato A, Fujiishi K et al (1962) A new detector for gaseous components using semiconductive thin films. Anal Chem 34:1502–1503
47. Iijima S (1991) Helical microtubules of graphitic carbon. Nature 353:56–58
48. Arafat MM, Dinan B, Akbar AA et al (2012) Gas sensors based on one dimensional nanostructured metal-oxides: a review. Sensors 12:7207–7258
49. Varghese OK, Gong D, Paulose M et al (2003) Hydrogen sensing using titania nanotubes. Sens Actuators B 93:338–344
50. Carney CM, Yoo S, Akbar SA (2005) TiO2–SnO2 nanostructures and their H_2 sensing behavior. Sens Actuators B 108:29–33
51. Lu HF, Li F, Liu G et al (2008) Amorphous TiO2 nanotube arrays for low-temperature oxygen sensors. Nanotechnology 19:405504
52. Lupan O, Chai G, Chow L (2008) Novel hydrogen gas sensor based on single ZnO nanorod. Microelect Eng 85:2220–2225
53. Rout CS, Kulkarni GU, Rao CNR (2007) Room temperature hydrogen and hydrocarbon sensors based on single nanowires of metal oxides. J Phys D: Appl Phys 40:2777–2782
54. Novoselov KS, Geim AK, Morozov SV et al (2004) Electric field-effect in atomically thin carbon film. Science 306(5696):666–669
55. Wang T, Huang D, Yang Z et al (2016) A review on graphene-based gas/vapor sensors with unique properties and potential applications. Nano-Micro Lett 8(2):9–119
56. Kostiuk D, Luby S, Siffalovic P et al (2019) Graphene Langmuir-Schaefer films decorated by Pd nanoparticles for NO_2 and H_2 gas sensors. Meas Sci Rev 19(2):64–69
57. Yang S, Jiang C, Wei SH (2017) Gas sensing in 2D materials. Appl Phys Rev 4:021304
58. Zanjani SMM, Holt M, Sadeghi MM et al (2017) 3D integrated monolayer graphene-Si CMOS RF gas sensor platform. npj 2D Mater Appl 1:36
59. Schedin F, Geim AK, Morozov SV et al (2007) Detection of individual gas molecules adsorbed on graphene. Nat Mat 6:653–655
60. Ilnicka A, Lukaszewicz JP (2020) Graphene-based hydrogen gas sensors: a review. Processes 8(5):633
61. Wu W, Liu Z, Jauregui LA et al (2010) Wafer/scale synthesis of graphene by chemical vapor deposition and its application in hydrogen sensing. Sens Actuators B 150:296–300
62. Lange U, Hirsch VM, Mirsky VM et al (2011) Hydrogen sensor based on graphene: palladium nanocomposite. Acta 56(10):3707–3712
63. LeBaron TW, Ohno K, Hancock JT (2023) The on/off history of hydrogen in medicine: will the interest persist this time around? Oxygen 3:143–162
64. LeBaron TW, Sharpe R, Ohno K (2022) Electrolyzed reduced water: review I. Molecular hydrogen is the exclusive agent responsible for the therapeutic effects. Int J Mol Sci 23:14750
65. Huang CS, Kawamura T, Toyoda Y et al (2010) Recent advances in hydrogen research as a therapeutic medical gas. Free Radical Res 44(9):971–982
66. Cai J, Kang Z, Liu WW et al (2008) Hydrogen therapy reduces apoptosis in neonatal hypoxia–ischemia rat model. Neurosci Lett 441:16–172
67. Vincent BB, Loeve S (2014) Metaphors in nanomedicine: the case of targeted drug delivery. NanoEthics 8:1–17

Chapter 26
The Emergence, Development, and Future Mission of Hydrogen Medicine and Biology

Shigeo Ohta

Abstract Molecular hydrogen (H_2) has emerged as a promising therapeutic and preventive medical gas. In 2007, we overturned the conventional concept that H_2 had no effect on mammalian cells, leading to the emergence of the field of "hydrogen medicine and biology". H_2 has numerous advantages, including high efficacy, lack of adverse effects, and multiple functions, such as anti-oxidation, anti-inflammation, anti-cell death, and energy metabolism stimulation. Moreover, H_2 is beneficial from healthy individuals to severe chronic and acute diseases. The target molecule of H_2 was identified as the oxidized form of porphyrins, subsequent hydride form acts as a catalyst to stimulate the selective reaction of H_2 with hydroxyl radical. Various species of porphyrins are widely and abundantly distributed throughout the body. Thus, repairing the oxidized porphyrins by H_2 may result in multiple benefits across various cells. H_2 suppresses the free radical chain reaction and modifies signaling mediators involved in lipid peroxides. H_2 indirectly regulates hormones and cytokines through various signal transduction pathways. H_2 has the potential to address a wide range of issues, including cardiac arrest, Alzheimer's-type dementia, metabolic syndrome, advanced-stage of cancer, inflammatory cytokine storms, healthcare, beauty, and agriculture. The mission of H_2 medicine is to overcome these unsolved ailments.

Keywords Anti-oxidation · Lipid peroxide · Molecular hydrogen · Multiple functions · No adverse effect · Porphyrin

---

S. Ohta (✉)
Department of Neurology Medicine, Juntendo University Graduate School of Medicine, 2-1-1, Hongo, Bunkyo-ku, Tokyo 113-8421, Japan
e-mail: ohta@nms.ac.jp

Institure for Advanced Medical Sciences, Nippon Medical University, 1-25-16 Nezu, Bunkyo-ku, Tokyo 113-8602, Japan

J. Slezak and B. Kura (eds.), *Molecular Hydrogen in Health and Disease*, Advances in Biochemistry in Health and Disease 27, https://doi.org/10.1007/978-3-031-47375-3_26

The Initiation of the Hydrogen Medicine and Biology

The field of hydrogen medicine and biology was initiated with the discovery of therapeutic and preventive benefits of molecular hydrogen (H_2). For a long time, H_2 was believed that it had no functions in mammalian cells due to its inactivity to biological compounds at body temperature without a catalyst. Although some bacteria can produce or metabolize H_2 using hydrogenase enzymes, mammalian cells lack the hydrogenase genes. However, in 2007, we published a groundbreaking article in Nature Medicine titled "Hydrogen acts as a therapeutic antioxidant by selectively reducing cytotoxic oxygen radicals" that overturned the previous concept and ignited the birth of the H_2 medicine and biology field [1]. H_2 stands out due to its excellent efficacies, lack of adverse effects, and multiple functions, making it applicable in diverse fields [2] (Fig. 26.1). The outstanding example is the efficacy of H_2 on the recovery from cardiopulmonary arrest as recently reported [3].

Furthermore, H_2 is beneficial not only to animals and humans, but also to higher plants. Therefore, H_2 has a strong impact on agriculture [4].

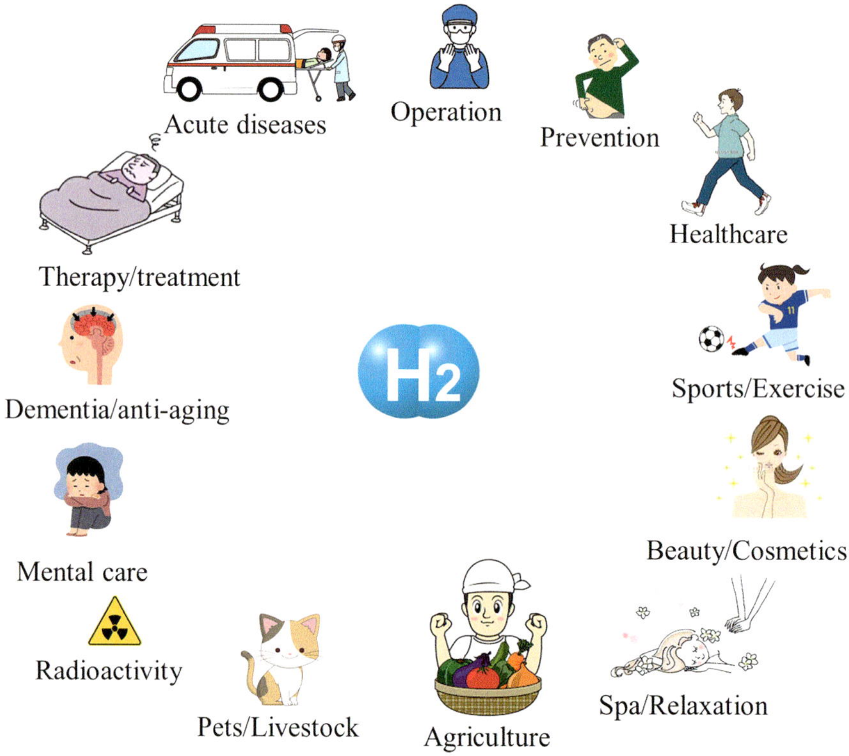

Fig. 26.1 Potential applications of molecular hydrogen in diverse fields. The fields in interest were selected from published clinical studies

From Mitochondrial Medicine to Hydrogen Medicine

In our laboratory, we have been investigating the pathogenesis of mitochondrial diseases from the aspect of molecular and cellular biology. Mitochondria are present in all cells except red blood cells, and their abnormalities cause a wide range of symptoms at any ages, in any organs, and with any types of inheritance. This notion contrasts with the conventional medicine, which tends to focus on organ specificity.

Mitochondrion has multiple functions, and its primary role is energy metabolism. As a byproduct of the energy metabolism, reactive oxygen species (ROS) are produced by the mitochondrial electron transport chain. Damage caused by ROS has various consequences, ranging from healthy individuals to serious illnesses.

The notion of diverse symptoms due to various abnormalities in mitochondrial pathologies is common with the diverse efficacies of H_2. Without our experience in mitochondrial medical research, we might not have arrived at the concept of H_2 medicine. Additionally, we have obtained experimental techniques related to ROS, such as specific detection of ROS and observation of their effects. Therefore, it was not so difficult for us to shift our focus to H_2 and continue our research.

Selective Reduction of Oxidative Stress by Molecular Hydrogen

Oxidative stress is caused by excessive production of ROS, including superoxide anion radicals ($^{\cdot}O_2^-$), hydrogen peroxide (H_2O_2), nitric oxide ($NO^{\cdot}$), and hydroxyl radical ($^{\cdot}OH$) [5]. Excessive ROS can cause a wide range of diseases by oxidizing and damaging biological substances. However, ROS such as H_2O_2, $^{\cdot}O_2^-$, and $NO^{\cdot}$ play crucial physiological roles in signaling cascades and biological processes such as cell proliferation, differentiation, apoptosis, and immunomodulation. Therefore, excessive antioxidant intake is therefore rather harmful and leads to increased mortality, as shown [6].

As reported by us, H_2 selectively reduces highly oxidative ROS such as $^{\cdot}OH$ and peroxynitrite ($ONOO^-$), but not $^{\cdot}O_2^-$, H_2O_2, or $NO^{\cdot}$ [1] (Fig. 26.2). $^{\cdot}OH$ is the most oxidative molecule to damage the cell components in a chaotic manner [7]. Experimental observations have repeatedly demonstrated that H_2 reduces $^{\cdot}OH$ not only in cultured cells but also in various tissues, as seen in testicular radioprotection, hematopoietic stem cell damage caused by total body irradiation, hyperoxia in cultured cells, lung hypoxia/reoxygenation, retinal ischemia–reperfusion, and retinal sonication [8]. Therefore, there is no doubt that H_2 reduces $^{\cdot}OH$ radicals in living cells and tissues.

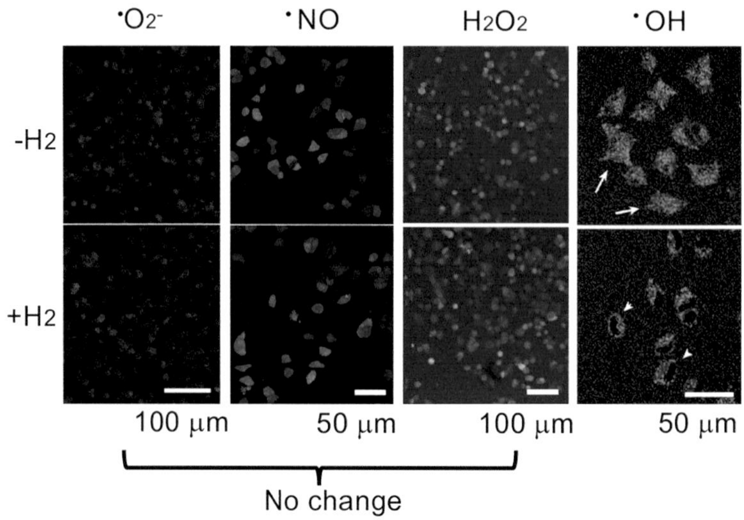

Fig. 26.2 Selective reduction of reactive oxygen species by molecular hydrogen. Each reactive oxygen species in cultured cells was visualized with its respective specific fluorescent indicator. Molecular hydrogen (indicated by $+H_2$) reduced the signal of the ·OH specific fluorescence only. These pictures were selected and rearranged from the previous publication. See [1] for the details

Pharmacokinetics of Molecular Hydrogen Ingested by Various Ways

There are no transporters or channels that transport H_2 into cells. Because H_2 is the smallest in the universe, non-ionic, non-polar and non-magnetic, its own physical properties allow it to pass through bio-membranes by diffusion. There are several ways to ingest H_2 into the body. Inhaling H_2 gas is a direct and effective method using a ventilator circuit, facemask, or nasal cannula. When inhaled, H_2 is absorbed in the lungs and distributed into the body via the arterial blood stream. H_2 is then delivered into most tissues through the blood vessels [9]. Inhalation may be the most suitable method to defend against acute oxidative stress, with a concentration of H_2 gas between 1–4% (v/v) being effective, resulting in blood concentrations of 8–32 μM.

H_2 is dissolved in water up to 0.8 mM (1.6 mg/L) at atmospheric pressure and room temperature, and does not alter pH. H_2-infused water (termed H_2 water or H_2-rich water) may also be beneficial because it is safe and easy to be drunk. Upon orally administration, H_2 was absorbed in the stomach and transported to the portal vein via the jejunal vein [10], where H_2 is released from the jejunal vein into the fluids within 10 min. and gradually returned to the portal vein. Importantly, the functional concentration of H_2 in the liver was maintained at 10–20 μM for one hour, which corresponds to inhaling 1.2–2.5% H_2 gas.

H_2 can also be administered intravenously or intraperitoneally as H_2 dissolved in saline (H_2-rich saline) with great efficacy in animal models [11]. H_2 delivery to cardiac grafts during cold preservation using an H_2-supplemented water bath efficiently ameliorated myocardial injury induced by ischemia and reperfusion. Alternatively, loaded eye drops containing H_2 were developed by dissolving H_2 in saline and administering it directly onto the ocular surface [12].

Discovery of the Target of H_2 that Resolves the Unresolved Discrepancy

Our group and others have demonstrated the reduction of $\cdot$OH by H_2 in living cells; however, the reaction rate of $\cdot$OH with H_2 in homogeneous aqueous solutions is much slower (0.35×10^{-8} M^{-1} s^{-1}) than the $\cdot$OH reactions with other antioxidants. For instance, $\cdot$OH reacts with glucose (15×10^{-8} M^{-1} s^{-1}), glutathione (230×10^{-8} M^{-1} s^{-1}), and other biomolecules much faster than with H_2 [13]. As mentioned above, H_2 cannot react with most molecules without a catalyst, but effective amounts of metals such as Cu, Fe, Ni, and Pt are unlikely to be present in living cells. Moreover, there is no report indicating the discovery of an organo-catalyst for H_2. The discrepancy between reactions in homogeneous aqueous solutions and living organisms has been debated for a long time.

Recently, Professor Qianjun He's group published a paper in Nano Research titled "Fe-porphyrins: redox-related biosensors of molecular hydrogen" [14], in which the oxidized form of Fe-porphyrins (termed "hematin") is identified as a molecular target/biosensor of H_2. Hematin is an oxidized form of Fe(III)-containing porphyrin (PrP) that was converted from Fe(II)-containing porphyrin (heme) (Fig. 26.3). They discovered a novel reaction showing that H of H_2 replaces the hydroxy group (−OH) of hematin with a hydrogen group (−H) attached to Fe(III). This H in the −H group should behave as a hydride (H^-) and, due to its high reactivity, $\cdot$OH should be rapidly converted to H_2O by this catalytic action.

It is shown that heme (PrP-Fe(II)) acts as a catalyst for the following reaction [15].

$$\text{PrP-Fe(II)} + {}^{\cdot}\text{OH} \rightarrow \text{PrP-Fe(III)-OH} \tag{26.1}$$

$$\text{PrP-Fe(III)-OH} + \text{H}_2 \rightarrow \text{PrP-Fe(III)-H} + \text{H}_2\text{O} \tag{26.2}$$

$$\text{PrP-Fe(III)-H} + {}^{\cdot}\text{OH} \rightarrow \text{PrP-Fe(II)} + \text{H}_2\text{O} \tag{26.3}$$

The overall formula indicates that heme (PrP-Fe(II)) catalyzed the following reaction:

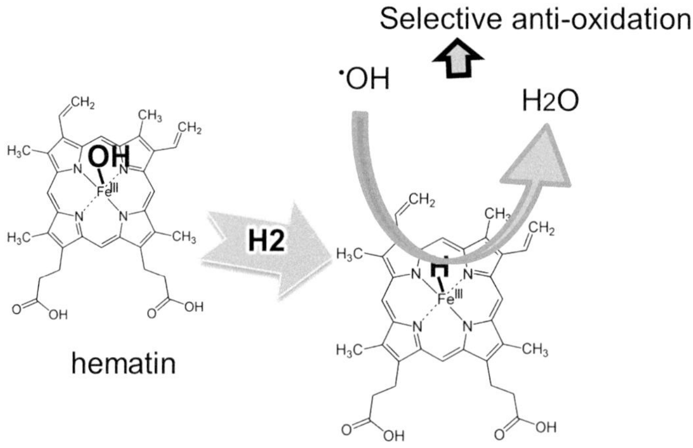

Fig. 26.3 Catalytic action of the oxidized form of a porphyrin (hematin) to stimulate the reaction of H_2 with ·OH. This illustration was drawn with reference to the previous publication. See [14] for the details

$$2^{\cdot}OH + H_2 \rightarrow 2H_2O \tag{26.4}$$

Thus, the discovery of the H_2 target resolved the discrepancy between living cells and homogeneous aqueous reaction kinetics because heme acts as a catalyst in the living cells.

Porphyrin-Involved Multiple Functions of Molecular Hydrogen

H_2 provides numerous beneficial effects by reducing oxidative stress in almost all organs [2]. In addition, H_2 has been demonstrated to possess several other functions, including anti-inflammatory, anti-apoptotic, and anti-allergic effects, anti-cell death, regulatory autophagy, and promote energy metabolism [2].

Since H_2 exerts various functions in a variety of types of cells, it was considered that the target molecules on which H_2 acts are diverse. However, it is possible that H_2 exerts various functions by acting on various types of the porphyrin group.

Porphyrin (heme in a broad sense) is abundantly distributed throughout the body, both inside and outside cells. Heme is present in hemoglobin in blood and myoglobin in muscles, and serves to deliver oxygen molecules (O_2) throughout the body. As a result, heme is exposed to O_2 or H_2O_2, and Fe(II) from the heme frequently catalyzes the formation of ·OH via the Fenton reaction or its mimetic reactions. Porphyrins are present as cytochromes in the electron transport chain of the inner mitochondrial membrane and facilitate electron transport by converting the reduced form to/from the oxidized form. In the intracellular cytosol, the antioxidant enzymes catalase and peroxidase, P450, and nitric oxide (NO) synthase have porphyrins as their essential components [15]. These Fe(II)-bearing porphyrins have a variety of functions, but a common function is that they act as mediators of redox reactions and are exposed to oxidative stress. As a result, porphyrins are frequently oxidized and rendered non-functional. In addition, several transcription factors contain each porphyrin, suggesting that H_2 is involved in various transcriptional regulations.

H_2 is capable of repairing various types of porphyrins, from their oxidized form, PrP-Fe(III)-OH, to their reduced functional form, PrP-Fe(II), through the following reactions:

$$\text{PrP-Fe(III)-OH} + H_2 \rightarrow \text{PrP-Fe(III)-H} + H_2O \quad (26.5)$$

$$\text{PrP-Fe(III)-H} + {}^{\cdot}X \rightarrow \text{PrP-Fe(II)} + \text{H-X}, \quad (26.6)$$

where ${}^{\cdot}X$ is an oxidative free radicals such as ·OH.

In pathological conditions, porphyrins are frequently damaged by oxidation, and H_2 can activate the porphyrin-containing enzymes by repairing their porphyrins. Thus, H_2 can be express many functions by repairing the various porphyrins. Importantly, H_2 exerts its function only on the porphyrins damaged by oxidation, and H_2 does not act on functional porphyrins. This mechanism may explain why there are no or little adverse effects of H_2. In other words, H_2 exerts its function only in pathological conditions.

Safety of Molecular Hydrogen Compared to Other Medical Gasses

Several medical gases are known to exhibit various efficacies in addition to H_2 gas. The other medical gases such as carbon monoxide (CO), hydrogen sulfide (H_2S), and nitric oxide ($NO^{\cdot}$) are highly toxic above a certain level of concentration, and only at low concentrations exhibit physiological roles as signaling molecules. These gases can affect pathological processes by acting directly on heme-proteins to exhibit their functions. In addition, these gases commonly act on K^+ channels to modulate the excitability of the cell membrane and promote the efflux of potassium ions within the cells, influencing various physiological processes. H_2 commonly interact with a K^+ channel.

While H_2 shares the target molecules with these toxic medical gases, H_2 has the advantage of being non-toxic even at high concentrations. High concentrations of H_2 gas have been used in deep-diving gas mixtures to prevent arterial gas thrombosis and decompression sickness, establishing its biological safety [16]. Moreover, no adverse effects have been confirmed in human studies related to the administration of H_2 therapy [17], and inhalation of H_2 gas was approved as safe in Phase I clinical trial [18]. Noteworthy, H_2 is not flammable at the biologically effective concentrations of 1–4%, which is an advantage from a safety point of view in practical uses.

Indirect Hormonal Regulation by Molecular Hydrogen

Hormonal regulation plays a key role to exhibit each specific function. In turn, H_2 modulates hormones and cytokines in specific tissues to reach target cells. Even when H_2 does not reach the necessary concentration for a target cell, H_2 can exert beneficial effects through the second messenger system under the hormonal regulation.

To exert multiple functions beyond antioxidant role, H_2 must influence a variety of signaling pathways, inducing the consequent expression of many genes. For example, H_2, which was drunk as H_2 water, is expelled from the lungs through the venous system, so it does not reach the brain; however, the H_2 can protect nerve cells by increasing the expression of the ghrelin hormone [19]. Ghrelin is produced in the stomach and reaches the brain, allowing H_2 to exert its effects on the brain.

In addition, H_2 stimulates energy metabolism by increasing the expression of fibroblast growth factor 21 (FGF21) [20]. FGF21 stimulates lipid metabolism, gluconeogenesis, and ketogenesis. Furthermore, H_2 has anti-inflammatory effects by reducing the levels of many pro-inflammatory cytokines [21].

Subsequent Signaling Involving Lipid Peroxides

Recent studies have delineated pathways through which H_2 exerts multiple functions (Fig. 26.4).

Lipid peroxides (LPOs) accumulate through free radical chain reactions. In this reaction, $^{\cdot}OH$ plays a major role in extracting H atoms from polyunsaturated fatty acids in the initiation step. The next step is amplified by the lipid free radical chain reaction. Antioxidants donate H atoms to lipid peroxy radical species in termination reactions, leading to peroxide accumulation [22]. The end products derived from LPOs often serve as second messengers in signal transduction pathways. For instance, 4-hydroxy-2-nonenal (4-HNE) functions as a significant second messenger in signal transduction. By decreasing the level of 4-HNE, H_2 can induce PGC-1α and the subsequent Akt/FoxO1 signaling pathway [23]. PGC-1α is a multi-functional co-activator that includes energy metabolism.

Alternatively, phospholipids containing polyunsaturated fatty acids are converted chemically or enzymatically, into oxidized mediators that modulate various signal transduction pathways. Oxidized phospholipids exert a wide range of biological effects on different cell types in diverse cellular responses, including inflammation, proliferation, and cell death, and in the development of several chronic diseases such as atherosclerosis. Oxidized phospholipids activated Ca^{2+} signaling, and H_2-modified oxidized phospholipid mediators suppressed Ca^{2+} signaling by acting as antagonists

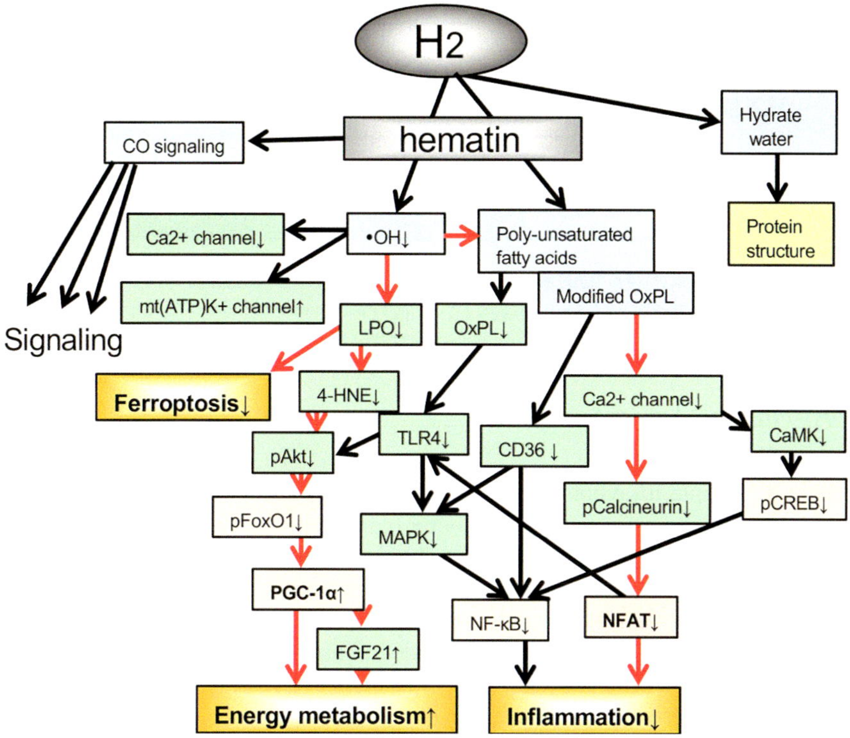

Fig. 26.4 Possible pathways for exerting molecularly hydrogen-induced anti-cell death, anti-inflammation, and energy metabolism-stimulation. This illustration was modified from the previous publication with the addition of the role of hematin. See [8] for the details

of Ca^{2+} channels. This decrease in cellular Ca^{2+} inactivates the transcription factors, nuclear factor of activated T cells (NFAT), and cAMP response element-binding protein (CREB). NFAT transcribes the following genes involved in inflammation: transcription factors (EGR-1, KLF2, ATF3, JUNB, and NFKB2), cytokines (TNF-α and IL-8), and enzymes (cyclooxygenase-2 (COX-2)) [24]. Therefore, H_2 can suppress inflammation by decreasing the expression of pro-inflammatory factors.

Future Mission of Hydrogen Medicine

The future mission of H_2 medicine is to apply H_2 for treatment, prevention, and quality of life across a wide range of fields. Furthermore, the use of H_2 in agriculture has the potential to produce safe, productive and high-quality foods. H_2 medicine will be able to be practiced with simple and low-cost treatments without adverse effects, and will contribute to many unsolved serious problems facing today's society.

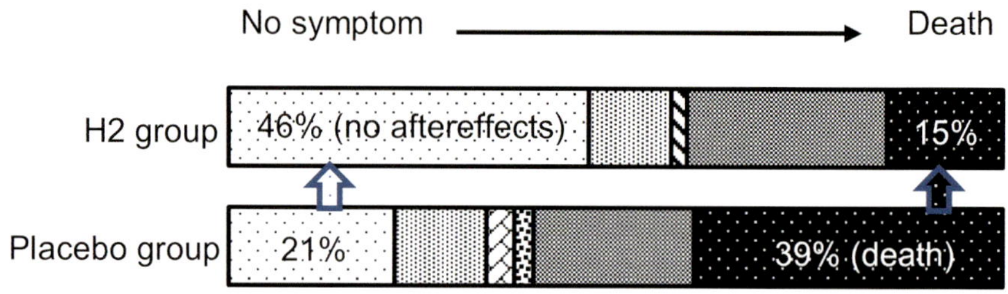

Fig. 26.5 Effects of H_2 gas inhalation on recovery after 90 days of cardiac arrest. The efficacies were assessed by a modified ranking scale. Hydrogen gas inhalation halved the mortality rate and doubled the rate of no sequelae. This graph was modified with the previous publication [3]

Post-cardiac Arrest Syndrome

The best example for emphasizing the efficacy of H_2 is the recovery after cardiac arrest. Cardiopulmonary resuscitation (CPR) can often save the lives of patients who experiences sudden cardiac arrest due to heart failure. Because blood is not circulating, the whole body, including the brain, suffers severe damage. Thermoregulatory therapy is currently being used, but its effectiveness is still uncertain. A multi-center, randomized, double-blind, placebo-controlled clinical trial was conducted in post-cardiac arrest syndrome patients. The patients inhaled 2% H_2 with O_2 in addition to thermoregulatory therapy, and mortality and neurological outcomes were investigated. Strikingly, the 90-day mortality was less than half of that receiving only the thermoregulatory (from 39% [without H_2] to 15% [with H_2]), and 46% of the H_2 group was recovered with no aftereffects (no sequelae), whereas only 21% of the control group was recovered with no aftereffects (Fig. 26.5). These results strongly suggest that practical use of hydrogen therapy will save many lives and livelihoods [3].

Alzheimer's Disease

Alzheimer's disease (AD) is a progressive and fatal neurodegenerative disorder that causes cognition, memory, and behavior deficits. Overcoming AD is one of the most important challenges in the today's aging society. Aging is the highest risk factor for AD, suggesting that multiple factors are involved in the complex AD pathogenesis. Therefore, the H_2's multiple functions may improve AD. In addition, H_2 has the advantage to cross the blood–brain barrier (BBB) by gaseous diffusion without a specific drug delivery system. Although it is impossible to revive dead cells in AD patients, the ideal strategy is to promote activity of surviving neurons with disease-modifying therapies, even after the onset of AD.

Extensive studies have shown that H_2 improves neurodegenerative disorder models with marked efficacies [25]. Drinking H_2 water prevented cognitive deficits

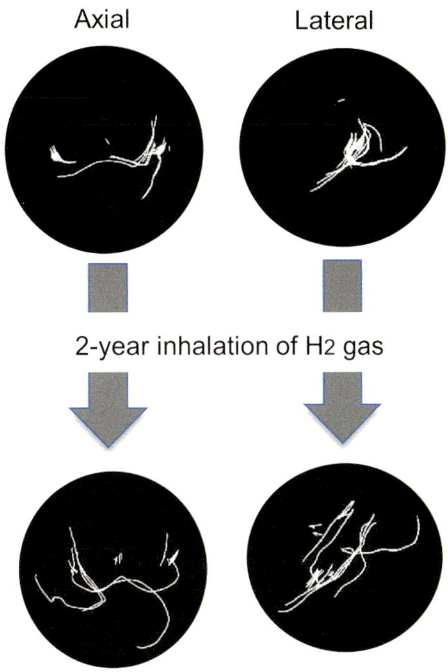

Fig. 26.6 Effects of long-term inhalation of H_2 gas on neurons of an Alzheimer's disease patient. The activated neurons that pass through the hippocampus were visualized from axial and lateral views by diffusion tensor imaging. Photographs were rearranged from the previous publication [28]. See that the active neurons increased after 2-year inhalation of hydrogen gas, even in the advances stage of Alzheimer's disease

induced by chronic physical restraint stress in mice with ameliorating the suppression of neural proliferation induced by the restraint stress in the dentate gyrus of the hippocampus [26].

In a double-blind, randomized clinical study, long-term (for one year) drinking of H_2 water improved cognition in subjects with mild cognitive impairment (MCI) who carry a genetic high risk of AD, the apoE4 genotype [27]. There is a case report in an advanced AD patient whose fecal incontinence improved with continuous inhalation of H_2 gas for 2 years [28] (Fig. 26.6). This study encouraged AD patients that the possibility of improvement cannot be ruled out even in patients with severe AD. As assessed clinically using ADAS-cog and objectively using Diffusion Tensor Imaging, clinical trials using H_2 inhalation for half a year showed marked improvements. Moreover, in a 1-year follow-up without inhaling H_2 gas, the improvement was maintained at least half year. This suggests that this treatment not only provided temporary symptom relief, but also had a disease-modifying effect [29].

Advanced Stage of Cancer

The application of H_2 gas therapy to patients with advanced cancer is under investigation [30, 31]. Inhalation of H_2 showed improvement of prognosis and quality-of-life, and reduced tumor-size in patients with advanced-stage of cancer.

Since PD-1 is transcribed by the transcription factor NFAT [24], H_2 reduced the PD-1 expression, enhancing immunotherapy, because PD-1 suppresses immunotherapy. In addition, H_2 promotes mitochondrial activation by increasing the PGC-1α expression through the 4-HNE/Akt/FoxO1 pathway [23], resulting in the enhancement of the immune activity of $CD8^+$ T cells.

In addition, H_2 was also shown to mitigate the adverse effects caused by chemotherapy and radiotherapy [32].

Metabolic Syndrome

Metabolic syndrome is characterized by excessive accumulation of visceral fat, elevated blood pressure, fasting hyperglycemia, and abnormal lipid levels [33]. Drinking H_2 water firstly showed to repress arteriosclerosis in model mice [34]. Long-term H_2 water intake showed significant decrease body and fat weights, as well as the levels of plasma glucose, insulin, and triglyceride [20]. More studies revealed that drinking H_2 water improved nonalcoholic steatohepatitis (NASH) in model mice.

Several clinical studies were conducted, and a meta-analysis with systematic review has revealed that drinking H_2 water can lead to a significant decrease in blood lipid levels [35].

Cytokine Storm Involved in Infection

Since 2019, the novel coronavirus disease (COVID-19) was world-widely expanded as a pandemic, resulting in a significant increase in human deaths by cytokine storm. The cytokine storm is induced by excessive inflammation. Similarly, the cytokine storm is induced in multiple organ failure, and H_2 prevents death by suppressing the cytokine storm [21]. H_2 inactivates the NFAT activity via modified oxidized phospholipids, as described above, suppressing the expression of inflammatory cytokines and COX-2, and thus alleviating inflammation. Inhalation therapy with H_2 gas was widely used to treat COVID-19 patients in China [36]. Additionally, inhalation of H_2 was shown to alleviate the COVID-19 sequelae.

Suppression of inflammation by H_2 should help treat many other infectious diseases, not just COVID-19.

Improving the Quality of Life and Beauty of Healthy People

Drinking H_2 water was shown to improve mood, anxiety, and autonomic function in usual daily life under no undue stress [37]. This study confirmed the beneficial effects of H_2 in the daily life of healthy people.

Many papers have shown that drinking H_2 water and inhaling H_2 gas are effective in improving exercise capacity and relieving fatigue [38]. As an example, drinking H_2 water reduced fatigue and improved endurance after moderate exercise on a bicycle ergometer [39].

It has been reported that H_2 improves skin condition and contributes to beauty [40].

Preventive medicine is important for healthy person's lives. However, if one drug or food ingredient would be effective in preventing only one disease, as is common sense in conventional medicine, we would have to continue taking many drugs. Also, it should not have any side effects. H_2 has multiple functions with no adverse effects, so it should be able to contribute greatly to prevention. Therefore, H_2 has a beneficial effect on healthy people.

Conclusion

H_2 has great potential in real-world applications for overcoming unsolved problems in severe chronic and acute diseases and for improving quality of life, in various medical fields. Recent studies have shed light on the fundamental molecular mechanisms underlying H_2. H_2 may be unusual or outstanding from the point of view of the conventional medicine. It is therefore important to receive official approval for the application of H_2 treatment.

Competing Interest The author declares that he is directors of H_2 affiliated companies, H2 WATER JAPAN, Inc. (Tokyo, Japan) and H2 Global Group s.r.o (Ostrava, Czech Republic).

References

1. Ohsawa I, Ishikawa M, Takahashi K, Watanabe M, Nishimaki K, Yamagata K, Katsura K, Katayama Y et al (2007) Hydrogen acts as a therapeutic antioxidant by selectively reducing cytotoxic oxygen radicals. Nat Med 13:688–694
2. Ohta S (2014) Molecular hydrogen as a preventive and therapeutic medical gas: initiation, development and potential of hydrogen medicine. Pharmacol Ther 144:1–11
3. Tamura T, Suzuki M, Homma K, Sano M, Group HIS (2023) Efficacy of inhaled hydrogen on neurological outcome following brain ischaemia during post-cardiac arrest care (HYBRID II): a multi-centre, randomised, double-blind, placebo-controlled trial. EClinicalMedicine 58:101907

4. Li L, Lou W, Kong L, Shen W (2021) Hydrogen Commonly Applicable from Medicine to Agriculture: From Molecular Mechanisms to the Field. Curr Pharm Des 27:747–759
5. Slezak J, Kura B, LeBaron TW, Singal PK, Buday J, Barancik M (2021) Oxidative stress and pathways of molecular hydrogen effects in medicine. Curr Pharm Des 27:610–625
6. Bjelakovic G, Nikolova D, Gluud LL, Simonetti RG, Gluud C (2007) Mortality in randomized trials of antioxidant supplements for primary and secondary prevention: systematic review and meta-analysis. JAMA 297:842–857
7. Lubec G (1996) The hydroxyl radical: from chemistry to human disease. J Investig Med 44:324–346
8. Ohta S (2021) Direct targets and subsequent pathways for molecular hydrogen to exert multiple functions: focusing on interventions in radical reactions. Curr Pharm Des 27:595–609
9. Sano M, Ichihara G, Katsumata Y, Hiraide T, Hirai A, Momoi M, Tamura T, Ohata S et al (2020) Pharmacokinetics of a single inhalation of hydrogen gas in pigs. PLoS ONE 15:e0234626
10. Ichihara G, Katsumata Y, Moriyama H, Kitakata H, Hirai A, Momoi M, Ko S, Shinya Y et al (2021) Pharmacokinetics of hydrogen after ingesting a hydrogen-rich solution: a study in pigs. Heliyon 7:e08359
11. Cai J, Kang Z, Liu K, Liu W, Li R, Zhang JH, Luo X, Sun X (2009) Neuroprotective effects of hydrogen saline in neonatal hypoxia-ischemia rat model. Brain Res 1256:129–137
12. Oharazawa H, Igarashi T, Yokota T, Fujii H, Suzuki H, Machide M, Takahashi H, Ohta S et al (2010) Protection of the retina by rapid diffusion of hydrogen: administration of hydrogen-loaded eye drops in retinal ischemia-reperfusion injury. Invest Ophthalmol Vis Sci 51:487–492
13. Wood KC, Gladwin MT (2007) The hydrogen highway to reperfusion therapy. Nat Med 13:673–674
14. Jin Z, Zhao P, Gong W, Ding W, He Q (2023) Fe-porphyrin: a redox-related biosensor of hydrogen molecule. Nano Res 16:2020–2025
15. Ohta S (2023) Molecular hydrogen may activate the transcription factor Nrf2 to alleviate oxidative stress through the hydrogen-targeted porphyrin. Aging Pathobiol Ther 5:25–32
16. Fontanari P, Badier M, Guillot C, Tomei C, Burnet H, Gardette B, Jammes Y (2000) Changes in maximal performance of inspiratory and skeletal muscles during and after the 7.1-MPa Hydra 10 record human dive. Eur J Appl Physiol 81:325–328
17. Ono H, Nishijima Y, Ohta S, Sakamoto M, Kinone K, Horikosi T, Tamaki M, Takeshita H et al (2017) Hydrogen gas inhalation treatment in acute cerebral infarction: a randomized controlled clinical study on safety and neuroprotection. J Stroke Cerebrovasc Dis 26:2587–2594
18. Cole AR, Sperotto F, DiNardo JA, Carlisle S, Rivkin MJ, Sleeper LA, Kheir JN (2021) Safety of prolonged inhalation of hydrogen gas in air in healthy adults. Crit Care Explor 3:e543
19. Matsumoto A, Yamafuji M, Tachibana T, Nakabeppu Y, Noda M, Nakaya H (2013) Oral 'hydrogen water' induces neuroprotective ghrelin secretion in mice. Sci Rep 3:3273
20. Kamimura N, Nishimaki K, Ohsawa I, Ohta S (2011) Molecular hydrogen improves obesity and diabetes by inducing hepatic FGF21 and stimulating energy metabolism in db/db mice. Obesity (Silver Spring) 19:1396–1403
21. Qi B, Yu Y, Wang Y, Wang Y, Yu Y, Xie K (2021) Perspective of molecular hydrogen in the treatment of sepsis. Curr Pharm Des 27:667–678
22. Iuchi K, Nishimaki K, Kamimura N, Ohta S (2019) Molecular hydrogen suppresses free-radical-induced cell death by mitigating fatty acid peroxidation and mitochondrial dysfunction. Can J Physiol Pharmacol 97:999–1005
23. Kamimura N, Ichimiya H, Iuchi K, Ohta S (2016) Molecular hydrogen stimulates the gene expression of transcriptional coactivator PGC-1alpha to enhance fatty acid metabolism. NPJ Aging Mech Dis 2:16008
24. Iuchi K, Imoto A, Kamimura N, Nishimaki K, Ichimiya H, Yokota T, Ohta S (2016) Molecular hydrogen regulates gene expression by modifying the free radical chain reaction-dependent generation of oxidized phospholipid mediators. Sci Rep 6:18971
25. Iketani M, Ohsawa I (2017) Molecular hydrogen as a neuroprotective agent. Curr Neuropharmacol 15:324–331

26. Nagata K, Nakashima-Kamimura N, Mikami T, Ohsawa I, Ohta S (2009) Consumption of molecular hydrogen prevents the stress-induced impairments in hippocampus-dependent learning tasks during chronic physical restraint in mice. Neuropsychopharmacol 34:501–508
27. Nishimaki K, Asada T, Ohsawa I, Nakajima E, Ikejima C, Yokota T, Kamimura N, Ohta S (2018) Effects of molecular hydrogen assessed by an animal model and a randomized clinical study on mild cognitive impairment. Curr Alzheimer Res 15:482–492
28. Ono H, Nishijima Y, Sakamoto M, Kitamura S, Naitoh Y, Suzuki K, Fujii N, Kikura R et al (2022) Long-term inhalation of hydrogen gas for patients with advanced Alzheimer's disease: a case report showing improvement in fecal incontinence. Med Res Arch. https://doi.org/10.18103/mra.v10i7.2951
29. Ono H, Nishijima Y, Ohta S (2023) Therapeutic inhalation of hydrogen gas for Alzheimer's disease patients and subsequent long-term follow-up as a disease-modifying treatment: an open label pilot study. Pharmaceuticals (Basel) 16:434
30. Akagi J, Baba H (2019) Hydrogen gas restores exhausted CD8+ T cells in patients with advanced colorectal cancer to improve prognosis. Oncol Rep 41:301–311
31. Chen JB, Kong XF, Mu F, Lu TY, Lu YY, Xu KC (2020) Hydrogen therapy can be used to control tumor progression and alleviate the adverse events of medications in patients with advanced non-small cell lung cancer. Med Gas Res 10:75–80
32. Nakashima-Kamimura N, Mori T, Ohsawa I, Asoh S, Ohta S (2009) Molecular hydrogen alleviates nephrotoxicity induced by an anti-cancer drug cisplatin without compromising anti-tumor activity in mice. Cancer Chemother Pharmacol 64:753–761
33. Qin S (2021) Role of hydrogen in atherosclerotic disease: from bench to bedside. Curr Pharm Des 27:713–722
34. Ohsawa I, Nishimaki K, Yamagata K, Ishikawa M, Ohta S (2008) Consumption of hydrogen water prevents atherosclerosis in apolipoprotein E knockout mice. Biochem Biophys Res Commun 377:1195–1198
35. Todorovic N, Fernández-Landa J, Santibañez A, Kura B, Stajer V, Korovljev D, Ostojic SM (2023) The effects of hydrogen-rich water on blood lipid profiles in clinical populations: a systematic review and meta-analysis. Pharmaceuticals 16:142
36. Guan WJ, Chen RC, Zhong NS (2020) Strategies for the prevention and management of coronavirus disease 2019. Eur Respir J 55:2000597
37. Mizuno K, Sasaki AT, Ebisu K, Tajima K, Kajimoto O, Nojima J, Kuratsune H, Hori H et al (2017) Hydrogen-rich water for improvements of mood, anxiety, and autonomic nerve function in daily life. Med Gas Res 7:247–255
38. Botek M, Krejci J, McKune A, Valenta M, Sladeckova B (2022) Hydrogen rich water consumption positively affects muscle performance, lactate response, and alleviates delayed onset of muscle soreness after resistance training. J Strength Cond Res 36:2792–2799
39. Mikami T, Tano K, Lee H, Lee H, Park J, Ohta F, LeBaron TW, Ohta S (2019) Drinking hydrogen water enhances endurance and relieves psychometric fatigue: a randomized, double-blind, placebo-controlled study. Can J Physiol Pharmacol 97:857–862
40. Bajgai J, Lee KJ, Rahman MH, Fadriquela A, Kim CS (2021) Role of molecular hydrogen in skin diseases and its impact in beauty. Curr Pharm Des 27:737–746

Printed by Printforce, the Netherlands